Neville on Concrete

An Examination of Issues
in Concrete Practice

Second Edition

Adam Neville

BOOKSURGE LLC

Copyright © 2006 Adam Neville
All Rights Reserved
ISBN: 1-4196-5207-9

To order additonal copies, please contact us.
BookSurge, LLC
www.booksurge.com
1-866-308-6235
orders@booksurge.com

To the memory of my parents
and to my wife, who has been my concrete inspiration over half a century.

Behind every successful man (even mildly so) stands a surprised wife.

Preface to the Second Edition

As the Preface to the First Edition indicates, that edition was published in the U.S. by the American Concrete Institute (ACI) and the exposure of the book to British and other European readers was limited because of ACI's marketing arrangements for this, the only ACI book. Up to the publication of the first edition, many of my articles and papers were published in the U.S. but, since 2003, I have not submitted many papers to American journals and none to the ACI. My activities shifted largely to the United Kingdom.

This book represents solely the views of the author, and some statements, and even classifications of materials, may not conform to the current orthodoxy. I have no illusions about my importance or influence, but I feel strongly that airing unorthodox views is important, if only to provoke discussion and hence ascertain what at present seems to be the correct view or the best interpretation.

Alas, the beginning of the 21st century appears to be the age of increasing political correctness. At the same time, we are in an age of more mixed societies and closer international relations, so that we should respect other people's views. But putting a taboo on criticism of many aspects of concrete and on gentle mockery is excessive and unhelpful to seeking, if not the truth, then the closest approximation to truth within the inevitably limited state of our knowledge. The cynical words by Voltaire put into the mouth of Dr. Pangloss "in the best of all possible worlds" surely do not apply to present-day concrete, and stifling criticism and publishing a glossy popular magazine is not conducive to progress. Hence, the need for publishing this edition in a more open and critical atmosphere.

In *Man and Superman*, George Bernard Shaw wrote: "The reasonable man adapts himself to the world: the unreasonable one persists in trying to adapt the world to himself. Therefore, all progress depends on the unreasonable man." I am all for being an unreasonable man.

Writing this Preface to the Second Edition of *Neville on Concrete* gives me an opportunity to promote two of my related books. One is *Concrete: Neville's Insights and Issues*, published in 2006 by Thomas Telford, which is similar in approach to the present book and complements it. *Concrete: Neville's Insights and Issues* is based on my investigations and publications subsequent to the publication of *Neville on Concrete*: there is thus no overlap and no duplication. I would submit therefore that, if you find *Neville on Concrete* to be useful, you are sure to find *Concrete: Neville's Insights and Issues* to be equally useful.

The other book which I would like to commend to you is *Properties of Concrete*, 4th Edition. This is an encyclopaedic tome, starting with a scientific approach to the various ingredients of concrete, and proceeding through properties of the various ingredients and properties of fresh and hardened concrete, and ending with mix selection (or mix design). For anyone seriously working in concrete, *Properties* is an essential item of education – 844 pages of it. *Properties* was first published in 1964, and the latest (and also final) edition was published in 1996 but partially updated in the 13 impressions produced up to 2006.

Good concrete and good concreting!

Adam Neville
London, England

October 2006

Preface to the First Edition

The purpose of this Preface is to present the rationale of this book—one more book about concrete. In other words, why is a collection of articles and papers being published in book form?

There exists already quite a number of books on concrete. Perhaps I should explain that by the term "books on concrete" I mean books dealing with concrete as a material relevant to construction, that is, engineering books, and not books dealing with purely scientific aspects of concrete. Strangely enough, the scientific books, normally concerned with the chemistry of cement and, to a lesser extent, of concrete, are more numerous than engineering books about concrete. My concern is with the latter category.

The books dealing with what we might call engineering concrete, or possibly concrete engineering, are of two types. In the first of these, a single author, or two co-authors, covers a wide range of topics relevant to concrete construction, and to the underlying properties of concrete. The number of books of this type is relatively small, and very few have been published in the last decade or so.

The second type of book on concrete engineering is a volume that is a collection of chapters, each chapter written by a different author or, in some cases, by several authors; I have seen as many as seven co-authors of a single chapter. This approach of stitching together a number of chapters, each providing information on one topic at a time, is not surprising in view of the very large amount of knowledge available on all aspects of concrete: it is increasingly difficult for one person to cover the whole field. So, while the stitching-together approach is becoming increasingly inevitable, it has the disadvantage of a lack of cohesion and of lacunae in the material covered. (I cannot resist saying that concrete is meant to be cohesive and free from lacunae, that is, voids.)

Computer search may help to provide the information required, but this works well only if the person searching knows exactly what he or she is searching for. In reality, an engineer may be looking for information on one aspect of concrete to be used under some novel set of circumstances, and he will find the relevant information. But he may well not have his attention drawn to some additional aspect because this additional aspect was not entered in the search formulation. It was not entered because the engineer was not aware of it, and the search vehicle is not configured to provide gratuitous information. To give an example, someone may look for data on drying shrinkage of concrete under different curing regimes; the relevant information will be provided, but it is unlikely to say: drying shrinkage is not the only thing that matters; small drying shrinkage can be accompanied by large autogenous shrinkage.

These are the realities of modern concrete engineering, and the present book will not solve the situation at a stroke. What this book can do, however, is to give a fairly comprehensive presentation on a number of individual topics, all in one volume. Each of these is presented as a section. Because all these sections have been written by a single author, the fundamental approach is similar in all cases, and this should make it easier for the reader to become familiar with the content. It is worth mentioning that almost all of the sections stem from practical problems encountered by me in consulting work or in litigation. It follows that the problems discussed are real. The content is a personal view, rather than a neutral review of everybody's work.

I have no illusions that the present book will do away with the need for a comprehensive tome on concrete. Indeed, for selfish reasons, I would be sorry were this to happen, and I hope my book *Properties of Concrete, Fourth Edition*, will continue to be used (and of course sold). The material in the present book is complementary to *Properties of Concrete*. Indeed, with one or two exceptions, all the papers and articles in this book were written in the years 1994-2002, so that they are more recent than *Properties*. This book is arranged in nine chapters, nearly every one containing five sections; in addition, Chapter 10 comprises the Epilogue. Each section is a paper or an article published in a journal or a magazine (except for one published in a volume of conference proceedings). Altogether, there are 42 Sections, the vast majority of which were originally published by ACI. This is not surprising, given the pre-eminent role of ACI on the world concrete scene. My only criticism of that organization, with its worldwide web of ACI Chapters, is that its name does not reflect its truly international character; alas, it is too late to coin a new name, and it is not easy to think of words that would convey the American, as well as worldwide, character of ACI.

Of the nine chapters in this book, five deal with cement and concrete, always bearing in mind the use of concrete in construction; two chapters deal more specifically with construction and design. One chapter deals with research in so far as it contributes to better concrete structures; advancement in scientific knowledge without an indication of how it can lead to better concrete structures is not within my purview. The final chapter deals with litigation about concrete structures; this is the least desirable aspect of construction and is often quite unnecessary. Nevertheless, it is useful to be aware of the real, often cruel and unjust, world in which we live.

The preface to a book is its starting point, the opening gambit. The easiest way to start is by stating the obvious: concrete is a splendid material. However, we have to learn more about concrete, concrete technology, and concreting techniques in order to avoid problems that crop up from time to time. I believe that few of these problems are caused by poor materials or by an inherent inadequacy of concrete. Most of the problems are due to the people involved in concreting, in the selection of the materials, or in the design that is not sufficiently rooted in the knowledge of concrete. As Reginald Heber, Bishop of Calcutta, said 200 years ago: "Though every prospect pleases, and only man is vile."

To improve our performance in building concrete structures, we need to learn more about concrete. My best way of expressing this is by plagiarizing the near-final words in *Properties of Concrete*: we require "*experience*, coupled with the *knowledge* of the influence of various factors upon the properties of concrete; this knowledge must be based on an *understanding* of the behavior of concrete." All three, experience, knowledge, and understanding, are necessary.

This book represents solely the views of the author, and not those of ACI. Thus some statements, and even some of the classifications of materials, may not conform to the current ACI orthodoxy. This may upset purists but, if no dissent is allowed to be aired, and no changes can be even considered, we are unlikely to improve. Anyway, even ACI modifies its terminology from time to time, and both ACI and concrete change and move forward. My approach stems from my firm belief in freedom of scientific and technical expression but, nevertheless, I apologize to those who may dislike my rocking the (concrete) boat. I am not a crusader and I have no revolutionary zeal or mission.

Finally, I would like to say how pleased I am that it is ACI that has published this book because ACI has done more for concrete worldwide than any other organization. I consider it, therefore, a privilege that ACI is willing to broadcast my views, but they are my views and not those of ACI.

Adam Neville
London, England

March 2003

Acknowledgments

No book is written by one person without help from other people. This book is no exception. I do not mean that other people wrote parts of this book, except that Sections 4.2, 4.3, and 4.5 were written jointly by Pierre-Claude Aïtcin and me (with Section 4.5 having also a contribution from P. Acker). It is other forms of input that have taken place, and these should be acknowledged.

First and foremost, the idea of this book came from Bill Semioli, Editor-in-Chief of *Concrete International*. For this I am very grateful to him, as well as for making the complex arrangements required to convert a number of articles published in recent years into a beautifully produced book. This may be an appropriate place for me to thank him also for his encouragement and friendship over the years. If readers find this book of use and of value, they should recognize his forward thinking.

A book is not just an author's manuscript; much work of compilation and arrangement, of processing and checking, and of overseeing that the output is coherent and clear is required. Several people at ACI Headquarters have ensured all this, with Emily Lorenz, P.E., Engineering Editor of *Concrete International*, taking the lead and managing the whole process of making the book "happen." For all her very competent work, combining engineering and editorial skills, her patience, and friendly courtesy, I am truly grateful.

I am very appreciative, too, of the skill of those in the Production Department at ACI, and especially Bonnie Schmick, who converted my words into a physical (not to say, concrete) book.

I also owe considerable thanks to Terry Holland, President of ACI 2002-03, for providing the momentum that ensured the production of this book in time for the ACI Convention in Vancouver in April 2003, just following my reaching (I hope) four score years.

A technical book like this one is not just a product of one person's mind: numerous references need to be consulted, and these are listed at the end of the various sections as appropriate. But identifying these references and searching for them requires not only hard work and effort but also very specialized knowledge and skill. It is in this respect that I have a considerable debt of gratitude to Robert Thomas of the Library of the Institution of Civil Engineers, London: this work is in the background, but its fruits are up front.

Another person who contributed to all aspects of this book is Dr. M. H. Neville. That person (this avoids the use of the clumsy he or she) applied critical thinking to the content and arrangement of the book as well as to the production of the manuscript, and to ensuring its accuracy. This is the type of activity that brings no glory on the title page, but the input of this *eminence grise* to the grey subject matter—concrete—cannot be overestimated. All I can say is: I am most grateful.

Finally, I owe thanks to those publishers and organizations that permitted me to reproduce my original papers and articles in this book. These are as follows:

- RILEM, the publisher of *Materials and Structures*, France, for Sections 2.3, 3.1, 3.4, 3.5, and 4.3.
- *Materiales de Construcción*, Spain, for Section 8.5.
- Elsevier Science, U.S., for Section 8.3. Reprinted from *Cement and Concrete Research*, Vol. 22, "Concrete Research on Micro- and Macro-Scale," pp. 1067-1076, © 1992.
- The American Society of Civil Engineers, for Section 4.1. Reproduced from *Advances in Cement and Concrete*, "Cement and Concrete: Their Interaction in Practice," pp. 11-14, © 1994.
- Thomas Telford Limited, publisher of *Magazine of Concrete Research*, for Section 9.4, "Guest Editorial" in Vol. 41, No. 148, 1989.
- The Concrete Society, publisher of *Concrete*, England, for Section 9.3. To acknowledge this in full, I should say: "This copyrighted material originally appeared in *CONCRETE*, January 2001, and is reproduced by permission of The Concrete Society. For further information, tel.: +44(0)1344-466-007: e-mail: watson@concrete.org.uk. Address: The Concrete Society, Century House, Telford Avenue, Crowthorne, Berkshire, RG45 6YS, UK."

A full list of the articles, papers, and letters included in this book, with details of their original publication, is given in the Appendix.

Neville on Concrete
Table of Contents

PREFACE .... **IV**

ACKNOWLEDGMENTS .... **IX**

CHAPTER 1: CEMENTITIOUS MATERIALS **1-1**
1.1 Cementitious Materials—A Different Viewpoint 1-4
1.2 There is More to Concrete than Cement 1-8
1.3 Silica Fume in a Concrete Specification: Prescribed, Permitted or Omitted? 1-12
1.4 Whither Expansive Cement? 1-20
1.5 A "New" Look at High-Alumina Cement 1-25

CHAPTER 2: WATER AND CONCRETE **2-1**
2.1 Water and Concrete: A Love-Hate Relationship 2-3
2.2 Water—Cinderella Ingredient of Concrete 2-14
2.3 Effect of Cement Paste on Drinking Water 2-26
2.4 Seawater in the Mixture 2-37
2.5 How Useful is the Water-Cement Ratio? 2-50

CHAPTER 3: DURABILITY **3-1**
3.1 Consideration of Durability of Concrete Structures: Past, Present, and Future 3-4
3.2 Maintenance and Durability of Concrete Structures 3-13
3.3 The Question of Concrete Durability: We Can Make Good Concrete Today 3-22
3.4 Good Reinforced Concrete in the Arabian Gulf 3-34
3.5 Chloride Attack of Reinforced Concrete: An Overview ... 3-53

CHAPTER 4: SPECIAL ASPECTS **4-1**
4.1 Cement and Concrete: Their Interaction in Practice 4-3
4.2 High-Performance Concrete Demystified 4-15
4.3 High Performance Concrete—An Overview 4-29
4.4 Aggregate Bond and Modulus of Elasticity 4-42
4.5 Integrated View of Shrinkage Deformation 4-54

CHAPTER 5: CONSTRUCTION **5-1**
5.1 Concrete Cover to Reinforcement or Cover-Up? 5-4
5.2 Specifying Concrete for Slipforming 5-17
5.3 Efflorescence—Surface Blemish or Internal Problem? Part 1: The Knowledge 5-23

5.4 Efflorescence—Surface Blemish or Internal Problem? Part 2: Situation in Practice 5-30
5.5 Autogenous Healing—A Concrete Miracle? 5-36

CHAPTER 6: TECHNOLOGY AND DESIGN 6-1
6.1 Concrete Technology—An Essential Element of Structural Design 6-5
6.2 What Everyone Who is In Concrete Should Know About Concrete 6-12
6.3 A Challenge to Concretors 6-22
6.4 Concrete Technology and Design—The Twin Supports of Structures 6-32
6.5 Creep of Concrete and Behavior of Structures: Part I: Problems 6-46
6.6 Creep of Concrete and Behavior of Structures: Part II: Dealing with Problems 6-59

CHAPTER 7: TESTING 7-1
7.1 Standard Test Methods: Avoid the Free-For-All 7-4
7.2 Core Tests: Easy to Perform, Not Easy to Interpret 7-26

CHAPTER 8: RESEARCH 8-1
8.1 Is Our Research Likely to Improve Concrete? 8-4
8.2 Suggestions of Research Areas Likely to Improve Concrete 8-9
8.3 Concrete Research on a Micro- and a Macro-Scale 8-23
8.4 Reviewing Publications: An Insider's View 8-32
8.5 Objectivity in Citing References 8-43

CHAPTER 9: LITIGATION 9-1
9.1 Litigation—A Growing Concrete Industry 9-4
9.2 An Expert Witness—As Reliable as Concrete? 9-14
9.3 An Expert Witness as Seen by an Expert Witness 9-24
9.4 The Influence of Litigation on Research 9-29

CHAPTER 10: EPILOGUE 10-1

APPENDIX: DETAILS OF ORIGINAL PUBLICATIONS 10-5

INDEX 10-9

Chapter 1

Cementitious Materials

Concrete is a composite material, the number of phases in the composite depending on the scale at which the observation is made. At the gross level of observation, we recognize the aggregate, large and small, and the hydrated cement paste. Clearly, the two phases have to be intimately linked, so that in reality there is, in addition, also an interface zone. This zone has a significant influence on the behavior of concrete and is considered in Sections 1.2 and 4.4.

Sometimes, it is argued that the main material is the aggregate, and that the hydrated cement paste is only the "glue" to hold the aggregate particles together. Other people take the view that the "real" construction material is the hydrated cement paste, and that the aggregate particles are there only as a cheap diluent. The second of these opinions has little validity because, even if cement were much cheaper than aggregate, we would be unwise to use neat cement as a construction material. The reasons for this lie in the harmful features of hydrated cement paste, such as shrinkage, creep, thermal movement, and vulnerability to physical and chemical attack. Some of these topics are discussed in Section 6.2.

What is relevant in the present chapter is that cement is essential to achieve what we call concrete. The cement has to be hydraulic cement, that is, a cement that sets and hardens under water by virtue of a chemical reaction with it. Because we cannot make concrete without cement, it is appropriate that Chapter 1 deals with cementitious materials.

The transition in terminology from "cement" to "cementitious materials" calls for an explanation. The word cement is used to denote

Chapter 1: Cementitious Materials

portland cement, which is the original hydraulic cement. For more than a century since the patent for "Portland cement" was taken out by Joseph Aspdin in England, concrete was made with that material alone.

With time, it was found that other extremely fine powders could be included in the mixture, in addition to portland cement, with beneficial effects. These materials, all inorganic in nature, include the following: ground-granulated blast-furnace slag; pozzolans of various kinds (natural and artificial); a particularly common pozzolan, fly ash (siliceous and high-lime); and silica fume. Even chemically inert, but physically active, fillers are included in these materials.

As these materials arrived on the concrete scene in a haphazard manner, their nomenclature was also haphazard. For example, fly ash is known in the United Kingdom as pulverized fuel ash. Even the generic name of these materials varies from the United States to Canada and to the United Kingdom. A proposal to introduce a rational categorization is made in Section 1.1, but I have no illusions about an immediate acceptance of my nomenclature. Nevertheless, the reader may find Section 1.1 useful in clearing the confusion.

ACI has espoused the term cementitious material in that the old water-cement ratio (w/c) is now water-cementitious material ratio (w/cm) regardless of whether the cementitious material contains anything in addition to portland cement. For most purposes, this is satisfactory but, on occasion, for example, to establish a very early strength, it may be useful to know the ratio of water to portland cement, that is, the old w/c. Thus, in my opinion, both w/cm and w/c are useful as appropriate. The concept of the water-cement ratio is specifically dealt with in Section 2.5. In any case, it may be worth pointing out that portland cement is an essential component of concrete. The only minor exception of a mixture without portland cement is blast-furnace slag with an activator used in masonry mortar and some other construction, but this is not concrete.

A recent arrival on the scene is silica fume. This material is more expensive on a unit mass basis than other cementitious materials, but it has important beneficial effects. Silica fume has been standardized in terms of its properties but there is no agreed approach to its categorization in specifications for concrete. I was confronted in a lawsuit by the consequences of the uncertain classification, and this has led to Section 1.3.

Although portland cement is by far the most common cement used in construction, there exists also high-alumina cement. This cement has a number of specialist applications, including its use in refractory concrete, but it is not a cement used in structural concrete, following problems and failures that occurred a quarter of a century ago. Its inclusion in the present book is due to some fairly recent attempts to resuscitate high-alumina cement as a cement for structural use, using a new name, calcium aluminate cement. Section 1.5 presents a current review of the applicability of high-alumina cement in construction, and is followed by letters to the

Editor sent by readers after the original publication of my article in *Concrete International*.

The moves to revive or reconsider the structural use of high-alumina cement can be by way of selection of exclusively positive references. This is discussed in Section 8.5. The moves can also be explicit.

My original article on high-alumina cement refers to a European standard, whose current status is pre-Standard prENV 197-10. I realize that European Standards are of limited interest in the United States but there is no ASTM standard on high-alumina cement, and construction is becoming ever more international. The pre-Standard still contains the long Annex, called "informative" about the use of high-alumina cement in construction works; construction is manifestly outside the scope of a standard titled "Cement—Composition, specifications and conformity criteria." And yet the Annex refers even to large beams, 350 by 1800 mm in cross section (14 in. by 6 ft).

What is even more surprising is a statement in the Annex that says: "If concrete is made in accordance with the principles given in the Annex, it does not imply any conformity with national or international codes for design." European Standards are mandatory in European countries and, to my way of thinking, they should not contain "principles" that are contrary to "national or international codes for design."

This chapter contains Section 1.4 on expansive cements. These cements, consisting of portland cement and an expansive agent, are useful in counteracting problems caused by drying shrinkage. Despite the long history of expansive cements, their use is very limited, and Section 1.4 attempts to revitalize that material.

Section 1.1: Cementitious Materials—A Different Viewpoint

The need to categorize the various cementitious and allied materials used in concrete has led to a certain classification. While it is convenient to have such a classification and a uniformity in nomenclature, I feel that the present system does not correspond to the actual usage of the various materials and is, in fact, inimical to their more widespread use. This section presents therefore a general discussion of the cementitious materials used in concrete but without describing the various types of portland cement.

Originally, concrete consisted of only three materials: cement, aggregate, and water; almost invariably, the cement was portland cement. Later on, in order to improve some of the properties of concrete, very small quantities of chemical products were added into the mixture. These are the chemical admixtures, often simply called admixtures.

Later on, other materials, inorganic in nature, were introduced into the concrete mixture. The original reasons for using these materials were usually economic: they were cheaper than portland cement, sometimes because they existed as natural deposits requiring little or no processing, sometimes because they were a by-product or waste from industrial processes. A further spur to the incorporation of these supplementary materials in the concrete mixture was given by the sharp increase in the cost of energy in the 1970s, and we know that the cost of energy represents a major proportion of the cost of the production of cement.

Further encouragement for the use of some of the supplementary materials was provided by the ecological concerns about opening of pits and quarries for the raw materials required for the production of portland cement on the one hand and, on the other, about the means of disposal of the industrial waste materials such as blast-furnace slag, fly ash, or silica fume. Moreover, the manufacture of portland cement in itself is ecologically harmful in that the production of one tonne of cement results in about one tonne of carbon dioxide being discharged into the atmosphere.

It would be incorrect to infer from the previous, historical account that the supplementary materials were introduced into concrete solely by the push of their availability. These materials also bestow various desirable properties on concrete, sometimes in the fresh state, but more often in the hardened state. This pull, combined with the push, has resulted in a situation such that, in very many countries, a high proportion of concrete contains one or more of these supplementary materials. It is therefore inappropriate to consider them, as was sometimes done in the past, as cement replacement materials or as extenders.

If, as just stated, the materials which we had hitherto described as supplementary are, in their own right, proper components of the cementitious materials used in making concrete, then a new terminology has to be sought. No single terminology has been agreed on or accepted on a worldwide basis, and it may be useful to briefly discuss the nomenclature used in various publications.

Insofar as concrete is concerned, the cementitious material always contains portland cement of the traditional variety, that is, pure portland cement; the exceptions are quantitatively negligible. Therefore, when other materials are also included, it is possible to refer to the ensemble of the cementitious materials used as *portland composite cements*. This is a logical term, as is *blended portland cements*. The new unified European approach of the Standard EN 197-1:2000 is to use the term *CEM cement*, which requires the presence of the portland cement component by implication (in that it excludes high alumina cement). In my view, the name CEM cement is not explicit or of general appeal.

The current American approach is given in ASTM Standard C 1157-00, which covers *hydraulic cements* for both general and special applications. A blended hydraulic cement is defined as follows: "A hydraulic cement consisting of two or more inorganic constituents which contribute to the strength-gaining properties of the cement, with or without other constituents, processing additions and functional additions."

This terminology is sound except that the term "inorganic constituent" is difficult to relate to the actual materials incorporated in concrete, typically natural or industrially produced pozzolans, fly ash, silica fume, or ground-granulated blast-furnace slag. Moreover, emphasis on the term "hydraulic" may conjure up a wrong image in the eyes of the general users of cement. Furthermore, the ASTM terminology is not used by ACI, even though both are American bodies concerned with concrete and both have a worldwide following.

The preceding, rather lengthy, discussion explains the difficulty of classifying and categorizing the different materials involved. The situation is not helped by a lack of international nomenclature. Indeed, more than one approach is possible but the difficulty is exacerbated by the fact that some of the divisions are not mutually exclusive.

In an effort to reflect the present-day usage of the various materials, I propose the following terminology:

- A cement consisting of portland cement with no more than five percent of another inorganic material will be referred to as *portland cement*. We should recall that prior to 1991, portland cements were generally expected to be pure, that is, not to contain minor additions other than gypsum or grinding aids.
- A cement consisting of portland cement and one or more appropriate inorganic materials will be called *blended cement*. This term is close to that used in ASTM Standard C 1157-00. Like ASTM, I use the term blended to include both the results of blending the separate powders and of intergrinding the parent materials, for example, portland cement clinker with ground-granulated blast-furnace slag.

Chapter 1: Cementitious Materials

Table 1.1.1—Cementitious nature of materials

Material	Cementitious nature
Portland cement clinker	fully cementitious (hydraulic)
Ground-granulated blast-furnace slag	latent hydraulic, sometimes hydraulic
Natural pozzolans (Class N)	latent hydraulic with portland cement
Low-lime (siliceous) fly ash (Class F)	latent hydraulic with portland cement
High-lime fly ash (Class C)	latent hydraulic with portland cement or hydraulic
Silica fume	latent hydraulic with portland cement but also physical in action
Calcareous filler	physical in action but with slight latent hydraulic action with portland cement
Other fillers	chemically inert; only physical in action

There is some difficulty in choosing the term for the components that make up a blended cement. The terms constituent and component run the risk of confusion with the chemical compounds in portland cement. What all the materials with which we are concerned have in common is that, in the words of ASTM C 1157-00, they "contribute to the strength-gaining properties of the cement." In actual fact, some of these materials are cementitious in themselves, some have latent cementitious properties, yet others contribute to the strength of concrete primarily through their physical behavior. I propose to refer to all these materials as *cementitious materials*. Purists might criticize this choice, but it has the important merits of simplicity and clarity.

Table 1.1.1 describes the relevant properties of the individual cementitious materials. It can be seen that there are no clear-cut divisions with respect to hydraulic, that is truly cementitious, properties.

All the cementitious materials as just defined have one property in common: they are at least as fine as the particles of portland cement, and sometimes much finer. Their other features, however, are diverse. This applies to their origin, chemical composition, and physical characteristics such as shape, surfaces texture, or specific gravity.

There are several ways of preparing a blended cement. One way is to intergrind the other cementitious materials with the portland cement clinker so that an integral blended cement is produced. The second way is for two or, more rarely, three materials in their final forms to be truly blended. Alternatively, portland cement and one or more cementitious materials can separately, but simultaneously or nearly so, be fed into the concrete mixer.

Furthermore, the relative amounts of portland cement and of the other cementitious materials in the concrete mixture vary widely: sometimes the proportion of the other cementitious materials is low, while in other mixtures it constitutes a major part of the blended cement. Thus, I propose that the term *cementitious material* be used for all the powder material, other than that which forms the finest particles of aggregate, provided that one of the powder materials is cement. With very few exceptions, the cement is portland cement. Thus, the cementitious material may be portland cement alone or it may comprise portland cement and one or more other cementitious materials.

A given cementitious material may be hydraulic in nature; that is, it may undergo hydration on its own and contribute to the strength of the concrete. Alternatively, it may have latent hydraulic properties; that is, it may exhibit hydraulic activity only in consequence of chemical reaction with some other compounds such as the products of hydration of portland cement which co-exists in the mixture. Yet a third possibility is for the cementitious material to be largely chemically inert but to have a catalytic effect on the hydration of other materials, for example, by fostering nucleation, or to have a physical effect on the properties of the fresh concrete. Materials in this category are called fillers.

I feel quite strongly that the term mineral admixtures used by ACI to describe the nonhydraulic supplementary materials should not be used. The word admixture conjures up a minor component, something added to the main mixture, and yet some of the so-called supplementary materials are present in large proportions and are not there just to supplement the alleged main constituent.

I realize that my proposals fly in the face of the received wisdom. They are not meant to represent an idiosyncratic view or just a personal whim. I believe that they represent the present-day usage of the various materials more closely than the other nomenclatures. The proposals also recognize that the various cementitious materials are important in their own right and that their use continues to grow. Acceptance of these proposals, perhaps only after a passage of time, would contribute to a better recognition of the position of the various cementitious materials and to their greater use in the future.

Section 1.2: There is More to Concrete than Cement

The difference between concrete and cement is well known, but we still sometimes hear of a "cement mixer" or a "cement floor." On the other hand, I have heard, in litigation, of Type V concrete. On occasion, I have seen papers with a title containing the word "concrete," but then this is followed by a description of a laboratory experiment on neat cement paste or, at best, mortar.

However, the purpose of this section is not to ridicule and correct such simple, albeit significant, mistakes. Rather, I want to comment on the importance of the various components of concrete in its performance in service. I am referring, in particular, to the situations when concrete is, or may be, subjected to aggressive agents. In the case of existing concrete in service, there may be a dispute about the properties of concrete being, or not being, appropriate for the particular conditions of exposure.

Cement and aggregate

Several points of view may be advanced. One of these is to regard the type of cement as being of primary importance; this may be argued when there is the possibility of alkali-aggregate reaction or sulfate attack. Another view is to concentrate on the composition of the cement paste, especially the water-cement ratio. Both these approaches ignore the fact that concrete is not just cement that has reacted with water, that is, hydrated cement paste.

At this point, the reader's reaction is likely to be, "Of course, there is also aggregate in the mixture." In reality, however, little attention is paid to aggregate, provided it conforms to ASTM C 33 and provided no alkali-aggregate reaction is suspected. This is probably a reasonable attitude given that, in most cases, local aggregate has to be used, as transport over long distances is uneconomical. So, I am not arguing for the use of "better" aggregate, whatever that may be.

Interface zone

My point is that concrete is not just cement paste plus aggregate. First of all, the two components of concrete have to act together so that there must be adequate bond between them.[1] But there is more to the contact between the cement paste and the aggregate: the interface zone. This is that part of the hydrated cement paste which is in contact with the surface of the aggregate particles. By this I mean not only the coarse aggregate but also the fine aggregate down to the smallest particle size.[2]

The reason for the existence of the interface zone is that the particles of aggregate are much larger than the typical particles of cement, whose median size is about 30 μm. When we imagine these cement particles covering the surface of the aggregate, it is easy to realize that there will be

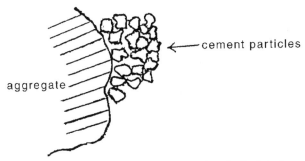

Fig. 1.2.1—Schematic representation of the wall effect in the interface zone

present tiny voids, with a resulting high porosity. This is the so-called wall effect: the "packing" of the particles is not as dense as further away from the surface of the aggregate where there are only cement particles next to one another (Fig. 1.2.1). The term "voids" means absence of solids as, in well-compacted concrete, the voids contain water.

The thickness of the zone having different packing is typically up to 50 μm around large aggregate particles in a given volume of concrete. When we think how large is the number of aggregate particles, we realize that the total volume of the interface zone occupies a large proportion of the volume of the hydrated cement paste. Given that the thickness of the cement paste between adjacent particles is between 80 and 150 μm, this proportion may be of the order of 30%.[3] Thus, taking a typical concrete mixture, whose volume consists of 75% aggregate of all sizes, the interface zone represents about 0.3×0.25, or 8% of the volume of concrete.

The importance of the presence of the interface zone is that its chemical and physical properties are different from the bulk cement paste. This article is not concerned with laboratory science but two facts should be noted. First, the interface zone contains calcium hydroxide—one of the main products of hydration of portland cement—in a greater proportion and in the form of larger crystals than is the case in the bulk cement paste. Second, the physical structure of the less well-packed interface zone, with its higher water-cement ratio, is probably conducive to easier passage of water, which may carry aggressive agents. There is still some uncertainty on this last point because the interface zone is not a continuous component of concrete whereas the bulk cement paste is continuous; indeed it is the only continuous component.

So, what I am trying to say is that, if an investigation of a possible, or suspected, inadequacy of concrete centers on the hardened cement paste, it is important to realize that this paste has two components: bulk paste and interface zone. In other words, concrete is more complicated than it seemed at the beginning of this section: we started with cement paste and aggregate but we now have two types of cement paste. Any investigation of the quality of the hydrated cement paste, usually referred to as its microstructure, should include consideration of the properties of the interface zone.

The preceding discussion dealt with concrete containing portland cement alone or with fly ash. There exist, however, cementitious materials much finer than portland cement, notably silica fume. Particles of silica fume can fit in between the cement particles and especially in the spaces at the surface of the aggregate. This is one way in which silica fume confers beneficial properties on concrete with respect to durability and strength.

If bleed water is trapped underneath coarse aggregate particles, it also represents a component of concrete. This influences the behavior of concrete, especially with respect to permeability. Because the bleed voids are induced by gravity, their orientation is anisotropic.

Influence of aggregate on interface zone

Without a scientific consideration, it is easy to think of several properties of aggregate that influence the interface zone, at least in qualitative terms. The grading, especially the maximum size of aggregate particles, influences the total surface area of aggregate upon which the interface zone is formed. The shape of the aggregate particles—rounded, angular, or flaky—also influences the total surface area, at a given grading. The texture of the aggregate particles is another factor because it influences the ease of packing of cement particles and the depth over which this packing is affected.

Quality of concrete

All this makes it clear that, while it is easy to make concrete, it is not easy to determine its properties in the case of a dispute about its quality. It is all too convenient for an interested party to focus on a component of the concrete to prove a point.

But there is even more to concrete than its components considered in a scientific way. What is of interest is not the mixture proportioned by a computer or made in the laboratory, but concrete in the actual structure. The factors affecting the quality of that concrete are not limited to mixture proportions, but they include also the quality of mixing, transport, placing, compaction, finishing, and curing. Each of these deserves careful consideration. In reality, this is not always the case. And yet, any one of these may be a true determinant of the quality of concrete and the life of the structure. This is what we are interested in in practice.

Conclusions

I started by pointing out that in assessing the quality of concrete, especially in disputes, the cement, albeit very important, is not the sole factor. In applying research on cement to concrete, and especially concrete in service, we should beware of the direct translation from one material to the other: concrete is not the same as cement, nor is it a "diluted" cement. Concrete is a material with three components: cement paste, aggregate, and the interface between the two. Aggregate deserves more attention than it is often given. Between the aggregate and the cement paste, there is a very important interface zone. Even this is not the end of the story: mixing and the various activities that follow right up to the finished product all have an important influence on the quality of the concrete in place. These operations may be low-tech but good concrete is not a low-tech material.

References

1. Neville, A. M., "Aggregate Bond and Modulus of Elasticity," *ACI Materials Journal*, V. 94. No. 1, Jan.-Feb. 1997, pp. 71-74 [Section 4.4].

2. Monteiro, P. J. M.; Maso, J. C.; and Ollivier, J. P., "The Aggregate Mortar Interface," *Cement and Concrete Research*, V. 15, No. 6, 1985, pp. 953-958.

3. Larbi, J. A., "Microstructure of the Interface Zone Around Aggregate Particles in Concrete," *Heron*, V. 38, No. 1, 1993, 69 pp.

Section 1.3: Silica Fume in a Concrete Specification: Prescribed, Permitted, or Omitted?

The genesis of this article is the difficulty sometimes encountered when interpreting a specification for concrete, particularly the meaning of terms used and the intent of the specification itself. This section is not about standards or specifications for silica fume; nor is it, except marginally, about the fundamental benefits of the incorporation of silica fume in a concrete mixture. Rather, I intend to discuss the treatment of silica fume in a specification that seeks to achieve the production of concrete with exacting properties, but without losing sight of the principle of economy in design. My article has been prompted by a case in which I was involved: it was alleged that silica fume was not properly dealt with in the concrete specification, and this resulted in a multi-million-dollar claim. Arising from this, I am trying to proffer some ideas on what to do and what not to do.

At this stage, it may be worth pointing out that, in general, the designers' aim is not to produce the best structure possible. Their task is to design a structure that fulfills all the owner's requirements at an economic cost. To give an example: if a farmer wishes to have a bridge that will allow a small herd of cattle to cross an obstacle, the engineer should provide a bridge wide enough, strong enough, and rigid enough to carry the cows in a single file, not a bridge capable of carrying a tank transporter. This approach is well-known by experienced practitioners but rarely explained clearly enough to students.

To underline my point, let me quote an epigraph in a book by a well-known novelist and story writer who was an aeronautical engineer, Nevil Shute (a more readable Neville), called *Trustee from the Toolroom*: "An engineer is a man who can do for five bob (two bits) what any bloody fool can do for a quid (a dollar)." This is not the only distinction between an engineer and a fool!

Concrete specifications

Advice on specifying concrete that will contain silica fume is given in ACI 234, which says in Section 8.4: "Silica-fume concrete has usually been specified as a separate section ... of a project specification."[1] This, of course, is predicated on the assumption that the designer "prescribes" the inclusion of silica fume in the mixture. However, in recent years, we have been increasingly using performance-type specifications that do not prescribe the mixture ingredients and proportions in detail, unlike in the past. Nevertheless, usually the end-product properties do not stand alone but are supplemented by other clauses or sections. These may prescribe the use of specific materials and methods, or list those from which the contractors may choose, either as they wish or only with the engineer's permission. My particular concern is with silica fume and also with the form in which it must, or can, be used. Thus, for concrete in a given structure, it may be desirable to use silica fume in the mixture, but it may not be absolutely necessary to do so in order to achieve concrete that has the specified properties. How does a specification writer go about it?

Regardless of the type of specification used, it is important that it reflect the state of the art with respect to concrete ingredients, mixing, transporting, and placing practices. As Jo Coke, then President of ACI, quotes in her memo in the September 1999 issue of *Concrete International*: "All too often ... we are asked to produce concrete for that state-of-the-art structure; then comes paragraph two and our hands are tied due to unrealistic or, as often is the case, outdated specifications."

Now, silica fume is one of the state-of-the-art materials, and this brings me to the theme of this section: how to treat the silica fume in the concrete specification. I assume, of course, that we are dealing with concrete in which the presence of silica fume "may" be beneficial. I am clearly not looking at a case where silica fume is essential; if it is, then it should be mandated, and that is the end of the story. However, many other cases exist where silica fume may be beneficial, but is not essential, because there exist other means of achieving a satisfactory concrete mixture.

The second genesis of this section is the confusion in the terminology applied to what may be called "active fine powders" in concrete. Section 1.1 presents an article on this topic,[2] exhorting a consistent use of terms to describe portland cement, pozzolan, fly ash, ground-granulated blast-furnace slag (or slag, for short), silica fume, metakaolin, processed rice husks, and fillers. I do not propose to repeat my ideas, but the fact is that a variety of terms continue to be used to describe all these materials, or even any one of them.

What is silica fume?

Silica fume is a by-product of the manufacture of silicon and ferrosilicon alloys from high-purity quartz and coal in a submerged-arc electric furnace. This amorphous, and therefore highly reactive, silica is in the form of extremely fine spherical particles. The reactivity with calcium hydroxide means that silica fume satisfies the definition of pozzolan, and a highly reactive one at that.

However, the fineness of the silica fume particles (about 40 times finer than portland cement) enables them to fit in between the cement particles and especially in the voids at the surface of the aggregate particles where the cement particles cannot fully cover the surface of the aggregate and fill all the space available. This is the so-called interface zone, whose properties influence some properties of concrete[3] (see Section 1.2).

Silica fume is a relatively recent arrival among the ingredients of concrete. Originally, it was just a waste material, extremely difficult to dispose of. In the 1970s, it began to be used as a pozzolan, but in the next decade the great value of its physical properties, that is, its fineness and shape, became appreciated. It is only fair to add that, like many good things in life and like some important drugs, it is expensive.

The combination of high reactivity and extreme fineness possessed by silica fume results in the possibility of producing more dense concrete with a very low porosity, the pores being small and discontinuous, and therefore with a high strength and a low penetrability by liquids. To quote Young[4]: "The use of silica fume to make strong, durable concretes is now a well-established technology around the world." He refers to the "less well known ... ability of the silica fume particles (0.1 to 0.3 mm) to pack between the larger cement particles, as well as undergo a rapid pozzolanic reaction."[4]

Standards for silica fume

Despite its benefits, standards for silica fume have been slow to be produced. As far as I can find out, the earliest American standard specification for silica fume was that produced by AASHTO under the designation M 307, published in 1990; curiously enough for an American body, the term "silica fume" is not used, the title being "Microsilica for Use in Concrete and Mortar." The ASTM Standard Specification C 1240 first appeared in 1993; the current edition is dated 1998. The first ACI Guide for the Use of Silica Fume in Concrete[1] was published in 1996, after 9 years of committee work and almost as many draft reports, the first of which was published in the ACI *Materials Journal* in March 1987. In 1976, the use of silica fume in blended cement was included in a Norwegian standard. Probably the first specification for silica fume in an English-speaking country (as well as French-speaking) was published in Canada in July 1986 as part of CAN/CSA A23.5-M86, titled "Supplementary Cementing Materials."

There is no British standard covering silica fume, but only a European Standard EN 13263-2001 titled "Silica Fume for Concrete—Definitions, Requirements and Conformity Control." This document may, in the fullness of time, become a British standard as well as a standard in other European countries. It is worth noting that in Europe, in addition to standards and codes, there exists another way of introducing a product or method into the marketplace, namely by way of certificates based on testing under supervision of a body called the Board of Agrément. The first such certificate for silica fume from a single producer was issued in the U.K. in November 1985.

What is relevant to the theme of this section is that silica fume is a highly valuable ingredient of concrete for some purposes, but the standards and guides are only recent; moreover, they differ greatly in the terminology they use to describe the role of silica fume in concrete.

I would like to point out that the ingredients of concrete do not necessarily have to comply with existing established standards. Always to do so would hinder progress and development, but, clearly, the use of any "nonstandard" materials should be only with the approval of the design engineer or the design engineer's representative on site.

This approach may raise the eyebrows of people who always like to follow a conventional path. My answer to such a point of view is as follows, although I realize that it might come as a bit of a shock to young engineers who may view codes as a source of all wisdom. Standards and codes are not, and are not meant to be, leaders in introducing new materials or new design methods. It is generally not possible to prepare a standard until a body of experience with the given material or method has been accumulated, and until there are a number of people with relevant experience who are able to draft the standard. To put it another way, it is not enough for a product manufacturer to say to the standard-producing organization: here is my new product; please prepare a standard for it so that the material can be used.

Indeed, silica fume was first used in the U.S. long before the ASTM standard appeared and even longer before the ACI Guide. The latter informs us that, as reported by Holland et al., "the first publicly-bid project using silica-fume concrete was done in late 1983."[1]

How to categorize silica fume

The standards and specifications discussed above fit the silica fume into different categories. For example, the Canadian specification includes silica fume in "supplementary cementing materials." AASHTO calls it a "mineral admixture." ACI Guide 234[1] describes silica fume as "a concrete property-enhancing material" and as "a partial replacement for portland cement." ASTM Standard Specification C 1240 refers to "silica fume for use as a mineral admixture." The British Agrément certificate describes silica fume as "a mineral addition to enhance the properties and applications of concretes," but it also says that "it has been assessed as a cement replacement."

For what it is worth, my view is that silica fume is not an admixture. It certainly is not a chemical admixture because it has no formulated chemical composition; nor does it have a special chemical action distinct from pozzolans. I do not view silica fume as a mineral admixture on a par with fly ash or slag because an important function of silica fume is, in the words of ACI 234, to act as "a concrete property-enhancing material."[1] Alas, there is no such separate category of mixture ingredients; clearly, mineral admixtures are also beneficial, but they are separately categorized. Some other rather unspecific terms, such as "additives" and "additions," are occasionally used; the latter term is used in the European Standard. The term "supplementary" material has, to my ear, an almost deprecating sound about it, such as supplementing income by doing odd jobs.

Some people have recently begun to use the term "binder" as an umbrella word for what I call cementitious materials. My objection is that this term is very broad. Binders used in construction include hydraulic lime and, indeed, polymers and glues; however, these materials are not ingredients of concrete. Some polymers are, but they are a class apart. What

I call cementitious materials always require the use of portland cement as an essential ingredient. They all belong to the class of hydraulic cements. For the sake of completeness, I should mention that silica fume is unlikely to be useful in concrete made with high-alumina cement because it does not produce calcium hydroxide that could react with silica fume. High-alumina cement, although a hydraulic cement, is not a portland cement and is not a material used in structural concrete in most countries.

Even the name of "silica fume" is not universally agreed upon or used. For example, in the United Kingdom, it is known as "microsilica." However, the term silica fume is the most common one and, to avoid confusion, we should stick with it.

The purpose of the preceding discussion was not just to encourage a clear and consistent terminology, but to ensure an unambiguous interpretation of concrete specifications. In many parts of the world, construction involves a designer from one country and a contractor that may be a consortium from two other countries. What is customary in one country, and therefore in the mind of the party from it, may be misinterpreted by parties from elsewhere. Moreover, what is customary in the country of the specification writer may be so obvious to that writer that it is not even given in writing; how is the other party to realize this?

Specifying silica fume

The lack of uniform categorization of silica fume may well lead to problems, especially when silica fume is not explicitly mentioned in the specification. As I said earlier, I was involved in a case of a very major construction project where silica fume was not mentioned by name in the concrete specification. In the section on cements, there was a requirement that the cement comply with one of the listed standards, none of which included silica fume. In the section on admixtures, there was a like list of standards; there was also a provision for the use of other admixtures "with the approval of the engineer." However, the term "admixture" might not be considered to include silica fume, at least in some countries. In any case, at the pre-tender stage, the contractor might be uncertain whether silica fume would be considered to be an admixture, although the contractor could, or perhaps should, seek clarification from the designer.

A modern concrete specification is likely to contain various performance requirements with respect to durability. Taking all these into account, contractors may envisage the use of silica fume and its price, or they may not do so. Their view may not be the same as the owner's view. Bearing in mind that a concrete specification forms part of the contract between the owner and the contractor, ambiguity must be avoided at all cost; otherwise, there may be a high cost of resolving the problem. So my plea is for clarity in the concrete specification with respect to all materials and methods, not just silica fume.

In my opinion, if the quality of the concrete required is such that the inclusion of silica fume is likely to be necessary, or if it may be preferred by the contractor, then an inclusion of an appropriate reference to it in the specification is desirable. There are several possibilities.

One possibility is to put an explicit clause or section on the lines: "the use of silica fume is permitted," followed by a reference to a standard with which the silica fume must comply; examples of such standards are ASTM C 1240, AASHTO M 307, Canadian CAN/CSA A23.5-M86, or even the British Agrément certificate. The last has a possible drawback of referring only to a single producer of silica fume.

It is arguable that such a separate mention of silica fume may give the impression of favoring the inclusion of silica fume in the mixture. Such presumed bias would be avoided by including the reference to silica fume in the clause or section dealing with cementitious materials. A possible wording would be that any one of these materials must comply with, say, ASTM C 150 (portland cement), or C 595 (blended hydraulic cements), or C 989 (slag), or C 1240 (silica fume). The list may include standards other than ASTM, for example AASHTO M 307.

Both the above approaches, that is, dealing with silica fume in a separate clause or including it under the umbrella of cementitious materials, provide for the situation where the use of silica fume is to be "permitted." When the use is to be "prescribed," the approach is simple: the specification says that silica fume shall be used; this may be followed by clauses dealing with the proportion of silica fume in the total cementitious material. Careful wording is necessary with respect to what is meant by cementitious material; for example, does its mass include the silica fume?

In my opinion, silica fume should be included in the total cementitious material whose mass is counted for the purpose of establishing the "cement" content of the concrete and its water-cement ratio $(w/c)^5$ (see Section 2.5). If the specification contains a clause laying down a minimum or maximum content of cementitious material per cubic meter (or cubic yard) of concrete, it has to be remembered that the specific gravity of silica fume is about 2.20, compared with 3.15 for portland cement.

It is also essential to state whether silica fume can be used with other materials such as slag or fly ash and, if so, in what proportions. Again, a clear statement with respect to the method of calculating the "cement" content and w/c or the water-cementitious materials ratio (w/cm) is necessary.

The specification should contain some statement about the form of the silica fume: as-produced; densified; in the form of micropellets; or in a slurry form, possibly also indicating the concentration in the slurry. Not all forms need be permitted. It may also be prudent to make an explicit statement about the use of portland cement blended with a fixed proportion of silica fume. Furthermore, at least in the United Kingdom, there exist ready-mixed concrete suppliers of concrete containing silica fume whose amount is not disclosed; the cement content may also not be stated. It may be wise to consider, either positively or negatively, the possibility of using concrete from such a source.

When silica fume is used, it is important that it be dispersed and uniformly distributed in the mixture. This may require a longer mixing time than usual (especially when the silica fume is not in a slurry form), and the specification should make an appropriate reference to this. A longer mixing time means a lower output of the mixer per hour, and this costs money. The contractor should be clear at the tender stage about these financial implications.

If it is intended that silica fume is not to be used in the concrete, it is best to say so in the specification in an explicit manner. Simply remaining silent on the topic may unwittingly mislead contractors into thinking that they may choose to use silica fume, even if only with the engineer's approval. It is best if the contractors are clear about the situation at the tender stage.

Conclusions

This section suggests approaches to clear wording in a specification with respect to the use of silica fume for all possible situations: when silica fume is prescribed; when it is permitted; or when its use is forbidden.

If none of these approaches is used, then there is a risk of a misunderstanding, as occurred on the recent very major construction project mentioned earlier. Because a misunderstanding is likely to have financial implications (particularly where an expensive material such as silica fume is involved), it may turn into a dispute. And a dispute may lead to litigation. And litigation is "a growing concrete industry,"[6] but it is an industry we can do without (see Section 9.1). Prevention is better than cure; hence the recommendations in this section.

Finally, I would like to offer a speculative generalization. This section is about one material and its use where knowledge is still moving from the domain of research to general application. But in many countries, this knowledge is not yet widespread enough among designers and materials technologists to produce clear specifications and unambiguous terminology. The problems with specifications for the use of silica fume could be repeated in other areas of new knowledge and technology.

We live in a world of proliferation of new materials and methods; while we should not stifle their use, we have to beware of an excessive "open door" policy that would allow an uncontrolled introduction of different materials by individual contractors. Thus, there is a general need for carefully worded specifications; it is not just silica fume that may be at issue.

References

1. ACI Committee 234, "Guide for the Use of Silica Fume in Concrete (ACI 234R-96)," American Concrete Institute, Farmington Hills, Mich., 1996, 51 pp.

2. Neville, A. M., "Cementitious Materials — A Different Viewpoint,"

Concrete International, V. 16, No. 7, July 1994, pp. 32-33 [Section 1.1].

3. Neville, A. M., "There is More to Concrete than Cement," *Concrete International*, V. 22, No. 1, Jan. 2000, pp. 73-74 [Section 1.2].

4. Young, J. F., "Densified Cement Pastes and Mortars," *Bulletin of Center for Advanced Cement-Based Materials*, V. 9, No. 2, 1997, pp. 4-5.

5. Neville, A. M., "How Useful is the Water-Cement Ratio?," *Concrete International*, V. 21, No. 9, Sept. 1999, pp. 69-70 [Section 2.5].

6. Neville, A. M., "Litigation—A Growing Concrete Industry," *Concrete International*, V. 22, No. 3, Mar. 2000, pp. 64-66 [Section 9.1].

Section 1.4: Whither Expansive Cement?

Whenever a new material or technique is developed, enthusiasts expect it to displace the existing practice. This was the case when prestressed concrete first seriously entered the practical arena. Likewise, when expansive cements were first developed half a century ago, it was thought that, because their use would obviate the problems of shrinkage cracking, the benefits would lead to their widespread use.

This has not come to pass. At most, one percent of concrete structures are made with expansive cement.[1] In view of this development, or lack of development, it may be of interest to take stock of the situation.

Shrinkage compensation by the use of expansive cements

Shrinkage is probably one of the least desirable properties of concrete. When shrinkage is restrained, it may lead to shrinkage cracking. This mars the appearance of concrete and makes it more vulnerable to attack by external agents, thus adversely affecting durability. But even unrestrained shrinkage is harmful; adjacent concrete elements shrink away from one another, thus opening "external cracks." Shrinkage is also responsible for a part of the loss of the initial stress in the tendons in prestressed concrete.

It is not surprising, therefore, that many attempts have been made to develop a cement which, on hydration, would counteract the deformation induced by shrinkage. In special cases, even a net expansion of concrete on hardening may be advantageous. Concrete containing such an expansive cement expands in the first few days of its life, and a form of prestress is obtained by restraining this expansion with steel reinforcement: steel is put in tension and concrete in compression. Restraint by external means is also possible. Such concrete is known as shrinkage-compensating concrete.

It is also possible to use expansive cement in order to produce self-stressing concrete in which the restrained expansion, remaining after most of the shrinkage has occurred, is high enough to induce a significant compressive stress in concrete (up to about 7 MPa [1000 psi]).[2] Expansive cement, although considerably more expensive than portland cement, is valuable in concrete structures in which a reduction in cracking is of importance, for instance, bridge decks, pavement slabs, and liquid storage tanks.

It is worth making it clear that the use of expansive cement does not prevent the development of shrinkage. What happens is that the restrained early expansion approximately balances the subsequent, normal shrinkage; this is shown in Fig. 1.4.1. Usually, a small residual expansion is aimed at because as long as some compressive stress in concrete is retained, shrinkage cracking will not develop.

Types of expansive cement

Early development of expansive cement took place in Russia and in

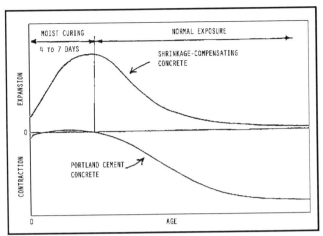

Fig. 1.4.1—Diagrammatic representation of length changes of shrinkage-compensating and portland cement concretes (from ACI 223R-90)

France, where Lossier used a mixture of portland cement, an expanding agent, and a stabilizer.[3] The expanding agent was obtained by burning a mixture of gypsum, bauxite, and chalk, which form calcium sulfate and calcium aluminate (mainly C_5A_3). In the presence of water, these compounds react to form calcium sulfoaluminate hydrate (ettringite), with an accompanying expansion of the cement paste. The stabilizer, which is blast-furnace slag, slowly takes up the excess calcium sulfate and brings expansion to an end.

Nowadays, three main types of expansive cement are produced, but only one, Type K, is commercially available in the United States. ASTM C 845-96 classifies expansive cements, collectively referred to as Type E-1, according to the expansive agent used with portland cement and calcium sulfate. In each case, the agent is a source of reactive aluminate which combines with the sulfates in the portland cement to form expansive ettringite. For instance, in Type K cement, the reaction is:

$$4CaO.3Al_2O_3.SO_3 + 8[CaO.SO_3.2H_2O] + 6CaO.H_2O + 74H_2O$$
$$\rightarrow 3[3CaO.Al_2O_3.3CaSO_4.32H_2O]$$

The resulting compound is known as ettringite.

Calcium sulfate reacts rapidly with $4CaO.3Al_2O_3.SO_3$ because it exists in a separate form,[4] unlike C_3A which is part of the portland cement clinker. While the formation of ettringite in mature concrete may be harmful, a controlled formation of ettringite in the early days after placing of concrete is used to achieve a shrinkage-compensating effect.

The three types of expansive cement recognized by ACI 223-98[5] and by ASTM C 845-96 are:

> Type K, which contains $4CaO.3Al_2O_3.SO_3$ and uncombined CaO,
> Type M, which contains calcium aluminates CA and $C_{12}A_7$, and
> Type S, which contains C_3A in excess of the amount normally present in portland cement.

In addition, in Japan, there is produced an expansive cement which uses specially processed calcium oxide to produce free-lime expansion.[6] This cement is called Type O.

Type K cement is produced by integral burning of the components or by intergrinding. It is also possible, as is done in Japan, to add the expansive component at the concrete batching plant.[6]

Shrinkage-compensating concrete

The expansion of cement paste resulting from the formation of ettringite begins as soon as water has been added to the mixture, but only *restrained* expansion is beneficial, and no restraint is offered while concrete is in the plastic state or while it has negligible strength. For this reason, prolonged mixing and delay before placing concrete containing expansive cement should be avoided.[7]

On the other hand, delayed expansion in concrete in service may prove disruptive, as is the case with external sulfate attack. It is therefore important that ettringite formation ceases after several days, and this happens when either SO_3 or Al_2O_3 has become exhausted.

ASTM C 845-96 prescribes a 7-day expansion of mortar of between 400 and 1000×10^{-6}: the 28-day expansion must not exceed the 7-day expansion by more than 15%. The latter value is a check on delayed expansion.

Because the formation of ettringite requires a large amount of water, wet curing of concrete made with expansive cement is necessary for full benefits of the use of such cement to be reaped.[8]

Information on the use of expansive cements so as to obtain shrinkage-compensating concrete is given in ACI 223-98[5] but some features of this type of concrete merit mention here. Its water requirement is about 15% higher than when portland cement only is used. However, as some of this additional water becomes combined very early, the strength of concrete is little affected.[5] Another way of representing the situation is to say that, at the same water-cement ratio, concrete made with Type K expansive cement has a 28-day compressive strength some 25% higher than concrete made with portland cement only.[4, 9] At a given water content, the workability of expansive cement concrete is lower and the slump loss is greater.[7]

The usual admixtures can be used in shrinkage-compensating concrete but trial mixtures are necessary because some admixtures, especially air-entraining ones, may not be compatible with certain expansive cements.[4, 7]

Because expansive cement has a large content of calcium sulfate, which is softer than portland cement clinker, the cement has a high specific surface, typically 430 kg/m². Excessive fineness, by promoting rapid hydration, may lead to premature expansion,[5] which is ineffective because very young concrete is unable to offer restraint. The expansion is greater the higher the cement content of the concrete and the higher the modulus of elasticity of the aggregate because the aggregate offers restraint to the expansion of the cement paste.[2] ASTM 878-95a prescribes a test method for restrained expansion of shrinkage-compensating concrete. This test can be used to study the effects of various factors on expansion.

Silica fume can be incorporated into shrinkage-compensating concrete in order to control excessive expansion.[1] Tests on Type K cement paste[10] have shown that silica fume in the mixture accelerates expansion, but the expansion stops before $CaO.3Al_2O_3.SO_3$ has been used up, probably due to a lowering of the pH. The absence of long-term expansion is desirable, and shortening the wet-curing period to four days is convenient.

If, following the expansive reactions, the cement is undersulfated, the concrete is vulnerable to sulfate attack. This may be the case with Type M and Type S cements.[9]

The future?

Finally, it can be noted that expansive cement is about 20% more expensive than portland cement; but the cost of preventing shrinkage cracking, dealing with it, or repairing it outweighs the initial outlay. The early problems of erratic behavior of expansive cements have long since been resolved. Admittedly, there is a need for a strict control of concrete production and, above all, of reliable wet curing, but surely quality concrete production is not unduly onerous and is highly desirable whatever the cement used. So what are the reasons for *not* using expansive cements in appropriate structures? If it is more than inertia and unwillingness to change, ACI readers might like to write and tell the membership at large.

References

1. Cohen, M. D.; Olek, J.; and Mather, B., "Silica Fume Improves Expansive Cement Concrete," *Concrete International*, V. 13, No. 3, Mar. 1991, pp. 31-37.

2. Polivka, M., "Factors Influencing Expansion of Expansive Cement Concretes," SP-38, *Klein Symposium on Expansive Cement Concretes*, American Concrete Institute, Farmington Hills, MI, 1973, pp. 239-250.

3. Lossier, H., "Cements with Controlled Expansions and Their Applications to Prestressed Concrete," *The Structural Engineer*, V. 24, No. 10, 1946, pp. 505-534.

4. Hoff, G. C., and Mather, K., "A Look at Type K Shrinkage-Compensating Cement Production and Specifications," SP-64, *Cedric Willson Symposium on Expansive Cement*, American Concrete Institute, Farmington Hills, MI, 1977, pp. 153-180.

5. ACI 223-98, "Standard Practice for the Use of Shrinkage-Compensating Concrete," *ACI Manual of Concrete Practice*, Part 1, 1996, 26 pp.

6. Kokubu, M., "Use of Expansive Components for Concrete in Japan," SP-38, *Klein Symposium on Expansive Cement Concretes*, American Concrete Institute, Farmington Hills, MI, 1973, pp. 353-378.

7. Cusick, R.W., and Kesler, C.E., "Behavior of Shrinkage-Compensating Concretes Suitable for Use in Bridge Decks," SP-64, *Cedric Willson Symposium on Expansive Cement*, American Concrete Institute, Farmington Hills, MI, 1977, pp. 293-310.

8. Mather, B., "Curing of Concrete," SP-104, *Tuthill Symposium on Concrete and Concrete Construction*, American Concrete Institute, Farmington Hills, MI, 1987, pp. 145-159.

9. Polivka, M., and Willson, C., "Properties of Shrinkage-Compensating Concretes," SP-38, *Klein Symposium on Expansive Cement Concretes*, American Concrete Institute, Farmington Hills, MI, 1973, pp. 227-237.

10. Lobo, C., and Cohen, M.D., "Hydration of Type K Expansive Cement Paste and the Effect of Silica Fume: II. Pore Solution Analysis and Proposed Hydration Mechanism," *Cement and Concrete Research*, V. 23, No. 1, 1993, pp. 104-114.

Section 1.5: A "New" Look at High-Alumina Cement

This section comprises an article that is a sequel to one published in the July 1998 issue of *Concrete International*,[1] whose theme was that concrete technology is an essential element of structural design and that decoupling what is perceived to be two separate disciplines can lead to unsatisfactory design (see Section 6.1). An investigation of the deterioration of a no-fines concrete slab led to the comment: "The case serves as a useful example of the importance of proper understanding of materials technology in civil engineering construction."[2] Another example, involving a large number of structures made of concrete containing high-alumina cement, is the subject of the present article.

But there is a more specific reason for writing about high-alumina cement just now, and that is the two recent attempts at restoring the structural use of this cement. Before discussing the merits or demerits of a renewed structural use of high-alumina cement, I would like to comment on the time interval since high-alumina cement was last used in structures in the United Kingdom and in most other countries in Europe. Such use was never widespread in the United States, but there, too, exist some important and sensitive structures built in the 1970s and before, containing high-alumina cement.

Well, the time interval is about a quarter of a century, and this is a significant length of time. The science weekly *Nature*, in the issue dated August 15, 1991, stated: "Publications more than 25 years old are likely to be forgotten, which is a shameful waste." In the case of high-alumina cement, the waste may indeed be shameful, if the new generation of engineers who have no knowledge of design and construction more than 25 years old were to be seduced by the "new" high-alumina cement presented as a satisfactory structural material. The word "new" is appropriate because, although the properties of the cement are exactly the same as before, it has been rebaptized "calcium aluminate cement." If the change of name is not governed by marketing but by chemical precision, then portland cement should be renamed calcium silicate cement. In both cases, I intend to stick to the traditional names.

Author's credentials

Why is it I who is writing on this subject? Because my qualifications to do so date back 40 years, I should restate them very briefly. Through an investigation of the deterioration of a floor in a dairy where milk churns were being washed with very hot water, I became aware of the adverse effects of high temperature on the strength of high-alumina cement concrete. What happens is that temperature accelerates the natural and inevitable conversion of calcium aluminate hydrates (the main product of hydration of high-alumina cement) from the unstable hexagonal crystal form to cubic crystals; hence the term, conversion of high-alumina cement.

The latter crystals have a higher density so that their formation results in an increase in the porosity of the hardened cement paste by a factor of about 2. In broad terms, the higher the porosity, the lower the strength. In addition, converted high-alumina cement is much more vulnerable to chemical attack. A particularly dangerous type of attack is that by alkalies derived from portland cement if water can travel through the latter into adjacent high-alumina cement.

My studies included laboratory tests on the effect of warm storage, continuous or intermittent,[3] and on the effects of humidity. My work culminated in a major paper. Despite pecuniary pressure, this paper was published in the *Proceedings*, Journal of the Institution of Civil Engineers.[4] The paper, in addition to further laboratory studies, contained an extensive review of the behavior of structures made with high-alumina cement concrete in the United Kingdom, France, Germany, Iraq, Iran, New Zealand, and Austria.

An argument against high-alumina cement

My conclusions, as I saw it, were cautious and moderate. I said:[4]

High-alumina cement concrete with a water-cement ratio (w/c) higher than about 0.5 will lose a considerable proportion of its strength under ordinary conditions existing outdoors in Europe, including England, over a period of 20 to 30 years. Many structures are expected to have a life in excess of this period. The loss of strength is accelerated by a rise in temperature and humidity, and the cement should not be used in buildings where a high humidity and considerable warmth, say in excess of 25°C (77°F) are expected. Such conditions may sometimes not be anticipated at the time of construction but may occur at a later date. For example, when an industrial building is put to a different use from that for which it was originally designed, this may require a temperature of, say, 30°C (86°F), or the use of hot water or spray over the floors or walls. While the structural strength, as far as the superimposed loads are concerned, would normally be checked, the new user may not even be aware of the fact that high-alumina cement has been used in construction. Even short periods of higher temperature, especially coupled with a rise in humidity, are detrimental because their effect is cumulative and irreparable. This danger of loss of strength at some time in the future is an argument against the use of high-alumina cement.[4]

I also wrote that the use of high-alumina cement in prestressed concrete is not recommended.

The written discussion[5] of my paper showed the great irritation of the producers of high-alumina cement (three from one company) and of seven manufacturers of precast concrete units made with high-alumina cement. Even chauvinistic criticism of Germans (who had long since forsaken high-alumina cement) was invoked, it being said: "That the main offenders were German contractors and manufacturers who, with their proverbial precision, applied religiously to high-alumina cement concrete the provisions of various DINs covering the use of portland cement, strength alone being the basis for comparison."

Collapses

In 1963, I moved to Canada and, for the next 10 years, my interest turned away from high-alumina cement. The cement manufacturers persisted in marketing high-alumina cement for structural use. As late as July 1973, the data sheet of Lafarge Aluminous Cement Co. Ltd. said: "If concrete is cured with water the concrete is cooled and the concrete will not convert unless the section is very thick." Ironically, on June 12 of the same year, the roof over the reading room in the University of Leicester, England, collapsed. By a strange coincidence, one day later, the roof over the assembly hall in the Camden School in London, England, collapsed. On February 8, 1974, the roof over a swimming pool in Stepney School in London, England, collapsed.

These collapses were extensively investigated.[6] In addition to the reduced strength of the high-alumina cement concrete, the investigators also identified some flaws in the original design and construction. One of these was the inadequate bearing area of the roof beams. I mention this because those proposing a revival in the structural use of high-alumina cement emphasize these contributory causes of collapse, and they may be missing a vital point.

The roofs that collapsed were 9, 10, and 19 years old, and there had been no change in the applied load. Thus, as long as the concrete in the roof beams had a constant strength, the bearing, though unsatisfactory, continued to support the beams. What changed was not the bearing area but the strength of the concrete, which was significantly reduced. I am making this point because it illustrates admirably the contention in my earlier article that the designer must be familiar with concrete as a material[1] (see Section 6.1). The complementary requirement is that the materials specialist, attached to a cement plant or a precast concrete factory, must understand structural action and behavior.

Reintroducing high-alumina cement for structural use

The first of the two proposals is a European Pre-Standard "Cement—Composition, Specifications and Conformity Criteria—Part 10: Calcium Aluminate Cement."[7] The old name, high-alumina cement, is not used probably because of its adverse associations. It is perfectly sensible to produce a European standard on high-alumina cement because it is a cement with many uses.

However, there is a nine-page Annex to the Pre-Standard, entitled "Essential Principles for use of Calcium Aluminate Cement in Construction

Works." The words "construction works" clearly suggest the use of high-alumina cement in structures. In the Annex, there are sections on protection of reinforcement, on curing, on concreting in cold and hot weather.

These topics, and indeed the whole content of the Annex, are appropriate for a building code or for a structural design code but, in my opinion, the Annex is outside the remit of a Standard for cement. The Annex, almost as long as the main body of the Pre-Standard, looks rather like a way of insinuating the structural use of high-alumina cement by cement specialists, as opposed to structural engineers. Maybe the cement specialists are trespassing on the structural domain because the structural engineers showed themselves in the past to be inadequately knowledgeable about cement.

Whatever the reasons for including the Annex, it is not objective in that neither in the text nor the bibliography are there any references to past failures of concrete structures built with high-alumina cement. It is as if there had never been any serious structural or durability problems.

The second "revival" document is a report of the (British) Concrete Society Working Party,[8] which concludes "that a safe basis for the use of high-alumina cement, including in structures, is a reasonable and desirable aim, to take advantage of the beneficial properties of the cement in applications where other materials may have disadvantages." The report refers to the recommendations of the European Pre-Standard[7] and specifically to the maximum total w/c of 0.4, which follows the approach in France. This reference stems from the fact that the French manufacturer of high-alumina cement is a member of the Concrete Society Working Party, as well as of the committee that drafted the European Pre-Standard.

To illustrate my point about the dichotomy between materials specialists and structural engineers, there is no doubt that the former are very knowledgeable about cement and concrete. But it is inevitable that structural behavior and even construction practice are not their strong points. Of the nine members of the Working Party, only one is a chartered (equivalent to professional) civil engineer, and there is not a chartered structural engineer.

The issue of the maximum w/c

Let us consider the practicality of ensuring that the total w/c does not exceed 0.4. First of all, because the value is given to one decimal place, a w/c of 0.44, and possibly even of 0.45, could be claimed not to exceed 0.4. But let us assume that the upper limit is 0.40. This requirement can be readily achieved in the laboratory, or possibly in a ready-mixed concrete plant. However, ready-mixed concrete producers do not supply concrete containing high-alumina cement because mixtures of portland cement and high-alumina cement, which could occur accidentally in the equipment, might exhibit false set.[9]

We have to consider, therefore, site batching and mixing. Here, some

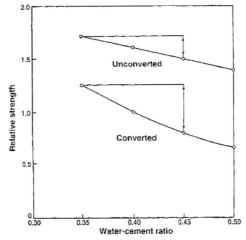

Fig. 1.5.1—Influence of the *w/c* on the strength of high-alumina cement concrete, before and after conversion, relative to the strength after conversion of concrete with a *w/c* of 0.40 (data derived from Reference 10)

variation in the *total* water content of concrete is inevitable, if only because the water content of aggregate varies and is not determined on a batch-to-batch basis with adequate precision. It can be pointed out that the same is true when portland cement is used. There is, however, a difference in the consequences of the *w/c* being occasionally, say, 0.45 instead of 0.40. Let us look at the difference in strength between concretes made with *w/c* of 0.35 and 0.45. With portland cement, the reduction in strength would be about 10 MPa (1450 psi). On the other hand, with high-alumina cement the difference could be as large as 17 MPa (2470 psi) because the loss of strength on conversion of high-alumina cement increases significantly with an increase in the *w/c*. This situation is illustrated in Fig. 1.5.1, based on the data of the cement manufacturers.[10]

The crucial point is that it is not possible to guarantee that the *w/c* (based on total water in the mixture) on site will never exceed 0.40. There is an analogy to some medications, which are beneficial at the right dose but toxic at higher doses. But then medications are dispensed by a pharmacist and not by a concretor working under difficult conditions in the field.

It is also important to point out that strength is not the sole criterion of good concrete: durability is equally important. The requirements for maintenance and durability of concrete structures were recently discussed in *Concrete International* [11] (see Section 3.2). The position of high-alumina cement in this respect will be discussed in the next section.

Why do this?

At the risk of being accused of indulging in polemics, I quote from the comments of the Concrete Society Working Party[8] on the earlier collapses in the United Kingdom: "These collapses, in a very small number of the estimated 50,000 total number of buildings incorporating the material, received unprecedented publicity." And later on, it is said: "Only a very small number of buildings (with high-alumina cement elements) have required remedial work."[8] May an aircraft designer say: Only a very small number of my structures collapse? Surely, the designer of structures connected to the ground is in the same position.

One further question can be asked about the suggested revival of high-alumina cement concrete as a structural material. Why do this, bearing in mind not only the strength and durability aspects but also the high price of high-alumina cement, when other materials are available to provide early high strength, low permeability, and good durability? The Strategic Highway Research Program in the United States decided in 1991 not to consider high-alumina cement concrete because of the consequences of conversion.[12] My own view is this: In all design, there is some uncertainty and therefore some risk, but no avoidable risk should be taken; high-alumina cement represents a risk that is avoidable by choosing portland cement concrete.

Recent problems with high-alumina cement use

The problems with high-alumina cement in structures discussed earlier in this paper cannot be consigned to history because they did not stop with the collapses in the 1970s. In the United Kingdom, I and other engineers[13] continue to monitor and assess structurally numerous buildings containing high-alumina cement. In some cases, especially roofs, remedial work and strengthening are required.[14]

In other countries, too, structures built prior to the ban on the use of high-alumina cement continue to create problems. The failure, around 1986, of a large part of a 30-year-old hall/building in Czechoslovakia was reported.[15] It is interesting that the other part of the hall, made with portland cement, was not damaged. The collapse was sudden and without "any symptoms of defects." The conclusion was that "high-alumina cement concrete cannot be, with respect to consequences, applied in principle in supporting structures."[15]

In Spain, some isolated cases of failure of high-alumina cement concrete structures in the 1970s were followed, in 1990, by a failure which caused one fatality and some injuries.[16] It is estimated that there are between 19,000 and 104,000 apartments built with high-alumina cement concrete, admittedly some of it of poor quality.[17] Various advisory meetings, to some of which I was invited, have been held. A range of remedial measures, including demolition, strengthening, and a reduction in the applied load, has been used to resolve the physical problem. In addition, the legal problem of liability and the social problem of housing a great many people are also being dealt with.

Experience with structures in some other countries is available only locally and there may be some language difficulties. For example, the

English-language abstract of a Hungarian paper[18] reports investigations over a period of more than 28 years up to 1967 and states: "Decomposed products (of high-alumina cement) are of decreased strength or carry no strength at all. These processes are unavoidable; therefore concretes made with high-alumina cements must not be used as structural units, irrespective of the manufacturing technology of cement." This last point is of interest because it is sometimes argued that only "foreign" cements lead to trouble.[5]

There is little point in adducing more evidence of problems with high-alumina cement. The structural use of this cement is either explicitly prohibited or the cement is not listed as permitted in numerous countries, including United Kingdom, United States, Canada, Denmark, Norway, Sweden, Germany, Czech Republic, Switzerland, Belgium, The Netherlands, Hungary, Spain, Japan, New Zealand, and Australia.

The only exception is France, which is the birthplace of high-alumina cement and a large producer and exporter (for example, to the United Kingdom where the manufacture of high-alumina cement ceased a long time ago). There has never been a French precast/prestressed concrete industry based on high-alumina cement, but the cement is used under strictly prescribed conditions in place in various structures and, to my knowledge, successfully. The French requirements include a minimum cement content of 400 kg/m^3 (675 lb/yd^3) and a maximum w/c of 0.40; the water considered is the total water in the mixture.

Durability of high-alumina cement concrete

High-alumina cement was developed at the beginning of the 1900s as a solution to the problem of decomposition of portland cement under sulfate attack. There is no doubt that concrete containing high-alumina cement which has not undergone significant conversion is resistant to sulfates and also to many dilute acids. However, conversion of high-alumina cement increases the permeability of the hardened cement paste and has, therefore, an adverse effect on durability. A particular problem is the attack by alkalies, mentioned earlier.

There is also an unfortunate synergy in that chemical attack of high-alumina cement has also a negative effect on its strength. Tests at a British government agency[19] have shown that wet conditions, or even occasional wetting, are deleterious as strength may be 10 to 15 MPa (1450 to 2180 psi) lower than that of concrete of the same composition and age kept under dry conditions. Wetting may occur accidentally, for example, by an overflowing bath or by condensation in a kitchen, or through fire fighting, or on a change of use of the structure. Similarly to the discussion on temperature effects, the new user may not even be aware of the presence of high-alumina cement, and hence runs an unexpected risk.

Another durability aspect is that of corrosion of steel reinforcement. Despite its somewhat lower pH than portland cement, unconverted high-alumina cement offers good protection to steel.

1-32   Chapter 1: Cementitious Materials

Conversion changes the situation. An investigation by a British government agency of 14 buildings in the United Kingdom, between 20 and 35 years old, reported in 1993, has shown that, in the majority of them, high-alumina cement concrete has undergone carbonation to the depth of the steel.[20] When such concrete becomes wet for long periods, the risk of corrosion is considerably increased.[20] The same paper states that there is "the possibility that corrosion may be present in buildings more than 25 years old, even in internal environments."[20] And furthermore, as "high-alumina cement becomes more porous as it converts, the potential for corrosion in the future is probably higher than in other forms of concrete construction."[20]

In a later paper, the same authors[21] state: "There are a number of considerations for the management of high-alumina cement concrete structures which are not material-dependent but are relevant to the form of construction in which high-alumina cement is commonly used." This statement is an excellent example of the importance of understanding structural behavior as well as the material used. The Secretary of the (British) Standing Committee on Structural Safety[22] wrote in 1995: "Although rare, errors in the advice given in authoritative guidance documents do arise, for example, the pre-1977 advice on use of calcium chloride in concrete and also on the structural use of high-alumina cement concrete."

Conclusion

It could be asked: Why is this topic discussed in *Concrete International*? The first answer is that *Concrete International*, and indeed ACI, is international. So is concrete, as neither designers nor contractors and cement manufacturers are constrained by international boundaries. Knowledge certainly is not.

Moreover, the United States is not entirely free from high-alumina cement problems. But as I know from my professional experience, as is the case all over the world, failures or difficulties are not publicized, and professional involvement is strictly confidential.

But the main reason for this article is that it contains a lesson. Structural engineers must have an adequate knowledge of construction materials if their designs are to be safe, serviceable, and durable. This does not necessarily mean a wide and detailed personal knowledge, but at least an in-house materials specialist or specialists, who have made an effort to understand the structural aspects of construction. The engineers who rely on external materials specialists who are often highly knowledgeable in their own narrow field but nothing else, run a risk of serious problems, usually at some later date. If they rely solely on the advice of manufacturers and suppliers, they do so at their peril.

Finally, high-alumina cement is a highly valuable material for use in refractory concrete and for other special purposes; this is not disputed. But as for structures, we have learned a lesson, and let the next generation of engineers learn from our mistakes. We have had a new look at high-alumina cement—the picture is the same.

References

1. Neville, A., "Concrete Technology: An Essential Element of Structural Design," *Concrete International,* V. 20, No 7, July 1998, pp. 39-41 [Section 6.1].

2. Alexander, M. G., "Dissolution of No-Fines Concrete Slab Due to Soft-Water Attack," *Journal of Materials in Civil Engineering,* V. 5, No. 4, Nov. 1993, pp. 427-435.

3. Neville, A. M., "The Effect of Warm Storage Conditions on the Strength of Concrete Made with High-Alumina Cement," *Proceedings,* Journal of the Institution of Civil Engineers, V. 10, June 1958, pp. 185-192.

4. Neville, A. M., "A Study of Deterioration of Structural Concrete Made with High-Alumina Cement," *Proceedings,* Journal of the Institution of Civil Engineers, V. 25, July 1963, pp. 287-324.

5. Discussion of reference 4, *Proceedings,* Journal of the Institution of Civil Engineers, V. 28, May 1964, pp. 57-84.

6. Neville, A. M., *High-Alumina Cement Concrete,* The Construction Press, Lancaster, U.K., 201 pp.

7. European Pre-Standard, prENV 197-10, "Cement—Composition, Specifications and Conformity Criteria—Part 10: Calcium Aluminate Cement," *European Committee for Standardization,* May 1996, 21 pp.

8. Concrete Society Technical Report 46, *Calcium Aluminate Cements in Construction: A Re-assessment,* The Concrete Society, Slough, U.K., 1997, 63 pp.

9. Neville, A. M., *Properties of Concrete,* Longman, Harlow, and John Wiley & Sons, Inc., New York, 1996, 844 pp.

10. George, C. M., "Manufacture and Performance of Aluminous Cement: A New Perspective," *Calcium Aluminate Cements,* E&FN Spon, London, 1990, pp. 181-207.

11. Neville A. M., "Maintenance and Durability of Concrete Structures," *Concrete International,* V. 19, No. 11, Nov. 1997, pp. 52-56 [Section 3.2].

12. Strategic Highway Research Program, *High-Performance Concretes: A State-of-the-Art Report,* SHRP-C/FR-91-10, NRC, Washington DC, 1991, 233 pp.

13. Mathieson, G., "Letters to the Editor: HAC Uncertainty," *New Civil Engineer,* June 12, 1997, p. 14.

14. Hill, R. N., "Are Old High-Alumina Cement Concrete (HACC) Roof Structures Still a Problem?" *The Structural Engineer,* V. 75, No. 23, Dec. 1997, pp. 421-422.

15. Lach, V., "The Deterioration of Alumina Cement Concrete," SP-100, *Concrete Durability,* American Concrete Institute, Farmington Hills, Mich.,

1987, pp. 1903-1914.

16. Goñi, J., et al., "A New Insight on Alkaline Hydrolysis of Calcium Aluminate Cement Concrete: Part 1. Fundamentals," *Journal of Materials Research,* V. 11, No. 7, July 1996, pp. 1748-1753.

17. Macias, F., "The Consequences of the Use of Concretes with High-Alumina Cement in Madrid," *Informes de la Construcción,* V. 44, No. 422, November/December, 1992, pp. 71-77.

18. Talaber, J., "Long-time Studies with High-Alumina Cements," *Epitoanyag,* V. 19, 1967, pp. 1-8.

19. Collins, R. J., and Gutt, W., "Research on Long-Term Properties of High-Alumina Cement Concrete," *Magazine of Concrete Research,* V. 40, No. 145, Dec. 1988, pp. 195-208.

20. Crammond, N. J., and Currie, R. J., "Survey of Condition of Precast High-Alumina Cement Concrete Components in Internal Locations in 14 Existing Buildings," *Magazine of Concrete Research,* V. 45, No. 165, December 1993, pp. 275-279.

21. Currie, R. J., and Crammond, N. J., "Assessment of Existing High-Alumina Cement Constructions in the U.K.," *Engineering Structures and Buildings, Proceedings,* Journal of the Institution of Civil Engineers, V. 104, Feb. 1994, pp. 83-92.

22. Menzies, J. B., "Hazards, Risks, and Structural Safety," *The Structural Engineer,* V. 73, No. 21, Nov. 1995, pp. 357-363.

Letter to the Editor No. 1

I am responding to the article by Adam Neville, "A 'New' Look at High-Alumina Cement," published in the August 1998 issue, pp. 51-55.

Adam Neville has contributed much to concrete technology in an illustrious career. The views he expresses are long held and I am sure genuinely meant; however, the approach and tone he adopts in the article are misleading and may cause offense to the integrity and professionalism of some to whom he refers.

I should state at the outset that many of the general aspects of the properties and performance of aluminate cement concretes are common between us. There are, however, significant differences in detail, interpretation, and conclusions drawn. Fundamentally, it is well known that under warm and moist conditions, aluminate cement concretes are likely to undergo a microstructural transformation—termed "conversion"—which has implications for strength and durability. This was incorporated into United Kingdom (UK) design codes long before Adam Neville's reported work in the early 1960s.

I particularly seek through this response to redress the balance of argument on behalf of the UK Concrete Society Working Party, whose work, published as Concrete Society Technical Report 46, in 1997, Adam Neville comments on. Readers should not be misled into believing that a concerted campaign is underway to establish calcium aluminate cements as a primary and mainstream structural material. This was not the intention of the Working Party nor is it the conclusion. As chairman of the Working Party, I am, I believe, well placed to present *Concrete International* readers a balanced view of our deliberations. This response does not attempt an unnecessary point by point rebuttal of Neville's case. This may be for another time and another place but it is important to cover the essential points of disagreement.

The Working Party started from a position that a family of materials exists—calcium aluminate cements (CAC)—with a considerable current market, and with many examples of good and long service performance, but which also had suffered in-service failures, some outwardly dramatic. We were interested to discover whether sufficient information now exists to give wider guidance to designers.

Our report is a wide ranging review of the data and experience with calcium aluminate cements. It considers service performance, laboratory assessments, and the treatment in design codes. It reviews behavior in many chemical environments and the behavior of structures. It gives an unbiased review of that data and gives clear guidance where that is possible. Uncertainty and aspects where future research is needed are highlighted.

The Concrete Society report expresses a view that, based on all the evidence reviewed, an approach exists which can form the basis of the safe use of the material in structures. This approach has some commonality with the European Standard also discussed in Neville's article but takes a more cautious and questioning view of the determination of a safe basis.

There is an implication, made by Neville, that because a representative of Lafarge Aluminates was on the Working Party, then it was inevitable that the recommendations of a proposed new European Standard would be followed unquestioningly. This implication is both untrue in fact and is not supported by the text of our report. I must add, however, that to have contemplated such a study and report on aluminate cements, without involving the major manufacturer and their considerable library of knowledge, would have been extraordinarily limiting. Furthermore, through members of the Working Party, other UK suppliers of aluminate cement were given the opportunity to input and comment.

At this point, I would like to give reassurance on the credibility and competence of the Concrete Society Working Party. It is suggested that, of its nine members, only one is a chartered civil engineer. (In fact, there are two.) This should not, however, be taken to accept that the skills and expertise of structural/civil engineers have been ignored. All Concrete Society reports are reviewed and approved by a range of experienced people including materials, structural, and civil engineers. In this particular case, because of the sensitivity of the subject, extreme care was taken to bring a very extensive consultation.

The Working Party itself comprised a considerable array of expertise in the nature of materials and was supported by the consultations undertaken by the Working Party members and by the even wider range of expertise in the society's committees reviewing the report.

Two contributions should be particularly highlighted in regard to the balance of materials and engineering inputs and of research and real practice. The first is that the Working Party membership contained three specialists from the UK Building Research Establishment (BRE)—the former "government agency" much quoted by Neville. In the consultation and reviewing of drafts all relevant sections of that organization, including materials and structural engineers, were extensively involved. This is particularly important because of the central role played by BRE in investigating the nature and behavior of aluminate cement concrete structures after the 1970s problems and their long-standing involvement in research into the material.

I, myself, drew heavily upon the experience of my engineering qualified colleagues in Ove Arup & Partners, an international engineering design consultancy of considerable resource and experience. Some of these colleagues are deeply involved with drafting the UK and European design codes; others were principal investigators in one of the "headline" structural collapses of 1973.

The evidence presented by Neville on one of these collapses should be corrected. The published conclusions of the investigation of one of the collapses was that the failure of some beams had principally been caused by inadequate resistance of support on nibs to the roof beams against horizontal tensile forces due to rotational thermal movement. There had been a cycling load which eventually induced the nib failure. It was the loading regime rather than the degradation of strength which was the primary cause of failure. Indeed it is relevant to point out that the remaining beams did not fall and were retained in the rebuilt structure, albeit with a revised support structure at beam ends. The beams remain in service to this day.

In producing the report, I, as chairman of the Working Party, devoted much time and energy to achieving a balance of views and conclusions that, on occasion, were quite divergent within the group. I was able to do so because I am in the fortunate position to fulfill the role that Neville applauds in his conclusions! I am by training a materials scientist and have spent the last 20 years working very closely with structural and civil engineers. I thus have a good understanding of structural design and construction practice, and find it somewhat ironic that Neville suggests an implicit shortcoming in materials specialists not understanding such matters while structural/civil engineers are free to become specialists in materials without question!

It is inevitable, because of the dramatic nature of collapse failures, that attention focuses upon them, irrespective of material. It is true that the UK events of the 1970s were a considerable shock to the engineering community. The important issue with failures is to understand the nature of the material, and of the situation, so that lessons can be learned and repeated failures avoided. This is true for any structural material—steel, portland cement concrete, masonry, timber—all of which have suffered dramatic failure without being universally "banned." Indeed, the same might also be said of aircraft which are not unknown to have failed in service but we still fly them. I do not believe any engineer, of any discipline, designs with materials that he expects to fail.

That production of aluminate cement concrete needs good levels of control to assure satisfactory performance is not in dispute. This is implicit in suggesting that the material will only find use in specialist applications, including perhaps those with a structural requirement, where such control is possible. That control and acceptable performance are achievable is acknowledged by Neville when he states, in relation to French practice, "...but the cement is used under strictly prescribed conditions in situ in various structures and, to my knowledge, successfully."

Neville, rightly, recognizes and agrees with the Working Party view that conversion has implications for durability as well as structural performance. Rather than just considering permeability of the concrete, it is important to consider and understand the chemical differences in the various converted and unconverted hydrates. For example, the apparently better performance of converted aluminate cement concrete in acidic conditions, despite its higher permeability, can be explained by the difference in inherent chemical resistance of the different hydrates. Each situation or application must be considered in the light of the requirements and the conditions. The Concrete Society assessment is a comprehensive start, not a finish, in bringing this information together for a wider audience, experienced or new.

Perhaps I can bring this response to a close by restating some of the concluding remarks from the Concrete Society Report:

> "A primary conclusion of this report is that the conversion reaction and the attainment of low but stable long-term strengths in CAC concretes is virtually inevitable. Any design basis for the use of CAC concrete must recognize this."

> "It is doubtful whether, even if a basis for the safe use of CAC concretes can be formalized and agreed, they would supplant portland cement concrete for the mainstream of structural applications whether in situ or precast. The cost penalties of CAC materials and the improved technical base of portland cements would both go against such a replacement."

> "...further coherent, detailed, and independent guidance should be developed of a safe basis for determining the minimum in situ strength of calcium aluminate cement concrete for particular structures."

> "...research should be undertaken and guidance developed, which is devoted to understanding more fully the nature and behavior of CAC concrete in aggressive service conditions. In particular, the role of the

various cement hydrates and microstructure in influencing performance should be examined further. This should include examples of both good and bad performance."

It is fundamentally incorrect to imply that the Concrete Society Technical Report 46 is solely attempting to restore or revive the structural use of aluminate cement concrete. Indeed, we are not the first to propose a design basis focused on establishing a safe stable long-term strength. Fundamental to the Concrete Society report was the belief that structural designers need to have lucid information on the nature and performance of the materials. The combined application of materials and engineering expertise is the appropriate route to achieve that. Any implication that the report suffers from a lack of credibility has been demonstrated to be untrue.

To take a "new" look at evidence and experience, it is necessary to have open eyes and minds. This the Concrete Society's Working Party set out to do and, I believe, succeeded.

Bob Cather
Chairman, Concrete Society Working Party on CAC
Ove Arup & Partners
London, United Kingdom

Letter to the Editor No. 2

Adam Neville's timely article on high-alumina cement concrete is important for the teacher and for us all as perpetual students. It is the reminder, that as Robert Stephenson (president of the Institution of Civil Engineers in 1856) said: "Nothing is so instructive to the younger members of the profession as the records of failures, far more valuable than description, of the most successful works."

Neville's crucial hypothesis is, however, that in the design process, an intimate knowledge of the properties and behavior of the chosen material is as essential as understanding the complexities of structural behavior. I hope this message is taken in by teachers because for most engineers their first feel of concrete is in the university laboratory and the first steps in design, in the college drawing office. Neville rightly says that these must not be seen as two separate functions but there is a danger that the technology of the concrete and the design of concrete structures will be taught by different people. This "decoupling," to use Neville's word, sends the wrong message to the student who should be taught how to exploit the best in both the material and in the structural form.

Structures are somehow seen to become alive and we are apt to forget that concrete does not become dead after a few months but is a living material affected by loading, age, and environment.

One cannot expect every professional engineer to know every detail of every material as well as every analytical equation and structural mechanism in the book. But one can expect, as Neville pleads, that the two specialists must not work in isolation but as real partners to progress the best and safest use of concrete in all its facets.

Rowland Morgan
Department of Civil Engineering
University of Bristol
Bristol, United Kingdom

Neville's response

It is gratifying to receive comments both from an engineer and a materials scientist. Professor Morgan's plea for the professional engineer and the materials specialist not to work in isolation but as "real partners" has my full support.

Mr. Cather's letter concentrates on my brief comments about the Concrete Society Working Party report on high-alumina cement. In reply, I would like to point out that my original work on the deterioration of high-alumina cement concrete was performed not in the early 1960s as Mr. Cather says, but in the 1950s.[1-3] Further, the implications of the conversion of high-alumina cement were not "incorporated into UK design codes...long before the 1960s."

The 1965 British Code for Precast Concrete referred only to "hot, wet conditions...which lead in some circumstances to a marked decrease in strength." The 1969 British Code for Reinforced Concrete used stresses based on a factored 1-day strength of high-alumina cement concrete, that is before degradation. The 1969 British Code for Prestressed Concrete said: "Concretes made with high-alumina cement are sometimes unsatisfactory in warm moist conditions. This cement should be used only in accordance with the manufacturer's recommendations." Presumably it was, and failures and extensive deterioration occurred in a many countries.

Even the 1972 British Code for the structural use of concrete recommended the use of unconverted strength of high-alumina cement when concrete is "cool and moist during the first 24 hours." And further: "At temperatures above 30°C (86°F), high alumina cement should be used on the basis of the manufacturer's advice." And yet, even nowadays, the manufacturers advertise the use of high-alumina cement in hot climates such as Mexico for bridges, canals, jetties, tunnels, and the like. A strength of 50 MPa (7300 psi) at 24 hours is listed without any mention of the fact that after conversion this would be reduced to 20 MPa (3000 psi), as stated in the 1972 British Code.

What is important is that conversion occurs not only under "hot, wet" or "warm moist" conditions. Normal temperatures and exposure conditions lead to conversion and a loss of strength. A 1994 paper refers to "the realization that under normal service conditions, the compressive strength of high-alumina cement concrete could be halved over a long period of time."[4] The Building Research Establishment (BRE) warned in 1988 of the possibility "that dry HAC concrete which has reached its minimum strength may, if subsequently subjected to wet conditions, undergo some further loss in strength to reach a minimum strength that would have been reached if subjected to wet conditions initially."[5] In 1994, BRE said: "The amount of strength loss depends on a number of complex interactions, including conditions at the time of casting and subsequent service conditions." The word "subsequent" is of importance.[6]

A general concern is that the change of use of a building may involve more humid conditions, and the new owner may not even necessarily be aware of the presence of high-alumina cement in the structure. When he becomes aware, he may have to resort to monitoring the behavior of the structure with the concomitant expense and economic blight.

BRE, which continues to monitor numerous buildings,[7] also expressed concern about the effects of carbonation of concrete on corrosion of prestressing steel.[4] It found "that the majority of the floor and roof components are carbonated to the depth of the steel."[7] There had been no disruption, but "the future risk from corrosion in such beams is still unclear, even in nominally dry internal conditions."[7] The 1994 view of BRE is that, following carbonation, "the reduced quality of the concrete cover often means that the reinforcement is no longer protected from corrosion which may develop even in some internal locations."[6] And further: "the risk of corrosion to reinforcement is increasingly important in the assessment of high alumina cement components."[6]

In any case, reliance on the residual strength of high-alumina cement (HAC) ignores the fact that converted cement has a greatly reduced resistance to chemical attack. The *BRE Digest* says about the chemical attack: "in the presence of alkali or sulfate ions and excessive and persistent dampness, it can lead to progressive deterioration." This causes a further loss of strength beyond that assumed by conversion.[6] The *Digest* says further: "Not all cases of chemical attack in wet conditions have resulted in disruption or weakening of the concrete but where disruption has occurred the effects on strength have been very severe."[6]

Moreover, Mr. Cather himself acknowledges that "we are at present still developing procedures for predicting values for long-term minimum strength in situ."[8] So, perhaps it is not surprising that the vast majority of European countries as well as the U. S. and Canada do not provide in their design codes for structural use of high-alumina cement.

What is actual industrial practice? In the U.S., the Strategic Highway Research Program lists under Technology-Related Exclusion: "High-alumina cement based concretes are not considered because of the conversion of the hydrated aluminate compounds that occur over time, resulting in dramatic increases in porosity and permeability and significant strength loss."[9] In the United Kingdom, a 1996 industry survey states: "the material has a tendency to lose strength under certain conditions during service which may in the extreme case lead to structural collapse."[10]

Mr. Cather tells us that he "drew heavily upon the experience of engineering qualified colleagues" and that he has "spent the last 20 years working very closely with structural and civil engineers," and concludes: "I thus have a good understanding of structural design and construction practice." I accept this fully but in most countries (although not in the U.K.) design must be undertaken by an engineer qualified professionally, who has passed the requisite examination and who holds a certificate as a registered or professional engineer. My good friend is an anaesthetist who has worked very closely with surgeons, but I still would not allow him to repair my heart.

It follows that a materials scientist cannot design a structure. On the other hand, a structural engineer who does not have an adequate knowledge of materials designs at his peril; this was the topic of my recent paper in *Concrete International*.[11] In my opinion, therefore, the Working Party should have included structural designers. It should also have included some members whose views are critical of the structural use of high-alumina cement. Reconciliation should take place within the Working Party if its views are to be seen and accepted as truly balanced and objective.

In this connection, it is worth noting that ASTM requires a balanced membership of producers, users, and independent members in the various committees. ACI also jealously guards the need for a balanced membership of its committees. Being neither a producer, supplier, or user of high-alumina cement, I have no vested interest in its future use in structures; I have no vested interest in portland cement either. But I believe that offering a warning about potential hazards is a professional engineer's duty; this was highlighted at a conference of the Royal Academy of Engineering in 1991.[12]

I would like to interject at this stage a recent elaboration of this topic. The ethical duty of engineers to issue warnings was confirmed in a Lloyd's Register Lecture by John Uff, who is Nash Professor of Engineering Law at King's College, London and a Queen's Counsel as well as a chartered civil engineer. The lecture was given to the Royal Academy of Engineering on April 22, 2002 and it has been published under the title "Do engineers owe duties to the public?" The relevant words are: "The papers (edited by E. Hambly) contain other examples of warning delivered, but acted on too late, including the case of High Alumina Cement (HAC) on which Professor Adam Neville CBE FREng FRSE had given clear published warnings in the 1960s. Despite this, a series of collapses in the 1970s still took the industry and the public by surprise."

Returning to Mr. Cather's letter, he says that the degradation of strength of high-alumina cement concrete was not the primary cause of failure in one of the collapses referred to in my article. Even if the loss was a contributory cause, should not *any* cause be avoided? His comments on the inadequate support of roof beams ignore the fact that it was only when the strength of the high-alumina cement concrete was reduced that the bearing pressure became excessive.

Of course, I have no quarrel with many of the Working Party's conclusions, and recommendations as quoted by Mr. Cather but I cannot see that, as Mr. Cather says, "an approach exists which can form the basis of safe use of the material in structures." As he says, alas, the Draft European Standard takes an even less "cautious and questioning view of the determination of a safe basis." As Mr. Cather points out, portland cement concrete and steel structures have been known to fail, but that was not due to intrinsic and foreseeable degradation. The basis question remains: why build structures using a cement which presents risks and offers no technical benefit but costs more than the usual cements?

From the structural engineer's point of view, it is not sound practice to use a material known to deteriorate under some circumstances which cannot be foreseen or controlled. For special purposes, high-alumina cement is valuable, and this is not disputed.

Mr. Cather quotes selectively some conclusions from his report. For the benefit of the readers, it is essential to set the record straight. Mr. Cather fails to quote the very first recommendation of his report: "Specifiers, users, and clients should be encouraged to consider applications where calcium aluminate cements would have technical and commercial benefits either in conventional concrete form or as specialist proprietary products." So, the Working Party encourages the use of HAC "in conventional concrete form." This is far more than Mr. Cather admits in his letter.

Finally, the purpose of my article was not to debate with the Working Party. Rather, I wanted to present what I believe to be a balanced and impartial view of the structural use of high-alumina cement. Such a view is necessary for the health of the structural engineering profession and for the safety of the public.

Adam Neville

References

1. Neville, A. M., "The Effect on Concrete Strength of Drying During Fixing of Electrical Resistance Strain Gauges," *RILEM Bulletin*, No. 38, 1957, pp. 95-96.

2. Neville, A. M., "The Effect of Warm Storage Conditions on the Strength of Concrete Made with High-Alumina Cement," *Proceedings of the Institution of Civil Engineers*, V. 10, June 1958, pp. 185-192.

3. Neville, A. M., "Tests on the Strength of High Alumina Cement Concrete," *Journal of the New Zealand Institution of Engineers*, New Zealand Engineering, V. 14, No. 3, July 1959, pp. 73-76.

4. Currie, R. J., and Crammond, N. J., "Assessment of Existing High Alumina Cement Construction in the UK," *Engineering, Structures and Buildings, Proceedings*, Journal of the Institution of Civil Engineers, V. 104, Feb. 1994, pp. 83-92.

5. Collins, R. J., and Gutt, W., "Research on Long-Term Properties of

High-Alumina Cement Concrete," *Magazine of Concrete Research*, V. 40, No. 145, Dec. 1988, pp. 195-208.

6. Building Research Establishment, "Assessment of Existing High Alumina Cement Concrete Construction in the UK," *BRE Digest*, No. 392, 1994, 12 pp.

7. Crammond, N. J., and Currie, R. J., "A Survey of Conditions of Precast High-Alumina Cement Concrete Components in Internal Locations in 14 Existing Buildings," *Magazine of Concrete Research*, V. 45, No. 165, Dec. 1993, pp. 275-279.

8. Cather, R., "Calcium Aluminate Cements," *Magazine of Concrete Research*, V. 49, No. 179, 1997, pp. 79-80.

9. Strategic Highway Research Program, "High-Performance Concretes: A State-of-the-Art Report," *SHRP-C/FR-91-10, NRC*, Washington, D.C., 1991, 233 pp.

10. Construction Industry Research and Information Association (CIRIA), "Excluded Materials: An Industrial Survey," *Project Report 47*, London, 1996, p. 16.

11. Neville, A. M., "Concrete Technology—An Essential Element of Structural Design," *Concrete International*, V. 20, No. 7, July 1998, pp. 39-41.

12. Hambly, E. C., ed., "Preventing Disasters," *Proceedings of Conference on Warnings of Preventable Disasters*, The Fellowship of Engineering, London, 1991, 138 pp.

Chapter 2

Water and Concrete

Traditionally, the topic of water in concrete has been regarded as of little importance because it is just … water. I am referring to the quality of water, and not to its quantity. The significance of quantity has always been considered: as the mass in liters per cubic meter (or gallons per cubic yard) because it is relevant to workability; and as the ratio of the mass of water to the mass of cement because it is relevant to strength.

And yet water is involved in the whole life of concrete: during construction, in service, and under special circumstances when electrical resistivity or thermal insulation are of importance. Section 2.1 reviews all these circumstances and effects; I believe this is the first such wide-ranging review in a single article.

It is known that water to be used in a concrete mixture must be "good," but this is an ill-defined and unspecific term. Usually, recourse is made to the use of tap water, that is, to water fit for drinking. While such an approach generally errs on the safe side, it is not practicable in those parts of the world where there is no distribution of potable water or in areas where there is no piped water whatever, and other sources have to used. Alas, we know very little about what are the minimum acceptable properties of water fit for use in concrete; indeed, there are no standards, although some guidelines may be found. Section 2.2 reviews the situation with respect both to mixing water and to water used for curing. The same section discusses the use of wash water from ready-mixed concrete plants and of domestic wastewater.

I am convinced that in the future it will become more necessary to use

Chapter 2: Water and Concrete

water that is not obviously potable. Moreover, for economic reasons, we shall be obliged to consider the use of "other" waters even when potable water is available. There exist already many parts of the world where water is scarce, and the situation is likely to become worse; even in the U.S., there are areas where water supply is becoming problematic. All the relevant aspects of the quality of water are discussed in Section 2.2.

Use of potable water in the mixture precludes the use of seawater. And yet, seawater has, on occasion, been used as mixing water, and the prohibition of the use of seawater under *any* circumstances is not an open-and-shut case. This topic is considered in Section 2.4. That section discusses also the use of seawater in curing.

A particularly important consequence of using seawater in the mixture is the corrosion of reinforcement. This topic is discussed in Section 2.4 and also in Section 3.5, which forms part of Chapter 3; the latter section looks at the broad topic of chloride attack of reinforced concrete.

While the influence if mixing water on concrete is known, and is discussed in Sections 2.1, 2.2, and 2.4, there exist almost no publications on possible effects of leachate from concrete or mortar upon the quality of drinking water. These effects occur only when the contact between water and hydrated cement paste is prolonged, but this does happen in domestic water supply. The main effects, discussed in Section 2.3, are an increase in pH and in the content of calcium carbonate. When high-alumina cement is used, there is also an increase in the content of aluminum. An understanding of the phenomena involved will be helpful with respect to the continuing use of concrete conduits and mortar-lined pipes in the distribution of drinking water.

The final section of this chapter, Section 2.5, deals with the water-cement ratio. The use of this ratio as an important parameter in concrete is 90 years old, but periodically questions are raised about the usefulness of this ratio. It is the usefulness that is the subject of Section 2.5. With the change from the term water-cement ratio (w/c) to the term water-cementitious material ratio (w/cm) introduced by ACI, it is important to bear in mind the difference between considering portland cement alone and the totality of cementitious materials; this is emphasized in Section 2.5.

The original publication of Section 2.5 produced some lively letters, which are reproduced following the text of Section 2.5. My closing words in my reply express my views but readers might like to look at both sides of the problem.

Section 2.1: Water and Concrete: A Love-Hate Relationship

Papers about concrete, when they refer to water, usually consider only one relationship at a time. The most common of these is the relation between water-cement ratio (w/c) and strength. Another example is water content and its influence on workability. At the more scientific end of the scale, much work has been done on the hydration reactions of cement and the state of water in the resulting products of hydration: chemically combined, physically adsorbed, and so-called free water.

But water is not just a liquid used to make concrete: it is involved in the whole life of concrete, for good or evil. Concrete in the environment is usually in contact, permanent or intermittent, with water, in liquid or in vapor form. Most actions on concrete in service, other than loading, involve water, either pure or carrying salts or solids. There are numerous relations between water and concrete, and even multifaceted interactions between these two materials.

I do not recall ever seeing a discussion, or even enumeration, of all of these relations in a single paper. And yet, this would be appropriate because water and concrete are the two materials most used by mankind: water in the first place and concrete in the second. I shall give the relevant quantities at the end of this section, which will briefly explore the connection, or indeed the numerous connections, between concrete and water. This should dispel the notion that concrete is a dry subject!

Part I: During construction

The opening observation, and perhaps the most obvious one, is that you cannot make concrete, as usually defined, without water. Thus, the first topic is mixing water.

Mixing water

There are three aspects of mixing water that should be considered. The first of these is its quality. The water should be "good," such as city supply water but, if it is not, what are the effects of impurities on the properties of the resulting concrete: setting time, gain of strength, discoloration, and long-term durability? Relatively little is known about the limitations on the suitability of water for mixing.

And yet, with an increasing scarcity of water in many parts of the world and ecological considerations of impure water, in making concrete we have to consider the use of treated domestic sewage and industrial waste waters, as well as water used to wash out concrete mixers and trucks, and ready-mix concrete yards. Such waters must not be discharged into the natural surface waters. The obvious solution is to recycle the wash water by reusing it as a part of the mixing water. As the wash water contains alkalies from portland cement and also solid material in suspension, we need to know what is acceptable. The entire issue of suitability of water for mixing and curing was discussed in a previous article[1] (see Section 2.2).

The second aspect is the quantity of water in a unit volume of concrete. This can be expressed in liters per cubic meter of concrete (or gallons per cubic yard). This is the so-called water content, which greatly influences the workability of the resulting mixture.

The third aspect is that it is also possible to express the quantity of water in the mixture as the *w/c*, nowadays always by mass, in decimal form (except by the Japanese, who prefer percentages). Fortunately for those who work outside the U.S., gallons per sack (which is an expression using inconsistent units) is just a thing of the past. That the *w/c* greatly affects the strength of concrete has been well known since the pioneer work of Duff Abrams more than 80 years ago. I have recently written a short article on the usefulness of the w/c^2 (see Section 2.5).

Water of hydration

When concrete is mixed, the cement and water do not long stay simply as neighbors. Cement powder is hydrophilic; chemical reactions of hydration take place and various products of hydration are formed. They all contain water, but it does not all exist in the same form. Some of the water becomes chemically combined, that is, it becomes a part of the compound, such as calcium hydroxide. Some of the water is physically adsorbed on the internal surfaces of those products of hydration that are in the form of gel. It would not be appropriate in this section to discuss in detail the various forms of water in hydrated cement paste; suffice it to say that the energy of binding of water varies. It is therefore convenient to distinguish evaporable water and nonevaporable water, but the division is not absolute and depends on the method of "evaporating" the water from a sample.

Not all the space in the mixture as placed becomes filled with solid products of hydration (the term "solid" including gel water). The excess space forms capillary pores that, at least to begin with, are full of so-called free water. It is called that because free water can be fairly easily removed from the capillary pores; the pores can also be refilled. The presence of at least some water in the capillary pores is essential for the hydration of the hitherto unreacted cement to take place. The minimum vapor pressure is 80%.

It is useful to note that the products of hydration of cement have a very low solubility in water; indeed, it is the essence, as well as the origin of the name, of hydraulic cement (and portland cement is its prime example) that it is stable in water. Thus, at this stage, hydrated cement paste and water exist side by side, at least to a large extent.

Bleed water

In addition to the various forms of water of hydration, in concrete as distinct from cement paste, there may also be trapped bleed water. This water appears to travel upwards through the concrete mass but, in reality, it is the solid particles, which are heavier than water, that are subject to sedimentation by gravity. If the bleed water reaches the top surface of a concrete element, it may evaporate. The rate of evaporation and the rate of "supply" of bleed water, between them, influence the development of plastic shrinkage and possibly plastic-shrinkage cracking.

If the bleed water, on the way up, is trapped underneath large aggregate particles, this forms voids that may adversely affect the durability of concrete in later life by providing a preferential path for the ingress of aggressive agents. The bleed voids are originally water-filled but, with time, the bleed water may become used up in continuing hydration of cement or migrates outwards. But even when it stays as water, it provides a preferential path in the case of ingress of aggressive agents into the concrete.

Whether a given mixture bleeds more or less depends on some properties of the mixture, but whether the bleed water becomes trapped depends on the aggregate size and, above all, shape. Flaky particles are much more likely to intercept water in its apparently upward motion than equidimensional particles. Thus, we have an interaction between aggregate shape and water in the mixture.

If the bleed water readily reaches the top surface of a concrete element and evaporation takes place, this lowers the temperature of the concrete near the surface, which has to supply the relatively large heat of evaporation of water. This may be beneficial in a hot climate, especially as the temperature of the concrete at the time of setting affects its longer-term strength development. The temperature at the time of setting affects also the development of thermal cracking: if a concrete element adopts its final dimensions at a higher temperature, then subsequent lowering of temperature may induce cracking.

Water curing

The temperature of the concrete, once it has set, at its surface is affected also by the application of externally supplied water, that is, curing water. If this water evaporates, heat is abstracted from the concrete. Moreover, if the curing water is cold, the concrete is cooled by heat transfer. This may be beneficial in a hot climate. Conversely, curing with cold water can be dangerous if the temperature drops and there is a risk of frost. Generally, water curing should start very early and be continuous and prolonged.

It is obvious that curing water must not contain ions that can attack concrete or the reinforcement within; aggressive waters are discussed in a later section. This precludes the use of seawater for curing reinforced concrete elements.

It is worth observing that the requirements for mixing water and curing water are not identical. For example, pure water, such as distilled water or other desalinated water, is perfectly suitable for mixing, but it would attack hardened concrete, as discussed later on. Conversely, many waters containing some organic compounds that might interfere with the hydration of cement are harmless when used for curing. The only exception is the case when these compounds may cause discoloration of the surface of concrete that will be exposed to view.

In passing, I should mention that, in the past, when high-alumina cement was used in construction, strong cooling by the application of water was recommended by the cement manufacturers: the argument was that prevention of a large temperature rise on hydration of cement would obviate the process of so-called conversion of the products of hydration with the concomitant loss of strength. We now know that conversion cannot be prevented, and in most countries high-alumina cement is no longer used in construction.

In considering the cooling of the surface of a concrete element by curing water, we should pay attention to a possible differential in temperature between the surface and the interior: if the differential induces a large temperature gradient, there may develop excessive thermal stresses and possibly cracking.

It is well known that the supply of water to concrete by means of curing is essential for the progress of hydration of cement in mixtures with medium and low water-cement ratios. In most cases, this aspect of water curing overshadows the temperature aspects, but both show a clear interaction between curing water and the quality of resulting concrete.

Part II: In service

The important role of water in concrete continues in service, and there are many aspects of this role.

Drying shrinkage

When discussing the loss of water from fresh concrete, that is, before its setting, I mentioned plastic shrinkage. Now, the loss of water from hardened concrete may also lead to drying shrinkage. It is not, however, all the water whose loss leads to shrinkage: the loss of free water does not result in shrinkage. It is only when, following the departure of free water, adsorbed water is lost to the ambient medium that contraction takes place. Under some circumstances, drying shrinkage can lead to shrinkage cracking, which, in my opinion, is the greatest shortcoming of concrete and the most common cause of problems in concrete structures exposed to air; this means all structures except those under water or fully embedded in wet soil. Drying shrinkage thus represents a particularly important relationship between water in hardened concrete and its deformation and deterioration.

Creep

Creep of concrete, which is a time-dependent deformation under steady load, also involves movement of water, although this is more complicated and less well established than in the case of shrinkage. Nevertheless, we know that when concrete is drying while under a sustained load, the contraction known as drying creep occurs. This drying creep is larger than when no water movement in or out of concrete under load takes place; the deformation under the latter conditions is called basic creep. Thus, water plays a significant role in the time-dependent deformation of concrete under sustained load. This is quite separate from the role of water in shrinkage when no load is involved.

The mechanisms of the two types of creep are still not well understood. Alas, there has been only limited progress in this respect since the last major book on creep of concrete (as distinct from a collection of symposium papers) was published in 1983.[3] Creep continues to be studied, but not so much in order to understand its mechanism as to develop mathematical models for prediction, none of which has consensual support.[4] In my view, an understanding of the mechanism of creep, including the role of water, should precede reliance on complicated mathematical expressions (see Sections 6.5 and 6.6).

Wetting and drying

Dry concrete in contact with water will rapidly imbibe it. The water gradually penetrates deeper and deeper into the concrete element: quite rapidly, total saturation is possible. The process is reversible, but not at the same rate: drying is extremely slow, so that those parts of a concrete element that are more than, say, 500 mm (20 in.) from a drying surface may never become completely dry, or at least not in the lifespan of the concretor! Wetting and drying is a means of ingress of salts because a solution enters, but only water evaporates.

Water in high-performance concrete

In high-performance concrete, when the w/c is extremely low, autogenous shrinkage may develop, which has only recently become recognized as a significant factor in the behavior of concrete.[5] Autogenous shrinkage, like drying shrinkage, is due to the loss of water from hydrated cement paste, but the loss is caused by the water being used up in chemical reactions of continuing hydration, and not by movement to the atmosphere.

High-performance concrete also presents a special problem as far as water in the hydrated cement paste is concerned in the case of fire. The very low permeability of such concrete (due to the extremely low w/c) means that water still present within the hardened paste cannot readily escape. The high temperature during a fire causes vaporization and the concomitant increase in the volume of the original liquid water. The change in phase and a high-vapor pressure can lead to bursting of the outer zone of the concrete. In consequence, under particular circumstances, what was intended to be a high-performance concrete may result in a performance inferior to that of ordinary concrete.

Chapter 2: Water and Concrete

When fire is fought with cold water, the concrete whose surface temperature has become very high in the fire, finds itself suddenly subjected to quenching. This may have serious consequences for the serviceability and strength of concrete.

The preceding brief discussion shows that, although its w/c is very low, high-performance concrete exhibits a large sensitivity to water.

Autogenous healing

It is well known that the presence of cracks in concrete is virtually unavoidable. However, when concrete at the surface of a narrow crack is in contact with water, static or slow-moving, healing of the crack is possible. This occurs by hydration of the hitherto unhydrated cement or by the formation of calcium carbonate from the leached-out calcium hydroxide, if carbonation takes place. Autogenous healing is discussed in Section 5.5.

Aggressive waters

Concrete in service may be exposed to aggressive waters. In many countries, the most common deleterious ion is sulfate, combined with one of several cations. In other waters, acids and chemical by-products from industrial processes may be present. In some locations, the water in contact with concrete is seawater or brackish water. The chlorides present in such waters, if carried into the interior of the concrete, often lead to corrosion of steel reinforcement.

It is important to remember that only salts in solution attack concrete: a heap of calcium sulfate sitting on a dry floor is harmless. So is a heap of common salt, that is, sodium chloride, except that it is deliquescent so that, if the air is humid, the salt will eventually produce a solution in which chloride ions will be present; they will then penetrate into the concrete, but they can do so only in water.

Even pure water is not good for concrete in service: in pipes and conduits, pure water leaches calcium hydroxide from the hydrated cement paste (see Section 2.3). Rain, at least away from a polluted atmosphere, is a reasonably pure water, and it also leaches calcium hydroxide. If rain is followed by exposure of the concrete surface to the sun, the leachate forms efflorescence that, although not harmful in itself, is unsightly and mars the appearance of the concrete (see Sections 5.3 and 5.4).

The other extreme of rainwater is the so-called acid rain, which is a highly harmful consequence of some industrial exit gases dissolved in water in the atmosphere and carried by wind. Acid rain contains mainly sulfuric and nitric acids and can have a pH as low as 4.5 or even 4.0. These acids etch the surface of the concrete.

Corrosion of reinforcement

Corrosion of embedded steel, whether of the general type or chloride-induced, occurs only if the concrete is sufficiently wet so that the pore water acts as an electrolyte. This permits the development of an electrochemical cell. Thus, water is necessary for corrosion to proceed.

Conduits for drinking water

Drinking water is not an aggressive water in usual terms, but it causes leaching of concrete that, under some circumstances, may make the water harmful with respect to human consumption. This situation arises primarily in small-diameter conduits with a dead end, in which the rate of flow of water is low so that the contact between the water and the portland cement paste is prolonged. In consequence, with waters of a very low alkalinity, that is, a very low calcium carbonate content, the pH level of the water in the conduit may rise above 9.5. Also, the amount of aluminum entering the water may become excessive.

The amount of aluminum leached and carried by the water may be particularly high in the case of concrete or mortar made with high-alumina cement. This cement could be used in repairs or linings when rapid rehabilitation of corroded water mains is undertaken; however, such a use is questionable or inappropriate.

The topic of the use of concrete in conduits for drinking water is discussed in Section 2.3.

Water-resistant concrete

It is sometimes asked whether attack by aggressive waters can be prevented by using water-resistant concrete. I am not using the term "waterproof," which is not recommended in British Standards, if only because impermeable concrete is probably unattainable. What is possible is concrete that has a high resistance to water penetration.

Why is this? By its nature, concrete is a porous material. Porosity and permeability are not the same, but many pores in concrete are interconnected so that transport of water into and through the concrete is possible. Not only flow, but also other transport phenomena are involved. The better compacted the concrete and the fewer gross voids in it the more difficult it is for water to travel through the concrete. Furthermore, the exact structure of the hydrated cement paste determines the ease with which water can move through the concrete. Particularly important are the amount and interconnectivity of pores; these are usually assessed by their size distribution: the same volume of fine pores allows less movement of water than when the pores are few in number but large and interconnected.

Under many circumstances, the movement of water through concrete is very small, but concrete, as we make it, is not waterproof in the same way as a plastic membrane. In most cases, this situation creates no problems. However, if a virtually waterproof concrete is required, then we have to consider the use of waterproofing admixtures, possibly those that are also hydrophobic, or we may resort to special concretes such as polymer concrete.

It is worth remembering that much depends on the ambient conditions to which a given concrete element is exposed. For example, in the case of a basement wall with waterlogged ground on the outside and dry space inside, there will be some transport of water through the thickness of the wall. If someone believes that installing a powerful air conditioner in the basement will solve the problem, there will be disappointment: the greater the difference in relative humidity between the two sides of the concrete wall the more water will be transported. Likewise, the greater the head of water on the outside (that is, if the basement is very deep) the more water will be transported.

In any case, the exposure conditions and the quality of the concrete are not the sole determinant of watertightness. However carefully constructed, concrete walls and floors are rarely entirely monolithic and crack-free, and they may be subjected to differential strains. If we wish to build a water-retaining structure, we need, in addition to concrete of the right quality, appropriate detailing of reinforcement and possibly a provision of water stops. This is the domain of structural design and not just concrete technology. May I emphasize the term "watertight": this is not the same as impermeable.

Freezing and thawing

The deleterious action of repeated cycles of freezing and thawing on concrete is well known in many parts of the world. For the purpose of this section, it is sufficient to note that freezing and thawing involve water. The water may be that originally present in the mixture or it may be water that ingressed into the concrete; what is significant is that it is water that leads to damage, and it does so with a vengeance.

Damage by repeated cycles of freezing and thawing is quite distinct from damage of fresh concrete by the action of frost on fresh or very young concrete. Here, the expansion on conversion of liquid water into ice disrupts the concrete; generally, the damage is irreparable. So is, of course, damage by cyclic freezing and thawing.

What is relevant to the present article is that both frost action and freezing and thawing involve water: it is water that is the culprit.

Carbonation

Carbon dioxide in the atmosphere can react with some products of hydration of cement, notably calcium hydroxide. However, it is not gaseous carbon dioxide that reacts: the presence of water is necessary so that the actual agent is a weak carbonic acid. It is evident then that water is the essential element in carbonation of concrete, which reduces the alkalinity of the hydrated cement paste and can lead to corrosion of reinforcement.

Alkali-aggregate reaction

Two types of reaction exist: one involving silica, the other carbonate, either of these being present in the aggregate; the source of alkalies is the cement. It could be thought that consideration of this deleterious reaction is not within the scope of this article; however, this is not so: the alkali-aggregate reactions proceed only in the presence of water. Even if the reaction has developed to the point of causing some damage, drying out the concrete and maintaining it dry will arrest all further reaction; the damage is not reversible but it is certainly stoppable. Thus, once again, water is an essential element in the process of deterioration of concrete.

Cavitation and erosion

Although concrete is used extensively in hydraulic works, including spillways and closed conduits, flowing water can cause severe damage by cavitation. Cavitation can occur when the flow of water is not steady and not tangential to the surface of the concrete at all locations, but when there is a divergence between the direction of flow and the surface of the concrete. Under such circumstances, vapor bubbles can form when the local absolute pressure drops to the value of the ambient vapor pressure of water at the ambient temperature. These bubbles travel downstream and, on entering an area of higher pressure, they collapse with great impact. This collapse enables high-velocity water to enter previously vapor-occupied space and exert an extremely high pressure on a small area of concrete surface. Such repeated application of high pressure results in pitting of the surface, making it more uneven and rough, thus exacerbating further damage.

Erosion of concrete surface is another type of damage of concrete in contact with flowing water. The damage here is caused by solid particles carried by water, rather than by the water itself. Thus, water is no more than the transporting medium. Nevertheless, both in cavitation and in erosion, water is a factor in the development of damage.

Part III: Special properties

It may now be useful to consider some properties of concrete that are significantly affected by the presence of free water in the concrete.

Thermal insulation

Concrete is a relatively good insulator, and lightweight concrete is especially effective. Of particular interest is thermal conductivity; this is defined as the ratio of the flux of heat to the temperature gradient. Because the conductivity of air is lower than that of water, the conductivity of a given concrete depends on the degree of saturation with water of the voids in the concrete. The effect is particularly significant in lightweight concrete, which has a larger proportion of voids than ordinary weight concrete. For example, an increase in moisture content of 10% increases the conductivity of concrete by 50%. It follows that, if the insulating properties of concrete are of importance, it must be dried out and not allowed to become saturated. We can see thus that free water in concrete, either from the original mixture or ingressed later, has a considerable influence on the thermal insulation properties of concrete in service.

Electrical resistivity

Electrical properties of concrete are of importance in some applications. The resistivity of concrete is greatly affected by its degree of saturation. For example, air-dried concrete has a resistivity of the order of 10,000 ohm-m;[6] the resistivity of oven-dried concrete is about four orders of magnitude higher.[7] The topic is fairly complicated, but what is of essence is that electric current is conducted through moist concrete primarily by electrolytic means, that is, by ions in water in the capillary pores. Other things being equal, the less water in these pores, whatever its provenance, the higher the resistivity of the concrete.

Quantities of concrete and water used

In addition to everything discussed so far, there is another special relationship between concrete and water: as I mentioned at the outset of this section, they are the two materials most used by mankind, water being the leader and concrete the runner-up. The consumption of concrete, that is, the quantity of concrete placed per annum per head of world population— woman, man, child—is about 2½ Mg (2.8 tons).

The figure for water is less readily determined and depends of course on what uses of water represent consumption. Also, some of the water is treated and sold to households and industry; other water, for example, that used for cooling, is taken from a river and returned to it. It may be of interest that in England, the total consumption per annum per capita is about 130 Mg (143 tons), but there is a considerable variation worldwide.

Conclusions

I do not suppose that I have considered all the topics under the heading of "water and concrete," but I hope to have shown the intimate relationships between the two materials in the life of concrete. These relationships are numerous and they exist from the construction stage, right through life in service, including exposure to aggressive conditions. Sometimes the relationships are for good, sometimes for evil, but they are inevitable. I hope that the consideration of all of these relationships in the present article may help us in understanding the nature and behavior of concrete: such a wide-ranging approach is likely to be beneficial, as all too often studies are limited to a very narrow single relation under a single set of conditions.

Let me summarize in terms of the title of this section. On the one hand, it is not possible to make concrete without water: water is not just an "optional" ingredient. Furthermore, application of external water in curing and sometimes in cooling is highly beneficial to concrete. This is the love element of the relationship.

On the other hand, a great many mechanisms of deterioration and damage to concrete involve water as an essential factor. In those cases, water represents the hate element.

My wife of 50 years' standing, who has contributed a great deal to my "concrete thinking," would not allow me to liken this love-hate relationship to marriage. Let me just say that there is no doubt that, for the foreseeable future, we shall need not only water but also concrete: the two will have to live together.

References

1. Neville, A. M., "Water—Cinderella Ingredient of Concrete," *Concrete International*, V. 22, No. 9, Sept. 2000, pp. 66-71 [Section 2.2].

2. Neville, A. M., "How Useful is the Water-Cement Ratio?," *Concrete International*, V. 19, No. 9, Sept. 1999, pp. 69-70 [Section 2.5].

3. Neville, A. M.; Dilger, W.; and Brooks, J. J., *Creep of Plain and Structural Concrete*, Longman Group, London, 1983, 361 pp.

4. *Adam Neville Symposium: Creep and Shrinkage — Structural Design Effects*, SP-192, A. Al-Manaseer, ed., American Concrete Institute, Farmington Hills, Mich., 424 pp.

5. Aïtcin, P.-C., "Demystifying Autogenous Shrinkage," *Concrete International*, V. 21, No. 11, Nov. 1999, pp. 54-56.

6. Whittington, H. W.; McCarter, J.; and Forde, M. C., "The Conduction of Electricity Through Concrete," *Magazine of Concrete Research*, V. 33, No. 114, 1981, pp. 48-60.

7. Monfore, G. E., "The Electrical Resistivity of Concrete," *Journal of the Portland Cement Association*, V. 10, No. 2, 1968, pp. 35-48.

Section 2.2: Water—Cinderella Ingredient of Concrete

To say that water is a necessary accompaniment of cement is to state the obvious: you cannot make concrete or mortar without water; you cannot even make neat cement paste. And yet, of all the ingredients of concrete, water seems to have been treated like Cinderella: it has been the subject of the least amount of study, and hardly any of the studies are recent.

Not only is there a dearth of research data on the properties of water for making concrete, there is also an absence of standards or even serious guides on the properties of mixing water. This situation is the background to this section, the purpose of which is to collate our knowledge of the influence of mixing water on the properties of concrete and, even more importantly, to stimulate work in this area.

Strictly speaking, there are four uses of water that are of interest: water put into the mixture; water used in curing; water used to wash out mixers, agitators, and other equipment; and water used to wash aggregate. Although the various uses present some requirements in common, they are not identical. For the sake of clarity, I should add that mixing water includes crushed ice and ice shavings, and also the surface water on aggregate.

This section deals with water used in the mixture and for curing. Section 2.1 looks at the whole spectrum of relationships between water and concrete at various stages in the life of concrete and under many circumstances.

Standards for mixing water

A search for standards for mixing water has proved all but fruitless. There is no ACI document telling us what properties mixing water must have or must not have. There exists no British standard either. The nearest approach is an appendix to a British standard. Let me explain this rather feeble and timid effort.

There exists a document published by the British Standards Institution under the number BS 3148: 1980, titled "Methods of test for Water for making concrete (including notes on the suitability of the water)."[1] The wording as well as the use of capital letters are as in the original document, which is more than 20 years old. So nothing much has happened during the period: when the use of admixtures, especially high-range water-reducing admixtures (or superplasticizers), has burgeoned; when high-performance concrete as well as the use of silica fume has become firmly established; when ready-mixed concrete has become a dominant material; and when bigger, taller, and more complex concrete structures have been built.

The "notes" heralded in the title of the British document are contained in Appendix A (indeed, the only appendix).[1] A "note" has, to my mind, a distinctly noncommittal ring about it. Even though the British Standard puts the words "Guidance on the suitability of the water, based on the initial setting time test, is given in A.5" in parentheses, the wording lacks authority. "Guidance" is less definite than "guide," with the latter probably indicating good practice.

Although there are various ASTM standards for cements and cementitious materials, for aggregate, and for admixtures, there is no standard for mixing water for concrete in general. What exists is a section on water in ASTM C 94-00, "Standard Specification for Ready-Mixed Concrete." This section is brief and rather qualitative, being couched in terms like "The mixing water shall be clear and apparently clean" and expressions like "smell or taste unusual or objectionable." I shall consider these criteria in a general way in the next section. The only quantitative requirements in ASTM C 94-00 refer to compressive strength and setting time; again, these are subjects of a full discussion later in this section.

I am aware of only one English-language standard — Canadian Standard A23.1, whose latest version was published in 2000 (in French as well as in English). This standard contains a section on mixing water, but it is couched in very general terms about the presence of harmful material; for more specific information, reference is made to U.S. publications in the 1960s and 1970s.

As for standards and codes in languages other than English, they are not easy to find. All I have managed to establish is that neither the German nor the Swiss codes contains detailed requirements for the quality of mixing water. In any case, unless they have been translated into English and publicized, they are unlikely to be used in English-speaking countries or in international construction.

There exists a German memorandum on mixing water.[2] This is not a standard of the DIN (*Deutsche Norm*) type. This memorandum contains limit values on various ions and other substances in the water. The memorandum also refers to the test for the soundness of cement using the untried water; the need for this is not obvious.

What is unusual is that the German memorandum, as well as the German guide for the use of wash water,[3] gives detailed information on the possible presence in water of humins, which is that part of naturally occurring organic matter that is not alkali-soluble. The reason for avoiding an excessive amount of humins is that they interfere with the hydration of cement. As the term *humin* is not common, it may be helpful to quote its definition in *Webster's Dictionary*: "a bitter, brownish yellow, amorphous substance, extracted from vegetable mold, and also produced by the action of acids on certain sugars and carbohydrates."

The quantity of humins in water is acceptable if the color is lighter than yellowish brown and there is no smell of ammonia. (Ammonia would be the product of putrefaction of nitrogenous animal or vegetable matter.)

Almost as a curiosity, we can note that there exists a French standard (*Norme*) NF P 18-303, unaltered since 1941. It is limited to consideration of the quantity of solids and the quantity of dissolved salts without differentiation in their nature.

Since I wrote my original article, the European standard EN 1008:2002 on mixing water for concrete was published. The standard is fairly comprehensive.[4] It deals with a range of waters, including water from underground sources, natural surface water, and industrial waste water, as well as wash water and similar water from concrete processes. Sewage water is said to be "not suitable"; also seawater and brackish water are said to be "in general not suitable" for reinforced or prestressed concrete. In my opinion, the term "in general" used in a standard may give rise to confusion, if not dispute.

Preliminary assessment of an unknown water is made by: detecting oils and fats; foam stability; color; solids in suspension; smell; pH of at least 4; and humic (organic) matter. Limits on the content of chlorides, sulfates, sugars, phosphates, nitrates, lead, and zinc are laid down. Alkalies are limited if there is a risk of reaction with aggregate.

Failure to conform to any of the above requires tests on initial and final setting time of mortar, as compared with mortar made with distilled or de-ionized water. A percentage variation as well as an absolute minimum value of the time of initial setting and a maximum value of final setting are specified. Furthermore, the 7-day strength of mortar specimens must be not less than 90% of specimens made with distilled or de-ionized water. An annex to EN 1008, which is non-mandatory, describes a testing scheme.

ASTM advice on mixing water

Although I seem to deplore the absence of standards with which mixing water must comply, I realize that we do not live by standards and codes alone. Good, reliable, and up-to-date scientific information would also enable us to make correct and prudent decisions about the quality of an untried mixing water. Before reviewing the relevant publications, I would like to comment on what is probably the most easily accessible handbook-style information: a chapter in ASTM STP 169 C, published in 1994.[5] This is the fourth version of ASTM STP 169, the first one having been published in 1956. In that year, the chapter on mixing water was written by Walter J. McCoy, who updated the information for the 1956 edition. We are told that he made only minor changes for the 1978 edition.

Now, the chapter in the 1994 edition of ASTM STP 169C was written by James S. Pierce,[5] who says about it: "This current version is essentially Mr. McCoy's (now retired) chapter with minimal updating. There has been very little new technology published regarding mixing and curing water for concrete." In this, probably the most recent, major paper on mixing water, Pierce[5] says that "most references appear to be outdated"; indeed, they are all more than a quarter of a century old.

I hasten to say that just because a publication is old, it does not mean that its content is not valid. But what McCoy wrote in 1956, as well as the very extensive test results of Duff Abrams published by ACI in 1924,[6] was based on cements and mixtures of the day. Those cements were substantially different from modern portland cements in their chemical composition, fineness, and setting characteristics. Moreover, nowadays there exists a whole range of different cementitious materials. The mixture proportions, especially the water content and the values of the water-cement ratio (w/c), have changed, too. Furthermore, the parts of the world in which large-scale concrete construction takes place have extended enormously: in the various climates, and various topographic areas, water is not the same as what comes out of a tap in Farmington Hills!

Received wisdom on mixing water

Despite all this, the received wisdom with respect to mixing water remains as it was, and yet water is not just a condiment added to the dish: in every cubic meter of concrete, there are 130 to 200 L of water. Spending money on water of the right quality is sometimes frowned upon as extravagance. I have experience of construction in a remote part of the world where it was necessary to import cement, reinforcement, and some of the aggregate, but when it came to water, economy prevailed, the attitude being that "any water will do." It did not "do," and the consequences were very costly. I shall discuss this in some more detail in Section 2.4 dealing specifically with seawater in the mixture. Here, I would like to emphasize that water is a proper ingredient of concrete, and it should be factored into costs.

This is distilled (no pun intended) in a Portland Cement Association (PCA) publication by Kosmatka and Panarese,[7] in the opening words of Chapter 4: "Almost any natural water that is drinkable and has no pronounced taste or odor can be used as mixing water for making concrete. However, some waters that are not fit for drinking may be suitable for concrete." These statements may be true, but they are a rough-and-ready approach.

The opening word "almost" is wisely inserted. For example, mineral waters from springs, good for our health, can be harmful in concrete. Specifically, some natural mineral waters may contain alkali carbonates or bicarbonates that could be conducive to alkali-aggregate reaction.

As far as piped water is concerned, on one of the Galapagos Islands, I found the local drinking water to be distinctly brackish and quite undrinkable (I had to make do with beer!) even though the local people, who had no choice, routinely drank the available water. I have also heard about a case in the desert in Africa, where the specification laid down "drinking water," which, although drunk locally, turned out to be bordering on saline and unsuited for concrete-making. Likewise, water containing quite small quantities of sugar may be potable, but sugar would upset the time of setting.

At the other extreme, waters not fit for drinking because they are foul-smelling or have a disgusting taste may, in some cases, be perfectly satisfactory as mixing water. In other cases, if used to make concrete, they may create serious problems.

So there is no quick fix by determining the color or odor of an untried water, or even by establishing the presence of impurities: reliance on comparative tests is necessary. Whereas strength and the time of setting can be readily compared, longer-term effects are not easily established. In any case, a purely comparative case-by-case approach is rather primitive and not necessarily reliable. In particular, "impure" waters may vary from time to time.

We definitely need to have an objective way of determining whether the available water is suitable as mixing water or whether water has to be imported. This information in a reliable form is needed at the tender stage as the cost differential to the contractor can be significant.

My remarks are by no means a criticism of the PCA. It is simply that neither that organization, nor the rest of us, knows any better. So let me review what we do know or, more correctly, the views that have been published. I am using the word "views" advisedly because the relevant papers or parts of books rarely give test results demonstrating the limits of properties of water that have proved satisfactory or actual experience with waters demonstrating that exceeding some limits has proved to be unsatisfactory.

Paradoxically, water that attacks hardened concrete may, in some cases, be satisfactory as mixing water. For example, pure water, if flowing over a surface of concrete, would leach calcium hydroxide, but is good as mixing water.

Criteria for acceptance of mixing water

These criteria are of two kinds: performance requirements, and physical and chemical requirements. The performance requirements, which are purely comparative in nature, are the time of setting and compressive strength. The physical and chemical requirements refer to dissolved salts and solids in suspension.

Time of setting

An untried water can affect the time of setting through salts dissolved in it or through other impurities. ASTM C 94-00 (and we should recall that it applies to ready-mixed concrete only) requires the time of setting of cement made with the untried water to be not more than 60 min earlier and not more than 90 min later than when the test is performed on the same cement using distilled water or tap water. Appendix A to BS 3148:1980 (and we should remember that this is only a note) has more stringent limits: 30 min earlier or 30 min later.

The German guide[3] has different limits. With untried water, the time of setting should not be less than 1 h or more than 12 h, and should not differ from the time of setting of cement mixed with "good water" by more than 25% of that value.

Table 2.2.1—Extreme values of selected impurities in city water[7]

Ion	Content, ppm	
	Minimum	Maximum
Calcium	1.3	96
Magnesium	0.3	27
Sodium	1.4	183
Potassium	0.2	18
Bicarbonate	4.1	334
Sulfate	2.6	121
Chloride	1.0	280
Nitrate	0.0	2
Total dissolved salts	19.0	983

It is interesting to note that when the times of setting vary more than is acceptable, the German memorandum on mixing water recommends that the causes of this behavior be established.[2] Such an approach is more rational than a rejection of suspect water out of hand.

Compressive strength

ASTM C 94-00 requires the 7-day compressive strength of standard mortar cubes made with the untried water to be not less than 90% of cubes made with distilled water or tap water. The German guide[3] contains the same requirement. The British note also gives 90%, but the test is performed on concrete cubes at the age of 28 days.[1] The note also states that water that results in a strength reduction up to 20% can be acceptable, but the mixture proportions should be adjusted as appropriate. Given that we are not dealing with mandatory documents, this makes good sense.

Chemical tests

These tests are concerned with elements and ions present in an untried water. There is no difficulty in determining the quantities of these by standard chemical methods. It is, however, not obvious how much can be tolerated. One approach is to compare the results for an untried water with values for waters supplied in American cities. Unfortunately, the values readily available to concrete specialists (as distinct from city water engineers) date back to 1944.[7]

Moreover, in some cases, the range is enormous. As an example, some actual extreme values of ions in city waters in towns with a population of more than 20,000 are given in Table 2.2.1, extracted from Reference 7. It is, therefore, very difficult to say how much is too much, and to put specific values as requirements in a specification. Also, while any one solute at its maximum value may be satisfactory, when several of them are present in large quantities, they may interact with one another; we simply do not know what limits are reliable and truly required, but not unnecessarily overcautious.

Furthermore, untried waters that require chemical tests are likely to be either natural waters from wells in less accessible areas, or industrial or mine waste waters. In either case, the quality of the water is likely to vary from time to time, being affected by rainfall (wet or dry season) and other climatic changes, or by industrial operations that may influence the resulting waste water. It follows that periodic testing is likely to be necessary. It is important to remember that water is a powerful solvent and therefore is unlikely to be pure.

An easy test is to determine the pH of the water, and it is generally thought that this should not be outside the range 6.0 to 8.0. Test methods for this, as well as for chloride and sulfate content and also the content of inorganic and organic material, are prescribed by AASHTO T26-79.[8] The lower limit on the value of pH is prescribed not because of concern with hydrogen but rather because of some of the associated anions, such as chloride or sulfate. Also, an extremely low value of pH can have health effects on workers.

Overall, with respect to chemical aspects of water, I cannot do better than to direct readers to References 4 and 5. There is, however, a specific comment that I would like to make about the presence of chloride ions in an untried water. We know that chloride ions in the mixture, when present above a certain threshold, are conducive to corrosion of reinforcement or embedded metal. Alas, there is no universally agreed value of the threshold, except that in prestressed concrete, an extremely low value is necessary: 0.06% of water-soluble chloride ion by mass of cement.

The difficulty in limiting the chloride ion content in water is that water is by no means the only source of chlorides in the mixture. Chlorides can be present on the surface of the aggregate particles and also in portland cement (although the latter quantity is very small). What matters in practice is the total quantity of chlorides per cubic meter of concrete, whatever their provenance. It follows that a certain water may be acceptable for one kind of concrete but not for another. I shall discuss the progress in our learning about the use of seawater (and this, of course, contains chlorides) in the mixture in Section 2.4. At this stage, I shall point out only the fact that seawater in the mixture can cause efflorescence or dampness on the surface; these would mar the appearance or even upset an applied finish.

To give a sense of proportion to the content of dissolved salts in mixing water, it is useful to look at an example of the quantities involved. If the content of a given salt is 2000 ppm and the water content is 150 L/m^3 (30 gal./yd^3) of concrete, the content of the dissolved salt is 300 g/m^3 (8 oz./yd^3) of concrete.

Solids in suspension and algae

Some waters contain silt or clay particles in suspension. These very fine particles can affect some properties of concrete, possibly aggregate bond. The adverse effect of excessive amounts of very fine particles upon compressive strength is small, but they can increase the water demand. The use of settling basins is an effective way of reducing the content of such particles, but it requires appropriate space and pumping facilities. There remains, of course, the question of disposal of the waste, and modern legislation on waste disposal generally precludes dumping.

Algae may entrain air, with a concomitant reduction in strength, but quantitative data are lacking. This is not surprising, given the variety of algae and the difficulty of quantifying their presence.

Wash water

In the past, when mixers or agitators were washed out, wash water was just water down the drain, so to speak, finding its way into a stream or a lake. Nobody cared about the ultimate disposal of wash water and, at the same time, fresh water was plentiful and cheap. In the last two decades, the situation has changed dramatically, and there now exists, in many parts of the world, legislation controlling the discharge of polluting waters, including wash water.

Moreover, when water is scarce or expensive—and there is no doubt that it will be increasingly so in the future—there is an economic incentive to reuse the wash water. Such water can generally be used as mixing water, preferably if blended with fresh water. In this manner, the limit on total solids of 50,000 ppm, optionally recommended by ASTM C 94-00, can be satisfied. The water should not produce films or coatings on the surface of aggregate that could interfere with its bond, nor should the water contain admixtures. It can be useful to point out that the wash water has to be of good quality in the first place.

The section in ASTM C 94-00 on the use of wash water as mixing water, while helpful, is not highly prescriptive. The German Committee on Reinforced Concrete[3] offers much more detailed guidelines. They deal with wash water arising from various operations connected with fresh concrete, as well as sawing and grinding hardened concrete. The guidelines deal also with wash water from cement-lime mortar because of the consequences of introducing sulfates from that source. The sulfates originating from lime can affect setting and hardening; furthermore, by their presence in the hardened concrete, they can influence the durability of concrete.

Although this section does not purport to give detailed information on assessing the suitability of an untried water as mixing water, some of the German guidelines[3] are worth quoting. In addition to considering color (at most, pale yellowish), they refer to oil and grease (traces only); oil need not be harmful, but it can affect the bond of aggregate. The guidelines also consider chlorides and detergents (small, unstable foam formation); they can entrain air and therefore reduce strength.

If the wash water fails to satisfy the various criteria, it can still be used

Chapter 2: Water and Concrete

when blended with "good" water to such an extent that the resulting water conforms to the values in the guidelines.[3] Specific rules on the frequency of testing are given.[3]

The German guidelines[3] contain an ingenious, yet simple, approach to determining the content of solids: the density of the wash water. To facilitate the calculation of the blend proportions, there is a table that interprets the density of wash water in terms of the solids content as a function of the amount of wash water per cubic meter of concrete. When blending wash water with "good" water, it is important that the solids are uniformly dispersed.

As wash water contains alkalies from the washed-out cement, the German guidelines[3] give specific advice on the use of wash water with potentially alkali-reactive aggregates. When the cement content is higher than 400 kg/m^3 (675 lb/yd^3), it seems best not to use wash water at all.

If any cleansing agents are added to the wash water, it is essential that they do not interfere with setting or hardening.[3]

There is one other respect in which the use of wash water is of particular importance: the influence of hardness of water on the efficacy of air entrainment. Water is said to be hard when it contains a significant amount of calcium ions. Pierce[5] states that there is no effect, but the data were published in 1946 and have not been verified since then. More recently, Gaynor[9] advised that air-entraining admixtures should not be allowed to come into contact with hard water, so that they should be batched into the mixer with clean water or with sand; my view is that the latter procedure may not lead to a uniform dispersal of the admixture. The German guidelines[3] simply forbid using wash water in air-entrained concrete. This seems unduly restrictive for American usage.

Hardness of water is also of potential interest because of a possible interaction with some admixtures—something that is not easily determined *a priori*. I believe that the issue of the influence of hard water on air entrainment should be properly investigated.

Use of domestic waste water

Domestic waste water need not just be water wasted. Admittedly, biological treatment to reduce the content of pathogenic bacteria and viruses is essential before the water can be handled by humans, and a reduction in the content of organic material is also desirable. Such treated water is used in various parts of the world to irrigate trees and grass, and also for street washing. Warnings are posted to prevent accidental ingestion of this water, especially by children. There is little doubt that an appropriate use of domestic waste water will continue to grow. Will this include mixing water for concrete?

So far, there is very little experimental evidence on the use of biologically treated domestic waste water as mixing water. Cebeci and Saatci[10] showed in laboratory tests that such use is feasible. Some support for this was found in laboratory tests in Qatar.[11]

At the present time, a generalization is not possible, especially in view of a variability in the domestic wastewater in consequence of differences in water consumption (and therefore dilution of sewage) and of the actual water treatment used. Nevertheless, there is bound to be an increased economic and ecological pressure to reuse more wastewater. Compromises are possible by way of a more appropriate water treatment. In addition, as shown in the laboratory,[11] blending biologically treated domestic wastewater with good water can be a way forward. In my opinion, the use of recycled domestic waste water is bound to increase not only in the production of concrete, but within a few years, also for human consumption.

Water for curing

It might be thought that the requirements for mixing water should also apply to curing water and that water suitable as mixing water is also satisfactory for curing purposes. However, this is not always so. For example, pure water is entirely satisfactory as mixing water but, as already mentioned, if flowing over or in prolonged contact with concrete, such water would leach some of the calcium hydroxide. Conversely, water containing excessive solid matter or water containing alkali carbonates or a residue of admixtures is harmless when used for curing.

Moving away from a comparison with the requirements for mixing water, it is obvious that curing water should be free from substances that attack hardened concrete or embedded reinforcement.

There is one warning that I would like to add on the basis of my experience. If a reinforced or prestressed concrete structure intended for immersion in seawater is built onshore, it is vital that the concrete is not cured by seawater until it has matured substantially. Premature exposure to seawater, especially if the concrete has been allowed to dry out, will result in imbibition of seawater with the associated risk of corrosion of the reinforcement.

A specific requirement for curing water is that it does not contain substances that will cause staining or discoloration of surfaces to be exposed or to be subsequently treated and visible. A performance test for the purpose of assessing the risk of staining is prescribed by the U.S. Army Corps of Engineers.[12]

Conclusions

This section was not intended, and does not purport, to be a state-of-the-art guide on the requirements for the quality of mixing water. Relying on what we know today, it is not possible to establish safe, but not uneconomic, limits on the properties of mixing water: much work needs to be done.

My aim was to indicate the relevant criteria and say enough about them in order to show how tentative and fragmentary our knowledge is. Worse than that: much of it is old, and it has been handed down the years and quoted again and again. But, in the intervening years, the world of concrete has changed quite significantly: not only have the properties of portland cement evolved, but we use a number of other cementitious materials as well. The range of water-cementitious materials ratios (w/cm) has been extended downwards, and the use of powerful water-reducing admixtures (that is, superplasticizers) has greatly altered the volumetric proportion of mixing water in concrete.

We could ask: what improvements in our knowledge of the required or desirable properties of mixing water have taken place? The answer is: almost none. At first sight, this situation is surprising in its contrast with advances in our knowledge of all the other ingredients of concrete, but there is an explanation. It is this: no single firm, nor even a trade association, has a commercial interest in tackling the problem and in preparing a standard, or even a serious guide, for mixing water. To achieve this would require undertaking extensive testing and research evaluation. No one has to gain by it, so no one does it. Amongst all the advances in concrete, improving our knowledge of requirements for mixing water is treated like Cinderella.

Establishing the requisite knowledge will make it possible to widen the safe use of wastewater, both domestic and industrial. This will contribute to the economy of making concrete and thus to extending its use.

In the meantime, it seems well worthwhile for ACI to publish (with an appropriate permission) an English translation of the German documents listed in References 2 and 3; together they are only 16 pages long but their content is really valuable.

By remaining largely ignorant, we are likely to be losers through not being able to use the most economic sources of water at one extreme, and at the other, avoiding problems with concrete that, unbeknownst to us, can be the consequence of some undesirable property of water that has been used as "good mixing water."

Realizing this situation is perhaps not quite half the battle, let me be realistic and say: it is a quarter of the battle. The remainder of the battle still to be won is for someone to start the appropriate research. That someone could be the PCA, which 80 years ago sponsored the historic and splendid work of Duff Abrams.[6] It could be a cooperative undertaking, which would be particularly appropriate for studying various waters and various cementitious materials as well as types of admixtures. I do not believe that the work will be very difficult, but it will be time-consuming. With luck, the work will be accomplished before water shortage begins to bite in many parts of the world and before fiscal measures impinge seriously on a profligate use of high-quality pure water (needed to accompany whisky). Let battle commence!

References

1. British Standard BS 3148: 1980, "Methods of Test for Water for Making Concrete (including Notes on the Suitability of the Water)," British

Standards Institution, London, 1980, 3 pp.

2. Deutscher Beton-Verein, "Zugabewasser für Beton," Merkblatt, 1996, 12 pp.

3. Deutscher Ausschuss für Stahlbeton, "Richtlinie für Herstellung von Beton unter Verwendung von Restwasser, Restbeton und Restmörtel," 1995, 6 pp.

4. Plumat, M., "L'eau de Gâchage," *Les Bétons,* Eyrolles, Paris, 1996, pp.143-150.

5. Pierce, J. S., "Mixing and Curing Water for Concrete," *Significance of Tests and Properties of Concrete and Concrete-Making Materials,* ASTM STP 169C, West Conshohocken, Pa., 1994, pp. 473-477.

6. Abrams, D. A., "Tests of Impure Waters for Mixing Concrete," ACI *Journal, Proceedings,* V. 20, 1924, 44 pp.

7. Kosmatka, S. H., and Panarese, W. C., "Mixing Water for Concrete," *Design and Control of Concrete Mixtures,* 6th Canadian Edition, Portland Cement Association, 1995, pp. 32-35.

8. AASHTO T 26-79, "Standard Method of Test for Quality of Water to be Used in Concrete," AASHTO, 1979, p. 27.

9. Gaynor, R. D., "Ready-Mixed Concrete," *Significance of Tests and Properties of Concrete and Concrete-Making Materials,* ASTM STP 169C, West Conshohocken, Pa., 1994, pp. 511-521.

10. Cebeci, O. Z., and Saatci, A. M., "Domestic Sewage as Mixing Water in Concrete," *ACI Materials Journal,* V. 86, No. 5, Sept.-Oct. 1989, pp. 503-506.

11. El-Nawawy, O. A., "Use of Treated Effluent in Concrete Mixing in an Arid Climate," *Cement and Concrete Composites,* V. 13, 1991, pp. 137-141.

12. U.S. Army Corps of Engineers, "Method of Test for the Staining Properties of Water," CRD-C 401, Vicksburg, Miss., 1975.

Section 2.3: Effect of Cement Paste on Drinking Water

The title of this section has not mistakenly reversed the words used. Admittedly, when considering concrete or mortar in contact with water, the common concern is the attack upon hardened cement paste by water, usually containing aggressive agents. In Section 2.1, I reviewed the broad topic of the numerous relationships between concrete and water.[1] That review included the action of aggressive waters on concrete, but recognized that pure water is also aggressive in that it can leach calcium hydroxide from the hydrated cement paste. These reactions are not surprising because water is a powerful solvent and an all pervasive agent.

There is, in addition, one other situation in which there may be a significant interaction between water and concrete or mortar. It is the action of hardened cement paste on water, and such action may come as a surprise to some readers. The problem is real when the water is destined for human consumption as drinking water. This is the subject of the present section.

The practical importance of this problem lies in the fact that leaching of hardened cement paste adversely affects the quality of drinking water, and regulations are being introduced to limit the allowable amount of the leachate. This is likely to impinge on the use of portland cement in conduits for drinking water unless the problem is resolved so that water is not "attacked" by cement.

Why is there a problem?

The quality of drinking water must be satisfactory not only at the exit from the water treatment plant but also at the point of delivery to the user. Hence, any effects of conduits on the water being transported may be of importance.

For more than a century, conduits were made of ferrous metal. With time, problems arose owing to the corrosion by electrochemical action, leading to scaling or leakage. Thus the usual concern was with the deterioration of the pipe by reaction between the metal and the aggressive chemical species in the water, rather than with the effect on the quality of water in consequence of the reaction between the pipe material and the water being transported.

Since the mid-1980s, in the United Kingdom, the rehabilitation of deteriorated conduits has often involved lining the interior of the metal pipes with cement mortar, thereby providing an electrically inert material between the metal and the water.[2,3] This lining is placed by inserting a plug of cement mortar and passing a mandrel to distribute the mortar on the surface of the pipe and to ensure a smooth finish.

Whereas the use of cement mortar lining prevents the corrosion of the ferrous metal pipe, the reaction between the hydrated cement and water

Fig. 2.3.1: A 7.6 m internal diameter pipe; crack repairs visible

leads to another type of "corrosion," that is, attack by cement on water.

More recently, steel or ductile-iron pipe, usually ranging in diameter between 80 mm and 1.6 m, has been manufactured with mortar applied in the factory as a part of the production process. The mortar is centrifuged onto the inner surface of the pipe and then heat cured, using a maximum temperature of about 55°C. Thus the manufacture of the pipes can be completed in one day, and thereafter the pipes are stored outdoors.[4]

The lining thickness ranges between 1.5 and 3 mm, being greater in larger pipes.[3] Sometimes, an asphaltic seal coat is applied to the water-facing surface of the mortar lining, thus preventing contact between the hardened cement paste and the water. However, such a seal coat may contain extractable organic compounds;[3] the problem of these compounds entering the water is not considered in this section.

In addition to pipes lined with cement mortar, there are used also concrete pipes, including large-diameter prestressed concrete pipes. These carry water both for agricultural purposes and for domestic use. I was involved in some investigations of prestressed concrete water pipes, 7.6 m in internal diameter, extending over several thousand kilometers. Such a size is not easy to appreciate (Fig. 2.3.1 and 2.3.2); each segment weighs 90 tonnes. A very ambitious, albeit somewhat smaller, project in Europe is the proposed pipeline, 3 m in diameter, to transport water over a distance of 520 km, from the river Rhone near Montpellier, in France, to Cardedeu near Barcelona, in Spain.

As far as the effect of hardened cement paste on the quality of drinking water is concerned, there is no difference between the use of mortar in lining applied in the manufacture of new pipes, or in the rehabilitation of old ones, or when the pipe is made of concrete. What is crucial is the

Fig. 2.3.2: Pipes ready for the construction of a 2000 km pipeline

presence of the hardened cement paste, and this is why I have used the word "cement" in the title of this section. What we want to know is the kind of chemical reactions involved and their effect on the quality of the drinking water. Such information can lead to guidelines on the use of cementitious materials in water conduits.

One further clarification may be useful. I am using the term "drinking water" to include not only water for drinking but also for cooking, washing, and for food production. The generic term used in the European Union documentation is "water intended for human consumption." The conduits of concern are not only water-supply pipes but also service reservoirs, water towers, and various ducts. The conduits may be under the control of public utility companies or they may belong to industrial or domestic users. The latter two categories are likely to be much less aware of potential problems than public bodies, which are directly controlled by authorities enforcing health requirements. A recent, but important, authority is an organ of the European Union.

It is worth noting that public supply of drinking water does not exist on a world-wide basis. For example, in Taiwan and Turkey, the water supplied by the public network is generally not intended to be used for human consumption. A separate system of providing containers of such water is in operation. With an increasing shortage, in many parts of the world, of water that can be treated to a degree fit for human consumption, the use of a dual provision of water—for drinking and for other uses—is likely to grow.

What is the problem?

The problem of the attack on water by hardened cement paste can be approached from two angles: cement chemistry and concrete technology. I am sure cement chemists know a great deal about leaching of hydrated cement paste, and have extensive data about the solubility of the various compounds present in the paste. My concern is with the influence on the quality of drinking water of the various ions and species that enter the water in the conduit in consequence of leaching.

The main problematic issues arising from the attack by the hardened cement paste are the raised pH of the water and its increased content of $CaCO_3$, referred to as carbonate alkalinity or water hardness. The increase in $CaCO_3$ is induced by carbon dioxide dissolved in the water and a reaction with calcium hydroxide in the hydrated cement paste. There may also be a concern with an increased content of aluminum, as well as of calcium, sodium, and potassium, all in consequence of leaching by the water.[2] Furthermore, the admixtures used in mortar or concrete may also enter the water; this last topic will be considered in a separate subsection.

Where does the problem occur?

It is important to bear in mind that the extent of leaching by water in contact with mortar or concrete depends on two major factors: the volume of water involved in relation to the internal surface of the lined pipe, and the length of time during which the water is in contact with the hardened cement paste in the lining (or with the pipe itself if it is made of concrete). The first factor is geometrical, and could be expressed as the volume-surface ratio. Thus, in a large reservoir or tank, where the volume-surface ratio is large, the potential for leaching afforded to any unit volume of water is small; consequently, in such situations the extent of leaching is not of importance unless the period of storage is very long.

The time factor means that water flowing continuously in a conduit has only a short period during which leaching can take place so that, again, the extent of leaching is not of importance. In other words, there are no problems when the pipes have a large diameter and the flow is fast and continuous. However, when the pipe has a small diameter or when the flow stops at times or when its velocity becomes very low, the situation allows a considerable amount of leaching to take place.

The notion of small-diameter pipes containing nearly static water may seem unusual. This is, however, the case when users of small amounts of water are distant from one another. Even when there are many domestic users in close vicinity, there occur periods with a very low flow velocity near dead ends of pipes. This is also the case in residential areas where water consumption during the night is minimal; the same situation can occur in some industrial areas at night or at weekends. I can vouch for the occurrence of very low flow rates: in my apartment, shower taps of unusual design that have been fully turned off in daytime sometimes drip at night. This is due to an increased hydrostatic pressure in the absence of friction losses when a very small amount of water flows through the supply pipes; in other words, the water in the pipe is nearly static.

Does leaching matter?

The potential problems in conduits in which hydrated cement paste comes into contact with drinking water have been known for some time, but a serious consideration of these problems, including specifications for concrete or mortar, is recent, at least in the United Kingdom and in many other European countries. I believe there are two reasons for this. First, in the past, there did not seem to exist much established knowledge of the levels of the various chemical species present in the hydrated cement paste giving rise to health hazards. Providing the water was biologically harmless, and did not have an unpleasant smell or taste, or an excessive turbidity, it was accepted as it stands (or flows). It is only when water is added to whisky that its taste and pH matter: when drinking a single malt whisky, cognoscenti like to add water from the same source as was used in the malting.

The second reason is that in the past there existed few rules or regulations with respect to water at the point of delivery as distinct from the exit point from the water treatment plant.

Excessive values of pH in concrete pipe came to notice in the United Kingdom in the 1980s, but investigations of this phenomenon started only around 1990, largely owing to health directives from the European Union.[2]

I should add that the European Union rules on construction materials in contact with water—a varied and complex situation—are still being evolved. This is not surprising, given that the European Union itself is still very much in the process of development and evolution. Thus, the European Approval Scheme for water conduits will come into being in the year 2005 at the earliest.[5] As from the year 2000, in the United Kingdom, there exist some regulations requiring approval of the use of concrete pipe or in-situ applied cement mortar linings. However, it will be quite some time before there is a serious effect on the actual level of leachates in drinking water.

Specifically, the introduction of new British regulations or European Union directives does not necessarily alleviate the existing problems because products (such as cement mortar linings) that have been in use for at least one year prior to the new regulations are considered to be "traditional" and, consequently, are exempt from the new requirements. More than that: there is no need to test the old products and their effect may remain undetected for a long time.[2] I dare say this approach is fair in not hurting the existing pipe manufacturers, but there remains the question of risk to consumers. It is not obvious where the right balance lies.

While this section is concerned only with conduits for water at ambient temperature, I should add that, in some countries, the rules about materials in contact with drinking water are applied also to hot water because people may consume (by design or by accident) water from hot systems.[5]

Although solubility is affected by temperature, perhaps this is a case of excessive caution.

Why does leaching vary?

The attack of hardened cement paste on water is very much dependent on the hardness of the water in the conduit, referred to as carbonate alkalinity. In the United Kingdom, many public utilities supply water with carbonate alkalinity of less then 55 parts per million (ppm) of calcium carbonate. It is in waters with a low carbonate alkalinity that leaching leads to a high value of pH of the water and to a higher content of aluminum.

It could be argued that, to prevent excessive leaching in conduits, the hardness of water should be deliberately raised in the water treatment plant. However, there are health concerns about ingestion of hard water, as well as practical consequences of scaling in domestic pipes, kettles, water tanks, and water heaters. In my opinion, therefore, it is not the water but the cementitious material that has to be adapted.

What is tolerable?

I have mentioned the fact that one important, perhaps the most important, effect of leaching is a rise in the value of the pH. In water treatment plants, the value of the pH is lowered to purify the water bacteriologically, then raised, and finally the drinking water enters the distribution conduits with a value of the pH typically somewhat above 7, that is, very slightly alkaline. It is generally accepted that, for health reasons, the value of the pH at the point of delivery should not exceed 9.5, but values well in excess if 10, or sometimes even more, have been found. Occasionally, there have been complaints about skin irritation developed by contact with such water.[3]

Excessive carbonate hardness, as I have already mentioned, is undesirable with respect to water installations and is probably not good for human health. However, hard waters are supplied in many areas so that, when hard water is the consequence of attack on naturally soft water (see section on tests) by hydrated cement paste, the situation is no worse that what occurs naturally elsewhere.

Leaching can introduce excessive amounts of different species into the drinking water. Limiting values of noxious levels have not been established, and cannot be easily determined, because much depends on the amount of water drunk by an individual and on whether the given water is drunk regularly or only occasionally. Moreover, testing humans up to the noxious limit is hardly a sensible approach.

In consequence, at least for the time being, all we have is a series of guidelines, rules, and directives promulgated by national governments and by the European Union.[4,6] Some limiting values for different species, extracted (no pun intended) from the relevant documents, are given in Table 2.3.1 in parts per million (ppm) of the water.

Table 2.3.1—Limiting values of some species in drinking water[4,6]

Species	Guide level: ppm	Maximum admissible level: ppm
Aluminum	0.05	0.2
Potassium	10	12
Sodium	20	150
Calcium	100	—
Nitrate	25	50
Sulfate	25	250
Arsenic	—	0.05
Mercury	—	0.001

What do tests show?

The nature of the problem is such that it is difficult to obtain data on the actual quantities of the various species in pipes containing stationary or slowly moving water. It is not practicable to take samples of water in such a way that the contact time between the water and the hydrated cement paste is known. Moreover, the characteristics of the supply water vary between different supply plants. For example, in the United Kingdom, the content of calcium carbonate ranges between 8 ppm and 160 ppm. This complicates an interpretation of any actual data, and yet information about the input characteristics of water and its "exposure" conditions is necessary in order to establish the best, and the worst, mortar mixtures for waters of various types.

Because of these difficulties, laboratory tests have been performed under clearly defined conditions. So far, the results available are meager. The question is: what kind of cementitious material should be used in order to minimize deleterious leaching? As this section is intended for a wide readership, I shall limit myself to broad brush statements.

It is important to establish the circumstances under which excessive leaching can take place. An investigation was performed by the British Department of the Environment.[7] Field tests were performed using in place mortar lining made with Type I portland cement in various parts of the United Kingdom so that waters of different quality were involved. Typically, the water-cement ratio was between 0.29 and 0.32, and the sand-cement ratio between 1 and 1.5; the sand was siliceous.[4]

The critical parameter was found to be the calcium carbonate alkalinity of the water in the conduit. This influences the period during which leaching may cause a high value of pH. Three levels of alkalinity of the water being transported in the conduit were distinguished; very low alkalinity (about 10 ppm of calcium carbonate); low alkalinity (about 35 ppm of calcium carbonate); and high alkalinity (more than about 55 ppm of calcium carbonate).

In the last-mentioned case, the adverse effect of hydrated Type I cement paste disappears after about one week, so that there are no problems in practice.[7] With low-alkalinity waters, pH rises above 9.5 but this lasts no more than one month. It is only very-low-alkalinity waters that create serious problems: pH can stay above 9.5 for several years. Furthermore, the amount of aluminum leached and present in the water can exceed 0.2 ppm during a period of one or two months,[7] and can even reach 0.3 ppm for a shorter time.[4]

These data lead us to consider whether cements other than Type I perform better. From laboratory tests that followed the field tests, it appears that a blended cement containing about 65% of ground-granulated blast-furnace slag is much more satisfactory than Type I cement alone.[7] Field trials have confirmed that the use of slag is beneficial in the case of high alkalinity waters; with low and very low alkalinity waters, the reduction in pH was small and took several weeks.[4] Indeed, with very low alkalinity waters in the conduit, Type I cement alone should not be used.[7]

Other tests have confirmed that the use of blended cement containing 65% ground-granulated blast-furnace slag results in a lower rise in the pH, but only after a period of several months.[3] However, this is not always the case, and different test conditions yield different results so that Douglas et al. concluded "that performance at one site does not necessarily translate to other sites."[3] This statement, albeit true, is unhelpful, as is also Douglas's recommendation to "regularly flush low-flow sections (of pipe) to prevent buildup of components causing adverse water quality effects."[3]

Thus, it seems that the validity and reproducibility of small-scale laboratory tests are not yet satisfactory.[2] In my opinion, if we are interested in continuing use of portland cement in water conduits, we need more tests in order to obtain a better understanding of the underlying problem of attack on water.

In passing, we should note that the use of a blended cement containing ground-granulated blast-furnace cement slag in mortar linings is permitted in the United Kingdom, provided the carbonate alkalinity of water is not less than 40 ppm.[2]

The behavior of slags originating from heavy metals was investigated, and it was found that there was no significant leaching of these heavy metals.[2]

Inclusion of fly ash in the mixture at a level of 35% of the cementitious material did not result in an improved performance.[4]

Can admixtures be used?

As it is common to include one or more admixtures in mortar or concrete, any possible effect of admixtures on drinking water should be examined.

The British Water Inspectorate issued, in May 2000, a draft list of authorized admixtures.[8] Most of the usual chemical admixtures are permitted; these include accelerators, retarders, air-entraining admixtures, as well as butyl stearate water repellent, and superplasticizers, both naphthalene- and melamine-based. They can be single-acting or acting in combination. However, corrosion inhibitors are expressly excluded because tests have demonstrated that corrosion inhibitors exhibited prolonged leaching and concentrations that could present a risk to health.[8] The letter from the Inspectorate adds that it "has received advice that corrosion protection for reinforcing bars is unnecessary in concrete mixtures intended for use in water retaining structures."[8] I, for one, would not subscribe to such a blanket opinion.

The European Union rules, still to be promulgated, will also deal with curing compounds and formwork release agents, as these can persist on the surface of the concrete and eventually enter the water in contact with that surface.[4] Organic fibres included in concrete should also be assessed for any possible interaction with water.

There still remains an uncertainty about the effects of organic compounds used as grinding aids in the production of portland cement, as well as of some organic plasticizers used in concrete. Given that pore water in hydrated cement paste contains these species,[4] their possible effect on the quality of water may need further study.

Mortar containing organic admixtures or plastic additives as bonding agents has been observed to lead to increased microbial growth, thus possibly adversely affecting the quality of the water.[9]

Can high-alumina cement be used?

As a possible means of avoiding the rise in the value of pH, it was suggested in France that high-alumina cement be used instead of portland cement.[10] However, high-alumina cement allows leaching of aluminum products in excess of values recommended by the European Union, namely, 0.2 ppm of aluminum[2,10] (see Table 2.3.1). While tests using portland cement showed aluminum contents in water up to 0.3 ppm,[2] Lawrence reported Conroy's test on mortar with high-alumina cement, which showed that, during early contact between stationary deionized water and mortar, the aluminum contents were 40 times higher than with portland cement.[4] It is therefore not surprising that the use of high-alumina cement is not recommended by the Concrete Society.[11]

On the other hand, the recommendations of the manufacturers of high-alumina cement are less clear.[12] They recognize that "in large pipes with a high volume of water flow the aluminate concentrations in the water will not be significantly affected and will remain well below acceptable levels."[12] That, of course, is well-known and was stated at the outset of this section. However, the manufacturers of high-alumina cement say that in waters with a low calcium carbonate content, the high-alumina cement "can provide a solution"[12] but qualify that statement by saying that "careful assessment should be made of the possible impact of aluminate leaching."[12]

In my opinion, given the actual test data of independent investigators reviewed in the preceding paragraphs, it should be accepted a priori that high-alumina cement is unacceptable for use in lining-mortar because of the high level of aluminum leached by the water in the conduit.

Are there any conclusions?

This section is not of the kind that lends itself readily to a summary of clear findings that can be converted into conclusions. Rather, the section identifies a problem that is not well known and has so far not been tackled adequately. For these reasons, the best conclusion is by way of some recommendations.

The review of the problem and the discussion have shown that the wide range of interactions between concrete and the ambient water considered in the past[1] is even wider than is generally realized. The problems reviewed in this section are somewhat specialized and, for this reason, probably not well known. But the avoidance of attack by concrete or mortar on water is important if the market for the concrete conduit or mortar-lined pipe is not to be lost. Further studies are required, and several directions need to be pursued: What levels of the various chemical species are reached under *practical* conditions? What levels are *really* higher than tolerable? Which cementitious materials minimize leaching of harmful species? Are there any preferable layouts and methods of operation (for example, flushing) of pipes?

Clearly, testing is required. However, because of the dependence of the attack on the properties of the water involved, any tests should be performed using waters simulating those likely to be present in the actual conduits. If, say, deionized water is used, all portland cements are likely to fail.[2]

It is important to establish the various parameters from the engineering standpoint before health agencies impose rules and limits that are unnecessarily stringent, just to be on the safe side. Such an approach might harm the concrete market.

Unlike the common concern about the attack of various waters on concrete, in this section, I have reviewed the rather unusual situation where water is the "victim" and the concrete is the "attacker." The concrete remains in good health but the water is adversely affected. This is reminiscent of Oliver Goldsmith's "Elegy on the Death of a Mad Dog" where he says:

"The man recover'd of the bite,
The dog it was that died."

We must ensure that both the mortar-lined or concrete conduits and the water in them remain healthy.

References

1. Neville, A. M., "Water and Concrete: A Love-Hate Relationship," *Concrete International*, V. 21, No. 12, Dec. 2000, pp. 64-68 [Section 2.1].

2. Conroy, P.; Fielding, M.; and Wilson, I., "Investigating the Effect of Pipelining Materials on Water Quality," *Water Supply*, V.11, No. 3-4, Berlin, 1993, pp. 343-354.

3. Douglas, B. D.; Merrill, D. T.; and Catlin, J. O., "Water Quality

Deterioration from Corrosion of Cement-Mortar Linings," *Journal of American Water Works Association*, V. 88, No. 7, 1996, pp. 99-107.

4. Lawrence, C. D., "International Review of the Composition of Cement Pastes, Mortars, Concretes and Aggregates Likely to be Used in Water Retaining Structures," Department of the Environment, London, England, 1994.

5. Taylor, M. G., "Concrete in Contact with Drinking Water – Developments in Europe," British Cement Association, May 2000.

6. British Standard BS 6920: Part 1: 1990, "Suitability of Non-Metallic Products for Use in Contact with Water Intended for Human Consumption with Regard to their Effect on the Quality of Water: Specification," British Standards Institution, 1990.

7. Conroy, P. J., "Deterioration of Water Quality in Distribution Systems: The Effect of Water Quality Arising from In Situ Cement Lining," Water Research Centre, Swindon, England, 1991.

8. Drinking Water Inspectorate, "List of Authorised Cement Admixture Components," Department of the Environment, Transport and the Regions, London, May, 2000.

9. Schoenen, D., "Influence of Materials on the Microbiological Colonization of Drinking Water," in *Microbiology in Civil Engineering*, P. Howson, ed., Spon, 1990.

10. Wagner, I., "Internal Corrosion of Pipes in Public Water Distribution Networks," International Report, DVGW Research Institute, Karlsruhe, Germany, 1998.

11. Concrete Society Working Party, "Calcium Aluminate Cements in Construction – A Re-Assessment," *Technical Report*, 46, The Concrete Society, England, 1997.

12. Scrivener, K. L. and Capmas, A., "Calcium Aluminate Cements," *Lea's Chemistry of Cement and Concrete*, 4th Edition, Arnold, 1998, pp. 709-778.

Section 2.4: Seawater in the Mixture

In Section 2.2, which comprises a recent article[1] I discussed our knowledge, still imperfect, of the required quality of mixing water, but I did not include the consideration of seawater. My reason for this was that to most engineers practicing nowadays in Western countries, it is almost axiomatic that seawater must never be used in reinforced concrete; never, never be used in prestressed concrete; and preferably not be used in plain concrete, which often contains some steel. The crucial reason for this is the corrosion of steel (not only reinforcement but also inserts such as anchor bolts) owing to the presence of seawater in the mixture.

For the sake of completeness, I should add that, in addition to the water actually poured into the mixer, seawater may enter the mixture via the aggregate if this is sea-dredged or won from the seashore and not washed. Also, even clean aggregate could be sprayed by seawater if the aggregate is transported by a barge without protection. Cooling the aggregate by spraying it with seawater would have the same result.

Previous positive attitude to seawater

The attitude described in the preceding paragraphs is actually fairly recent, and I thought it might be instructive to see how our knowledge has developed and improved.

Earlier on, there was a qualified approval of the use of seawater as mixing water; for example, in 1970, Taneja wrote: "With proper mix design, quality, and workmanship it may be appropriate to use seawater for making reinforced concrete."[2] A permissive approach was adopted in the *Overseas Building Note* of the Building Research Establishment, published in 1971: "In arid countries it is often necessary to employ seawater or brackish well water."[3]

To my surprise, I found that even in recent years papers were published under titles such as "Seawater for Concreting?"[4] in 1988 and "Suitability of Seawater for Mixing Structural Concrete Exposed to a Marine Environment"[5] in 1995. Furthermore, in 1991, Ehlert published an article reviewing the long-term behavior of a particular reinforced concrete structure made with seawater that was in a good state.[6]

At the other extreme, in 1994, O'Connor published a paper[7] that gave the impression that, if seawater was used in the mixture, it was the seawater alone that was responsible for all deterioration of the concrete in service. I shall discuss References 6 and 7 in more detail later in this section. However, the continuing publication of somewhat conflicting information has led me to the view that a brief review of how we have learned about the use of water in the mixture may be of interest.

Chlorides in seawater

What distinguishes seawater from other waters in the mixture is the presence of chlorides; the other salts are less significant. For clarity, I should explain that I am using the term seawater to subsume brackish water, that is, water whose immediate provenance is not the sea but that contains a significant amount of dissolved salts, especially sodium chloride. The increased salt content is usually due to the following sequence of events. First, there is evaporation of seawater from previously inundated areas, leaving behind inactive salt deposits. Subsequently, there is movement of fresh water, often of pluvial origin or due to human activities, which dissolves the highly soluble chlorides. In consequence, the salt content of brackish water is unpredictable and variable with time. Brackish water is usually less saline than seawater, but in some desert areas the brackish ground water contains more salts than seawater.

A further clarification concerns the amount of salts in seawater, as there are considerable variations among different bodies of water. Typically, seawater has a total salinity of 3.5%, of which 78% is sodium chloride, and 15% magnesium chloride and magnesium sulfate taken together. In passing, we can note that 10 g (0.4 oz.) of sodium chloride contains 6 g (0.2 oz.) of chloride. The North Sea has a salinity of 3.3%; the Atlantic and Indian Oceans have a salt content of 3.9%; and the Red Sea, 4%. Some seas have a much lower salinity; for example, the Baltic Sea has only 0.7%. At the other extreme is the Arabian Gulf with 4.3%. The Dead Sea, which, being landlocked, is not really a sea, has a salinity of 31.5%, but a very low content of sulfates.

Other aspects of seawater

Two other aspects of seawater should be mentioned. First, the main cation associated with chlorides is sodium, which of course is an alkali in a reactive state. Thus, the risk of alkali-aggregate reaction with a susceptible aggregate is increased. Second, seawater tends to cause persistent dampness of the surface of concrete and efflorescence (see Sections 5.3 and 5.4). Thus, even in plain concrete, seawater should not be used as mixing water when appearance is of importance or where a plaster finish is to be applied.[8]

Calcium chloride in the mixture

Because it is the presence of chlorides that characterizes the seawater for the purposes of the present discussion, it is useful to remember that calcium chloride used to be deliberately introduced into concrete as an effective and cheap accelerator that would promote the hydration of calcium silicates in portland cement and thus lead to a more rapid development of early strength. After many decades of use, various national specifications banned the use of chloride-based accelerators and introduced limits on the total content of chlorides in the mixture. The reason for this is that, under many circumstances, chloride ions in concrete lead to the corrosion of reinforcement.

With respect to concrete itself, as would be expected from our knowledge of the use of calcium chloride as an accelerator, the presence of chlorides in seawater must affect the early stages of hydration of portland cement and possibly the long-term development of strength. These and other effects of chlorides in the mixture are discussed elsewhere.[8] However, recent laboratory studies have indicated that "the quality and quantity of phases formed after one year in hydrated (Type I portland cement) pastes are unaffected by the water type (including seawater)."[9]

This does not apply to high-alumina cement whose setting and hardening are adversely affected, possibly because of the formation of chloroaluminates.[8]

Limits on chlorides in concrete

Although this section is not concerned with safe limits of chlorides in reinforced concrete (usually expressed as a percentage of the mass of cement) it is useful to note that these limits have not been reliably established; a partial reason for this is that the initiation of corrosion of steel in the presence of chloride ions and the progress of corrosion depend on the exact conditions of a given reinforced concrete member insofar as an electrochemical cell is established.

Laboratory tests on corrosion provide information on the behavior under specific conditions only, whereas on site the range of possibilities is large and also variable in time. It is not surprising, therefore, that Hansson wrote that there have been almost as many estimates of the threshold value of chloride concentration for corrosion to occur as there have been tests performed.[10]

An extreme case was reported by Lukas, who wrote: "Remarkable was that in the majority of the investigated 15-20 year old bridge structures and/or road surfacings corrosion had occurred only in cases where the chloride content in the area of steel parts was above 1.8% (by mass of cement)."[11]

The preceding value should be viewed against the limit for reinforced concrete, in the British and European standards, of a total chloride content of 0.40% by mass of cement. For reinforced concrete that will not be exposed to chloride in service, a maximum content of water-soluble chlorides of 0.30% by mass of cement is given in the ACI Building Code 318-99. The ratio of water-soluble chlorides to the total, that is, acid-soluble chloride content, is not fixed, but we can say that generally the American limits are more strict than the European ones. In all codes, the limits for prestressed concrete are more strict than for reinforced concrete. It is useful to add that for reinforced concrete that will be dry or protected from moisture in service, ACI 318-99 prescribes a limit of 1.00%.

How do those limits compare with the chloride content of seawater when used as mixing water? By way of an example, let us consider a mixture with a cement content of 320 kg/m^3 (540 lb/yd^3) and a water-cement ratio of 0.50. For 1 kg (2.2 lb) of cement, we have 0.5 kg (1.1 lb) of water with a typical content of 3.5% salts, of which chloride ion concentration is usually 2.0%. Thus 0.5 kg of water contains about 10 g (0.4 oz.) of chlorides. This represents 1.0% of chlorides by mass of cement. We should bear in mind that in concrete there may also be present chlorides from other sources, such as aggregate. Portland cement may contain 0.01% of chlorides by mass of cement; this is not much, but it should not be ignored in calculating the total chloride content per unit mass of cement.

Many codes preclude the use of seawater by dint of limits on the chloride content in the mixture, as discussed above. The German Standard DIN 4030 is explicit about not allowing the use of seawater as mixing water for reinforced concrete. The European Standard follows suit. From the preceding, it follows that, from considerations of corrosion of steel, seawater must not be used as mixing water for reinforced concrete, even when it is expected to be dry in service.

Is seawater the sole cause of deterioration?

I referred earlier to the paper by O'Connor[7] that described extensive corrosion of reinforced concrete structures made with seawater in the mixture. He concluded that "operational and environmental influences were, with some exceptions, minor contributors to deterioration." As it happens, I spent considerable time investigating the structures described by O'Connor, and I feel that it is misleading to negate the environmental influences on the deterioration of concrete, whether made with seawater or not.

In my field investigation, I found that the extent of the deterioration depended very much on local exposure, that is, on the microclimate: not only on the proximity to the sea, but also on the exposure to the prevailing wind; on air current eddies, which are influenced by the structure shape and by adjacent structures; on the frequency and duration of wetting with seawater, be it the tide, sea spray, or wash down; on exposure to solar radiation; on aeration of seawater; and on the internal conditions in a reinforced concrete member insofar as they lead to the formation of electrochemical cells. O'Connor denies these influences in his conclusions, and yet admits the "wide variations in the conditions of concrete" when describing the observed state of the structures. In the body of his paper, he says that there was "a wide variation of corrosion, from light to severe, and from uniform to localized." Such a disagreement between the content of the paper and its conclusions could be the consequence of a biased attitude.

While I subscribe to the opinion that seawater must not be used as mixing water in reinforced concrete, it is misleading to ignore all the other factors influencing the deterioration of concrete. Indeed, when investigating the structures described by O'Connor, I found that some structures made with seawater were seriously damaged owing to poor detailing of reinforcement, especially with respect to shrinkage. It is only when wide shrinkage cracks developed that the ingress of chlorides from outside led to rapid corrosion of the steel.

Conversely, in the same location, some structures made with fresh water also suffered damage owing to the ingress of chlorides. Again, O'Connor mentions that chloride profiles in the concrete clearly show the ingress of chlorides, but he ignores this fact in his analysis.

The reason for my discussing this paper is to point out the need for scrupulous reporting and analysis of site behavior; this is essential to improve our understanding of the physical and chemical phenomena involved.

One other investigation of reinforced concrete made with seawater merits discussion because it illustrates the influence of the exposure conditions on the deterioration of such reinforced concrete in service. This investigation is described in a paper by Ehlert, published in 1991, describing the contemporary condition of three military structures built in 1953 on Bikini Atoll in the Marshall Islands in the central Pacific Ocean, using coral aggregate and seawater in the mixture.[6] The state of the structures varied greatly. The structure that was partially buried below grade and was protected on three sides by earth berms, as well as surrounded by coconut trees, was found, 28 years after construction, to be in "excellent condition."[6] Specifically, Ehlert wrote: "Virtually no concrete cracking or spalling has occurred in the walls, and only minor cracking is visible in the concrete roof.... Interior concrete wall surfaces show no deterioration at all and exterior surfaces show no signs of reinforcing steel rusting or concrete spalling."[6]

Conversely, the other two structures exhibited serious damage, and Ehlert concluded that "coral aggregates may not be primary sources of rapid structural deterioration of reinforced concrete. Factors such as degree of atmospheric exposure, concrete cover to reinforcing steel, and surface cracking are of greater importance."[6]

We must accept Ehlert's field observations, but I am dubious about accepting, as advice for the future, his optimistic conclusions. My view is that we should not be led into a false sense of security. The various factors mentioned by Ehlert are important, but the consequences of using seawater as mixing water should not be belittled. The other factors may or may not be controlled in service, depending on circumstances. For example, the exposure conditions may change when other structures that originally provided protection are removed at a later date. A similar situation has been encountered in snow loading when a hitherto protected structure becomes exposed to snow drifts once its neighbors have been removed. I am, therefore, firmly of the opinion that the use of seawater must be avoided. This is my message.

I am saying the above even though I am personally familiar with another structure (having visited it in 1990) made of reinforced concrete containing seawater: this is a jetty on the island of St. Eustatius in the Netherlands Antilles. The structure performed well until there was damage to the concrete cover in a number of places due to ships' impact and, later, to two hurricanes. No repairs were affected and serious corrosion ensued in a number of areas, other areas remaining in a good condition (Fig. 2.4.1).

I would like to interject here a general comment: it is important to

2-42 Chapter 2: Water and Concrete

Fig. 2.4.1(a)—Overall view of jetty at St. Eustatius

Fig. 2.4.1(b)—Initial mechanical damage; corrosion followed at jetty at St. Eustatius

beware of generalized conclusions based on a single investigation, particularly when its objective is to prove a point. My remark implies bias, which need not be a deliberate desire to support one's employer (although sometimes it is) but may simply be an enthusiastic interpretation in accordance with the investigator's belief and expectation. It is recalled that the great philosopher Bertrand Russell referred to the "delusive support of subjective certainty."

Some current views on seawater in the mixture

Is simply not using seawater the end of the story? I would have thought so, but recent publications continue to offer less definite opinions, thus keeping ajar the door to the use of seawater. For example, a 1992 paper from South Africa[12] concludes: "Some caution is necessary when using seawater as mixing water in view of the increased risk of reinforcement corrosion, alkali-aggregate reaction, and sulfate attack." To ask for "some caution" is pretty feeble.

Pierce cites several instances of the use of seawater in reinforced concrete without harmful effects and recommends "extreme caution."[13] This is somewhat more definite. The Portland Cement Association, in its 1995 Bulletin on Design and Control of Concrete Mixtures, is even more definite: "Seawater is not suitable for use in making steel-reinforced concrete.... particularly in warm and humid environments."[14] In the 1998 edition of his concrete construction book, Dobrowolski says: "Seawater may reduce strength only slightly but is not recommended for reinforced concrete because of the corrosive effect on the steel. Nevertheless, it is sometimes used in some areas because of a scarcity of fresh water."[15] This is a compromise between the devil (corrosion) and the deep blue sea (water).

The above views temper science with kindness; and kindness leads to problems later. The trouble is that it is precisely in the areas where fresh water is scarce, like the Arabian Gulf area and other arid regions, that the exposure conditions are harsh and very severe so that a tolerant attitude to the use of seawater just tips the balance the wrong way (see Section 3.4).

Curing with seawater

The purpose of wet curing is to make water ingress into concrete. If that water is seawater, and therefore contains chlorides, chlorides would be carried into the interior of the concrete and towards the reinforcement. If at the start of curing the concrete is partially dry, then the chlorides in the seawater will move by flow or absorption at a rapid rate. If the concrete has not dried out significantly prior to the commencement of curing, then the chlorides will move into the originally chloride-free water in the capillary pores by diffusion.

It follows that curing concrete at an early age with seawater must not be practiced. Fresh water should be used or, if this is uneconomic, membrane curing should be used. Curing in seawater must not be permitted even if the concrete is intended for service by way of being submerged in seawater. If the submersion takes place when the concrete is more mature, it will have a denser structure owing to more advanced hydration so that the ingress of seawater will be limited.

To paraphrase from the poem "The Rime of the Ancient Mariner" by Samuel Taylor Coleridge:

> Water, water, every where.
> Nor any drop to cure!

Recommendations

The preceding discussion leads me strongly to recommending that

Chapter 2: Water and Concrete

seawater or brackish water not be used as mixing water for reinforced concrete. Escape statements such as "do not use seawater unless this is unavoidable" are unhelpful, if not outright dangerous, and the design cannot claim to follow such codes as the ACI, British, or German. Of course, if something is unavoidable, you can do it, but this would be Robinson Crusoe's concrete, and not concrete as we know it. If you are dying of thirst, drink the water from a brackish well, but do not call it drinking water.

So, in my opinion, we must not use seawater in the mixture and call the result reinforced concrete on a par with structures built according to standards and codes of good practice. There are only two possible exceptions. The first is when it is known for certain that the concrete made with seawater will always remain under dry conditions; the word *always* precludes a change in use. Even then, the ACI 318-99 limit of 1.00% of water-soluble chloride by mass of cement should be complied with. The second exception is when the structure will be permanently submerged in seawater, or any water for that matter, that contains very little dissolved oxygen; this would not be the case in a water bed with highly turbulent flow. The absence of oxygen is important but sometimes forgotten.

My recommendation is to have an unequivocal rule, even in arid parts of the world, that seawater not be used as mixing water in reinforced concrete or for curing such concrete; we should not be beguiled by "optimistic" publications. This rule would not apply to plain concrete without any steel inserts. With such an approach we shall avoid problems in the future.

References

1. Neville, A. M., "Water—Cinderella Ingredient of Concrete," *Concrete International*, V. 22, No. 9, Sept. 2000, pp. 66-71 [Section 2.1].

2. Taneja, C. A., "Seawater for Making Concrete," *Cement and Concrete*, 1970, New Delhi.

3. Harrison, W. H., "Concrete and Soluble Salts in Arid Climates," *Overseas Building Note*, No. 139, 1971, Building Research Establishment.

4. Anonymous, "Seawater for Concreting?" *Indian Concrete Journal*, V. 62, No. 10, 1988, pp. 505-506.

5. Kaushik, S. K., and Islam, S., "Suitability of Seawater for Mixing Structural Concrete Exposed to a Marine Environment," *Cement and Concrete Composites*, V. 17, 1995, pp. 177-185.

6. Ehlert, R. A., "Coral Concrete at Bikini Atoll," *Concrete International*, V. 13, No. 1, Jan. 1991, pp. 19-24.

7. O'Connor, J. P., "Middle Eastern Concrete Deterioration: Unusual Case History," *Journal of Performance of Constructed Facilities*, V. 8, No. 3, 1994, pp. 201-212.

8. Neville, A. M., *Properties of Concrete*, Fourth Edition, Longman, Harlow, and John Wiley, N.Y., 1995, 844 pp.

9. Ghorab, H. Y.; Hilal, M. S.; and Kishar, E. A., "Effect of Mixing and Curing Waters on the Behaviour of Cement Pastes and Concrete," *Cement and Concrete Research*, V. 19, No. 6, 1989, pp. 868-878.

10. Hansson, C. M., "The Effect of Cement Type on the Chloride Induced Corrosion of Steel in Concrete," *Tenth Scandinavian Corrosion Congress*, 1986, Stockholm.

11. Lukas, D. W., "Relationship Between Chloride Content in Concrete and Corrosion in Untensioned Reinforcement on Austrian Bridges and Concrete Road Surfacings,"*Betonwerk & Fertigteil-Technik*, No. 11, 1985, pp. 730-734.

12. Sephton, S. S., and Ballim, Y., "A Preliminary Investigation into the Effects of Using Seawater as Mixing Water on the Properties of Concrete," *Concrete Beton*, South Africa, Aug. 1992, pp. 14-22.

13. Pierce, J. S., "Mixing and Curing Water for Concrete," ASTM STP 169C, 1984, pp. 473-477.

14. Kosmatka, S. H., and Panarese, W. C., "Mixing Water for Concrete," *Design and Control of Concrete Mixtures*, 6th Canadian Edition, Portland Cement Association, pp. 32-35.

15. Dobrowolski, J. A., *Concrete Construction Handbook*, Fourth Edition, McGraw Hill, 1998, Section 12.2.2.

Letter to the Editor

Professor Neville's article comes to the correct conclusion that one should not use seawater as mixing water for hydraulic-cement concrete containing corrodible embedded iron or steel except under very special, very rare circumstances. By rather strong implication, it seems to me he is saying that if fresh water is very scarce and you are "dying of thirst," you can use brackish water but don't call it "mixing water," and don't call the concrete "specification concrete" in compliance with a proper building code—because it isn't!

So far, I am in complete agreement. But what about the case where it is presumed to be impossible to prevent chlorides from being available in large quantities over a long service life, as in a reinforced concrete bridge deck or a parking garage heavily exposed to chloride deicing salts, where it has been determined to use presumably non-corrodible reinforcement. I say "presumably" non-corrodible reinforcement because the first two candidates for this title—hot-dipped galvanized (zinc-coated) steel and epoxy-resin coated steel—had fairly inconclusive claims to the title of non-corrodible.

Some years ago, I saw some small lengths of nickel-coated large diameter reinforcing steel in the offices of the engineering firm responsible for design of a nuclear power plant on the Atlantic Coast. The plant would use seawater for cooling and would circulate it out of and back into the ocean through concrete conduits reinforced with such reinforcement. Since then I have seen both stainless steel and stainless-steel clad reinforcement. Perhaps if this was the reinforcing and there was a dire shortage of fresh water, seawater might be employed.

Then there is the case of the concrete that very soon after 28 days becomes very dry (internal RH below 80%) and stays that way forever. To the caveats that Adam raised to this rearrangement of the neighborhood, so to speak, I would add that even in buildings that are nominally very dry there are always restrooms and frequently kitchens that are not.

When I was called upon to review the state of the art of making concrete to use in structures at the oceanic shore and on the effects of seawater on the durability of concrete(not involving embedded metal), the references I encountered told me that the chemical composition of seawater throughout the world was remarkably uniform and that all the chloride was associated with sodium except for a very small amount of potassium. All the sulfate was associated with magnesium. From the standpoint of chemical effects of seawater on unreinforced concrete, it was the sulfate that was viewed with the greatest suspicion.

Later work, especially by Browne,[1] showed that the need for using even moderate sulfate-resisting cement was greatly reduced by employing concrete of presumably lower water-cement ratio (w/c). As Verbeck[2] had suggested, it's advantageous in concrete containing corrodible metal to use cement with as much C_3A as can be tolerated without incurring sulfate attack. The more C_3A in the cement, the more chloride ion will be intercepted by aluminate (precipitated as non-detrimental calcium chloroaluminate), taking longer for the ions to build up at the surface of the steel.

Professor Neville properly discussed the relationship of the phenomenology of chloride ion in concrete resulting from the intentional use of calcium chloride as an accelerating admixture, and the chloride ion so present, from the intentional use of seawater as a mixing water. There are still many specifications that allow up to 1%—or in some cases 2%—calcium chloride by mass of cement when there is not embedded ferrous metal, no aluminum conduit, no stay-in-place metal forms, and the like. In many cases one other restriction was added that forbade employment of calcium chloride when sulfate-resisting cement (Type V) was being used.

Some years ago I received a phone call from a very anxious government resident engineer who realised, after the fact, that he had mistakenly given permission on a very cold day to use calcium chloride even though Type V cement was being used. He wanted to know what would happen. After some discussion I told him not to worry. After reviewing the literature, I concluded that the sulfate-resistance of the concrete would not be reduced by the incorporation of calcium chloride. I published the results. To me it was interesting that for all sorts of cements that were not highly sulfate-resisting, the use of $CaCl_2$ did reduce the sulfate resistance, but not when Type V was employed. So I concluded that one should be permitted to use $CaCl_2$ as an accelerator in cold weather with Type V cement to the same limited extent as with other cements. I recall one almost hysterical response from an engineer who regarded chloride in concrete as poison that should never be permitted in any amount or for any reason.

Professor Neville speaks of the use of water in curing concrete as intended to penetrate the concrete. Of course, this is quite often the case, but it shouldn't be. If steps are taken to prevent loss of water from the concrete, and these steps are moderately or highly successful, no added water will be needed as part of curing. In the report of ACI Committee 308, Curing Concrete, issued when I was its chair, we concluded that if loss of water were prevented, no adding of water was needed except in two circumstances: (1) when the w/c was less than 0.4; and (2) when the concrete was made using expansive cement.

The committee withdrew the second caveat when it was explained by a major producer of shrinkage-compensating expansive cement that he knew just as well as we that half or more of the expansive potential of his cement would be lost if no added water was available. He also requested we please not say that added water was mandatory since a lot of such cement was being used in paving when water curing was economically impractical and membrane curing compounds were satisfactory for the concrete to expand enough to serve its purpose.

In guidance to the U.S. Army Corps of Engineers Beach Erosion Board,[3] the available literature had been reviewed. I concluded with a quotation from Carver[4]: "Experienced engineers are practically unanimous in the belief that seawater should not be used for mixing in reinforced concrete work, particularly in the tropics."

Bryant Mather
Director, Emeritus
U.S. Army Corps of Engineers
Structures Laboratory
Vicksburg, Miss.

References

1. Browne, R.D., "Mechanisms of Corrosion of Steel in Concrete in Relation to Design, Inspection, and Repair of Offshore and Coastal Structures," *Performance of Concrete in a Marine Environment*, SP-65, V. M. Malhotra, ed., American Concrete Institute, Farmington Hills, Mich., 1980, pp. 169-204.

2. Verbeck, G. J., "Field and Laboratory Studies of Sulphate Resistance of Concrete," *Performance of Concrete: Resistance of Concrete to Sulphate and Other Environmental conditions*, E. G. Swenson, ed., University of Toronto Press, 1968, pp. 113-124.

3. Mather B., "Factors Affecting Durability of Concrete in Coastal Structures," U.S. Army Corps of Engineers, Beach Erosion Board, 7M 96, 1957.

4. Carver, G.D., "Concrete Viaducts Over the Key West Extension of the Florida East Coast Railway," *Engineering Record*, Oct. 20, 1906.

Chapter 2: Water and Concrete

Neville's response

Bryant Mather's sharp analysis coupled with a prodigious memory mean that his letter enhances the original article. For this I am naturally grateful to him.

It is on the issue of whether curing water is needed to ingress into the concrete that we do not see quite eye-to-eye. Theoretically, Bryant is right in that, provided the w/c is not above about 0.4, there is enough water in the mixture for hydration to proceed to completion. I have used the qualification "theoretically" because it is necessary for the water to be uniformly distributed throughout the mass of concrete. This is possible in a laboratory specimen under controlled conditions.

On the other hand, in full-size structural members, there is inevitably some loss of water by evaporation from the surface. Consequently, hydration may proceed in the interior of the member but, near the surface, there is an inadequate amount of water in the capillaries so that penetration by curing water is highly desirable. However, if the water used for curing is seawater, chloride ions enter the surface zone and from there move inwards by diffusion.

My views are influenced by my experience in the Persian/Arabian Gulf area where the removal of formwork at a high temperature and a low relative humidity (RH) of the ambient air leads to rapid drying of the near-surface zone of the concrete. In the case of marine structures cast on land but destined for immersion in the sea, the risk of imbibition of seawater is high, unless thorough curing with fresh water has previously taken place.

I cannot resist adding that, from the standpoint of durability, it is the near-surface concrete that is much more important than the concrete in the interior of the mass: many durability problems start at the surface or through attack progressing form the surface inwards.

I am sure Bryant is fully aware of it but, for the sake of completeness, it is useful to note that even at a w/c of 0.475 there exist empty capillary pores. This is illustrated in Fig. 2.4.2.[1]

Finally, Bryant is right that the chemical composition of seawater is remarkably uniform. Specifically, of the total salts in seawater, the various ions represent approximately the following percentages: chloride, 51.3; sulfate, 7.2; sodium, 28.5; magnesium, 3.6; calcium, 1.3; and potassium, 1.0. What varies widely, however, is the *total* amount of any ion; values for some bodies of water are given in the article. The consequence is that, for a given mass of seawater that ingresses into the concrete, the amount of any given ion is proportional to the salinity of the seawater.

Bryant's closing quotation is particularly valuable as it pre-dates both him and me: if Carver's words of 1906 had been heeded, the use of seawater in the mixture would have ceased long since, and there would have been no need for my article. I have shown that the necessity to point out the risk associated with the use of seawater still

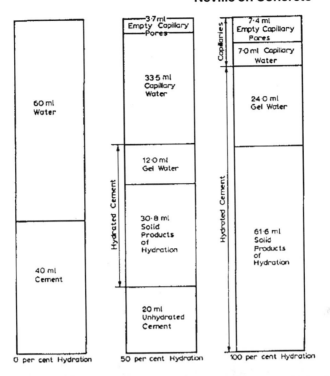

Fig. 2.4.2—Diagrammatic representation of the volumetric proportions within cement paste made with a *w/c* of 0.475 by mass and allowed to hydrate in a sealer container[1]

continues but, with luck, my article and this correspondence will put an end to such use.

Adam Neville

Reference

1. Neville, A. M., *Properties of Concrete*, Fourth edition, Longman, Harlow, and John Wiley, N.Y., 1996, 844 pp.

Section 2.5: How Useful is the Water-Cement Ratio?

The concept of the water-cement ratio (w/c) is a part of probably the oldest "law" in the field of concrete. The term invariably appears in specifications and in numerous ACI Standards. So why do I have the temerity to question the usefulness of the w/c?

I am not the first one to do so. In 1996 Bernie Erlin wrote a note on a discussion on this topic but, as is the case with most discussions, no positive conclusions were reached.

Definition of w/c

Let me start by expressing the opinion that the term "w/c" is unlikely to disappear from concrete technology, but we need to know exactly what it means and how it can be used. First of all, we have to recognize that the use of portland cement alone in concrete is on the way out: more and more mixtures will contain additional cementitious materials or other fine powders. To use the term "water-cementitious material ratio" is clumsy; to use "cementitious" as a noun is poor English.

The best solution is probably to use the term "cement" as inclusive of portland cement, fly ash, ground-granulated blast-furnace slag, silica fume, metakaolin, and even fillers (see Section 1.1). The only exception to this may be necessary when establishing the strength or other properties of concrete at very early ages, at which stage some of the "other" ingredients of "cement" may not yet have undergone significant chemical reactions. In such a case, it is the w/c based on the mass of portland cement alone that is relevant. At later ages, all the cementitious materials should be included in the calculation of the w/c. In any case, it is best, in addition to giving the w/c, to state also what kind of cement is being used.

This may be an appropriate place to point out that the liquid content of superplasticizers should be included in the total mass of water in the mixture.

As for the "cement," we should remember that the various cementitious materials have different specific gravities; for example, the specific gravity of fly ash is slightly above 2.0, as against 3.15 for portland cement. Some of these cementitious materials react with calcium hydroxide produced by the hydration of portland cement. In consequence, the long-term volume of solid products of hydration will be affected by the exact ingredients of the "cement." We should remember that the microstructure of concrete, which influences its strength and durability, depends on the volumetric proportions of the various components of hardened concrete and not on proportions by mass.

One other fact should be noted. Except when concrete is permanently submerged in water, some of the original cement will remain unhydrated. This may be of consequence when hardened concrete is being examined to determine the *w/c* of the original mixture.

Use of w/c

The traditional role of the *w/c* is as a determinant of strength. Additionally, strength used to be interpreted as an indicator of durability, and is still used by some people, although I do not think this is sound. Nowadays, the situation is more complex.

In the past, the water content of the mixture with a given workability varied little with variation in strength. The explanation is as follows. With an increase in the cement content, and at a constant water content, the *w/c* decreased and the strength increased. Now, with the advent of powerful water-reducing admixtures it is possible to vary the water content of the mixture very significantly. In consequence, at a given workability, the *w/c* can vary widely. And, as the *w/c* governs the porosity of the hydrated cement paste, the value of the *w/c* is relevant to many aspects of durability.

But the *w/c* alone does not determine the durability or even the permeability of the concrete. First of all, what matters is the voids in the concrete as a whole and not in the cement paste alone. To take a simplistic example, keeping the *w/c* constant, a mixture that has twice the cement content of another mixture will have twice the volume of pores. Secondly, it is not only porosity that influences durability but also the extent of connectivity of the pores that determines the penetrability with respect to aggressive agents. So, undue reliance on the *w/c* as an indicator of durability is not justified.

An attempt has been made to split the denominator because, with the various cementitious materials used nowadays, the influence of the *w/c* on various properties of concrete is not the same. The *w/c* splits into two terms: the mass of portland cement component, c plus a factored term for the other cementitious materials. For example, with fly ash whose mass is f, the denominator would become:

$$c + kf,$$

where k accounts for the difference between the influence of portland cement and of fly ash. The trouble is that the value of k is different, not only for different fly ashes, but also for different ages of the concrete. By the age of several months, I would not expect any difference between fly ash and portland cement. More importantly, the same value of k would not satisfy the relation between the *w/c* and strength as well as between the *w/c* and permeability. At least one reason for this is the variation in the absolute volume of the different cementitious materials of a given mass.

Determination of w/c

The issue of the *w/c* of concrete becomes particularly important in disputes about the quality of the concrete in an existing structure. At that stage, attempts are sometimes made to determine the w/c of the hardened concrete. There are two problems with this.

First, although there exist some petrographic and chemical methods of measuring the w/c, their precision is low because they require a number of assumptions to be made about their interpretation. Moreover, given that the w/c varies within a structure, the value of the w/c determined by tests on a few samples is unlikely to be better than within ±0.1 or, at best, ±0.05 of the "true" value. For example, a joint report of the Concrete Society and Society of Chemical Industry states: "In favorable circumstances, with reliable analysts the result is likely to be within 0.1 of the actual w/c ratio."[1] Even with a precision of ±0.05, this means that, if the question is, "Did the mix have a w/c ratio of 0.50 as specified?" the tests indicate that the w/c was between 0.45 and 0.55. This is not good enough to settle a legal dispute.

I have seen some maverick claims that a "trained eye" can establish the w/c of hardened concrete within ±0.02, or even ±0.01, by using vacuum impregnation with a fluorescent epoxy. Such treatment may induce spurious cracks which will become filled with the epoxy. Moreover, without strictly relevant reference samples of concretes made with the same ingredients at known values of the w/c, estimates are unreliable.

Furthermore, in a sample taken from an actual structure, what appears to be the original w/c is influenced by the degree of hydration of cement: less hydrated cement has more voids which were originally occupied by water. In the test, these voids became filled with fluorescent epoxy, but this does not mean that the w/c is higher than that of a well-hydrated companion sample. Thus, reference samples with the same degree of hydration as in the structure should be used. But the degree of hydration varies depending on the conditions of exposure of the concrete in the various parts of the structure during its entire life so that the actual degree of hydration is rarely known. Therefore, comparisons even with reference samples may be unsatisfactory.

Overall, in the absence of ASTM or international standards, it is wise to remain skeptical. Such an approach is recommended in other fields as well. In discussing genetically modified food, the President of the Royal Society said recently: "I would stress that premature, partial or selective release, or misrepresentation, of unsubstantiated research only serves to mislead the general public in a complex area."[2] Hydrating cement paste is complex.

The second problem with determining the w/c is that this term has a proper meaning only at the time when the concrete has set and begun to harden. What was put into the mixer is a "statement of intent" with respect to the w/c. Much can influence the actual value. To begin with, the exact water content of the aggregate is not known at all times. Also, with ready-mixed concrete, it is possible that some water was added prior to discharge of the mixture; purists may criticize doing this but it is better than discharging concrete that cannot be properly compacted. Also, changes occur during transportation if water can evaporate; changes occur also during placing if bleeding occurs; and the process of placing itself and of compaction induces local changes.

Thus, when we look at hardened concrete, we are unable to establish its "original" w/c. Indeed, ACI 318 clearly states that, once concreting has started, the criterion for evaluation and acceptance of the concrete is the compressive strength of cylinders or cores. Likewise, a recent Swiss report states that the w/c can be estimated only on fresh concrete.[3]

This is not to deny that a reliable method of determining the w/c of hardened concrete would not be welcome. However, until it has been established, we should beware of premature claims of success.

Conclusions

Overall then, the w/c is of use in assessing the properties of a concrete mixture at the time when it is designed, bearing in mind the nature of all the cementitious materials being used. The usefulness of the w/c is smaller than in the old days of concrete made with portland cement only, but even then the type of cement influenced the relation between the w/c and strength at a given age. However, if you cannot get full information, this is not a reason for getting none, which would follow from abandoning the w/c altogether. So, we should continue to use the w/c but we should remember the limitations on its interpretation which did not exist in the past.

The future

Let me finish on a slightly provocative note. Most specifications prescribe the w/c either explicitly or by laying down the cement content of the mixture and its workability, as well as requiring a specific approval of admixtures. However, compliance is invariably determined by the strength of standard test cylinders. This may be pretty well inevitable but it is not logical. We are all aware of the burden of tests for the compressive strength of standard cylinders and of the nightmare of dealing with instances of a low strength, that is, of suspected non-compliance (see Section 7.2). Shouldn't we try harder to determine the w/c of fresh concrete sampled from concrete that has just been placed? Such apparatus for this purpose as exists is cumbersome and insufficiently precise, so we need something new. Who will offer a prize for an easy-to-use gadget that will "read" the w/c? And who will win it? Here is a thought for the magic year 2000.

References

1. Concrete Society, "Analysis of Hardened Concrete," *Technical Report,* No. 32, 1989, 110 pp.

2. *Royal Society News,* London, March 1999.

3. "Définition et Détermination du Rapport Eau sur Ciment," *Société Suisse des Ingénieurs et des Architectes,* No. 7, Apr. 1999, pp. 128-129.

Letter to the Editor No. 1

In his usual manner, "Justice" Adam Neville does a good job of

2-54   Chapter 2: Water and Concrete

presenting the salient facts but may have slipped up with the emphasis placed upon factors involved.

Attention is drawn to the importance of specific gravity (SG) of the various "cements." When dealing with cementitious components having such a broad range of SG values, it would seem to be time to consider expressing the ratio by volume (if at all) rather than by mass.

However expressed, the w/c ratio can give a false sense of security. Mr. Neville specifically mentions the problem of increasing pores in the concrete when cement content increases at fixed w/c. He also stresses the significance of pores and voids in general. What he does not address is that, in his example, where the cement content is double, the unit water content of the concrete has also doubled. Surely, with the current use of multiple-component cements and multiple admixtures, it is the total fluid content of the concrete by volume that is important. Fortunately, in metric terms, 200 kg/m^3 (340 lb/yd^3) of water equates to 200 L (50 gal.) or 20% by volume of concrete. When the w/c law was written, most concrete contained close to 200 kg/m^3 (340 lb/yd^3) of water. These days, with appropriate use of admixtures, water content below 140 kg/m^3 (240 lb/yd^3) is not uncommon.

This raises the question of how to treat admixtures. Mr. Neville refers to including the "liquid content" of superplasticizers. What does this mean? Commonly it is interpreted as including only the water component of the admixture. However, the total volume of all liquid admixtures contributes to the potential void space in the hardened concrete and hence should be included along with the water. Both drying shrinkage and durability are influenced by the total volume of free water in the fresh concrete.

Just as concrete mixtures are designed by volume, so should w/c (or better adjusted unit water content) be measured by volume. Similarly, water absorption expressed by volume gives a much better indication of its significance. An absorption of 2% by mass may be 5% by volume and a w/c of 0.5 by mass is about 1.5 by volume. These values are independent of the varying density of aggregates and "cements" between different mixtures and provide a valid basis of comparison.

Bob Philleo (ACI past president and Honorary Member) was a leading proponent of expressing these values by volume, before the introduction of high-performance concrete. The argument now has significantly more validity.

W. Barry Butler
ConSult International
Marmong Point, New South Wales, Australia

Letter to the Editor No. 2

It..(the ratio)..is very useful indeed and Mr. Neville has clearly outlined the types of high-performance concrete being used today to achieve low permeability, long-term durability, and minimal shrinkage and cracking. Success in achieving these qualities of concrete in the hardened state depends on many factors and certainly the water-cementitious ratio (w/cm) alone is not sufficient. However, the proper w/cm is a very important aspect of the success of these types of concrete. I regularly see mixes designed to fulfil specification requirements for the above-mentioned qualities, that is, parking structures, that are not accurate as to the water content listed. Traditionally, this inaccuracy was only discovered after the hardened concrete had failed.

However, the microwave test, AASHTO TP 23, "Proposed Standard Method of Test for Water Content of Freshly Mixed Concrete Using Microwave Oven Drying," is being currently employed with great success by the Port Authority of New York and New Jersey and the states of Minnesota, North Dakota, and New Hampshire, and by leading engineering firms whose concrete is required to exhibit long-term durability, impermeability, and minimum cracking and curling (important floors and slabs). Certainly compressive strength results cannot be a measure of the w/cm.

High-performance concrete in parking structures and bridge decks is achieved by the conformance to the following requirements:
- w/cm of 0.40 or less with low water content;
- Proper air content throughout the mass, particularly the top surface, of 4.5 to 7.5%;
- Prompt and proper curing to minimize cracking; and
- Good surface abrasion resistance.

Mr. Neville's article outlines the impact of supplementary cementitious materials, porosity, and varying degrees of hydration on hardened concrete performance. The microwave test allows us to verify the water content of the plastic concrete prior to placement. That is a tremendous benefit.

His article and questions should cause many engineers to review their specifications for accuracy with respect to the desired concrete qualities. It should also alert all members of the industry that 6 x 12 in. (150 x 300 mm) cylinder results are not what the owner is buying and compressive strength is not directly related to other important concrete qualities.

William S. Phelan
Senior Vice President
Euclid Chemical Co.
Cleveland, Ohio

Letter to the Editor No. 3

Reference is made to the article..(in the) September 1999 issue. After the first reading of the article, it appeared to me that it was a product of the author's preparation for expert testimony over a concrete dispute on behalf of the defense. While the author speaks with apparent expertise on the subject of concrete, some of his statements are incorrect and out of context

Chapter 2: Water and Concrete

and only serve to create confusion over a complex issue in the less-than-expert reader. It is difficult to determine whether the author suggests to discard the w/c ratio criterion or whether to find more accurate ways to measure it, therefore recognizing its importance and usefulness in a concealed fashion.

While strength of the concrete may be the only important consideration for many concrete building structures, there are other structures where the concrete material must meet many other property requirements beside strength. An expert concrete design is one which specifies requirements and limits for the constituent ingredients and admixtures to be used in the concrete, and at the same time specifies strength, workability, placement, consolidation, finishing, environmental, and curing requirements in order to achieve the desired structural adequacy, to minimize shrinkage, to limit the depth of penetration of aggressive agents in order to protect the reinforcing, and to assure durability and watertightness. Proper w/c ratio for each class of concrete used is only one of these specification requirements. Although other factors are also influential, strength, shrinkage, density, porosity, durability, and watertightness are significantly dependent on the w/c ratio. It is well known that the major concrete ingredient contributing to concrete shrinkage is the cement paste and its w/c ratio (and the resistance provided by the aggregate and the reinforcing), and it is also well known that porous concrete resulting from excessive mix water cannot be as durable and watertight as one with the proper w/c ratio. Therefore, we cannot discard its use as one of the many enforceable specification requirements. (The only exception that would make the w/c determination less meaningful is when porous, expanded and moisture-absorbing lightweight aggregates are used.)

It is absurd to suggest that the recommended 5 to 9 fl oz. (150 to 270 mL) of superplasticizer per 100 lbs (45 kg) of cement be added into the w/c calculation since this alters the w/c ratio by 0.005. What would be important, and not mentioned by the author, is the amount of water added to the batch by the moisture content of the fine aggregate (or latent water in the drum). This latter amount is about 10 to 40 times the amount added by the superplasticizer. The author's point about the lesser specific gravity of fly ash only serves to further confuse the non-expert reader.

Many of my firm's projects require extensive considerations for durability, density, shrinkage, abrasion resistance and watertightness, in addition to strength requirements. Our experience shows that adhering to our specification requirements as written and with the limits given, including w/c ratio requirements based on the types of ingredients and admixtures, accompanied by our design, reinforcing and detailing methods, will produce the desired results. The author may be surprised to hear this, but the w/c ratio is nowadays always defined as the weight of total water divided by the weight of the total permissible cementitious materials at the time of placement. (Mr. Neville believes that the industry still limits the w/c ratio to cement alone.) My preference, and that of many others, is to refer to it as "water-cementitious material" ratio (w/cm). Others, however, prefer to adhere to the old "water-cement" phrase which is fine if we know what is understood by it.

The author incorrectly states that "compliance...(with w/c ratio requirements)...is invariably determined by the strength of the standard test cylinders." Better specifications require many controls such as the submission of design mixes for all concrete proposed for a given project before any concrete is placed. These specifications also require that the record of the mix design as well as the amount of water used, percent of air entrained, amount of admixtures used, determination of the concrete's density, and slump tests be indicated on every concrete test report. These listed test results generally indicate whether the amount of water in the batch is greater than that of the approved design mix. I never heard of anyone who is able to determine the water content of a mix solely by the strength of a test cylinder.

As a reader of *Concrete International*, I strongly recommend that the editors in the future screen the submitted articles for technical correctness to assure that only appropriate and useful conclusions are drawn therefrom by the many readers. This article should not have gone past the "Letters" column as it merely expresses one man's opinion.

Miklos Peller
Peller & Associates, Inc.
Cleveland, Ohio

Neville's response to letters No. 1-3

Mr. Butler's comments are constructive and valuable. The significant point which I made and with which he agrees is the importance of remembering that different cementitious materials have different values of specific gravity. The idea of expressing water content by volume is a broader issue than the scope of my article. I can add that, inevitably, we do so with entrained air. The topic should be visited when the U.S. finally and seriously uses the SI systems of measurements.

I cannot agree with Mr. Butler's proposition that the water content must increase in proportion to the increase in the cement content. It used to be so, but nowadays superplasticizers allow us to make highly workable mixtures with a high cement content and a low water content.

I am grateful to Mr. Phelan for his letter, which is not only supportive of my views but also emphasizes the need for assessing fresh concrete. I shall refer to the method recommended by him later.

Mr. Peller concludes his letter by saying that my article "merely expresses one man's opinion." Of course it does, but it is individuals that advance new views, more robust and controversial than committee reports which present a consensus smoothed out of anything controversial. It was said long ago that a camel is a horse designed by a committee.

Chapter 2: Water and Concrete

Mr. Peller says that some of my statements "serve to confuse the less-than-expert reader." That risk exists when a novel approach is aired, and I should therefore explain more fully some of the points that he may have misunderstood.

Starting with the meaning of the word "cement" and therefore of "w/c," I clearly recommend in my article "to use the term 'cement' as inclusive of portland cement, fly ash, ground-granulated blast-furnace slag, silica fume, metakaolin, and even fillers." So, when he says that I "believe" otherwise, this is contrary to what I wrote.

Arguing against my article, Mr. Peller says that "we cannot discard its (w/c) use as one of the many enforceable specification requirements." This is precisely what I do say, namely: "We should continue to use the w/c but we should remember the limitations on its interpretation which did not exist in the past." More than that: there is a banner headline, ¼ in. high, proclaiming those very words.

He also says that the moisture content of the fine aggregate is "not mentioned by the author"; it is indeed mentioned in the words: "the exact water content of the aggregate is not known at all times." Conversely, Mr. Peller says that it is "absurd" to add the liquid part of a superplasticizer as water in the mixture. I do not think he is right because the superplasticizer dosage at low values of the w/c when silica fume is included in the mixture, can be as high as 20 L/m^3 (4 gal./yd^3) of concrete, the liquid content of the superplasticizer being up to 60%.[1] Thus, taking as an example, the mixture used in the Roize bridge in France,[1] the total water content was 143 L (37 gal.) and the total cement content was 495 kg/m^3 (835 lb/yd^3) of concrete. Including the liquid part of the superplasticizer, we have a w/c of 0.29. Now, ignoring the liquid part of the superplasticizer, the w/c is 143 minus 60/100 times 20 (that is 131) divided by 495, which gives a w/c of 0.26. This difference in the w/c, admittedly at a maximum dosage of superplasticizer, is 0.03, and not 0.005, as given by Mr. Peller.

Mr. Peller's approach to what he calls "expert concrete design" in which he "specifies requirements and limits for the constituent ingredients and admixtures to be used in the concrete, and, at the same time, specifies strength, workability, placement, consolidation, finishing, environmental and curing requirements" has its place. However, in recent years, there has been a strong shift toward performance specifications, in which it is the contractor that chooses (within limits or subject to the engineer's approval) the exact ingredients, their proportions, and the method of construction. This modern approach is conducive to economy and encourages progress; otherwise, every contractor would do the same as any other contractor.

Like Mr. Peller, I consider durability to be an important aspect of concrete, and I say that "the w/c alone does not determine the durability or even the permeability of the concrete." Mr. Phelan expresses the same opinion when he says that "the water-cementitious ratio alone is not sufficient" for success with respect to durability and other qualities of concrete.

When it comes to assessing hardened concrete, my point is the difficulty of establishing the value of the w/c. As Mr. Phelan points out, the determination should be made on fresh concrete, and his reference to the proposed AASHTO standard is of interest. This is a method for the determination of the water content by absorption of electro-magnetic radiation, known as the microwave-absorption method.

Mr. Phelan refers to the use of this method by various authorities. I would like to take this opportunity to add some very new data on the variability of the w/c of fresh concrete. Field test data from one of these authorities, namely the Port Authority of New York and New Jersey on about 1000 samples of some 100,000 yd^3 (76,000 m^3) of fresh concrete gave the following results.[2] When the batched w/c was between 0.41 and 0.43, the standard deviation was 0.03; smaller batches with a w/c of 0.39 gave results with a standard deviation of 0.02. The above values of the standard deviation are exactly the same as those found by R. D. Gaynor, reported in Reference 1.

Given that 95% of test results are expected to lie within plus-or-minus two standard deviations from the "true" mean, this means that, when the "true" mean is 0.42, 95% of the results lie between 0.36 and 0.48. If the Port Authority tests are reliable, and the above is indeed the variability of the w/c of samples of fresh concrete, then the variability of the w/c of hardened concrete, which has undergone construction treatment, must be even greater. Moreover, the Port Authority calculations assume that the cement content is exactly as specified; in reality, there must be some variability due to batching errors, and this would increase the variability of the w/c. It follows that the statements in my article on the variability of the w/c determined on hardened concrete are indirectly confirmed.

In my article I say that, whatever way a specification is constructed, "compliance is invariably determined by the strength of the standard test cylinders." Mr. Peller chooses to insert the words "with w/c ratio requirements" in the middle of my sentence. Not only is such a distortion unjustified, but also, as cited by me, ACI 318 clearly states in the Commentary of Section 5.6: "Once the mixture proportions have been selected and the job started, the criteria for evaluation and acceptance of the concrete can be obtained from 5.6." And Section 5.6 deals exclusively with compression strength testing, and does not provide for the determination of the w/c of hardened concrete. Further, the same section says; "An effort has been made in the Code to provide a clear-cut basis for judging the acceptability of the concrete...." Now, ACI 318 also says, in the Commentary on Section 4.0, that "it is difficult to accurately determine the water-cementitious materials ratio of concrete during production...."

Mr. Peller opens his letter with the words: "After first reading of the article, it appeared to me that it was a product of the author's preparation for expert testimony over a concrete dispute on behalf of the defense." He does not say defense of what. Nor does he say what he though after second reading. It is like saying "at first sight the man appeared dead" without adding that on closer examination he was found to be asleep.

To add a concordant note, I agree with Mr. Peller when he says: " I never heard of anyone who is able to determine the water content of a mix solely by the strength of a test cylinder." Of course, my article makes no such recommendation.

On Mr. Peller's closing comment about screening articles for "technical correctness," I wonder whether the reviewers or the editor should act as censors. I do not believe that the editor considers himself to be a protector of the correct concrete dogma from a heresy. In my opinion, there is no single "correct" view of concrete, just as there is no single correct way to mix a martini. I like them both, concrete as well as martinis.

Adam Neville

References

1. Neville, A.M., *Properties of Concrete*, John Wiley & Son, N.Y., and Longman, Harlow, England, 1996, 844 pp.

2. Port Authority of New York and New Jersey, Federal Aviation Authority, "Aeronautical Concrete Pavement Designs: Field Testing Data Analysis," 1997, 15 pp.

Letter to the Editor No. 4

Dr. Neville states in his paper that "the w/c governs the porosity of the hydrated cement paste." I would like to express my agreement with this statement by presenting the numerical form of this relation for fresh cement pastes:[1]

$$p_o = \frac{(w/c)(1 - 0.01a)}{w/c + 1/G} + 0.01a$$

where

p_o = initial total porosity in the fresh cement paste; that is, capillary porosity filled with water + air content immediately after compaction;
w/c = water-cement ratio, by mass;
a = air content, %; and
G = specific gravity of the cement.

It is noteworthy that once the water-cement ratio has been increased, the created extra initial porosity in the fresh paste acts something like a birth defect. Concretes cannot overcome with age the produced strength reduction, although the strength difference decreases with age.

Dr. Neville's severe criticism of the petrographic method for testing hardened concrete may be valid, as far as the original water-cement ratio is concerned. However (sorry, Adam, there is always a "however" after I make a complimentary remark), for most of us, the ultimate goal of such testing is not the water-cement ratio, per se, but rather the estimation of the overall quality of the concrete, especially the strength. Fortunately, the argument Dr. Neville is using suggests that the results of a suitable petrographic method may correlate better directly with the strength of concrete than with the water-cement ratio because some of the interfering variables, age for instance, would be mitigated in the direct strength estimate.

Finally, I do not find surprising that Dr. Neville sees the improvement of the determination of w/c in fresh concrete as the problem to be solved in the future. Who would argue with him that the best time to check the strength potential of a concrete is either immediately after mixing or when the concrete has just been placed? It is too bad that Dr. Neville devoted only a brief paragraph to this important topic.

References

1. Popovics, S., "Strength and Related Properties of Concrete—A Quantitative Approach," John Wiley and Sons, Inc., N.Y., 1998, 535 pp.

Sander Popovics
Department of Civil and Architectural Engineering
Drexel University
Philadelphia, Pa.

Letter to the Editor No. 5

This is in response to my good friend Adam Neville who suggested that I comment on his article, "How Useful is the Water-Cement Ratio?"

Dr. Neville, in his article, referred to a review I wrote that was published in a 1995 issue of *Concrete International*. The review was about a session that I organized and chaired for ACI Committee 201 on the water-cement ratio. The session was held during the ACI spring convention in Salt Lake City. It is entitled, "There is an Enigma in Our Mixture." The enigma, of course, is the water-cement ratio. The subject questioned the traditional age-long w/c terminology, and raised the question of whether or not it is still needed in "modern times," particularly in light of current technology. If retained then, what should we, as an industry, call the water-cement ratio: w/c, w/cm, $w/c + m$, $w/(c + m)$, w/cp, w/m, or whatever else? ACI has dictated that it be w/cm; aside from the italicization, is that right or wrong? Is there still a need to use it, let alone rely upon it for whatever purpose? The answer to the latter depends upon the specific reliance you place upon it or your needs.

Dr. Neville, in his conclusions, says: We should keep using it; that its usefulness is now limited because of extensive use of supplementary cementing materials; that its usefulness is limited at early concrete ages; that is usefulness is better at long concrete ages after all of the cementitious materials have reacted; and that we must recognize the limitations incidental to its use. He says that getting some information by reference to the water-cement ratio is better than getting none. To all of this, I almost agree. But, perhaps, should we evaluate it further, and, if possible, extend its use so that it provides more meaningful information, and overcome some of the "limitations" commented on by Dr. Neville.

For example, we should know when either fly ash or ground-granulated blast-furnace slag is used in conjunction with portland cement because each modifies some early and late concrete properties, such as initial and final set, time for finishing; development and magnitude of bleeding, sensitivity to the formation of plastic shrinkage cracks, and strength gain at different temperatures. At low temperatures, fly ash will slow strength development, extend initial set, and prolong the time when proper finishing should begin. Ground-granulated blast-furnace slag at those same low temperatures can cause concrete to act somewhat similarly—but at high temperatures the slag will usually provide normal setting times and higher early and late strengths at earlier ages than most fly ashes. To call them both "*m*" in *w/cm* and classify them as similar cementitious materials does not allow an evaluation of more specific concrete properties when these mixtures are used in different environmental conditions during and shortly after concrete casting. That may not be of particular significance to many; however, it is important to investigators of concrete field problems, during discussions, and to reviewers of documents.

So, rather than being too general in the use of *w/cm*, a more specific symbolism is needed to let us know more exactly what the supplemental cementing material is. If just portland cement is used, then why not call it *w/c*? And use the following as appropriate: *w/cfa* for fly ash, *w/cs* for ground granulated blast-furnace slag, *w/csf* for silica fume, and *w/cmk* for metakaolin. Currently, that use is easy because there are a limited number of different categories of cementitious materials. The situation becomes somewhat complicated when two supplementary cementitious materials are used in conjunction with portland cement, such as fly ash and ground granulated blast-furnace slag, or even three when silica fume is also used. In those cases the symbolism could be *w/(c, fa, s)* and *w/(c, fa, sf)*. Who says life should be easy? For those who are more eager, *w/cfa* or the more complicated string of *cms* are more informative.

Although the *w/cm* (or however it is symbolized) does not alone reflect on all concrete properties and performance, as Dr. Neville said, it certainly does play a major role in interpreting many potential concrete properties.

As usual, the engineers have simplified data about concrete-making materials in order to plug the data into equations and formulas as, for example, the calculations of concrete mixture designs. And Dr. Neville has fallen into that same simplicity trap by stating, almost as gospel, that the specific gravity of fly ash is 2.9 (in other documents he says it is 2.35), and of portland cement, 3.15. In real life, the specific gravity of fly ash can vary from 2.4 to 2.9, and specific gravity of portland cement from 3.0 to 3.2.

With reference to the estimation of the water-cementitious materials ratio, I both concur and differ with Dr. Neville. Anyone can estimate the water-cementitious materials ratio using any method or technique, including assumptions incidental to their use. I concur that they are not reliable unless suitable background information and reference materials and standards are available, and even then those estimates are questionable.

But there is one technique that, although subjective, does not need to rely on anything but the deftness and experience of the individual performing the work—detailed petrographic examinations and evaluations of the petrographic data generated. Information relied upon for the estimation includes:

1. Color and color tone of the paste;
2. Size and number of residual and relict portland cement particles;
3. Degree of portland cement hydration;
4. Mineralogy of the portland cement;
5. Amount, degree of chemical reactions, and color of supplementary cementing materials;
6. Color, particle size, and amount of mineral admixtures;
7. Textures of surfaces of freshly fractured paste and lapped surfaces;
8. Hardness/softness of paste;
9. Size and morphology of the calcium hydroxide component of cement hydration; and
10. Others as may be decipherable and appropriate.

I don't know if Dr. Neville's reported "maverick claims" of close estimates of water-cementitious materials ratios (for example, ± 0.01 to 0.02) using a "trained eye" involves applying the full gamut of microscopical observations such as listed above. However, based upon their relationships, and in the hands of some experienced petrographers, relatively close estimates of water-cementitious materials ratios can be made. But perhaps I am prejudiced in the above approach, because I am a hands-on petrographer.

Dr. Neville has aptly criticized, but also shorted, the yellow-green fluorescent thin section technique (Nordic method) for estimating water-cementitious materials ratios. He certainly is correct in concluding that it is not suitable for most field concrete because the same concrete-making materials are usually not available and the degree of cement reactions cannot be duplicated in reference samples. It does, however, have utility for new concrete when reference samples can be made using the same concrete-making materials and curing that duplicates the degree of hydration of the portland cement and chemical reactions of other cementitious materials. For example, on projects where it is included in quality control and quality assurance programs, the concrete-making materials are available and suitable reference standards can be prepared. The method is sensitive to a number of things and, assuming suitable care has been exercised in accommodating them, there should be relatively good accuracy in the estimation of water-cementitious materials ratio.

The recipient of a water-cementitious materials ratio estimate done using any method of analysis should be aware that the estimate is for the specific piece of concrete examined and that piece of concrete may vary from mixture design because of: 1) variations of proportions of the concrete-making materials in different locations within the mixer and after placement; 2) (as commented on by Dr. Neville) water may have been lost due to absorption by the subbase and to evaporation into the atmosphere during mixing, handling, and after placement when the concrete is still plastic; and 3) bleeding that can create variable water-cement ratios from top to bottom of concrete lifts.

I agree with Dr. Neville that we should continue to use the water-cementitious materials ratio acronym. However, in addition to recognizing its limitations (as he concludes) when we use it, we should make it more versatile by having it more specifically reference the actual cementitious materials involved.

Bernard Erlin
The Erlin Co.
Latrobe, Pa.

Neville's response to letters No. 4 and 5

I am grateful to Dr. Popovics for adding a numerical expression to my verbal description of the relation between the porosity of the hydrated cement paste and the water-cement ratio. I am also pleased to have his support for the view that the best time to determine the w/c of concrete is when it is fresh.

The letter from Bernie Erlin, the doyen of concrete petrographers, complements my article that presents an engineer's view. He aptly deals with the "limitations" on the use of the w/c that I could not discuss in the space available, and his letter should be read by all who seek to understand the petrography of concrete.

Mr. Erlin's detailed description of his approach to the determination of the w/c of hardened concrete is impressive. Such a painstaking procedure in the hands (or is the eyes?) of a competent and experienced petrographer is likely to yield valuable results but, as Mr. Erlin admits, it is subjective. This cannot be avoided, at least so far.

Mr. Erlin concurs with me when he says that estimates of the w/c "are not reliable unless suitable background information and reference materials and standards are available, and even then those estimates are questionable." This qualification should be carefully borne in mind when reading the preceding words: "Anyone can estimate the water-cementitious materials ratio using any method or technique." Likewise, anyone can diagnose my ailments, but I would prefer not to rely on a witch doctor using lizard's entrails.

The same applies to a "maverick" determination of the w/c of hardened concrete which relies solely on a rapid assessment of color of a thin section specimen impregnated with epoxy. I am glad that Mr. Erlin says: "He (Neville) is certainly correct in concluding that it (the method) is not suitable for field concrete (because the same concrete materials are usually not available and the degree of cement reactions cannot be duplicated in reference samples)." Mr. Erlin says that I "aptly criticized, but also shorted" the Nordic method; I did not "short" it because, as he rightly points out, the method requires reference samples. Using the method without proper safeguards is perverting its intention. Alas, there seems to be an increasing misuse of standard test methods in order to provide a convenient point in a dispute. Surely this can only bring the scientific approach into disrepute as well as being potentially misleading (see Section 7.1).

I realize the value of establishing by petrographic methods the various parameters of hardened concrete, including its w/c, when investigating their influence on the behavior of concrete. Such scientific study for the purpose of improved understanding of the behavior of concrete, and therefore making better concrete in the future, is undoubtedly useful.

However, we should ask what is the "need" for the determination of the w/c of hardened concrete in a dispute over an existing structure in which no problems are observed. What we want to know is whether the concrete is sound or whether it has been damaged or has deteriorated. If it has not, then the "true" w/c of the concrete is of limited interest to engineers. Likewise, if a person walks away from a car accident entirely unscathed, what matters is his state of health. If he is well, it is of little interest to that person (as distinct from the garage) whether the seat belts were properly fitted. If he is well, all is well.

And all is well with using the w/c, subject to a proper understating of mixture ingredients and hydration of cement; none of the letters disagreed with this. My article sought to stimulate discussion and, in this, it has succeeded. I am therefore grateful to those who have contributed.

Adam Neville

Chapter 3

Durability

The durability aspect of concrete has been gaining importance in the last 20 years or so, but it was not totally ignored in the preceding years.

It may be instructive, as well as amusing, to look at the British Standard Code of Practice CP3–Chapter IX, headed Durability, published in 1950. This code deals with buildings in general, and is not restricted to concrete. The definition of durability is as follows: "The quality of maintaining a satisfactory appearance and satisfactory performance of required functions." While the second part of the definition is still valid today, it is revealing that it is listed after appearance. At the same time, it is a little sad to reflect that very little attention is paid to sustained good appearance of concrete structures; indeed, their appearance often suffers with the passage of years.

The 1950 Code of Practice CP3 recognizes four classes of durability: 100, 40, 10, and less than 10 years, but it gives no advice on how to achieve these values of satisfactory life. An Appendix to the Code lists numerous causes of deterioration. The main ones relevant to concrete are: soil and ground water action, electrolytic action resulting in corrosion of metals, and "unsound" materials, such as "furnace clinker concrete containing an excessive amount of unburnt or partially burnt coal…"

With respect to what the 1950 Code CP3 labels "cement products," there is a mention of frost (but not of repeated freezing and thawing), moisture and thermal changes, and of chemical attack by sulfates. No advice is given on ensuring durability of concrete structures.

Of course, 1950 was early days for concrete, and much progress has been made since then, but the changes in thinking, let alone guidelines, have been slow to come. The progress is traced in Section 3.1, and it is instructive to note the development in our thinking.

Whereas acceptance of the durability as a specific design requirement, and not merely as a concomitant of compressive strength of concrete, was slow to come, the recognition of the need to provide a systematic maintenance of concrete was even slower. This is largely due to the oft-heard view that concrete is maintenance-free, unlike steel, which needs to be painted and re-painted. The anecdote of a permanent team of painters who started to paint the Sydney Harbour Bridge from one end as soon as they had finished the other end was repeatedly used in promoting the "naturally durable" concrete. This may have been an innocent error or an expression of commercial interest.

Section 3.2 emphasizes the need for maintenance, and I commend to all those responsible for concrete structures the provision of regular preventive maintenance, and not just trouble shooting and repair at the stage when substantial deterioration has already taken place.

Nowadays, we are knowledgeable about ensuring durability, and Section 3.3 is emphatic about making and achieving good concrete. My views are neatly presented and emphasized, as well as extended, in the comment by Bill Semioli, the Editor-in-Chief of *Concrete International* in the July 2000 issue of the journal. He says:

> An old question you may have seen examined a number of times in *CI*: "Who should be knowledgeable about concrete." Contributor Adam Neville within his "In My Judgment" piece examines this question, among other items. He notes that in the past, great structural designers and engineers were "expert in structural analysis as well as in the materials they used." Neville points to past great personages, veritable giants in concrete of international repute: Torroja, of Spain; Rusch, in Germany; Philleo, the United States of America; Glanville, from the United Kingdom; and of course, Freyssinet, of France; and Nervi, of Italy. Particularly with regard to the "modern education" in the United States and Great Britain, he emphasizes that, "The teaching of concrete material is negligible." The author adds that compensation for this is the employment by consultants of concrete technologists, but who in turn are not knowledgeable about structural matters. There is brief comment on problems this can cause. It bears repeating then that the most important ingredient for any durable concrete is the human one. Insufficiency here will certainly and quickly neutralize obtaining the maximum from even the very best of the inert materials of construction.

Of course, exposure conditions vary in different parts of the world, and the severity of attack on concrete is probably greatest in the Arabian Gulf area. Special effort and stringent controls are needed there and in other areas with a similar climate. Given that what is done on site is, at best, what is specified by the designer, it is very important that the designer be aware of the conditions of exposure of the structure to be built. If the designer's experience is limited to temperate climates and he or she relies on codes of practice for those conditions, the consequences are likely to be catastrophic. In the past, they often were, but Section 3.4, as well as various publications emanating in the Arabian Gulf area, should be of help.

There is one particular form of attack that occurs not in arid climates but in temperate and cold areas: repeated cycles of freezing and thawing, and the associated action of deicing salts. The damaging action is that of chlorides, which lead to corrosion of reinforcement. The topic of chloride attack is reviewed in Section 3.5, which concludes that it is possible to design successful reinforced concrete structures that will withstand chloride attack.

Nevertheless, Section 3.5 ends with the recognition of extreme conditions of exposure involving wetting by seawater alternating with drying, combined with severe changes in temperature; under such conditions, the use of reinforced concrete is highly questionable. This is not a cowardly admission of defeat: reinforced concrete is a marvellous construction material, but even it has its limitations.

Section 3.1: Consideration of Durability of Concrete Structures: Past, Present, and Future

Many an author, especially advanced in years, likes to impress his readers by telling them that he has long ago discovered the right idea, and that it is other people who have followed the wrong approach. I propose to do the opposite, and to tell you that, for many years, I was wrong about consideration of durability in design and construction. I should make it clear that this paper is about the attitudes to ensuring durability and not about specific measures and actions.

I will then give you my current ideas. As for the future, it is included in my title because it sounds attractive and intriguing, but all I can offer is to make some recommendations and to hazard a few guesses. Being old, I do not have to worry about the day when I am shown to be wrong because I shall not be there to be told that I had been wrong. In structural design, my parallel favorite remark is the following definition of the factor of safety: the number of years to retirement plus two.

What is durability?

Before talking about my past views, and they were views held by most people at the time, I should define what I mean by durability of concrete structures. Durability means that the given concrete structure will continue to perform its intended functions, that is, to maintain its required strength and serviceability, during the specified or traditionally expected service life. It is worth analyzing this definition with some care. First of all, I referred to "a given structure" because there is no such thing as a durable concrete in the general sense: if I want to build a floor slab for a shed for gardening tools, which I intend to pull down in a year's time, a certain quality of concrete will be "durable." But the same concrete will not be durable for a major bridge, for a tunnel, or for an important dam. So, it is wrong to talk about "durable concrete" or "not durable concrete." This is obvious but, acting as an expert in many legal disputes, I hear lawyers talk about concrete which is "not durable," without any reference to the specific conditions of exposure foreseen at the design stage.

In fact, there are two qualifications of the term "durability." I have already mentioned the first one, namely, the expected service life. The expectation may be specified as a given number of years, for example 120 years for a major tunnel, or may be simply traditional, like 25 to 50 years for an industrial building, or 50 to 70 years for the foundations for a single family timber home in a housing development.

The second qualification concerns the processes of deterioration to which a given concrete can be expected to be exposed. In some cases, like a cow shed, it may be the carbon dioxide breathed out by the cows; in other cases, like a bridge in a temperate climate, it may be the action of cyclic freezing and thawing. The point that I am laboring again is that concrete which is durable under one set of conditions of exposure may not be durable under another set of conditions. So again, there is no such thing as an inherently durable concrete.

Once spelled out, all this is obvious, but it is not always recognized, especially in legal disputes where there is sometimes a tendency to indulge in broad statements and even emotional pleas like "my client's concrete is not durable." Occasionally, the concrete is not durable because a change of use has occurred; for example, a floor designed to store paper is now subjected to spilling of noxious chemicals.

All that I have said so far was simply scene setting, but perhaps a little more should be said about the causes of deterioration consequent upon inadequate durability. Broadly speaking, the causes can be external to the concrete structure or they may be internal causes within the concrete itself. The external causes acting on the surface of the concrete are generally mechanical in action, namely, impact, abrasion, erosion, and cavitation. I do not propose to discuss these topics further.

The causes that I wish to consider are chemical or physical in action; sometimes, they are external in origin, in other cases, they are internal. The internal causes include the alkali-silica reaction and, in some parts of the world, the alkali-carbonate reaction, both of which are chemical actions. The most common chemical forms of attack arise from outside of the concrete, being the action of aggressive ions, such as sulfates and natural or industrial liquids. In addition, chlorides and carbon dioxide in the form of a mild carbonic acid are conducive to the corrosion of reinforcement. Corrosion requires also the ingress of oxygen at the cathode.

Physical action includes repeated cycles of freezing and thawing and the associated action of deicing salts, and also temperature effects. These can be a high temperature of the concrete at the time of placing or a high temperature differential between different parts of a concrete member. A large differential between the coefficient of thermal expansion of the aggregate and of the hardened cement paste can be destructive if there are numerous cycles of temperature variation.

I have not presented an exhaustive list of factors affecting durability, nor a matrix of external and internal factors, and internal and external agents because this is a presentation of a broad view, and not a detailed technical paper. What I think is important to realize is that the various agents can act in a synergistic manner. Moreover, it is worth noting that the deterioration of concrete is rarely due to a single isolated cause. Concrete is a patient material, which can perform satisfactorily despite some adverse conditions but, with an additional adverse factor, deterioration can take place. So it behooves us to make as good a concrete as possible, especially with respect to minimizing the ease with which aggressive agents can penetrate into the concrete. As the aggressive agents act in solution, it is the penetrability of concrete by liquids that matters most. I am using the term "penetrability" to cover the various mechanisms of ingress of fluids into concrete. These are: permeability, which is flow under a pressure differential, usually a head of water; diffusion, which is transport by a concentration gradient; and sorption, which is capillary movement in pores open to the ambient liquid.

Past attitude to durability

Following this rather lengthy introduction, let me look at the past attitude to durability. A good example is offered by what I said in the first edition of my book *Properties of Concrete*, which appeared as far back as 1963. I discussed various deleterious actions but the general impression given was that, at the design stage, there was no need to consider durability in an explicit manner. In the chart for procedures in mixture design, included in the book, I indicated that the liability to chemical attack influenced the choice of the type of cement, but not that it influenced directly the choice of the water-cement ratio.

Although, in my book, I discussed some specific forms of attack, I wrote: "The usual primary requirement of a good concrete in its hardened state is a satisfactory compressive strength. This is aimed at not only so as to ensure that the concrete can withstand a prescribed compressive stress but also because many other desired properties of concrete are concomitant with high strength." That was then the generally held view. Indeed, in 1969, that is, six years after the publication of the first edition of *Properties of Concrete*, the British Code of Practice for Reinforced Concrete in Buildings, CP 114, said in a very sweeping way: "The greater the severity of the exposure the higher the quality of concrete required."

In the second edition of my book, *Properties of Concrete*, which appeared in 1973, I did better. I wrote: "Concrete of reasonable strength, properly placed, is durable under ordinary conditions but when high strength is not necessary and the conditions are such that high durability is vital, it is the durability requirement that will determine the water-cement ratio to be used." This is not incorrect, but it still gives the impression that adequate strength and adequate durability run hand-in-hand. This probably actually worked with the portland cements used around the year 1970. A little later, the properties of cements manufactured in modern plants changed, and this had an adverse effect on durability.

Before I deal with that topic, I would like to emphasize the words in the second edition that I have just quoted, namely "properly placed." Of course, it is obvious to anyone experienced in concreting that as full compaction (consolidation) as possible must be achieved. It follows that if a particular batch of concrete with the specified water-cement ratio is delivered on site, and the concrete is too stiff to be compacted by the means available, it is necessary to add more water, so that the mixture has the appropriate workability. It is no use adhering to the specified water-cement ratio so that the capillary porosity is low, and producing concrete in place with large voids. If we do this, the theoretical penetrability of the hardened cement paste is low but the reality is that the large voids provide easy access for aggressive fluids. Of course, if the workability of the mixture does not match the compacting equipment and manpower, then the mixture needs to be re-designed.

In the third edition of *Properties of Concrete*, which appeared in 1981, I did not advance much, but soon afterwards I realized the effect of the changes in the properties of modern cements upon the development of strength and thus upon the microstructure of the hardened cement paste, which is a controlling factor with respect to the penetrability of concrete and thus its durability under given conditions. I would like to develop this topic.

It was actually in 1985 that I commented on the fact that concrete placed in the 1970s was more vulnerable to carbonation than older concretes, on an age-for-age basis. These concretes made with the "new" cements also exhibited a higher penetrability. And yet, the "new" cements were, according to their manufacturers, stronger; the implication was that the cement manufacturers were giving us a better product. In my opinion, this was definitely not so, or at least not so because the engineers did not fully understand the change. Let me elaborate.

It was customary, and it often still is, to specify a concrete mixture by the 28-day strength of standard test specimens. In, say, the year 1960 and the year 1980, the structural designer specified the same 28-day strength and he or she was under the impression that he was obtaining the same concrete at all times. The prescriptive specification of mixture proportions went out of fashion as being uneconomical, and rightly so. The performance specification, which replaced it, was on the lines: tell the ready-mixed concrete supplier what 28-day strength you want, and leave the mixture proportioning to him.

So the specified 28-day strength did not change, but the composition of the mixture did. Why? The answer lies in the change in the properties of the then modern cements. Not all cement plants in all countries changed at once, so it was even more difficult to know what was happening. First of all, the cements became somewhat more finely ground, so that they hydrated more rapidly and therefore developed a given strength at an earlier age. More importantly, the chemical composition changed in that there was much more tricalcium silicate and less dicalcium silicate; this, too, led to a more rapid hydration and a more rapid development of strength. In particular, this resulted in a higher strength at the age of 28 days and a much lower *increase* in strength at later ages. We could describe the situation by saying that the two strength–age curves, that is, one for the "old" cement, the other for the "new" cement, pass through the same value of strength at 28 days, and of course they both pass through the origin. However, the curve for the "old" cement rises more slowly up to the value at 28 days, but continues to increase beyond that age, increasing by perhaps 25 or 30% of the 28-day value at the age of one year. On the other hand, the

3-8   Chapter 3: Durability

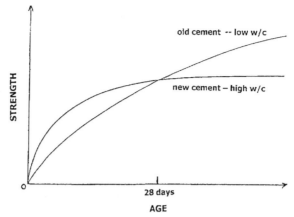

Fig. 3.1.1—The development of strength of old and new cements

curve for the "new" cement rises steeply at early ages and flattens out near the 28-day value so that there is little increase in strength later on (Fig. 3.1.1).

The shape of the curve for the "new" cement, that is, its early steep rise, can be exploited by using a higher water-cement ratio than before and still achieving the specified 28-day strength. In practical terms, for the ready-mixed concrete supplier, it means that less cement per cubic meter can be used, and this of course saves money, while the specification for the 28-day strength is satisfied.

What is the significance of all this with respect to durability? At a higher water-cement ratio, the hardened cement paste has more capillary pores and a more connected system of pores. In consequence, the penetrability is greater so that aggressive agents can enter the concrete more readily.

The remarkable thing is that this change occurred without those concerned knowing much about it. In a way, no one is to blame. The cement manufacturers improved the burning process in the kiln and generally increased the efficiency of cement production. The ready-mixed concrete suppliers increased their profits or lowered their price per cubic meter while adhering to the strength specification. And so it went on. All this could happen because, in most countries, there was no upper limit on the content of tricalcium silicate or on fineness. In actual fact, excessive fineness was rarely problematic because the high cost of grinding clinker discouraged the cement manufacturers from making excessively finely ground cement.

It may be worth adding that there was one other consequence of the higher rate of gain of strength of concrete at early ages. We know that in order to remove formwork we need a particular minimum value of strength. If this is achieved earlier, the contribution of the formwork to water retention within the concrete ceases. Unless active wet curing or membrane curing is applied, and this is rarely done on vertical surfaces, the external surface of the concrete is more porous and less durable. And it is this part of concrete that is exposed to any possible aggressive agents. On the other hand, the contractor likes the "new" cement because it permits a more rapid rate of construction.

The present view of durability

The present is, by definition, ephemeral because in a day or in a month the present will have become the past, and the future will become the present before this paper has been published. Thus, the boundaries between the past, the present, and the future, as listed in the title of this paper, are not clear-cut and writ on tablets of concrete. Nevertheless, I am able to say that the durability of concrete with a given water-cement ratio is now worse than it was in the past.

It could be argued that we should remedy the situation by specifying a lower water-cement ratio than is necessary to achieve the strength required by structural design. However, if we do that, we should not, at the same time, also specify the level of strength required by the structural designer. My reason for saying this is that it is confusing to the ready-mixed concrete supplier to be confronted by two incongruous items in the specification: a relatively low strength and an unconnected low value of the water-cement ratio. There is a practical consequence of such a pair of specified values: strength is routinely checked by standard test specimens crushed at the age of 28 days, so that the ready-mixed concrete supplier knows that he must not make concrete with too low a strength. On the other hand, once the materials have been batched into the mixer, the water-cement ratio cannot be readily verified. So there is a temptation not to pay too much attention to the specified maximum value of the water-cement ratio. It follows that if we really want to specify a certain maximum value of the water-cement ratio, we should specify a corresponding value of strength, even if the structural designer does not need that strength. But, of course, if we specify a high value of strength, there is no point in specifying the water-cement ratio as well. Moreover, it would be logical to establish the higher value of strength required from durability considerations right at the outset of the design, so that the structural designer can possibly take advantage of it.

The second possibility is to specify the cement content per cubic meter of concrete. This approach is sometimes favored by cement manufacturers, but there is no intrinsic merit in it. The cement content is a measure of the amount of cement in a given volume of concrete, but it is only the amount of cement in a given volume of cement paste that affects the durability, the amount of water being controlled by the required workability. It follows that, if the volume of aggregate in a cubic meter of concrete is larger, then the cement content in a cubic meter of concrete may be lower. Such an approach is used in the British and French codes by means of factors for cement content allowing for the maximum size of aggregate. The rationale of this is that, with larger aggregate, there is a higher total aggregate content in the mixture. I am not discussing the properties of aggregate because we have adequate knowledge to ensure the use of good aggregate.

It seems to me that the approach discussed in the preceding two paragraphs is workable but only if the mixture contains exclusively portland cement. Durability is not controlled directly if other cementitious materials are included in the mixture. These days, in many parts of the world, what I call "serious" concrete contains more than just portland cement. In consequence, the approaches mentioned previously are inadequate.

As for my published views, I recognized the past error of my ways in the fourth edition of *Properties of Concrete*. I wrote: "in many situations, durability is of paramount importance." And later on: "It is now known that, for many conditions of exposure of concrete structures, both strength and durability have to be considered explicitly at the design stage. The emphasis is on the word 'both' because it would be a mistake to replace overemphasis on strength by overemphasis on durability."

Future improvements in durability

This brings us to the future. I am not looking to spectacular developments arising from revolutionary research. I do not believe that we shall see such developments in the next few years. I am also not convinced that we need them. We have at our disposal a whole gamut of good cementitious materials, as well as admixtures: what we need is a judicious and intelligent use of these materials and good quality execution of the concreting operations.

In my opinion, the preceding statement about an appropriate choice of the ingredients of the concrete and about a high quality of concrete in the structure tells us all we need to know in order to ensure durable concrete in the future. The materials mean principally the cementitious materials because it is generally the hardened cement paste that undergoes deterioration. Even in those cases where the aggregate or the steel is involved in deleterious reactions, it is still the hardened cement paste that enables the aggressive agents, ions or carbon dioxide, to travel from outside into the interior of the concrete. In cases when no outside chemical agent is involved, as for example in the case of the alkali-aggregate reaction, water is essential for the reaction to take place. If the concrete is dry and no water from outside can enter into the interior of the concrete, no further alkali-aggregate reaction will take place. So it is the low penetrability of concrete that is crucial to minimize deleterious chemical action.

However, I would like to emphasize my earlier choice of the expression "high quality of the concrete in the structure." Making good concrete in the laboratory is not all that difficult: the ambient temperature is kind to the concrete and to the laboratory worker. The mixture ingredients can be weighed with precision that would satisfy a pharmacist. The laboratory mixer is good enough for a top confectioner. The compaction is performed meticulously. Curing is effected at a closely controlled humidity.

And so on.

Not so on site. It is not necessary for me to describe the vagaries of the weather or the physical difficulties facing the worker because of the shape and position of a concrete member; contrast this with a factory worker under a roof in air-conditioned space. More than that: the place of work of the latter is always the same, and does not change every few months from one site to another. There is another difference at play. I do not know whether it is the consequence of work conditions or simply the consequence of historical development, but the worker on a construction site is much less well trained than his counterpart in a factory. If a worker does not understand why he should not move concrete horizontally in the form because this may result in segregation; if he does not understand the consequences of adding more water to the mixture so as to make it easier to compact; if he is not aware of consequences of using a sledge hammer to "adjust" the position of reinforcement; if he does not realize that he would displace the steel by standing on an unsupported bar; and so on; then why should he do better? Moreover, he knows that, once the concreting has been finished, his "sins" will be hidden.

In many countries there is no recognized trade of concretor requiring training, examinations, and certification. It is often thought that anyone standing idle can be directed to place or compact concrete. I am describing this situation for the purpose of drawing the conclusion that in the future we must have properly qualified personnel involved in concreting. I recognize that this cannot be achieved overnight, but we must move in the right direction without delay. One way forward is to put in the conditions of contract that, for example, in any concreting gang, there must be at least one person with specified trade qualifications; otherwise, the contractor is liable to a financial penalty. A year later, the contract may require that at least one-third of workers are so qualified; and progressively on until there is no unqualified concretor, just as there is no unqualified welder.

Such improvement in the training of concretors would make it possible to achieve the use of correct placing methods, compaction techniques, finishing operations, and curing. In my experience, achieving durable concrete requires all the operations of concreting to be as nearly perfect as possible. This does not sound like a high-tech recommendation. But what use is a high-tech approach to, say, mixture proportioning, if a perfect mixture is so badly compacted that the actual concrete in the structure is honeycombed? The answer is obvious.

Maintenance

Maintenance is not covered by the title of this paper, and yet maintenance is essential for concrete to continue to satisfy the durability requirements.

In the past, proponents of concrete, and especially cement manufacturers, prided themselves on the "fact" that concrete needs no maintenance, unlike steel, which needs repeatedly to be painted. As late as

1969, the British Code of Practice for Reinforced Concrete in Buildings, CP 114, stated: "No structural maintenance should be necessary for dense concrete constructed in accordance with this code."

We now know that this is not true and can never be true. Both corrective maintenance and preventive maintenance are essential. This may be unpalatable to some people but it is not really surprising: all construction materials, as well as all natural materials, which are subjected to moisture or to temperature variation deteriorate. In the fullness of time, rocks disintegrate and so will concrete, however good. Maintenance ensures that concrete does not do so in our lifetime or that of our children.

Concluding remarks

You may be surprised at the seemingly simple content of this paper. My views on ensuring the required durability can be summed up by: the use of appropriate ingredients, including the various cementitious materials and admixtures, by a good, or very good, execution from batching, through mixing, transporting, placing, compacting, finishing, up to curing. We must not assume that achieving the desired strength ensures an appropriate durability, but neither should we be preoccupied with durability to the exclusion of strength. Strength and durability are two separate aspects of concrete: neither guarantees the other.

Lest I am accused of not providing a recipe for concrete with an appropriate durability, I would like to remind you that this paper was intended to review the *consideration* of durability, in the past, at present, and in the future. This is what I have done. The present paper is not the place to give detailed recipes for concrete that will prove durable under various conditions of exposure. Once we have established the principles of making durable concrete, appropriate details are not difficult to determine.

Section 3.2: Maintenance and Durability of Concrete Structures

It could be thought by some that the issue of maintenance and durability of concrete structures is so well known that there is not a need to write about them. Alas, this is not so. Indeed, this is a topic that is of vital importance on a truly international and gigantic scale. International because concrete is undoubtedly the most widely used construction material all over the world, and yet the durability of concrete structures all too often leaves much to be desired. Gigantic because the cost of maintenance, repair, and rehabilitation, in money spent or which should be spent, runs annually into billions of dollars. For example, in the United Kingdom, 40% of the total expenditure on construction is devoted to repair and maintenance. This represents four percent of the Gross Domestic Product.

Past attitudes

A part of the difficulty in establishing the importance of my topic is due to the fact that the international engineering community was for many years under a considerable misapprehension about the durability of concrete structures and about the need for maintenance. I am as guilty in this respect as the next man, and I had better admit it. In the first edition of my book, *Properties of Concrete*, published in 1963, I wrote the following: "The usual primary requirement of a good concrete in its hardened state is a satisfactory compressive strength. This is aimed at not only so as to ensure that the concrete can withstand a prescribed compressive stress but also because many other desired properties of concrete are concomitant with high strength." This was too sweeping a statement, even though I was clearly aware of specific durability problems such as sulfate attack and deterioration due to cycles of freezing and thawing and by deicing salts, and I devoted one chapter out of ten to durability. But, in those days, emphasis was on strength and there was a belief that strong concrete is a durable concrete. For example, the British Code of Practice for Reinforced Concrete in Buildings, CP 114, published in 1969 (that is, six years after my book) said merely the following about durability: "The greater the severity of the exposure, the higher the quality of concrete required." The advice to the designer was encapsulated in the words: "It is essential that it (referring to cover) should be dense, impermeable and of a quality suitable for the conditions of exposure involved." Even though this was a British code, it was used indirectly in some other countries.

What I have just written explains why so many structures built in the 1960s and the 1970s have a poor durability and need much repair. The situation has changed very substantially, and modern codes place emphasis on durability, but not necessarily by using the right approach. At this juncture, I would like to sound a word of warning by quoting from the fourth edition of the same book of mine, *Properties of Concrete*, published in 1996: "It is now known that, for many conditions of exposure of concrete structures, both strength and durability have to be considered explicitly at the design stage. The emphasis is on the word 'both' because it would be a mistake to replace overemphasis on strength by overemphasis on durability."

Maintenance

In this respect, too, we erred in the past. Thirty years ago, the view, forcefully advanced by many cement manufacturers, was that concrete needed no maintenance, unlike steel that had to be repeatedly painted. Up to as late as 1969, the British Code of Practice CP 114 stated: "No structural maintenance should be necessary for dense concrete constructed in accordance with this code."

We now know that this is not true, but we are still very vague about the kind of maintenance that needs to be provided. I would like to consider this topic before returning to durability because, as I see it, in order to ensure durability it is not enough to make durable concrete, but it is also necessary to put into place a system of regular maintenance procedures. From what I have just said, it should be clear that I am referring to preventive maintenance and not to repair at the stage when substantial deterioration has already taken place.

Maintenance includes, as a first step, a periodic inspection of the structure. The inspector must be familiar not only with concrete as a material but also with structural action. For example, if cracking is observed, the inspector must be able to distinguish between, on the one hand, cracks due to overloading of a properly designed structure or caused by the structure being inadequately strong and, on the other hand, cracks induced by the corrosion of reinforcement or by chemical action, including alkali-aggregate reaction, or by the delayed formation of ettringite or by thermal effects. The inspector must also be very tactful so as not to alarm the occupants of offices and, above all, of apartments who may all-too-readily jump to the conclusion that their structure is in trouble.

The maintenance that follows the inspection may involve minor repair work. Repair means making good, but prior to any substantial, as distinct from cosmetic, repair work, it is essential fully to establish the extent and causes of the problem. Otherwise, the problem may recur so that repeated repair work will be necessary. So I cannot overemphasize the importance of understanding the causes of the apparent deterioration and damage. Let me give some examples.

It is possible that the observed cracks are due to changes in temperature or moisture, combined with restraint of deformation. As these changes will continue to occur in the future, simply filling the cracks with a rigid material would result in the opening of new cracks, probably just in the vicinity of the old ones. What is necessary is to use a flexible filling material that will be able to accommodate alternating contraction and expansion in the future. Moreover, because the deformation to be accommodated can be developed only over a certain minimum width, it may be beneficial actually to widen the crack prior to filling it.

Another example of reversible crack movement is when cracking is due to soil movement, which is itself reversible. This situation occurs, for example, in clays in consequence of a prolonged period of dry weather where large trees exist in the vicinity of a house. Filling the cracks would be harmful as this would prevent the cracks from closing when the clay becomes resaturated. This is a case when the best maintenance is to do nothing, but the inspector must be able to recognize the situation. That is, to realize the cause of cracking.

A further possibility is that the cracks are stable, old cracks due to earlier shrinkage or to initial thermal stresses. In such a case filling with a patch material is likely to be an adequate solution.

If the crack is straight and follows the reinforcement, the most likely reason for cracking is corrosion, and it is this that has to be dealt with, and not just the crack itself. In other words, repairs are useless if the causes of corrosion persist. By this I mean that, if chlorides are present near the surface of the reinforcing steel, corrosion will continue unless dry conditions can be induced. Even replacing the concrete in the cover zone is useless if chlorides can continue to ingress toward the surface of the steel. To determine whether such migration has been taking place, it is necessary to determine the profile of the chloride content over the depth of the concrete cover or even further. The shape of the profile will make it possible to determine the source of the chloride. For example, if the amount of chloride in concrete is approximately constant with depth, this would strongly suggest that the chloride was in the concrete at the time of placing. The source of the chlorides may have been the mixing water that was saline or brackish, or the aggregate that was contaminated with chlorides. If the profile is variable, generally decreasing from the surface of the concrete inwards, then the chloride ingressed after the concrete had been placed. I have seen, in the Middle East, cases where both sources of chlorides were present. Such a situation underlines the need for a careful and intelligent interpretation of test results. An open mind and an inquisitive attitude are essential requirements for the inspector in charge of maintenance.

Corrosion can also be due to the depassivation of the steel in consequence of carbonation of the hydrated cement paste. This is the most likely cause above ground level in inland areas. Determination of the depth of carbonation is very simple, and replacement of cover concrete by a suitable material is likely to provide a lasting solution.

Remedial measures

In deciding how to deal with a problem found in the course of inspection, it is vital that the action taken does not aggravate the problem. This is obvious, and yet, on more than one occasion, I saw the following sequence of events:

A flat roof sags so that a puddle of water is formed and may cause leakage. The presumed remedy is to put an additional layer of a more-or-less impermeable material. If this layer is 50 mm (2 in.) of concrete, this additional dead load increases the deflection, even if laid to a level surface. With time, a puddle is formed, and the "remedial" measure is repeated. The increased load may well induce cracking in the original roof slab, thus increasing leakage. There is no need to describe what follows.

The title of this article does not include the word "repairs" but, as I have already indicated, that topic cannot be totally divorced from maintenance. The consideration of repair methods must include the owner's expectations from the structure. For example, is the appearance of the structure important? I was once involved in the case of a high-rent office block in a prime location in Australia where the owner would not accept any local repair method that would detract from a fine and uniform appearance of the façade. Another question to be asked is: is serviceability important? For instance, is the use of a building such that the floors have to be quite level, or is strength the sole consideration?

Even more important is the question: what is the expected life of the structure? Is it enough to repair it so that it will stand up and serve for a few years, and then be demolished to make room for a new structure? Or is there a specific life, say 30 years, required, in which case the repairs have to be much more substantial. And, of course, substantial repairs may be so uneconomical that demolition and replacement offer the best solution. I shall come back to this shortly.

There are other questions, too. Is there a danger to the occupants or to the public if something goes wrong? Is a shutdown possible so that repairs can be effected all at once in the most convenient and economic way? This was certainly not the case with a natural gas processing plant that I was involved with, where the consequences of a shutdown for even a short period militated against the use of cheaper but slower repair methods. Convenience, by which I mean minimal interference with the use of a structure because it is needed as an abode or an office, may have a similar effect and point toward a more costly solution. This may lead to the fundamental question whether repair is preferable to building a new structure altogether. For example, building a new bridge alongside the deteriorated one, and then demolishing the latter, would not interfere with the traffic and might be the preferred solution.

All these considerations have to be taken into account in choosing the appropriate repair option. Broadly speaking, the options are as follows:
- Allow the deterioration to continue but inspect the structure sufficiently frequently to know when its useful and safe life is coming to an end;
  - Demolish the damaged part of the structure and build a new one in its place or elsewhere. It is worth bearing in mind that elsewhere

might indeed prove preferable;
- Repair the damaged parts but take no other measures; or
- Repair substantially so that no deterioration takes place in the future.

Any decision on repairs includes the choice of repair materials. This is a major topic, and I shall limit myself to a listing of the characteristics that the repair material should have: adequate compressive and shear strengths; the modulus of elasticity and the coefficient of thermal expansion similar to those of the original concrete; a low drying shrinkage; chemical compatibility with the substrate; and a sufficiently high rate of gain of strength. In specific cases, there may be additional requirements, such as those concerning creep.

I am emphasizing the importance of systematic inspection because I believe in it, but I have to admit that such practice is not widespread in the United Kingdom. Generally, an inspection is effected when there is reason to suspect that damage may have occurred or when the adequacy of the structure is suspect. An example of the latter is when concrete has been made with high-alumina cement, and I personally conduct a number of such inspections.

In some industrial plants, general maintenance procedures may include concrete. For instance, in one important and safety sensitive plant, there was an instruction to the effect that "the behavior of the concrete should continue to be observed as a routine maintenance item." It is worthwhile to quote more fully the relevant instructions: "The objectives of the maintenance operation are to ensure that the equipment installed can perform its designed duty for the desired period of time with a high degree of reliability, due regard being paid to the overall economics of the operation, and to safety of personnel and plant."

There are two basic methods by which these objectives can be met. The more usual one is to carry out repair and rehabilitation when an item fails or when it falls below an acceptable standard. This is corrective maintenance. The second method is to intervene in the life cycle of each item immediately before it can be expected to fail, and to restore it to an acceptable standard. This is preventive maintenance. Depending on the nature of the structure and on the consequences of failure in terms of life, health and money, one or the other type of maintenance should be used. It is worth noting that, even in the case of corrective maintenance, the critical expression is "acceptable standard" and not as good as new.

I feel I should explain why I have dealt with maintenance prior to considering durability. The reason is that maintenance applies to all structures, even to those that we think have been designed to be truly durable. Moreover, a vast number of structures already in existence are to a greater or lesser degree deficient with respect to durability, and the best we can do for them is to ensure proper and regular maintenance.

Ensuring durability

Looking to the future, we should ask: what should we do to ensure that concrete structures built in the future are more durable than those built in the last 30 years? I should add that the last five or seven years have shown considerable improvement but we still require to effect a further change. Broadly speaking, the durability of concrete structures depends on the choice of the concrete mixture and on the execution of the concreting operations. In addition, durability may be affected by the shape and form of the structure, which influence wetting and drying as well as the temperature variation. A particularly telling example is the difference in durability of jetties over seawater. Norwegian work has shown that flat slabs perform much better than beam and slab construction. Exposure to the sun and the pattern of wetting and drying may also be significant factors.

I shall now deal with the influence of the concreting operations on durability; fortunately, to ensure good construction practice we need no new knowledge and no new research. However, from the standpoint of ensuring durability, the construction of concrete structures presents problems generally absent in, for example, industrial manufacture. In terms of quality system standards, construction is a "special process." This means that a concrete structure is a product that cannot be verified by inspection and testing in its finished state: deficiencies may become apparent only after the passage of time. Hence, the enormous importance of adequate supervision so as to ensure that the execution of construction complies with the specification.

To be brutally frank, a lack of supervision is likely to result in poor quality of construction due to human nature or, worse still, due to ignorance, idleness, or greed of some of the people involved. A favorite example of mine is inadequate wet curing of concrete, and I should start by explaining why curing is so important in ensuring durability. Curing affects primarily the concrete in the cover to the reinforcement and, by definition, this is the concrete that protects the reinforcement from corrosion by the ingress of aggressive agents. It may be worth reminding ourselves that adequate cover, in thickness and in quality, is necessary also for other purposes—to transfer the forces in the reinforcement, to provide fire resistance to the steel, and to provide an alkaline environment at the surface of the steel.

Wet curing is specified on the majority of construction projects but it is rarely achieved. Like batching and mixing, curing requires close supervision, but in my experience, not only in the United Kingdom but also in desert countries, curing is left to the man with the water hose or to no one at all. I must emphasize that I am not suggesting a modification to the specifications but a change in practice. At present, curing is not paid for separately in the Bill of Quantities. In my view, curing should be detached from the general item of concrete placing, and paid for by the consumption of water, or by man-hours, or in some other way.

The reader may feel that curing is such a low-tech topic. But a low-tech failure can ruin a high-tech operation, and, anyway, concrete is a low-tech material in the sense that it is used in any part of the world, under any conditions, and in very large quantities.

I will not discuss any other aspects of the site operation, but simply repeat that proper supervision is essential to make certain that good concrete coming out of the mixer becomes good concrete in the structure in service.

Mixture selection

It is particularly important that we choose a concrete mixture that will result in a durable concrete structure. I shall start by explaining what I have hinted at before, namely, why it is that structures built prior to the 1980s are more durable than those built in the succeeding 20 years. There was a change in the properties of portland cement, at least in many European countries, in that period. The change of greatest practical interest was the increase in the 28-day strength at a fixed water-cement (w/c) ratio. The main reason for this was a large increase in the content of tricalcium silicate; on average, this was from about 47% in 1960 to about 54% in the 1970s in the United Kingdom. In France, the increase between the mid-1960s and 1989 was from 42 to 58%. This change was made possible by improvements in the manufacture of cement, but it was also driven by the benefits of using a "stronger" cement as perceived by the users. Namely, a reduction in the cement content for a given specified strength, earlier removal of formwork, and faster construction.

These benefits were, unfortunately, associated with disadvantages. For example, in the United Kingdom in 1970, concrete with a characteristic strength of cubes of 32.5 MPa (4700 psi) required a w/c of 0.50. In 1984, the same cube strength could be achieved using a w/c of 0.57. Designers continued to specify the same characteristic strength so that concrete producers could take an advantage of this change in the 28-day strength by changing the mixture proportions. They had to maintain the same workability and therefore the same water content in liters per cubic meter, but they could lower the cement content of the mixture. This could be reduced by 60 to 100 kg/m^3 (100 to 170 lb/yd^3) of concrete. The concomitant increase in the w/c was between 0.09 and 0.13.

Even if it was not widely known then, we all now know that concrete with a higher w/c and a lower cement content has a higher permeability and is therefore more liable to carbonation and to penetration by aggressive agents. In other words, even though the 28-day strength remained unaltered, the concretes made with the "new" cements had a lower durability. This brings us back to the point that I made at the beginning of this paper to the effect that compressive strength is not an adequate measure of durability.

Perhaps it is worth mentioning two additional negative effects on durability. First, the rapid early gain of strength of the "new" cements means that formwork could be removed earlier so that effective curing ceased at an early age. Secondly, the "new" cements exhibited a much lower increase in strength beyond the age of 28 days so that there was no long-term improvement in concrete that had reassured users in the past, even if such improvement was not taken into account in structural design.

In my opinion, some of the existing codes and standards are too lenient for extreme conditions of exposure. For instance, it would be unwise to rely on British Standard BS 5328:1991 for conditions more severe than in the United Kingdom and possibly even under certain conditions in that country. Alas, that standard still expresses the view that a satisfactory strength "will generally ensure that the limits on free w/c and cement content will be met without further checking." The range of cementitious materials available makes this assumption unwarranted. In particular, German work has shown that some cementitious materials increase the compressive strength of concrete but the higher strength does not necessarily contribute to the resistance to freezing and thawing or to carbonation. So, once again, I am expressing the view that strength alone cannot be used as an indicator of durability.

It is also important to note that the cement content as such does not control durability. It does so only in so far as it influences the w/c at a given workability. Moreover, considering reliance on a minimum cement content, it should be remembered that this value applies to a unit volume of concrete, whereas durability depends largely on the properties of the hydrated cement paste. Thus, it is the cement content of the paste that is relevant. Having been critical of BS 5328:1991, I have to admit that it recognizes that situation by recommending an adjustment to the required cement content as a function of the maximum size of aggregate because the relative volume of cement paste in concrete is smaller the larger the aggregate.

High-performance concrete

So much for an explanation of what has happened, not surreptitiously, but imperceptibly as far as design for durability is concerned. Now, we know better. Now, we can select mixture proportions so as to ensure the desired durability, although we are not yet able to produce charts or tables for an automatic selection of mixture proportions. In my opinion, the various computer programs on the market may be a convenience but are not a source of technical knowledge. One difficulty is that we are not able to describe exposure conditions, especially the microclimate, by a number. We can, nevertheless, select and produce so-called high-performance concrete. I am using this term because it has become accepted, but it is rather presumptuous. High-performance concrete is concrete selected so as to be fit for the purpose for which it is required. There is no mystery about it, no unusual ingredients are needed, and no special equipment has to be used. All we need is an understanding of the behavior of concrete and the will to produce a concrete mixture within closely controlled tolerances.

Let us look at the ingredients to be used in making high-performance concrete. We have already seen that the w/c alone does not determine the resistance of concrete to, for example, penetration by aggressive agents, but may do so with appropriate cementitious materials. In particular, concretes containing both ground-granulated blast-furnace slag and silica fume offer particularly good resistance. Silica fume reduces the permeability of the transition zone around the aggregate particles as well as the permeability of the bulk cement paste. In relative terms, the influence of silica fume on permeability is much greater than on compressive strength.

Reduced permeability of concrete containing silica fume results in a better resistance to ingress of aggressive ions, such as chlorides. In the case of sulfate attack, there is an additional benefit in that silica fume reduces the content of calcium hydroxide and alumina, which become incorporated in calcium silicate hydrate. Silica fume is also particularly effective in controlling the expansive alkali-silica reaction.

The essence of high-performance concrete is its very low penetrability due to a particularly dense structure of the hydrated cement paste, with a discontinuous capillary pore system. Usually, this requires a very high content of cementitious material, between 450 and 550 kg/m^3 (760 and 930 lb/yd^3) with silica fume representing 5 to 10% of this value. Sometimes fly ash or ground-granulated blast-furnace slag are included. A high dosage of superplasticizer is essential so as to reduce the water content by 45 to 75 L/m^3 (9 to 15 gal./yd^3) of concrete. The superplasticizer must be compatible with the actual portland cement used. Thus, a very low w/c is achieved: always below 0.35, often around 0.25, and occasionally even 0.20. Lest there are any worries that such concrete, even well-cured, which is vital, contains a significant proportion of unhydrated portland cement, let me say that such remnants of cement can be viewed as very fine "aggregate" particles, which are extremely well bonded to the hydrated cement paste. This situation contributes to a low penetrability.

Conclusions

This section has painted no more than a broad picture, but I hope that it has illustrated both the wide ramifications of the durability of concrete structures and its enormous importance. In essence, to ensure adequate durability, we must select an appropriate concrete mixture; this requires an up-to-date knowledge of the available cementitious materials, often producing so-called high-performance concrete. However good the mixture, the concrete in the structure will be good only if all the concreting operations have been well executed. This can be achieved only with a high level of supervision. Of course, the structure has to be properly designed, not only from the point of view of strength, but also with respect to exposure to the local conditions and microclimate. However well all this has been done, the vagaries of operations and circumstances, sometimes entirely unexpected, may have an adverse effect on durability. For this reason, and also because of the very nature of concrete, it is essential that structures are subject to a systematic maintenance. Let us strive to provide truly durable concrete in the future.

Section 3.3: The Question of Concrete Durability: We Can Make Good Concrete Today

It is stating the obvious to say that concrete is the most widely used building material in the world. Alas, a proportion of the concrete in service is not as good as it should be or could be, especially with respect to durability. Such a situation is not satisfactory, but not unavoidable, and this is why in the title of this article I have included the words, "We can make good concrete today."

So, I am suggesting that we have the knowledge and capability to make good concrete, but we do not always do so. I propose to discuss the reasons for this situation and the problems involved but, first of all, I should say what I mean by good concrete. I certainly do not mean concrete with a particularly high strength or with some special properties. I mean concrete fit for the purpose for which it is intended and for the expected life during which it is to remain in service.

By way of a simple example, if we are dealing with a slab on ground or with a domestic garage floor, then the stresses in service are extremely low and low-strength of concrete is adequate; nothing is to be gained from installing a 50 MPa (7200 psi) concrete. By way of analogy, if you are transporting gold, you need an armored vehicle, but if you are transporting copies of *Concrete International*, valuable though it is, you do not. As for the expected life, this is really a durability issue.

Cementitious materials for durability

I define durability by saying that every concrete structure should continue to perform its intended functions, that is, maintain its required strength and serviceability during the specified or traditionally expected life under conditions to which the structure is expected to be exposed. There are a number of words in my definition, but they all matter. Serviceability means, among other requirements, an absence of excessive deflection. The expected life is nowadays increasingly specified explicitly: for example, 120 years for major bridges or tunnels. In many other cases, however, there is a traditional life expectancy that is shorter for warehouses and industrial buildings than for offices and private dwellings. I discussed durability and the necessary maintenance of concrete structures in an article in *Concrete International* in November 1997[1] (See Section 3.2).

In a single article dealing with concrete in general, I cannot go into details, but my point is that we have an adequate knowledge of concrete to design (or select) durable mixtures. For many conditions of exposure, these mixtures should contain a number of cementitious materials and not just portland cement. Portland cement is an essential ingredient, but it is much less satisfactory alone than when other cementitious materials are present as well. These are: fly ash, ground-granulated blast-furnace slag, silica fume, and some other very fine powder materials, including calcareous, and even inert, fillers.

I discussed these various materials in an article in *Concrete International* in July 1994,[2] and silica fume in particular in a later article.[3] I will not rehearse my views, but I would like to make one plea: let us call all these materials "cementitious materials" even if, strictly speaking, some of them have only latent cementitious properties, or even if their contribution is largely, or even exclusively, physical in action. My view is that, as long as we use terms such as "replacement materials," "mineral admixtures," or "supplementary materials," we create an impression that they are inferior to portland cement. They are not. They are co-equal with portland cement. Indeed, in my opinion, concrete containing solely portland cement is appropriate for only a very limited range of uses.

In an article in *Concrete International* in April 1999,[4] I mentioned that some producers of portland cement are opposed to the use of other cementitious materials and admixtures. At one time, they used a slogan: "The best admixture is more portland cement." My view is that, if these various materials result in better concrete, concrete will continue to be used extensively without fear of serious competition. In commercial terms, it is in the interest not only of portland cement manufacturers, but also all of us "in concrete," to put less portland cement into every cubic meter of concrete, but to produce more cubic meters of good concrete. The editor-in-chief of *Concrete International* paid me a compliment in singling out this comment in his editorial in the same issue of *Concrete International*.[5]

Admixtures

In my opinion, admixtures are essential ingredients of most concrete mixtures, but their use requires an understanding of their actions and not a blind following of the advice of a salesperson. I shall deal with the need to understand concrete in a separate section.

The action and use of admixtures are large topics, and I shall limit myself to just one aspect. It is well known that any admixture to be used has to be compatible with the cement in the mixture. Now, with most ordinary admixtures of the water-reducing type, there is only infrequently a problem. It is not so in the case of superplasticizers, also known as high-range water-reducing admixtures (HRWRA), that are increasingly being used and are virtually necessary when silica fume is included in the mixture.

A possible problem arises from the fact that both the superplasticizers of the polysulfonate type and the sulfates in the portland cement can react with C_3A, which is always present in portland cement. The situation is probably as follows. On the one hand, a certain amount of superplasticizer is necessary during mixing to achieve an adequate workability; on the other hand, it is essential that the superplasticizer does not interfere unduly with the normal cement reactions. The superplasticizer molecules are needed to deflocculate and disperse the cement particles in the mixture. It is probable that a combination with C_3A occurs when the sulfate ions are not released fast enough to react with the C_3A. When they are released too slowly, the initial workability can be lost very rapidly, and the portland cement and the superplasticizer are said to be incompatible.

The above has not yet explained why there is a variable reactivity of calcium sulfates in portland cement. If they were always in the form of gypsum, that is, calcium sulfate dihydrate, we would know their solubility. However, in reality, in the manufacture of portland cement, there are used, and quite legitimately so, different forms of calcium sulfate such as gypsum, hemihydrate (which can be the result of a partial decomposition of gypsum during intergrinding with portland cement), anhydrite, or synthetic calcium sulfate (a by-product of some industries). Moreover, some of the sulfate in the cement (and it is really only the sulfate ions that we are concerned with) originates from the sulfur contained in the coal or oil used in firing the cement kiln. This sulfur reacts with volatile alkali oxides in the cement kiln and forms alkali sulfates that are highly soluble.

The essential issue is then that, depending on their origin, the sulfates can be more or less soluble, and the chemical composition of the cement does not inform us about this. The problem of compatibility can be readily resolved because, for each portland cement, there exists an optimum amount of soluble alkalies (that is, those existing as alkali sulfates), which ensures compatibility with a given superplasticizer. This can be established by a simple physical test on grout containing the superplasticizer and the actual cement to be used. Performing such a test is essential for mixtures with a very low water-cement ratio (w/c) because in those mixtures there is less water available to accept the sulfate ions. Of course, it is precisely in such mixtures that a superplasticizer is used. The whole question of compatibility has been admirably tackled by Aïtcin.[6] The important point to remember is that you cannot just buy *any* portland cement and *any* superplasticizer: you have to use a compatible combination of the two materials.

High-performance concrete

The title of the paper by Neville and Aïtcin to which I have just referred is "High-Performance Concrete—an Overview."[6] We used the term high-performance concrete because it was, and it continues to be, fashionable. Actually, high-performance concrete is just concrete fit for the specific purpose, sometimes possessing a high strength, occasionally a high modulus of elasticity, but more often, a good durability under particular conditions of exposure. I do not view high-performance concrete as a distinct sort of material, but rather as a part of the spectrum of concretes, each of which is fit for its particular purpose. I believe that, before long, we shall cease to use the distinctions in nomenclature and simply refer to "concrete" as a generic term. It should, of course, always be good concrete.

High-alumina cement

High-alumina cement is totally different from portland cement, and it is important to be aware of that fact. In some countries, it is marketed as "ciment fondu" without (at least in advertisements) a mention of the fact that it is not a portland cement. High-alumina cement is also marketed as "calcium aluminate cement," possibly to distance it from the tarnished image of high-alumina cement. While the new name is correct, portland cement should, in parallel, be called "calcium silicate cement," but then, why change the name of the reliable old portland cement?

In the past, high-alumina cement was used structurally in many European countries (very much so in Great Britain) and also in some sensitive structures in the U.S. An essential feature of this cement is that its hydrates in hardened concrete undergo chemical conversion, which results in a greatly reduced strength and also has adverse effects on durability. Consequently, in most European countries, except France, where the cement was invented, the use of high-alumina cement in structural concrete is not allowed or is severely limited. This situation has existed for a quarter of a century, but at the end of the 1990s, there started a move to re-introduce the structural use of this cement. This topic was discussed in *Concrete International* in August 1998,[7] and I shall not repeat my arguments (See Section 1.5).

I hasten to add that high-alumina cement has many valuable properties and is an excellent material for refractory purposes and for specialized uses, such as in rapid repairs or in nonstructural elements where resistance to some types of chemical attack or electrical properties are particularly appropriate. My concern here is solely with the structural use of this material.

The lively discussion of my article[8] and a further letter[9] touching on this subject make interesting reading with respect to the importance of the composition of committees that write standards or proffer technical advice.

Who should be knowledgeable about concrete?

The reason for my asking this question is that we have more problems with concrete in service than we ought to have. In Great Britain and in some other countries, these problems often arise from the split in the knowledge possessed by the people involved in the design of concrete structures.

Let me back up in time. The great designers and engineers in the past were expert in structural analysis as well as in the materials they used. Limiting myself to people no longer alive, I would mention Eduardo Torroja in Spain, Hubert Rüsch in Germany, Robert Philleo in the U.S., and Sir William Glanville in the U.K., all of whom I knew, or Pier Luigi Nervi in Italy and Eugène Freyssinet in France. Architects also married their knowledge of structural form with knowledge of the properties of the materials involved. Modern education of civil engineers in Great Britain and the U.S. concentrates on structural analysis, numerical methods, and use of computers. (I admit to having contributed in this field.[10]) The teaching of concrete as a material is negligible.

And yet, the behavior of concrete structures involves properties such as shrinkage, creep, modulus of elasticity, and of course, durability. For the physical and mechanical properties, the designer uses arbitrary values that may or may not be fulfilled by the actual concrete when the structure is actually built. With respect to durability, there is a need to balance the desiderata for resistance to aggressive agents and mixture proportions that will not lead to excessive temperature gradients. These, of course, are just two examples of the many factors existing in practice.

I would like to mention one other specification requirement that exists in the case of vertically slipformed concrete. This is the need for the concrete to remain plastic sufficiently long to avoid the formation of cold joints, to stiffen sufficiently and also to have an adequate strength soon enough to allow the formwork to travel upwards without the concrete sloughing off or bulging. This combination of rheological properties has to be achieved by a mixture that will not be faulty with respect to its long-term strength or durability or thermal gradients. This topic was discussed in *Concrete International* in November 1999[11] (See Section 5.2).

Let me return to the structural designer and his knowledge of concrete, or rather, lack of knowledge. To compensate for this lack, many consulting engineering firms now employ concrete technologists. Their background is often in geology, sometimes in physics. They become highly competent in concrete technology. However, inevitably, they are not knowledgeable about structural action, about loads and load paths, about load-induced deformation, about the effects of elastic deformation (as well as creep and shrinkage) on redistribution of stresses, about the effects of very early temporary loads during construction, or about practical exigencies and difficulties during construction. The resulting situation is that we have some designers who lack knowledge of materials and, at the same time, technologists who lack knowledge of structures and methods of construction. I must emphasize the word "some" because there are structural engineers with an excellent knowledge of concrete as a material—unfortunately, there are not enough of them.

We need many more engineers who are broadly educated and possess a wide knowledge. In the meantime, it may be useful to consider the consequences of this dichotomy between designers who are engineers and materials people who are not. The materials scientists often write the concrete part of the specification. It is then that a lack of understanding of structural action and behavior, and sometimes also a lack of knowledge of construction methods, may result in inconsistent elements of the specification and in requirements that cannot be achieved in practice. This, of course, gives the contractor a wonderful opportunity to exploit the contradictions, if not the impossibility of achieving what is specified; more money may be claimed for overcoming the problems.[12] I am not suggesting that this occurs very often, but, when it does, the sums involved may be large.

Specifying for durability

An interesting comment about inappropriate or outdated specifications was made in the ACI President's Memo in *Concrete International* in September 1999.[13] The situation is that, these days, we are increasingly using performance-type specifications rather than prescriptive instructions. It is possible that the water-cement ratio (w/c) is not expressly specified by a maximum value but the compressive strength is. Elsewhere in the specification, there is a requirement for durability.

Depending on exposure conditions, durability may require the use of a particular type of cement, and of an adequately dense concrete, with full compaction and a satisfactory pore structure, especially pore size, of the hydrated cement paste. One of the factors is the w/c, but this must be compatible with workability necessary to achieve full compaction (or consolidation). So the w/c must not be too low because full compaction is more important than a low w/c coupled with a poorly compacted concrete.

On the other hand, the criterion of pore size distribution in the hydrated cement paste may require a w/c significantly lower than the value necessary from strength considerations. The point I am trying to make is that specifying a relatively low compressive strength that can be achieved with a higher w/c than is necessary to ensure the required durability may be misleading to the contractor at the tender (or bidding) stage. It is wiser to specify a higher strength so that it is compatible with what is necessary for durability.

Doing this has an additional advantage. If the structural designer knows that he is going to have a higher strength, he may exploit it in his design. For example, if basement walls and the columns in the lower part of a building are made of stronger concrete than higher up, then they need not have a larger cross section to carry a higher load than the columns in the upper stones. This is advantageous not only in terms of useable space but also in terms of a lower load on foundations. A similar situation exists in the case of a bridge substructure.

A mismatch between a relatively low specified compressive strength and a high cement content specified for durability (as practiced in some countries) was fraudulently exploited by a ready-mixed concrete producer in the U.K.[14] He achieved the specified strength (which was routinely determined on site) but, by manipulating the software operating the batcher, he put less cement than on the computer printout (knowing that the cement content would not be verified on site).

Cover to reinforcement

I would like now to move to several more practical aspects of making good concrete. Because they are practical, some people, mainly academics, regard them to be unworthy of consideration or study; above all, these problems cannot be solved by an elaborate computer program.

The first of these is cover to reinforcement. Cover is the shortest distance between the surface of a concrete member and the nearest surface of the reinforcing steel. In a number of structures in service, I found the thickness of cover to be incorrect. In many of these, the cover was too small, with resulting corrosion of the steel.

There are several purposes of providing cover. The first of these is to put concrete around the reinforcing steel in a beam so that the strain in concrete in flexure is transferred to the steel, which can then develop a tensile force. This is obvious to every structural designer. What may not be obvious to the person on the building site is that, in a beam or a slab, too much cover may result in a reduced moment-carrying capacity of the structural member.

Cover is also important from the standpoint of shrinkage cracking. Unreinforced concrete, if restrained (which is nearly always the case) will allow concentrations of tensile strain to develop. This would lead to shrinkage cracking. The remedy is to provide reinforcement, spaced fairly closely, and located sufficiently near the exposed drying surface of the concrete member. The corollary of this is that the thickness of cover must not be excessive. We should remember that when we hear the occasional clamor for thicker and thicker cover by those seeking to provide protection of reinforcement from corrosion.

This brings me to the need for adequate cover for the protection of reinforcement from corrosion. Although the surface of the steel is passivated and protected by the alkaline environment of the pore liquid in hydrated cement paste, this protection may be destroyed by carbonation and by the ingress of aggressive ions so as to reach the surface of the steel. The most common ion is chloride, either from seawater, splashed or airborne, or from deicing agents on roads and bridges.

Another important reason for providing adequate cover to reinforcement is fire protection of the steel. Fire endurance is a complicated topic because it involves structural action but, in essence, design codes specify the minimum cover of various types of structural elements, such as beams, columns, floors, and ribs, necessary to ensure fire resistance during a certain number of hours.

All of the above demonstrates that achieving the appropriate cover, not too little and not too much, is of great importance. There are three reasons why, in practice, the thickness of cover may be unsatisfactory. First, the cover may be incorrectly specified. Second, the specification may be incorrectly formulated. Third, the actual cover as built may be different from what was specified. These three problems are discussed in *Concrete International* in November 1998[15] (See Section 5.1). Here, I want to consider the issue of the meaning of the term "cover" in the specification or on drawings.

To expect *exactly* the same cover along every reinforcing bar is unrealistic. There must be some tolerance. ACI specifies a negative tolerance, that is, too small a cover, of 10 or 13 mm (½ in.), depending on the depth of the member. The British tolerance is only 5 mm (¼ in.). Neither code tells us by how much the cover may be exceeded. And yet, as I mentioned earlier, excessive cover has an undesirable effect on the load-carrying capacity of a flexural member and on resistance to shrinkage cracking. In any case, does the tolerance mean that there cannot be a single location where the cover is incorrect? How are you going to verify this?

And even if the cover is correct prior to the actual placement of concrete, there is a considerable risk that some bars will be displaced during concreting. This may be due to the operations involved in compaction (that is, consolidation) if all the reinforcing steel is not rigidly fixed in space. Displacement may also be caused by the weight (or, strictly speaking, mass) of the operatives who, after all, have to stand on something! It is no use just shouting, "Do better!" The structure must be buildable. It follows that detailing the reinforcement must be done by someone with personal experience of construction.

The execution of steel bending must be carefully verified; "adjustment" of the position of the steel by a sledgehammer will not do. The chairs and spacers, as well as ties, must be fixed by a competent operative, and not just by someone with time on his hands. There are other precautions that can and should be taken.

Compaction

I have mentioned the corrosion protection of steel by having an adequate thickness of cover, but an adequate thickness provides adequate protection only if the concrete in the cover zone is fully compacted (that is, consolidated). I would go so far as to say that the quality of the concrete actually achieved in the cover zone is more important than the quality of concrete anywhere else in the structure.

And yet, how do we know that full compaction has been achieved? And what precisely does full compaction mean? Answers to these questions are not yet available, and I am unable to offer any definite opinions on these matters.

Mixing

I have come across a number of cases where the concrete in the actual structure appeared not to have been adequately mixed. The cause of this was usually found to be inadequate mixing time. Mixing time is specified in a general way by ACI 304.R.[16] Mixer manufacturers often specify the minimum number of revolutions of the mixer. There also exist methods of determining the homogeneity of the concrete as it is discharged from the mixer.[17]

In reality, there is a temptation on the part of the concrete supplier to reduce the mixing time as much as possible. This way, he achieves a larger output per hour from a given batching and mixing plant, and a bigger output means economy; that is, more money for the same capital outlay and possibly the same labor cost.

The consequences of inadequate mixing are serious: a variable aggregate content, a variable w/c, and here and there, concrete that cannot be properly compacted. The remedy is obvious: establish the necessary minimum mixing time, especially when using silica fume,[3] and make sure that you achieve it in practice in every batch of concrete, however much hurry there is.

Curing

Curing concrete is the lowest of low-tech operations. And even if it isn't, it is viewed as such. But is curing truly a matter of no importance?

Curing affects primarily the concrete in the cover to the reinforcement, and I have already pointed out that the quality of the cover concrete is of considerable importance with respect to durability. What is particularly significant is that curing affects the *actual* quality of concrete rather than its *potential* quality.

It is worth emphasizing that moist curing is essential for the cement to hydrate as much as possible. This should not be interpreted as requiring full hydration of all the cement. Not only is this not necessary but, at a low w/c (less than about 0.4), it is impossible.

Despite the importance of curing, it is a fact that curing is invariably specified, but satisfactory curing is rarely achieved. This topic is discussed in *Concrete International* in May 1996.[18] (See Section 8.2) It may be worth repeating from that article some reasons, admittedly not excusable, for the prevailing bad practice of inadequate curing.

First, curing is an operation that follows the end of the concreting operations; in consequence, there is a not surprising desire to move on to the next phase of work.

Second, curing is seen by many as a silly operation, a nonjob: just sprinkling water with nothing to show for it at the end of the day. In some climates, it can be argued that watering is not necessary because, any minute now, it will rain. In other climates it can be argued that water evaporates as soon as it has been applied to concrete, so that nothing is, and nothing can be, gained. Thus, for one reason or another, why bother?

The third reason for not curing is that most personnel on site, often including even supervisory staff, do not believe in their heart of hearts that curing serves a really useful purpose. Many of them have never applied, or supervised, proper curing in their lives, and they have been successful, so why change?

The fourth reason is the rather unkind argument that curing does not show. Who will know tomorrow whether the concrete was subjected to curing today? Perhaps they argue that what the eye does not see, the heart does not grieve over.

The fifth and last reason is the most compelling one, but it is also one that points toward a remedy: curing is not paid for as a separate item. To ensure good curing, it would be worthwhile to develop a method of payment for it. This could possibly be by the consumption of earmarked water, but a more clever method needs to be developed. In any case, curing should be a separate item in the bill of quantities.

If I emphasize ensuring curing, it is because curing can make all the difference between having good concrete at the end of the placing operation that becomes good concrete in service, on the one hand, and, on the other, having good concrete ruined by the lack of a small effort. The importance of curing is nowadays even greater than in the past for three reasons.

First, modern cements, with their higher rate of gain of strength, have unwittingly allowed a worsening in the curing practice. The explanation is as follows. Because strengths adequate for the removal of formwork or for trafficking the surface of concrete are reached very soon, there is an excuse to discontinue effective curing at a very early age.

The second reason is that lower values of w/c are used than was the case in the past and, to prevent excessive autogenous shrinkage and self-desiccation, early ingress of water into the concrete is necessary.[19]

Third, modern mixtures often contain fly ash and ground-granulated blast-furnace slag. These materials, especially the latter, react over longer periods of time, and consequently, need prolonged curing.[18]

Conclusions

The objective of this article is to show that it is possible to make good, durable concrete today, using existing knowledge, without extensive and expensive research and without clever high-tech methods. This does not mean that I view concrete as a low-tech material. It often is, but we can make it into a high-tech material by simply doing better and by operating correctly and properly on the basis of an understanding of factors influencing properties and behavior of concrete in actual structures.

I have discussed, perforce briefly, but with reference to more extensive published articles, several areas where we can easily do better. First of all, we should recognize the wisdom of including in the mixture a whole range of cementitious materials and not just portland cement. We should also use appropriate admixtures, which offer a way of tailoring the properties of concrete, mainly in the fresh state, to the needs of any specific construction. In the case of superplasticizers, it is essential to ensure a compatibility between the admixture and the portland cement actually being used.

Superplasticizers are necessary in making concrete with a very low w/c; such concretes are often referred to as high-performance concrete. In my opinion, this expression should be phased out—we should always make concrete that is fit for its intended purpose.

High-alumina cement should not be used in structural concrete. Nevertheless, there is commercial publicity for the material, and it is useful to be familiar with the behavior of that cement in structures in service. Indeed, it is vital for structural designers to be fully familiar with the properties and behavior of concrete that will be used in the structure. In other words, they must have a good knowledge of concrete and of the influence of its various ingredients. Some materials specialists have that knowledge but are not familiar with structural action and behavior in service. If those specialists write specifications, there is a risk of incompatible requirements in the various clauses.

A good specification is vital when durability under the expected conditions is to be ensured and, of course, it is essential that concrete be durable. The requirement of satisfactory strength must not be ignored: both durability and strength should be considered at the design stage and when writing the specification.

Finally, I have discussed four topics that some people regard as low-tech: cover to reinforcement, compaction (or consolidation) of concrete, adequate mixing time, and curing. They may be low-tech, but proper attention to all of them, as well as a good knowledge and understanding of concrete, are essential if concrete is no longer to be regarded as a cheap and low-tech material. We should always make good concrete with an appropriate durability. There are many answers to the question of durability, but we *can* make good concrete today.

References

1. Neville, A. M., "Maintenance and Durability of Concrete Structures," *Concrete International*, V. 19, No. 11, Nov. 1997, pp. 52-56. [Section 3.2]

2. Neville, A. M., "Cementitious Materials — A Different Viewpoint," *Concrete International*, V. 16, No. 7, July 1994, pp. 32-33. [Section 1.1]

3. Neville, A. M., "Silica Fume in a Concrete Specification: Prescribed, Permitted, or Omitted?" *Concrete International*, V. 23, No. 3, Mar. 2001, pp. 73-77. [Section 1.3]

4. Neville, A. M., "What Everyone Who is 'in' Concrete Should Know About Concrete," *Concrete International*, V. 21, No. 4, Apr. 1999, pp. 57-61. [Section 6.2]

5. Editor's Comment, *Concrete International*, V. 21, No. 4, Apr. 1999, p. 4.

6. Neville, A. M., and Aïtcin, P.-C., "High-Performance Concrete — An Overview," *Materials and Structures*, V. 31, Mar. 1998, pp. 111-117. [Section 4.3]

7. Neville, A. M., "A 'New' Look at High-Alumina Cement," *Concrete International*, V. 20, No. 8, Aug. 1998, pp. 1-5. [Section 1.5]

8. Letters to the Editor, Comment on Reference 7, *Concrete International*, V. 21, No. 3, Mar. 1999, pp. 7-10.

9. Letters to the Editor, Comment on Reference 7, *Concrete International*, V. 21, No. 5, May 1999, pp. 7-8.

10. Ghali, A. and Neville, A. M., Structural Analysis—A Unified Classical and Matrix Approach, Fifth Edition, E & FN Spon: London & N.Y., 2003, 831 pp.

11. Neville, A. M., "Specifying Concrete for Slipforming," *Concrete International*, V. 21, No. 11, Nov. 1999, pp. 61-63. [Section 5.2]

12. Neville, A.M., "Litigation—A Growing Concrete Industry," *Concrete International*, V. 22, No. 3, Mar. 2000, pp. 64-66. [Section 9.1]

13. Coke, J., "Keeping Pace with Technology," President's Memo, *Concrete International*, V. 21, No. 9, Sept. 1999, p. 5.

14. Parker, D., "Ready Mix Action on Software Scam," *New Civil Engineer*, Nov. 27, 1997, London, p. 3.

15. Neville, A. M., "Concrete Cover to Reinforcement—or Cover-Up?" *Concrete International*, V. 20, No. 11, Nov. 1998, pp. 25-29. [Section 5.1]

16. ACI Committee 304, "Guide for Measuring, Mixing, Transporting, and Placing Concrete," *ACI Manual of Concrete Practice Part 2: Construction Practices and Inspection Pavements*, American Concrete Institute, Farmington Hills, Mich., 2000, 49 pp.

17. Neville, A. M., *Properties of Concrete*, Fourth Edition, Longman, London, 1995, and John Wiley, N.Y., 1996, 844 pp.

18. Neville, A. M., "Suggestions of Research Areas Likely to Improve Concrete," *Concrete International*, V. 18, No. 5, May 1996, pp. 44-49. [Section 8.2]

19. Aïtcin, P.-C., "Demystifying Autogenous Shrinkage," *Concrete International*, V. 21, No. 11, Nov. 1999, pp. 54-56.

Section 3.4: Good Reinforced Concrete in the Arabian Gulf

In the past 30 years, there has been a large growth in construction in the Arabian Gulf Area, almost all of it using reinforced concrete. A great many structures there have suffered damage, which has been described in a number of articles containing alarming pictures. Even though I have been involved in studies and repair of some of the damaged structures, I do not intend to follow suit because, in my opinion, the time has come for some optimism in our outlook. I believe that much progress has been achieved in recent years. So, a new look, not wholly novel perhaps, but with a new eye (actually an old eye) may clear the air and give some new broad insights.

Part I: Conditions and Problems

In this part of my paper, I discuss the specific conditions of exposure in the Gulf Area and the problems in reinforced concrete structures caused by those conditions. Part II will deal with the response to the problems, and Part III with the requirements for reinforced concrete that will minimize the occurrence of problems.

What is special about the Gulf Area?

I am planning to refer to the causes of deterioration, but without indulging in horror stories, and then to follow this by what I consider good and sound practices that are necessary to produce good reinforced concrete. I must admit right up front that there is no single winning card and no single magic formula. My remark is a polite criticism of many a paper, in the past, in which the author demonstrates, at least to his or her entire satisfaction, that a particular product — an admixture, a surface coating of concrete or of steel, or a cathodic protection system — is the panacea: buy my product and all will be well! Alas, life is not that simple; if it were, all we would need is a single handbook of instructions on concreting and no one would buy my book *Properties of Concrete*[1] in order to study the influence of various factors on the complex structure and behavior of concrete.

Still continuing in terms of generalities, I subscribe to the view that the exposure conditions of concrete structures in the Gulf Area are more severe than almost anywhere else in the industrialized world. I am, of course, referring to climatic conditions affecting the ingress of chlorides and sulfates into concrete, and not to other severe conditions of exposure, such as freezing and thawing or abrasion and erosion. It follows that the conditions in the Gulf Area have to be considered quite apart from what happens in other parts of the world.

So the difference between the situation in the Gulf Area and in Europe or America is at the root of the "concrete problem" in the Gulf Area. It is well known that the original concrete practices were imported from

Western countries in a virtually unaltered form. This is not surprising as, at the time, the practices in those countries were the main source of the knowledge of concrete. Nobody is to blame for using a foreign starting point; indeed, it is usual and almost inevitable for engineers to rely on their existing knowledge and experience. Modification and adaptation follow, and this is what has been happening in the Gulf Area.

I would like to start by listing the main causes of deterioration of reinforced concrete and then look at the specific conditions that contribute to the severity of deterioration in the Gulf Area. In doing this, I should urge that we do not lose a sense of proportion: it is not only in the Gulf Area but *everywhere* that reinforced concrete deteriorates. That point of view was not always held; indeed, cement manufacturers used to insist that concrete was durable and needed no maintenance unlike the derided steel that required repeated painting. I have to admit that I, too, was a victim of such brainwashing. Since then I have recognized the error of my ways.

What is the fundamental cause of problems with concrete?

The answer to this question is very brief: portland cement. This statement may sound provocative but it is serious. Let me consider the most common problems with concrete in general, not just in the Gulf Area. The first on my list is shrinkage cracking. This is due to restrained excessive shrinkage of concrete. Although we talk about shrinkage of concrete, it is the hydrated cement paste that undergoes shrinkage; the aggregate is generally inert. Now, the hydrated cement paste is that phase of concrete that comprises portland cement.

Even though it is a much less common cause of problems, I would like to list next the creep of concrete. My choice is not entirely due to my many years of research on this topic,[2] but also because of close links between creep and shrinkage. Creep increases deflection, causes loss of prestress, and may also lead to time-dependent failure, for example in highly-stressed columns. I hasten to add that creep may be beneficial in redistributing loads in statically indeterminate structures. My point is that, while we talk about creep of concrete, it is the hydrated cement paste that undergoes creep; the aggregate does not. At this point it may be worth adding that, although the deleterious effects of shrinkage and creep depend on a number of factors, to a large extent, the higher the content of portland cement the greater the magnitude of shrinkage and creep and the larger the resulting problems.

As far as deformation-related problems are concerned, next on my list is thermal cracking. This is caused by too high a temperature gradient within a concrete member or by too high temperature-induced tensile stresses in a restrained member. Ignoring externally induced changes in temperature, we know that the main cause of a rise in temperature in the interior of a concrete member is the heat of hydration of portland cement. Here again, the higher the cement content the greater the problem.

Very many durability problems are caused by external agents that penetrate into concrete. These agents are ions carried by water ingressing into concrete: that is, sulfates and chlorides, which may attack the hydrated cement paste or travel through it to attack the reinforcing steel. Even pure water passing through concrete causes leaching of calcium hydroxide. These agents do not travel through the aggregate, so that their path is through hydrated cement paste. Even when no external chemical attack is involved, for example in the case of alkali-aggregate reaction, the presence of water is essential for the chemical reactions to take place. And the path taken by the water is through hydrated cement paste. So, once again, it is the permeable or penetrable hydrated cement paste that is the culprit.

Of course, you cannot make concrete without portland cement, but you should use it sparingly: other cementitious materials should be included in the mixture. I shall refer to their beneficial characteristics later on.

Climate

I am not qualified to discuss climate in a scientific manner, so I shall limit myself to repeating the oft-made statement that the climate in the Gulf Area is onerous with respect to concrete; the main positive feature is the absence of freezing and thawing.

Numerous papers give data on the extremes of temperature and relative humidity. The critical values depend on the actual location: on the coast or inland; they also depend on the elevation and exposure to wind. For some purposes, we should consider average daily or monthly maxima; for others, absolute values of peaks and troughs are relevant. For readers not familiar with the Gulf Area, it may be useful to give some general values of *absolute* extremes: temperature up to 50°C and down to 3°C. The relative humidity may be as high as 95% or as low as 5%.

These values are very high and very low (except for absence of freezing). More importantly, their range is very large so that a concrete member may have to undergo a considerable change in its exposure. The range of *average* monthly extremes is, of course, lower but still very large.

The above was a description of air temperature but the temperature of the concrete is affected by other factors as well. First of all, as is well known, the hydration of portland cement generates heat. If the ambient temperature is high so that the amount of heat lost to the ambient medium is small, then the temperature in the interior of a member can be 30 or perhaps even 40°C higher than the temperature at the exposed surface. If the ambient temperature is low, the maximum temperature in the interior of a concrete member will not be quite so high, but the temperature gradient between the interior and the exterior of the member may be very steep. In either case, thermal stresses are induced in the concrete and these can be so high as to cause cracking.

A higher temperature of fresh concrete increases the speed of hydration of cement and speeds up the stiffening of concrete prior to compaction, as well as increasing the moisture loss from the exposed surface and possibly the development of plastic shrinkage cracking. A high temperature at the time of placing has an adverse effect on the long-term strength. For these reasons, well known to those working in the Gulf Area and in other hot climates, the worst time to place concrete is in the afternoon, and the best is in the small hours of the night. The use of ice shavings as a partial replacement of mixing water, which may also be chilled, cooling the aggregate, and cooling the actual mixture by the use of liquid nitrogen—all of these are valuable and highly desirable.

The temperature itself affects the rate of the reactions involved in the processes of deterioration of concrete, and therefore in the life of the structure. Generally speaking, the rate of the various chemical reactions involved in the corrosion of steel increases with an increase in temperature. A rule of thumb is that an increase in temperature of 20°C speeds up these reactions by a factor between 2½ and 4. It follows that, if a certain extent of corrosion in England is reached after 25 years, the same level in the Gulf Area would be reached after about 6 to 10 years. As the ambient temperature cannot be modified, we need to concentrate on prevention of the initiation of corrosion and other deleterious actions.

There is one other influence of high temperature upon the quality of concrete structures: the physiological effect on the human body, and consequently on the performance of those who work physically. Those whose contribution to construction is mental are also affected, but they have a possibility of thinking in an air-conditioned hut. The first rational analysis of these effects was presented by Macmillan.[3] His findings are of considerable importance as they lessen the previously perceived view that the unskilled labor force was to blame for poor workmanship.

Local exposure

Changes in ambient relative humidity induce hygral movement, but in dense and well-cured concrete this effect extends probably no more than 20 to 50 mm into the interior. However, temperature influences the extent of drying and wetting. The temperature of the surface zone of a concrete member depends on exposure to the sun, quite apart from the influence of the ambient air temperature. Now, the surface temperature of concrete at a given location affects the speed of drying of the concrete and the depth of penetration of the dry "front." Wetting is less affected because it is always very fast, perhaps two orders of magnitude faster than drying. What I have just said explains why parts of a given structure facing in opposite directions are sometimes very differently affected by the climate, which should, more correctly, be called microclimate.

The actual situation is even more complicated by wind. Wind may affect the temperature but, even more importantly, it affects the relative humidity of the air in contact with the surface of the concrete. These effects can be calculated and taken into account in considering the durability of concrete. But even this is not enough: in addition to prevailing wind, there are local air currents and eddies. Eddies are affected by the shape of the given structure or even part of a structure. The effects are not readily amenable to analysis and are sometimes quite surprising. I remember a structure in the shape of a vertical cylinder near the coast: measurement of deposited chloride salt showed greater deposits on the leeward side than on the windward side.

Other configurations of structures and their juxtaposition may also have a large effect on the microclimate. Indeed, the microclimate at the site of a given structure can change by a later construction of another structure in the vicinity. My conclusion from this is that we cannot be very definite about the worst possible conditions at a specific location.

On the other hand, the shape of a structure affects local wetting and drying. For example, in a jetty, a flat slab does not have local, permanently wet pockets, which are present in a beam and slab construction; such differentials in humidity may contribute to the progress of corrosion of reinforcement.

Exposure conditions can change dramatically during the life of a structure owing to man-made effects. I remember being involved in an underground concrete duct system carrying electricity cables in the desert. The ducts became immersed in brackish water when the water table rose in consequence of residential and industrial development that generated wastewater.

Need for maintenance

I referred earlier to the fact that reinforced concrete structures require maintenance. The term "maintenance" is broad as several actions are involved.

Before discussing the different aspects of maintenance in some detail, it is useful to appreciate that virtually no material is inert if the ambient temperature varies or if the relative humidity is anything but extremely low. We know that rocks undergo weathering, that is, they disintegrate owing to the weather. Here, however, the time scale is very long. It is also very long for portland cement concrete which, just sitting quietly in the middle of nowhere, will on a geological scale disintegrate into the original mineral constituents of portland cement. This, however, gives no cause for concern. On the other hand, the 6- or 10-year life of a structure in the Gulf Area is much more worrying and virtually unacceptable.

Progress of deleterious reactions

Thus, we must accept that the various reactions of deterioration generally progress with time so that reinforced concrete structures undergo a progressive deterioration under steady conditions of exposure. However, the rate of progress of deterioration need not be constant. It may be constant in some cases such as erosion. But even there, because of the differential hardness of the aggregate and the hydrated cement paste, the formation of an uneven exposed surface can result in "plucking out" of large aggregate particles and a consequent acceleration of damage. Likewise, in the case of cavitation, because the surface of the concrete becomes jagged and pitted, it is possible for an initial period of small damage to be followed by a rapid deterioration, and then further damage at a slower rate.[1] Nevertheless, in general terms, chemical attack of concrete progressing from its surface under completely steady conditions of exposure, is likely to occur at a constant rate.

However, the rate will not be steady if concrete undergoes some changes, such as the development of cracks that open new paths for the ingress of the attacking medium. The cracks may be the consequence of the initial attack or they may be unrelated to it, but be caused by loading, especially overloading.

In yet other situations, less common, the deterioration may be self-limiting. For example, although magnesium sulfate in solution is very aggressive, when present in seawater it may result in the formation of brucite (magnesium hydroxide), which has a very low solubility and becomes deposited in the pores near the surface of the concrete, thus considerably slowing down further attack. However, if this "protective" layer becomes damaged or removed, for example by impact, the deterioration resumes.

Carbonation under steady exposure conditions also progresses at a decreasing rate because carbon dioxide has to diffuse into concrete through the pore system in the hydrated cement paste, including the already carbonated surface zone. Now, the product of carbonation of calcium hydroxide is calcium carbonate, which occupies a greater volume than the original calcium hydroxide. In consequence, the porosity of the already carbonated zone is reduced. Thus, the progress of carbonation is self-limiting. It follows that, if the carbonation front does not reach within a few millimeters of the surface of the steel, there are no deleterious effects on the structure.

As already mentioned, the situation described above obtains under steady conditions, but in real life there may be changes. For example, if cracks open, the carbonation front advances rapidly into the cracks and progresses from their tip. Moreover, the progress of carbonation is greatly affected by the relative humidity in the pores in the hydrated cement paste. Specifically, if they are filled with water, the rate of diffusion of carbon dioxide is four orders of magnitude slower than in air. On the other hand, gaseous carbon dioxide is not reactive with calcium hydroxide. The highest rate of carbonation occurs at a relative humidity of between 50 and 70%. It follows that the progress of carbonation varies from month to month or even day to day. On the other hand, variations in temperature have a small effect unless they affect the relative humidity. Washing down with rain significantly slows down the progress of carbonation. Washing down with fresh water is also beneficial when chloride salts are being deposited on the surface of concrete.[4]

Corrosion of reinforcing steel induced by the ingress of chlorides into the cover concrete occurs only after a certain time of exposure: this is the so-called initiation period, which is followed by deleterious corrosion. If this causes cracking, spalling or bursting of the cover concrete, the rate of damage increases.

The preceding lengthy discussion of the rate of progress of various reactions adversely affecting rein-forced concrete structures is of great relevance to the issue of maintenance, assessment of the state of the structure, and possibly associated repair.

Maintenance procedures

In an earlier paper,[5] I advanced arguments for a systematic maintenance of concrete structures (See Section 3.2). Zein Al-Abideen [6] puts forward the same approach when he recommends: "Routine inspections, regular maintenance, and repairs when required."

We can distinguish preventive maintenance, ad hoc maintenance and the associated minor repairs, and major repairs or replacement.

By preventive maintenance I mean periodic, preferably regular, inspection of the structure. The purpose of such inspection is to discover changes that are likely to lead to damage and to do so prior to the advent of damage This type of maintenance is invariably applied to aircraft and many other moving artifacts such as trains, buses, cars, ships, and machinery. Among structures connected to the ground, in the United Kingdom, it is probably only bridges and fun fairs that are routinely and regularly inspected. Regrettably, structures such as apartment blocks and office blocks are rarely subject to maintenance of the structure. Apart from cost, one reason for this situation is the psychological effect on inhabitants and office users. I have experience of monitoring the health of structures containing concrete made with high-alumina cement: inspection, for example by looking above a false ceiling, prompts questions of the type: is something wrong? are we safe to stay? In addition, of course, there is a certain amount of disruption, and inspection costs money.

Nevertheless, preventive maintenance of reinforced concrete structures is being increasingly introduced, at least in important and safety-sensitive plants. In one of these, I saw an instruction to the effect that "the behavior of the concrete should continue to be observed as a routine maintenance item." In the same plant, it was said:

"The objectives of a maintenance operation may be defined as:

To ensure that the equipment installed can perform its designed duty for the desired period of time with a high degree of reliability, due regard being paid to the overall economics of the operation, and to safety of personnel and plant.

There are two basic methods by which these objectives can be met:
1) to carry out repair and rehabilitation when an item falls below an acceptable standard (corrective maintenance);
2) to intervene in the life cycle of each item, immediately before it can be expected to fail, and restore it to an acceptable

standard (preventive maintenance)."

The above offers a clear distinction between preventive maintenance and repairs. Without elaborating, I want to emphasize the terms "overall economics" and "acceptable standard." They deserve a detailed discussion but not in this section.

As I intimated earlier, some deterioration of reinforced concrete structures is unavoidable, especially under very severe conditions, but it should not be allowed to progress too far or too rapidly. This is the rationale of inspection and maintenance, which should be undertaken however good the concrete. The frequency of inspection depends on the consequences of further progress of deterioration in terms of economics or of safety. This is the basis for the selection of corrective or preventive maintenance but, in my view, in the Gulf Area it should always be the latter.

Part II: Response to Problems

Having identified the problems and proper maintenance procedures, we should consider the appropriate response. Following this, in Part III, I intend to discuss the requirements for good concrete in the Gulf Area.

Action in response to problems

What was called corrective maintenance or ad hoc maintenance is action in response to observed real or suspected problems; this may involve minor repair work, which means making good. It may or it may not be possible to eliminate the causes of the deterioration. However, prior to any repair work, other than mainly cosmetic, it is essential to establish the extent and causes of the problem; knowing causes would make it possible to prognosticate future developments and to decide on the proper course of action. Without an understanding of causes, the problem may recur unnecessarily so that repeated repair work or repairs of repairs will be necessary.

If corrosion of reinforcement due to the ingress of chloride ions has been established, it is important to know whether the chlorides were present in the mixture at the time the concrete was placed or whether they have ingressed from outside. In the latter case, further ingress can possibly be prevented and adequate repairs would ensure safety of the structure. On the other hand, if the chlorides were present in the concrete from the start, repairs by way of replacing the concrete solely in the cover zone are less likely to be sufficient. The origin of the chlorides has to be established by determining the chloride profile from the surface of the concrete inwards: if the profile is more or less uniform, the chlorides were present already in the fresh concrete.

My second, equally obvious, example concerns the presence of cracking. First of all, the extent of cracking and the size of cracks have to be assessed. I have seen cases of a few hairline cracks alarming an inexperienced observer, particularly if he has very good eyesight and a powerful magnifying glass; a competent engineer would know that all concrete has some hairline cracks here and there.

Even if the cracks, or some of them, are wider, it is essential to determine the cause or causes of cracking. In *Properties of Concrete*,[1] I distinguish several causes of cracking: plastic settlement, plastic shrinkage, crazing, early thermal contraction, drying shrinkage, alkali-aggregate reaction, corrosion of reinforcement, and D-cracking due to frost damage by aggregate. Moreover, there are blisters due to trapped bleed water, pop-outs due to aggregate, salt weathering, and some more complicated surface damage. In addition to all these, there are cracks due to structural action: structural reinforced concrete members in flexure must have fine cracks on the tension side. Overloading would make these cracks wide, but removal of the overload would allow them to close again, usually no harm having been done. Furthermore, cracks may be due to bad detailing of reinforcement, especially at beam-column or slab-column connections. Such cracking may lead to durability problems under some conditions of exposure, but they are not a durability problem by themselves.

Not every cause of cracking can be unmistakably identified, but it is important to establish whether a crack is dead or live. If it is dead, it will not close up and is unlikely to grow. On the other hand, if the crack is live, it will respond to future changes in conditions by cyclic closing and opening. Repairing such a crack by filling it with a rigid material may cause more harm than good. I have seen cases of cracks developing in walls due to stress induced by differential settlement caused by clay movement beneath foundations during a severe drought. Filling such cracks would lead to subsequent cracking elsewhere in the structure when the soil movement is reversed on re-wetting of the clay under the foundations. Someone understanding the nature of the problem and the structural action involved would know that the best action is to do nothing. The same applies in the case of the more usual temporary overloading, mentioned in the preceding paragraph.

Cracks induced by temperature or moisture changes in concrete members subjected to restraint are also live cracks. Such cracks are likely to open wider at some times and, at other times, to close partially. Simply filling them with a rigid material would result in opening of new cracks, quite likely in the vicinity of the old ones. What is desirable is to use a flexible material that will be able to accommodate alternating movement in the future, but will prevent ingress of harmful agents into the interior of the concrete.

Another possibility is that the cracks are stable, old cracks due to earlier shrinkage or initial thermal stresses. In such a case, filling and patching is probably the right solution. But before deciding on this course of action, it is important to be sure that the crack, if straight and following a reinforcing bar, is not caused by its corrosion. The investigator should know enough about structural design to recognize such a possibility at appropriate locations. The remedy may lie in dealing with the causes of corrosion and not with cracks caused by corrosion.

Cracking can also be due to poor reinforcement detailing or to a lack of provision for movement of structural elements at some locations.[4] The investigator must be knowledgeable enough to consider these possibilities.

Competence of investigators

The reason for the preceding extensive discussion of the assessment of the problem during maintenance is to show the need for an understanding of structural action and behavior, in addition to the knowledge of concrete as a material. To put it bluntly, and my bluntness may offend some people, maintenance is a matter for structural engineers and not for "pure" materials scientists.[5,7] I have made this point in the past and so has Zein Al-Abideen.[8]

Zein Al-Abideen[8] and Rostam and Engelund[9] have also argued for intelligent testing; this topic deserves some elaboration. Just collecting a large number of test results without knowing beforehand what are the possible interpretations of the outcome is likely to be a waste of time and money at best; at worst, the test results will lead to controversy or dispute. I have seen, not in the Gulf Area, a wholesale determination of rebound hammer values and penetration resistance test results, which could not be compared with values for concrete known to be good or undamaged. Worse still, I have seen such determinations on surfaces that were not smooth, plane, or large enough. At the other extreme, I have seen the results of literally one or two proper tests used to opine that all concrete in a major structure is "bad."

I have a further criticism of some of the so-called testing specialists although of course many are highly competent. A person who is solely an "investigator" without keeping abreast of developments in codes of practice, design methods and construction practices is unlikely to be serving the owner of the structure well.

Decision on repairs

The knowledge of repair materials and methods of repair is another matter. I do not propose to deal with it because there exist numerous papers on these topics, but there are hardly any publications on what I have attempted to do, namely, to discuss the approach to taking a decision on repairs.

When the repairs are not very extensive or expensive it is easy to decide what to do: repair. There may, however, exist circumstances such that it is better to allow deterioration to continue but to inspect the structure periodically in order to ensure safety and to establish when its useful life is coming to an end.

However, when the repairs required to rehabilitate the structure are extensive, expensive, and cause a major upheaval, it is necessary to consider the alternative of replacing the structure, or even doing without it; after all, it is possible that the need for the structure, say a bridge, has diminished or altered significantly. It is also possible that an entirely new structure, even more costly, is economically preferable to a repaired or partly replaced one. The new structure may be able to carry heavier loads, have a higher headroom, provide a wider bridge or a larger unimpeded floor area, or have other new and better features.

At the other extreme, at the end of an investigation, it is legitimate to decide to do nothing but periodically inspect the structure until such time as it ceases to be useful or runs the risk of being unsafe; it should then be demolished. Under some circumstances, this may be the most economic solution. The drawback is that the owner, unless fully apprised of the reasoning behind the recommendation, may feel that he did not get his money's worth from the investigator, just as a patient who sees a doctor feels disappointed if not given a prescription. Ultimately, it is the owner who decides; the engineer's role is to provide the best and fullest information and unbiased advice.

Identifying causes of deterioration

Having considered inspection, maintenance, and the bases for taking decisions on repairs of reinforced concrete structures (and not just elements) I would now like to move very briefly to the causes of their deterioration, and then to making concrete that will resist those causes as much as possible.

The mechanisms of deterioration of reinforced concrete in the Gulf Area are well known so that there is no need to rehearse them. If we take as the principal causes the corrosion of reinforcement in consequence of chloride ingress and the disruption of concrete in consequence of sulfate attack, then we find that what they have in common is the ingress of salt-bearing solutions into the concrete. It follows that prevention of damage lies in minimizing such ingress.

There are several mechanisms by which ingress of these solutions takes place: flow, diffusion, and sorption.[1] What they have in common is that they depend on what I would like to call penetrability of concrete. Again, without going into scientific details, I would postulate that in order to minimize deterioration we should make the concrete with as low a penetrability as possible.

For the sake of completeness, I should add that deterioration of reinforced concrete can be caused also by salt weathering at the surface of the concrete, by cracking due to excessive thermal gradients and by cracking due to excessive plastic or drying shrinkage. It has recently been found that we should pay attention to autogenous shrinkage as well.[10] One factor that these mechanisms have in common is that the resulting cracking makes it easier for harmful agents in solution to penetrate into the concrete. It follows that we ought to try to minimize the development of cracking by these mechanisms.

It may be unpalatable to accept that it is virtually impossible to build structures in which the concrete will allow zero ingress of liquids. Admittedly, there exists the so-called reactive powder concrete, which has an extremely high density but, so far, this is not a generally practicable proposition for structures and buildings because of mainly economic reasons.

Thus, in practical terms, to achieve "the best" we have to concentrate on the selection of the concrete mixture, on processing the concrete between the batcher and the finished structure, and following by appropriate maintenance. I should add that, in some cases, the choice of an appropriate shape of members and of detailing can be very helpful. I now propose to consider these topics in some detail, but there is no simple and single recipe for success.

Part III: Requirements for Good Concrete
Cementitious materials

I have already discussed the problems arising from portland cement, but I have, of course, recognized the necessity for its inclusion in concrete. The solution lies in using as little portland cement as possible but, at the same time, including other cementitious materials: fly ash, ground-granulated blast-furnace slag (or slag, for brevity), and silica fume, as well as possibly a small amount of calcareous filler. With respect to the last-named material, there remains some uncertainty about its effects in concrete exposed to seawater. I realize that, generally speaking, these materials have to be imported into the Gulf Area. There may be a certain reluctance to do so, if only for reasons of cost, but I think the choice is simple: if you want durable reinforced concrete structures, either you pay more up front at the construction stage or you pay later by way of repairs and disruption. As a consolation, it is worth reflecting on the fact that some countries have abundant fly ash but they have no oil: the two are almost mutually exclusive!

I have not listed all the ingredients necessary: superplasticizers are virtually essential when silica fume is included in the mixture. Now, silica fume is not just an efficient pozzolana: the very high fineness of silica fume particles, coupled with their spherical shape, allows them to pack in-between the cement particles, especially at the aggregate-cement interface. A denser concrete is thus obtained. It is worth remembering that improved packing is obtained only up to a certain maximum content of silica fume: 8 or, at most, 10% of the total cementitious material. This is clearly not a case of "more is better." Other admixtures are also likely to be necessary. I should add that the use of silica fume requires care as it may prevent bleeding and therefore encourage plastic shrinkage cracking.

I realize that the owner of a structure or a building may be reluctant to spend more than absolutely necessary, possibly because he is awaiting income from the operation of the building or because he intends to sell the building soon and let somebody else worry. It may be drastic to say that the owner has to be forced to adopt the technically better choice even if it is more expensive. After all, we enforce a satisfactory structural choice in terms of resistance to rare loads such as earthquake, wind, snow or flood; codes of practice and building regulations do not allow the designer (acting on behalf of the owner) to say: "I shall take a chance — there won't be a severe earthquake in the near future." Admittedly, this approach is occasionally practiced in some countries but, when the earthquake or flood comes, there is a severe economic loss as well as a loss of life for which the society (and not just the original owner) pays and pays heavily.

Aggregate

Although the ingress of aggressive agents into concrete takes place primarily through the hydrated cement paste, the quality of aggregate is also of importance with respect to the durability of concrete. In some parts of the Gulf Area, the aggregates are of poor quality in terms of their mineralogical character and can have a relatively high absorption, which adversely affects the durability of the resulting concrete. I am not able to make any categorical statements about the limit of their acceptability; for this, specific studies are necessary, but there is a minimum quality below which it must not be said: "Well, that's the best there is."

There are other features of aggregate that have to be watched. Specifically, the grading of coarse aggregate and its shape have a considerable influence on the water demand of the mixture, and hence on the durability of the resulting concrete. If the aggregate is unduly friable, it is not enough that the grading is satisfactory at the time of batching: the mixing operation can result in the breakup of some particles so that the actual grading of the aggregate in the concrete as discharged from the mixer may be finer than it was to begin with, and also an excess of dust may be present. These changes in grading increase the water demand in an unexpected and variable manner.

It is worth adding that any increase in water demand that manifests itself after the discharge from the mixer must be satisfied. If the mixture has too low a workability, it is not possible to shrug one's shoulders and say: "Go on, place it." If the mixture cannot be fully compacted, then the resulting concrete in the structure is worse in terms of its durability than if more water had been added and full compaction had been achieved. To generalize: full compaction is more important than a low water-cement ratio.

Whereas the grading of the coarse aggregate can be carefully controlled by proper screening and possibly also finish screening, its shape is very much dependent on the crushing operations. I once saw, on an extremely large construction project, a crushing plant in which there were two crushers operating in parallel. Their output, after screening, was indiscriminately transported to the batching plant. Despite using fixed mixture proportions, the concrete coming out of the mixer had a variable workability that changed in a random manner. After a prolonged investigation, we found a bizarre explanation. One of the crushers was new and produced coarse aggregate with good shape characteristics. The other crusher was heavily worn and the resulting aggregate had a large proportion of elongated and flaky particles and of dust. It is not surprising that the resulting concrete had a high water demand. More than that, flaky particles can trap bleed water and thus create zones of high penetrability which compromise durability.

I should add that the crushing plant operator was not ignorant of what was happening but he was not willing to invest in a new crusher. This exemplifies the need for most careful supervision of all aspects of a construction project.

On the subject of dust on the surface of the aggregate particles, it is of course known that its removal is helped by washing. However, adequate washing, or even any washing, is not always practiced. Sometimes, the reason is the unavailability of water for the purpose or a very high price of water; at other times, the execution of washing is variable or simply poor because of inadequate supervision or even the belief that washing the aggregate is not an important task. But even on a well-regulated project, there may be problems. For example, the extent of adhesion of the dust to coarse aggregate particles varies with the weather. Adhering dust can become detached by abrasion during mixing. More importantly, the amount of adhering dust varies with the shape and texture of the aggregate particles and with the process of crushing, already mentioned.

Fine aggregate can also present problems. In some parts of the Gulf Area, the natural fine aggregate is too fine and consists of particles predominantly of one size. Blending in order to achieve good grading may be necessary. Some other fine aggregates are not sufficiently clean and may introduce chlorides and sulfates into the mixture; the finest particles are prone to a high salt content. Admittedly, the fine aggregate can be washed but, if fresh water is expensive, there may be a reluctance to wash thoroughly. Again, careful and continual, if not continuous, supervision is necessary.

Water

This may be an appropriate place to consider water. As far as mixing water is considered, with one major exception in whose investigation I was involved, I am not aware of instances in the Gulf Area of the use of water with too high a content of chlorides, sulfates, or solid matter. Nevertheless, I have heard of a case of construction in the desert (outside the Gulf Area) where the specification required the use of potable or tap water. Indeed, the mixing water was taken from the local supply. However, it became quickly evident that all was not well. On investigation, it was found that the local piped supply delivered brackish water, which was used in cooking by people accustomed to it. Reinforced concrete was less tolerant to the chlorides![11]

Water for curing also needs consideration. I came across a case in the Gulf Area where large caissons, made of high-quality concrete, were floated into the sea at a very early age. The reason for this was that it was necessary to vacate the casting area a soon as possible and someone thought that immersing the caissons in the sea would provide early curing. So it did, but the seawater allowed chlorides to penetrate into immature concrete. I suppose not everybody can think about everything all the time!

A review of the requirements for water for mixing and for curing has been published recently[11] (See Section 2.2).

Cover to reinforcement

As corrosion of steel reinforcement is one of the main causes of deterioration of concrete structures in the Gulf Area, it is important to consider how this reinforcement can be protected from the ingress of chlorides. The ingress takes place through the concrete in the cover zone; it follows that the quality of *that* concrete is what matters most. I would go so far as to say that, in many cases, but of course not all, the quality of concrete in the interior of a concrete member is relatively unimportant.

What is vital is to strive to achieve the best in cover concrete: "best" means excellent compaction without trapped air bubbles or bleed water; it also means not allowing escaping water locally to increase the water-cement ratio, the prevention of leakage of neat cement paste, and the avoidance of pockmarks and surface blemishes in general. Achieving all this requires a concentrated effort during placing and compaction as well as good quality formwork without blobs of form-release oil on its surface. In this respect, controlled permeability formwork seems to be beneficial, not only because it gives a better finish but also because it leads to less patching, which may be vulnerable to external attack.

The quality of the concrete at the instant of completion of placing is not enough: the concrete has to develop its full potential and this is achieved by curing. Curing is always important to maximize the quality of the concrete in the cover zone; in the Gulf Area, because of the climatic conditions, the importance of curing is even greater. I shall return to curing later on.

In addition to improving the quality of concrete surrounding the reinforcing steel, it is possible to make the steel itself less vulnerable. The usual steel can be replaced by stainless steel; alternatively, ordinary steel can be galvanized or epoxy coated. These topics require separate and extensive treatment, and here I shall limit myself to one remark. The use of these methods may be undertaken, but only *in addition* to ensuring good concrete in the cover zone and an adequate thickness of cover. In my opinion, to rely on, say, epoxy-coated reinforcement while the cover is too small or the quality of the concrete in it is poor, is very unwise.

This brings me to the thickness of cover, a topic which I recently discussed in some detail.[12] (See Section 5.1) I shall limit myself to saying that adequate cover is necessary, even if it means a greater volume of concrete being used and a greater self-weight. On the other hand, calls for thicker and thicker cover, made occasionally, must be resisted. Too thick a cover means an unreinforced thickness of concrete in which cracks can develop due to flexural tension or to drying shrinkage. While I avoid prescribing numerical values, I consider 80 mm (and very exceptionally 100 mm) to be a maximum thickness of cover.

An aspect of cover, minor but important and yet hardly ever mentioned, is the quality of cover blocks and chairs or other supports of the reinforcement. These items must not be a weak link providing easy access for the attacking medium. This essential point is sometimes forgotten so that making the cover blocks is left to the least skilled worker who happens to be around.

In some structures, there is a way of totally avoiding the issue of cover. Despite the fact that this paper is about reinforced concrete, I would like to make a gratuitous remark about the use of plain concrete: it is possible to use it in some types of structure, especially marine. That way, there are, at least, no corrosion problems.

Curing

All the above applies to wet curing. Unless wet curing is physically impossible, it is much preferable to membrane curing. This remark will make me lose my friends who manufacture curing materials, but what concrete needs is a supply of water from outside and not just prevention of loss of water from inside. I realize that there exist shapes of concrete elements that cannot be wet cured, but even with slipforming it is possible to drape wet hessian. I have seen this done, so don't believe those slipforming contractors who disagree!

When wet curing by suspending hessian above the concrete surface, make sure that the hessian is permanently wet, and not just prior to the inspector's visit. Make also sure that the space between the surface of the concrete and the hessian above does not create a wind tunnel through which dry air rushes; I saw that situation on an important and prestigious project in the Gulf Area.

Curing must be started as soon as practicable; this statement has to be interpreted intelligently. Let me give an example to the contrary. Some time ago, I was involved in investigating the cracking of precast concrete tunnel segments in a hot country. I found that the reason for the development of cracks was the interval of time between finishing the concreting operation and the start of wet curing. The explanation is amazing in its simplicity.

The specification laid down the following time schedule:

13.00 to 17.00 hours	Place concrete
18.00 hours	Apply damp hessian
08.00 to 15.00 hours	Demould and transport
15.00 hours onwards	Hose intermittently in daytime.

The contractor worked fast and some of the segments were cast by 15.00 hours. He then left them standing until 18.00 hours, that is, for three hours before starting wet curing, Clearly, this was not the intention of the specification writer, and presumably the contractor lacked an understanding of the phenomena involved. Anyway, a concrete surface left for three hours in the open, on a hot day, perhaps in a windy location, is liable to crazing, rapid plastic shrinkage development, and formation of small cracks. These cracks may not close on *subsequent* wet curing. Even if they cease to be visible during curing, they will re-open on subsequent drying and serve as crack initiators for the development of wide drying shrinkage cracks.

I have labored the topic of curing because I am convinced of its great importance, even though some scientifically-oriented researchers consider curing to be unworthy of consideration by a Doctor of Philosophy. Curing may be a low-tech operation but it is essential if you want to achieve high-quality concrete.

I wish to add three remarks about wet curing. First, it became even more important than it had been when we started to use cements that develop their strength more rapidly than, say, 20 years ago. The higher early strength allows an earlier removal of formwork, and therefore earlier loss of water from the concrete; this has to be counteracted by wet curing.

Second, the use of silica fume, fly ash, or slag requires particularly good and prolonged curing.

My third remark concerns the use of concretes with a very low water-cement ratio. Unless they are wet-cured from a very early age, they develop autogenous shrinkage.[10] The ill consequences of that can outweigh the benefits of a very low water-cement ratio. This may be an appropriate place to mention that water-cement ratios of 0.30 or even marginally lower are practicable and desirable. If anyone is worried that, even with the best wet curing, some of the cement will remain unhydrated, the answer is: remnants of cement grains are an excellent "aggregate," well bound to the hydrated cement paste for there is no interface zone.

Curing is notoriously performed in an unsatisfactory manner. I gave some reasons for this situation in an earlier paper,[13] but a widely-accepted remedy of the situation is still awaited. The solution most probably lies in separate payment for curing, possibly by a dedicated subcontractor. I realize, however, that he may interfere with the progress of the concreting subcontractor's work or be blamed for doing so even if he doesn't. Whatever the difficulties, early and good wet curing must be enforced.

Conclusions

Having discussed the various aspects of durable concrete in the Gulf Area, I should draw them together into conclusions or recommendations. This does not mean a single recipe: there is no panacea. Nor is there a single, or even major, cause of durability problems. I am saying this in the teeth of the oft-held view that the severity of exposure in the Gulf Area is the culprit. Admittedly, exposure is a major aggravating factor, but causes of deterioration include: inappropriate codes and specifications, bad design and detailing, poor ingredients of the mixture, lack of use of beneficial ingredients (cementitious materials and admixtures), poor workmanship and execution, especially lack of curing and, above all, lack of effective and continuous supervision of the entire process of construction. This list can be elaborated and extended. And the structure in service has to be regularly and systematically inspected and subjected to maintenance.

Strength and durability are separate aspects of concrete: one does not guarantee the other. To ensure durability it is not sufficient to specify a minimum compressive strength or a maximum water-cement ratio.

Concrete is a patient material that, under moderate conditions of exposure, can tolerate moderate imperfections. However, in the Gulf Area, every detail from design all the way to the finished structure has to be near-perfect. If that is achieved, we shall have good reinforced concrete in the Gulf Area. On the other hand, if there is a weakness or shortcoming, there will be deterioration. To avoid every weakness and every shortcoming is a tall order but, as I see it, it is the only solution.

The notion that a small sin of omission can lead to catastrophic results is not new. On the basis of a late fifteenth century French proverb, Benjamin Franklin wrote:

"A little neglect may breed mischief,...
For want of a nail, the shoe was lost,
For want of a shoe, the horse was lost,
For want of a horse, the rider was lost,
For want of a rider, the battle was lost.
For want of a battle, the kingdom was lost.
And all for the want of a horseshoe nail."

Having good reinforced concrete in the Gulf Area is too important to lose the "kingdom of concrete" through "a little neglect": this paper is intended to be a good starting point for seeking perfection in reinforced concrete construction.

References

1. Neville, A. M., *Properties of Concrete*, Fourth Edition, Longman London, 1995, and John Wiley N.Y., 1996, 844 pp.

2. Neville, A. M.; Dilger, W. H., and Brooks, J. J., *Creep of Plain and Structural Concrete*, Construction Press, London and N.Y., 1983.

3. Macmillan, G. L., "Ultra Hot Weather Concreting and the Ultra Hot Climate (UHC)—Its Influence on Workmanship," *Concrete Durability in the Arabian Gulf*, Supplement, Bahrain Society of Engineers, 1997, pp. 13-40.

4. Al-Rabiah, A. R., "Concrete Damage Assessment—Case Studies," *Deterioration and Repair of Reinforced Concrete in the Arabian Gulf*, Fifth International Conference, V. 2, Bahrain, 1997, pp. 693-705.

5. Neville, A., "Maintenance and Durability of Structures," *Concrete International*, V. 19, No. 11, Nov. 1997, pp. 1-5. [see Section 3.2]

6. Zein Al-Abideen, H. M., "Environmental Impact on Concrete Practice, Part 2," *Concrete International*, V. 20, No. 12, Dec. 1998, pp. 55-57.

7. Neville, A., "Concrete Technology—An Essential Element of

Structural Design," *Concrete International*, V. 20, No. 7, July 1998, pp. 39-41. [Section 6.1]

8. Zein Al-Abideen, H. M., "Expensive Errors in Buildings Assessment and Repair," *Deterioration and Repair of Reinforced Concrete in the Arabian Gulf*, Fifth International Conference, V. 2, Bahrain, 1997, pp. 595-622.

9. Rostam, S., and Engelund, S., "Reliability Based Framework for Economically Optimal Assessment of Structures and Design," *Deterioration and Repair of Reinforced Concrete in the Arabian Gulf*, Fifth International Conference, V. 2, Bahrain, 1997, pp. 707-731.

10. Aïtcin, P.-C., "Demystifying Autogenous Shrinkage," *Concrete International*, V. 21, No. 11, Nov. 1999, pp. 54-56.

11. Neville, A., "Water—Cinderella Ingredient of Concrete," *Concrete International*, V. 22, No. 9, Sept. 2000, pp. 66-71. [Section 2.2]

12. Neville, A., "Concrete Cover to Reinforcement—or Cover-Up," *Concrete International*, V. 20, No. 11, Nov. 1998, pp. 25-29. [Section 5.1]

13. Neville, A., "Why Usual Specifications Do Not Produce Durable Concrete," *Deterioration and Repair of Reinforced Concrete in the Arabian Gulf*, Third International Conference, V. 1, Bahrain, 1989, pp. 1-17.

Section 3.5: Chloride Attack of Reinforced Concrete: An Overview

In many parts of the world, reinforced concrete structures deteriorate, sometimes seriously, due to chloride attack. Numerous papers continue to be published; for the most part, these describe isolated experiments involving one or two variables. However, in practice, the variables interact so that a broad picture of the phenomena involved is difficult to discern. This paper attempts to provide such an overview.

A particular feature of chloride attack that distinguishes it from other mechanisms of deterioration of reinforced concrete is that the primary action of chlorides is to cause corrosion of steel reinforcement, and it is only as a consequence of this corrosion that the surrounding concrete is damaged. The topic of corrosion of steel is wide ranging, and this paper is limited to a consideration of those properties of concrete which influence corrosion, with emphasis on the transport of chloride ions through the concrete in the cover to the reinforcement. Nevertheless, a brief description of the mechanism of chloride-induced corrosion will be helpful in understanding the processes involved.

Mechanism of chloride-induced corrosion

Embedded steel develops a protective passivity layer on its surface. This layer, which is self-generated soon after the hydration of cement has started, consists of γ-Fe_2O_3 adhering tightly to the steel. As long as that oxide film is present, the steel remains intact. However, chloride ions destroy the film and, in the presence of water and oxygen, corrosion occurs. Chloride ions were described by Verbeck[1] as "a specific and unique destroyer."

It may be useful to add that, provided the surface of the reinforcing steel is free from loose rust (a condition which is always specified), the presence of rust at the time the steel is embedded in concrete does not influence corrosion.[2]

A brief description of the corrosion phenomenon is as follows. When there exists a difference in electrical potential along the steel in concrete, an electrochemical cell is set up: anodic and cathodic regions occur, connected by the electrolyte in the form of the pore water in the hardened cement paste. The positively charged ferrous ions Fe^{2+} at the anode pass into solution while the negatively charged free electrons e^- pass through the steel into the cathode where they are absorbed by the constituents of the electrolyte and combine with water and oxygen to form hydroxyl ions OH^-. These travel through the electrolyte and combine with the ferrous ions to form ferric hydroxide, which is converted by further oxidation to rust. The reactions involved are as follows:

anodic reaction
$$Fe \rightarrow Fe^{2+} + 2e^-$$
$$Fe^{2+} + 2OH^- \rightarrow Fe(OH)_2 \text{ (ferrous hydroxide)}$$
$$4Fe(OH)_2 + 2H_2O + O_2 \rightarrow 4Fe(OH)_3 \text{ (ferric hydroxide)}$$

cathodic reaction
$$4e^- + O_2 + 2H_2O \rightarrow 4OH^-$$
It can be seen that oxygen is consumed and water is regenerated, but it is needed for the process to continue. Thus there is no corrosion in dry concrete, probably below a relative humidity of 60%; nor is there corrosion in concrete immersed fully in water, except when water can entrain air, for example by wave action. The optimum relative humidity for corrosion is 70-80%. At higher relative humidities, the diffusion of oxygen through the concrete is reduced considerably.

The differences in electrochemical potential can arise from differences in the environment of the concrete, for example when a part of it is submerged permanently in seawater and a part is exposed to periodic wetting and drying. A similar situation can arise when there is a substantial difference in the thickness of cover to a steel system that is electrically connected. Electrochemical cells form also due to a variation in salt concentration in the pore water or due to a non-uniform access to oxygen.

For corrosion to be initiated, the passivity layer must be penetrated. Chloride ions activate the surface of the steel to form an anode, the passivated surface being the cathode. The reactions involved are as follows:
$$Fe^{2+} + 2Cl^- \rightarrow FeCl_2$$
$$FeCl_2 + 2H_2O \rightarrow Fe(OH)_2 + 2HCl.$$
Thus, Cl^- is regenerated so that the rust contains no chloride, although iron chloride is formed at the intermediate stage.

Because the electrochemical cell requires a connection between the anode and the cathode by the pore water, as well as by the reinforcing steel itself, the pore system in a hardened cement paste is a major factor influencing corrosion. In electrical terms, it is the resistance of the "connection" through the concrete that controls the flow of the current. The electrical resistivity of concrete is influenced greatly by its moisture content, by the ionic composition of the pore water, and by the continuity of the pore system in the hardened cement paste.

There are two consequences of the corrosion of steel. First, the products of corrosion occupy a volume several times larger than the original steel so that their formation results in cracking (characteristically parallel to the reinforcement), spalling or delamination of concrete. This makes it easier for aggressive agents to ingress towards the steel, with a consequent increase in the rate of corrosion. Second, the progress of corrosion at the anode reduces the cross-sectional area of the steel, thus reducing its load-carrying capacity. In this connection, it should be pointed out that chloride-induced corrosion is highly localized at a small anode, with pitting of the steel taking place.

Chlorides in the mixture

Chlorides can be present in concrete because they have been incorporated in the mixture through the use of contaminated aggregate or of seawater or brackish water, or through admixtures containing chlorides. None of these materials should be permitted, and standards generally prescribe strict limits on the chloride content of the concrete from all sources. For example, British Standard BS 8110: Part 1:1985 limits the *total* chloride-ion content in reinforced concrete to 0.40% by mass of cement. The same limit is prescribed by European Standard ENV 206: 1992. The approach of American Concrete Institute Building Code ACI 318-89 is to consider *water-soluble* chloride ions only. On that basis, the chloride-ion content of reinforced concrete is limited to 0.15% by mass of cement. The two values are not substantially different from one another because water-soluble chlorides are only a part of the total chloride content, namely, the free chlorides in pore water. The distinction between *free* and *bound* chlorides is considered in a later section, but, at this stage, it can be noted that the total chloride content is determined as the *acid-soluble* chloride content, using ASTM Standard C 1152-90 or British Standard BS 1881: Part 124: 1988. There exist several techniques for the determination of the content of water-soluble chlorides.

As a possible source of chlorides in the mixture, portland cement itself contains only a very small amount. However, ground-granulated blast-furnace slag may have a significant chloride content if its processing involved quenching with seawater. As far as aggregate is concerned, British Standard BS 882: 1992 gives guidance on the maximum total chloride ion content; compliance with this guidance is likely to satisfy the requirements for concrete of British Standard BS 5328: Part 1: 1991 and of BS 8110: Part 1: 1985. For reinforced concrete, the chloride content of the aggregate should not exceed 0.05% by mass of the total aggregate; this is reduced to 0.03% when sulfate-resisting cement is used. For prestressed concrete, the corresponding figure is 0.01%.

The various limits on chlorides referred to in the section are generally conservative, so that compliance with them should ensure no chloride-induced corrosion unless more chlorides ingress into the concrete in service. The view that the limits are conservative is disputed by Pfeifer.[3]

Ingress of chlorides

The problem of chloride attack arises usually when chloride ions ingress from outside. This can be caused by deicing salts, a topic not considered in this paper. Another, particularly important, source of chloride ions is seawater in contact with concrete. Chlorides can also be deposited on the surface of concrete in the form of airborne very fine droplets of seawater (raised from the sea by turbulence and carried by wind) or of airborne dust that subsequently becomes wetted by dew. It is useful to point out that airborne chlorides can travel substantial distances: 2 km has been reported,[4] but travel over even greater distances is possible, depending on wind and topography. The configuration of structures also affects the movement of airborne salts: when eddies occur in the air, salts can reach the landward faces of structures. Brackish groundwater in contact with concrete is also a source of chlorides.

Although it is a rare occurrence, it should be mentioned that chlorides can ingress into concrete from conflagration of organic materials containing chlorides. Hydrochloric acid is formed and deposited on the surface of concrete where it reacts with calcium ions in the pore water. Ingress of chloride ions can follow.[5]

Whatever their external origin, chlorides penetrate concrete by transport of water containing the chlorides, as well as by diffusion of the ions in the water, and by absorption. Prolonged or repeated ingress can, with time, result in a high concentration of chloride ions at the surface of the reinforcing steel.

When concrete is permanently submerged, chlorides ingress to a considerable depth but, unless oxygen is present at the cathode, there will be no corrosion. In concrete that is sometimes exposed to seawater and is sometimes dry, the ingress of chlorides is progressive. The following is a description of a situation often found in structures on the coast in a hot climate.

Dry concrete imbibes salt water by absorption and, under some conditions, may continue to do so until the concrete has become saturated. If the external conditions then change to dry, the direction of movement of water becomes reversed and water evaporates from the ends of capillary pores open to the ambient air. It is, however, only pure water that evaporates, the salts being left behind. Thus, the concentration of salts in the water left behind increases near the surface of the concrete. The concentration gradient thus established drives the salts in the water near the surface of the concrete towards the zones of lower concentration, that is, inwards; this is transport by diffusion. Depending on the external relative humidity and on the duration of the drying period, it is possible for most of the water in the outer zone of the concrete to evaporate so that the water remaining in the interior will become saturated with salt and the excess salt will precipitate out as crystals.

It can be seen thus that, in effect, the water moves outwards and the salt inwards. The next cycle of wetting with salt water will bring more salt present in solution into the capillary pores. The concentration gradient now decreases outwards from a peak value at a certain depth from the surface, and some salts may diffuse towards the surface of the concrete. If, however, the wetting period is short and drying restarts quickly, the ingress of salt water will carry the salts well into the interior of the concrete; subsequent drying will remove pure water, leaving the salts behind.

The exact extent of the movement of salt depends on the length of the wetting and drying periods. It should be remembered that wetting occurs very rapidly and drying is very much slower; the interior of the concrete never dries out. It should also be noted that the diffusion of ions during the wet periods is fairly slow.

It is apparent that a progressive ingress of salts towards the reinforcing

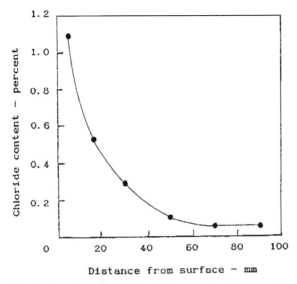

Fig. 3.5.1—A typical profile of total chloride ion content by mass of cement

steel takes place under alternating wetting and drying, and a chloride profile of the kind shown in Fig. 3.5.1 is established. The profile is determined by chemical analysis of dust samples obtained by drilling to various depths from the surface. Sometimes, there is a lower concentration of chlorides in the outermost 5 mm or so of the concrete where rapid movement of water takes place so that the salts are carried quickly a small distance inwards. The maximum chloride ion content in pore water can be in excess of the concentration in seawater; this was observed after 10 years exposure.[6] The crucial fact is that, with the passage of time, a sufficient quantity of chloride ions will reach the surface of the reinforcing steel. What constitutes a "sufficient" quantity will be discussed in the next section.

As just mentioned, the ingress of chlorides into concrete is influenced strongly by the exact sequence of wetting and drying. This sequence varies from location to location, depending on the movement of the sea and on the wind, on exposure to the sun, and on the usage of the structure. Thus, even different parts of the same structure may undergo a different pattern of wetting and drying; this explains why, sometimes, there is a considerable variation in the extent of corrosion damage in a single structure.

It is not only wetting and drying of the surface zone of the concrete that influences the ingress of chlorides: drying to a greater depth allows subsequent wetting to carry the chlorides well into the concrete, thus speeding up the ingress of chloride ions. For this reason, reinforced concrete in the tidal zone (where the period of drying is short) is less vulnerable to corrosion than concrete in the splash zone (where wetting may occur only when the sea is high or the wind is strong). The most vulnerable is the concrete wetted by seawater only occasionally, such as areas around bollards (where wet ropes are coiled) or in the vicinity of fire hydrants (using seawater), or in industrial areas subjected to periodic washdown with sea water, but at other times exposed to the drying effects of the sun and of a high temperature.

Threshold content of chloride ions

It was mentioned earlier that, for corrosion to be initiated, there has to be a certain minimum concentration of chloride ions at the surface of the steel. As far as chlorides incorporated in the original mixture are concerned, the threshold concentration was considered earlier. It is useful to add that the presence of a given excessive amount of chlorides in the original mixture results in a more aggressive action, and therefore a higher corrosion rate, than when the same amount of chlorides has ingressed into the concrete.[7]

As far as chlorides which have ingressed into the concrete are concerned, it is even more difficult to establish a threshold concentration of chloride ions below which there is no corrosion. This threshold depends on a number of factors, many of which are still imperfectly understood. Moreover, the distribution of chlorides within the hardened cement paste is not uniform, as found in chloride profiles in actual structures. For practical purposes, prevention of corrosion lies in controlling the ingress of chlorides by the thickness of cover and by the penetrability of the concrete in the cover.

While, under any given circumstances, there may be a threshold chloride content for corrosion to be initiated, its progress depends on the resistivity of the hardened cement paste, which varies with humidity, and on the availability of oxygen, which is affected by the immersion of concrete.

In any case, it is not the total chloride content that is relevant to corrosion. Some of the chlorides are chemically bound, being incorporated in the products of hydration of cement. Other chlorides are physically bound, being adsorbed on the surface of the gel pores. It is only the remaining chlorides, namely, free chlorides, that are available for the aggressive reaction with steel. However, the distribution of the chloride ions among the three forms is not permanent, since there is an equilibrium situation such that some free chloride ions are always present in the pore water. It follows that only the chloride ions in excess of those needed for this equilibrium can become bound.

Binding of chloride ions

The main form of binding of the chloride ions is by reaction with C_3A to form calcium chloroaluminate, $3CaO.Al_2O_3.CaCl_2.10H_2O$, sometimes referred to as Friedel's salt. A similar reaction with C_4AF results in calcium chloroferrite, $3CaO.Fe_2O_3.CaCl_2.10H_2O$. It follows that more chloride ions are bound when the C_3A content of the cement is higher, and also when the cement content of the mixture is higher. For this reason, it used to be thought that the use of cements with a high C_3A content is conducive to good resistance to corrosion.

This may be true when chloride ions are present at the time of mixing (a situation which should not be permitted) because they can react rapidly with C_3A. However, when chloride ions ingress into concrete, a smaller amount of chloroaluminates is formed, and, under some future circumstances, they may become dissociated, releasing chloride ions so as to replenish those removed from the pore water by transport to the surface of the steel.

A further factor in deciding on the desirable C_3A content of the cement is the possibility of sulfate attack on some parts of the given structure, other than those subject to the ingress of seawater. As is well known, sulfate resistance requires a low C_3A content in the cement. For these various reasons, it is nowadays thought that a moderately sulfate-resisting cement, ASTM Type II, offers the best compromise.

In the case of cements containing ground-granulated blast-furnace slag, it has been suggested that binding of chlorides takes place also by the aluminates in the slag, but this has not been confirmed fully.[8]

Concerning a possible use of cement with a high C_3A content, it should be remembered that a high C_3A content results in a higher early rate of heat evolution, and therefore a temperature rise. This behavior can be harmful in moderately large concrete masses often associated with structures exposed to the sea.[9]

Some standards, for example, British Standard BS 8110: Part 1: 1985, severely limit the chloride content when sulfate-resisting cement (Type V) is used, on the assumption that chlorides adversely affect sulfate resistance. This has now been proved not to be the case.[10] What happens is that sulfate attack results in a decomposition of calcium chloroaluminate, thus making some chloride ions available for corrosion; calcium sulfoaluminate is formed.[11]

Carbonation of hardened cement paste in which bound chlorides are present has a similar effect of freeing the bound chlorides and thus increasing the risk of corrosion. Ho and Lewis[12] quote Tuutti as having found an increased concentration of chloride ions in pore water to occur 15 mm in advance of the carbonation front. This harmful effect of carbonation is in addition to the lowering of the pH value of the pore water, so that severe corrosion may well follow. It has also been found in laboratory tests[13] that the presence of even a small amount of chloride ions in carbonated concrete enhances the rate of corrosion induced by the low alkalinity of carbonated concrete.

In considering both carbonation and ingress of chloride ions, it is important to remember that the optimum relative humidity for carbonation is between 50 and 70%, whereas corrosion progresses rapidly only at higher humidities. The occurrence of both of these relative humidities, one after another, is possible when concrete is exposed to long periods of alternating wetting and drying. Another occurrence of both chloride ingress and carbonation was observed by me in thin cladding panels of a building: airborne chlorides ingressed from outside and reached the reinforcing steel;

carbonation progressed from the relatively dry inside of the building.

Returning to the topic of the chloride ion concentration present in the pore water in an equilibrium situation, it should be noted that the chloride ion concentration depends on the other ions present in the pore water; for example, at a given total chloride ion content, the higher the hydroxyl (OH$^-$) concentration the more free chloride ions are present.[14] For this reason, the Cl$^-$/OH$^-$ ratio is considered to affect the progress of corrosion, but no generally valid statements can be made. It has also been found that, for a given amount of chloride ions in the mixture, there are significantly more free chloride ions with NaCl than with $CaCl_2$.[15]

Because of these various factors, the proportion of bound chloride ions varies from 80% to well below 50% of the total chloride ion content. Therefore, there may not exist a fixed and unique value of the total amount of chloride ions below which corrosion will not occur. Tests[14, 16] have shown that, in consequence of the various equilibrium requirements of the pore water, the mass of bound chlorides in relation to the mass of cement is independent of the water-cement ratio.

Influence of blended cements on corrosion

While the preceding discussion was concerned with the influence of the type of *portland* cement on the chemical aspects of chloride ions, it is also important, indeed more so, to consider the influence of the type of *blended* cement on the pore structure of the hardened cement paste and on its penetrability, as well as on resistivity. Those aspects of various cementitious materials that are particularly relevant to the movement of chloride ions will now be discussed. It should be added that the same properties of hardened cement paste, which influence the transport of chlorides, also influence the supply of oxygen and the availability of moisture, both of which are necessary for corrosion to occur. However, the locations on steel where chlorides are present and where oxygen is needed are different: the former at the anode, and the latter at the cathode.

The cementitious materials of interest are fly ash, ground-granulated blast-furnace slag, and silica fume. All three, when properly proportioned in the mixture, significantly reduce the penetrability of concrete and increase its resistivity, thereby reducing the rate of corrosion.[17-19] As far as silica fume is concerned, its positive effect is through improvement of the pore structure of hardened cement paste, which increases resistivity, even though silica fume reduces somewhat the pH value of the pore water as a consequence of its reaction with $Ca(OH)_2$.

It should be remembered that, because of its effect on workability, the use of silica fume usually involves the inclusion of a superplasticizer. Superplasticizers per se do not affect the pore structure and therefore do not alter the process of corrosion.

The beneficial effects of the various cementitious materials are so significant that their use in reinforced concrete liable to corrosion in hot climates is virtually necessary: portland cement alone should not be used.[20] Tests on chloride ion diffusion through mortar indicate that fillers do not affect the movement of chlorides.[21]

Chloride ions in concrete made with high alumina cement lead to a more aggressive situation than with portland cement,[22] the comparison being made at the same chloride ion content. It can be added that the pH value in high alumina cement concrete is lower than with portland cement so that the passive state of the steel may be less stable.[22]

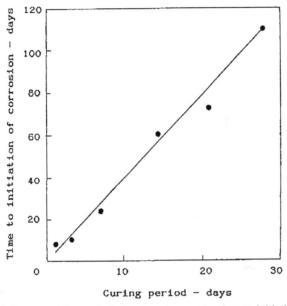

Fig. 3.5.2—Influence of the period of wet curing on the time to initiation of corrosion of reinforcing steel in concrete[23]

Further factors influencing corrosion

The preceding discussion of the influence of the composition of concrete upon its resistance to corrosion should be complemented by re-emphasizing the importance of good curing, the effect of which is primarily upon the concrete in the cover zone. The time to *initiation* of corrosion is reduced substantially by prolonged curing, as shown in Fig. 3.5.2 for concrete made with sulfate-resisting cement and a water-cement ratio of 0.5, partially immersed in a 5% solution of sodium chloride.[23] However, only fresh water must be used for curing because brackish water greatly increases the ingress of chlorides.[23]

Once the corrosion has been initiated, its continuation is not inevitable: the progress of corrosion is influenced by the resistivity of the concrete between the anode and the cathode and by the continuing supply of the oxygen at the cathode. On the one hand, it is very doubtful that the supply of oxygen can be stopped completely and reliably by the application of a membrane, although developments in this field continue. On the other hand, the resistivity of concrete is a function of its moisture condition so that drying out would halt the corrosion which, however, can re-start upon subsequent wetting.

Cracking of concrete in the cover facilitates the ingress of chlorides and therefore enhances corrosion. Although virtually all reinforced concrete in service exhibits some cracks, cracking can be controlled by appropriate structural design, detailing, and construction procedures. Cracks wider than about 0.2-0.4 mm are harmful. It may be worth mentioning that, although prestressed concrete is crack-free, the prestressing steel is more vulnerable to corrosion because of its nature; also, the small cross-sectional area of prestressing tendons means that pitting corrosion greatly reduces their load carrying capacity.

Higher temperatures have several effects on corrosion. First, the content of free chlorides in the pore water increases; the effect is more pronounced with cements having a high C_3A content and with lower chloride concentrations in the original mixture.[24] More importantly, the reactions of corrosion, like many chemical reactions, occur faster at higher temperatures. It is usually assumed that a rise in temperature of 10°C doubles the rate of reaction, but there is some evidence that the increase is only 1.6-fold.[25] Whatever the exact factor, the accelerating effect of temperature explains why there is so much more corrosion-damaged concrete in hot coastal areas than in temperate parts of the world.

It is known that initial hardening of concrete at high temperatures results in a coarser pore structure,[26] a consequence of which is a lower resistance to the diffusion of chloride ions.[27] The temperature differential between the surface of the concrete and its interior affects the diffusion; direct exposure to the sun can result in a significant rise in the temperature of the surface concrete above the ambient value.

Thickness of cover

The thickness of cover to reinforcement is an important factor controlling the transport of chloride ions: the greater the cover the longer the time interval before the chloride ion concentration at the surface of the steel reaches the threshold value. Thus, the quality of the concrete (in terms of its low penetrability) and the thickness of cover work together and can, therefore, to some extent, be traded off one against the other. For this reason, standards often specify combinations of cover and strength of concrete such that a lower thickness of cover requires a higher strength, and vice versa.

However, there are limitations to this approach. First of all, thick cover is of no avail if the concrete is highly penetrable. Moreover, the purpose of cover is not only to provide protection of reinforcement, but also to ensure composite structural action of steel and concrete, as well as, in some cases, to provide fire protection or resistance to abrasion. Unduly large thickness of cover would result in the presence of a considerable volume of concrete devoid of reinforcement. And yet, the presence of steel is required to control shrinkage and thermal stresses, and to prevent cracking due to those stresses. Were cracking to occur, the large thickness of cover would be proved to be detrimental. In practical terms, the cover thickness should not exceed 80-100 mm, but the decision on cover forms part of structural design.

Too small a thickness of cover should not be used either because, however low the penetrability of the concrete, cracking, for whatever reason, or local damage or misplaced reinforcement can result in a situation where chloride ions can be transported rapidly to the surface of the steel.

Tests for penetrability to chlorides

A rapid test for the penetrability of concrete to chloride ions is prescribed by ASTM Standard C 1202-97, which determines the electrical conductance, expressed as the total electrical charge in coulombs (ampere seconds) passed during a certain time interval through a concrete disc between solutions of sodium chloride and sodium hydroxide. The charge is related to the penetrability of the concrete to chloride ions, so that the test can be of help, in a comparative manner, in selecting a suitable concrete mixture. A somewhat similar test determines the AC impedance of specimens of various shapes.[28]

Tests of the kind just described do not necessarily replicate the transport of chloride ions in a real-life situation, nor do they have a sound scientific basis. Nevertheless, they are useful and certainly preferable to the assumption that resistance to chloride ion ingress is related simply to the strength of the concrete; this assumption has been shown not to be valid[29] except in the most general manner.

Prevention of corrosion

Simplified statements about methods of controlling or remedying corrosion that has been initiated could be unhelpful. All that should be stated here is that the progress of corrosion would be reduced by drying the concrete or by the prevention of oxygen supply using surface barriers. This is a specialized field, and ad hoc solutions may, in fact, prove harmful; for instance, applying a barrier at the anode (rather than the cathode) would increase the ratio of the size of the cathode to the anode, which would increase the rate of corrosion.

It is reasonable to raise the question of whether there exist integral corrosion inhibitors, that is, substances which, while not preventing ingress of chlorides into concrete, inhibit corrosion of steel. Nitrites of sodium[30] and calcium[31] have been found to be effective in laboratory tests. The action of the nitrite is to convert ferrous ions at the anode into a stable passive layer of Fe_2O_3, the nitrite ion reacting preferentially to the chloride ion. The concentration of nitrites must be sufficient to cope with a continuing ingress of chloride ions. Indeed, it is not certain that corrosion inhibitors are effective indefinitely; they may simply delay corrosion. The accelerating effect of the nitrites can be offset by the use of a retarding admixture, if need be. The search for other corrosion inhibitors continues.[32]

Being incorporated in the mixture, the inhibitors protect all the embedded steel. Nevertheless, inhibitors are no substitute for concrete of low penetrability: they are merely an additional safeguard. Moreover, sodium nitrite increases the hydroxyl ion concentration in the pore water, and this may increase the risk of alkali-aggregate reaction. Thus, the beneficial effect of an increased hydroxyl ion concentration upon the risk of corrosion of steel is accompanied by a negative effect on the risk of alkali-aggregate reaction. Of course, this is relevant only if the aggregate is susceptible to such a reaction in the first place.

A discussion of the prevention of corrosion of steel in concrete would be incomplete without a mention of the protection of steel by epoxy coating and by cathodic protection, which makes the entire steel surface cathodic. Epoxy coating of steel is a specialized technique that can be helpful in addition to an adequate thickness of cover concrete of low permeability. Cathodic protection has been shown to be effective in some applications but its use in a new structure is an admission of defeat in that the particular reinforced concrete structure is manifestly not durable.

A question that has to be faced occasionally is: can chloride ions be removed from the surface of the steel? There has been developed a technique for desalinating concrete, in which chloride is removed by passing a heavy direct current between the corroding reinforcing steel (now acting as a cathode) and an external anode in electrolytic contact with the concrete; chloride ions migrate towards the external anode, thus moving away from the surface of the reinforcement.[33] It seems that only about one-half of the chloride in the concrete can be removed and, with time, corrosion is likely to re-start. Some negative consequences of the process may follow.[34]

Concluding remarks

Understanding the phenomena involved in the corrosion of steel reinforcement in concrete and the factors influencing corrosion should make it possible to design successful reinforced concrete structures which will be exposed to chlorides. Nevertheless, there exist extreme conditions of exposure, in terms of temperature and the pattern of wetting by seawater alternating with drying, which put into question the appropriateness of reinforced concrete under those conditions. It could be postulated that there exist two excellent construction materials: concrete and steel. The combination of the two, one inside the other, may be catastrophic.

References

1. Verbeck, G. J., "Mechanisms of Corrosion in Concrete," *Corrosion*

of Metals in Concrete, SP-49, American Concrete Institute, Farmington Hills, Mich., 1975, pp. 21-38.

2. Al-Tayyib, A. J., et al., "Corrosion Behavior of Pre-Rusted Rebars After Placement in Concrete," *Cement Concr. Res.*, V. 20, 1990, pp. 955-960.

3. Pfeifer, D. W.; Perenchio, W. F.; and Hime, W. F., "A Critique of the ACI 318 Chloride Limits," *PCI J.*, V. 37, 1992, pp. 68-71.

4. Nireki, T., and Kabeya, H., "Monitoring and Analysis of Seawater Salt Content," Fourth International Conference on Durability of Building Materials and Structures, Singapore, 4-6 Nov. 1987, pp. 531-536.

5. Lammke, A., "Chloride-Absorption from Concrete Surfaces," *Evaluation and Repair of Fire Damage to Concrete*, SP-92, American Concrete Institute, Farmington Hills, Mich., 1986, pp. 197-209.

6. Nagataki, S., et al., "Condensation of Chloride Ion in Hardened Cement Matrix Materials and on Embedded Steel Bars," *ACI Materials Journal*, V. 90, No. 4, 1993, pp. 323-332.

7. Lambert, P.; Page, C. L.; and Vassie, P. R. W., "Investigations of Reinforcement Corrosion. 2. Electrochemical Monitoring of Steel in Chloride-Contaminated Concrete," *Mater. Struct.*, V. 24, 1991, pp. 351-358.

8. Bakker, R. F. M., "Initiation Period," *Corrosion of Steel in Concrete*, P. Schiessl, ed., RILEM Report of Technical Committee 60-CSC, Chapman & Hall, London, 1988, pp. 22-55.

9. Hoff, G. C., "Durability of Offshore and Marine Concrete Structures," *Durability of Concrete*, SP-126, American Concrete Institute, Farmington Hills, Mich., 1991, pp. 33-53.

10. Harrison, W. H., "Effect of Chloride in Mix Ingredients on Sulphate Resistance of Concrete," *Mag. Concr. Res.*, V. 42, 1990, pp. 113-126.

11. Mather, B., "Calcium Chloride in Type V-Cement Concrete," *Durability of Concrete*, SP-131, American Concrete Institute, Farmington Hills, Mich., 1992, pp. 169-176.

12. Ho, D. W. S., and Lewis, R. K., "The Specification of Concrete for Reinforcement Protection—Performance Criteria and Compliance by Strength," *Cement Concr. Res.*, V. 18, 1988, pp. 584-594.

13. Glass, G. K.; Page, C. L.; and Short, N. R., "Factors Affecting the Corrosion Rate of Steel in Carbonated Mortars," *Corrosion Sci.*, V. 32, 1991, pp. 1283-1294.

14. Tritthart, J. "Chloride Binding in Cement. II. The Influence of the Hydroxide Concentration in the Pore Solution of Hardened Cement Paste on Chloride Binding," *Cement Concr. Res.*, V. 19, 1989, pp. 683-691.

15. Al-Hussaini, M.-J., et al., "The Effect of Chloride Ion Source on the Free Chloride Ion Percentages of OPC Mortars," *Cement Concr. Res.*, V. 20, 1990, pp. 739-745.

16. Tang, L., and Nilsson, L.-O, "Chloride Binding Capacity and Binding Isotherms of OPC Pastes and Mortars," *Cement Concr. Res.*, V. 23, 1993, pp. 247-253.

17. Schiessl, P., and Raupach, N., "Influence of Blending Agents on

the Rate of Corrosion of Steel in Concrete," *Durability of Concrete: Aspects of Admixtures and Industrial By-Products*, 2nd International Seminar, Swedish Council for Building Research, June 1989, pp. 205-214.

18. Ellis, W. E., Jr.; Rigg, E. H.; and Butler, W. B., "Comparative Results of Utilization of Fly Ash, Silica Fume and GGBFS in Reducing the Chloride Permeability of Concrete," *Durability of Concrete*, SP-126, American Concrete Institute, Farmington Hills, Mich., 1991, pp. 443-458.

19. Al-Amoudi, O. S. B., et al., "Prediction of Long-Term Corrosion Resistance of Plain and Blended Cement Concretes," *ACI Materials Journal*, V. 90, 1993, pp. 564-571.

20. Stuvo, "Concrete in Hot Countries," Report STUVO, Dutch Member Group of FTP, The Netherlands, 1986.

21. Cochet, O., and Jesus, B., "Diffusion of Chloride Ions in Portland Cement-Filler Mortars," International Conference on Blended Cements in Construction, R. N. Swamy, ed., Elsevier Applied Science, Barking, 1991, pp. 365-376.

22. Goñi, S.; Andrade, C.; and Page, C. L., "Corrosion Behaviour of Steel in High Alumina Cement Mortar Samples: Effect of Chloride," *Cement Concr. Res.*, V. 21, 1991, pp. 635-646.

23. Rasheeduzzafar; Al-Gahtani, A. S.; and Al-Saadoun, S. S., "Influence of Construction Practices on Concrete Durability," *ACI Materials Journal*, V. 86, 1989, pp. 566-575.

24. Hussain, S. E., and Rasheeduzzafar, "Effect of Temperature on Pore Solution Composition in Plain Concrete," *Cement Concr. Res.*, V. 23, 1993, pp. 1357-1368.

25. Virmani, Y. P., "Cost Effective Rigid Concrete Construction and Rehabilitation in Adverse Environments," *Annual Progress Report*, Year Ending Sept. 30, 1982, U.S. Federal Highway Administration, 1982.

26. Neville, A. M., *Properties of Concrete*, Fourth Edition, Longman London, 1995, and John Wiley, N.Y., 1996, 844 pp.

27. Detwiler, R. J.; Kjellsen, K. O.; and Gjørv, O. E., "Resistance to Chloride Intrusion of Concrete Cured at Different Temperatures," *ACI Materials Journal*, V. 88, 1991, pp. 19-24.

28. SHRP-C-365, "Very High Performance Concretes," *Mechanical Behavior of High Performance Concretes*, Volume 5, Strategic Highway Research Program, National Research Council, Washington, DC, 1993.

29. Samaha, H. R., and Hover, K. C., "Influence of Microcracking on the Mass Transport Properties of Concrete," *ACI Materials Journal*, V. 89, 1992, pp. 416-424.

30. Alonso, C., and Andrade, C., "Effect of Nitrite as a Corrosion Inhibitor in Contaminated and Chloride-Free Carbonated Mortars," *ACI Materials Journal*, V. 87, 1990, pp. 130-137.

31. Berke, N. S., "Corrosion Inhibitors in Concrete," *Concrete International*, V. 13, 1991, pp. 24-27.

32. Nmai, C. K.; Farrington, S. A.; and Bobrowski, S., "Organic-Based

Corrosion-Inhibiting Admixture for Reinforced Concrete," *Concrete International*, V. 14, 1992, pp. 45-51.

33. SHRP-S-347, "Chloride Removal Implementation Guide," Strategic Highway Research Program, National Research Council, Washington, DC, 1993.

34. Tritthart, J.; Pettersson, K.; and Sorensen, B., "Electrochemical Removal of Chloride from Hardened Cement Paste," *Cement Concr. Res.*, V. 23, 1993, pp. 1095-1104.

Chapter 4

Special Aspects

This chapter has a rather broad title: Special Aspects. It is only fair to admit that the aspects covered are not focused on a single characteristic of concrete, but deal with several important issues.

Section 4.1 discusses the interaction in practice between cement and concrete. This is a topic that is not commonly addressed, and yet the objective of most studies on cement is to know and control the influence on the properties of concrete.

It may be useful to point out that occasionally the popular press confuses the terms cement and concrete, but the readers of this book are bound to be aware of the real distinction between these two materials. It was, I believe, Bryant Mather who said that cement is to concrete as flour is to fruit cake.

Nevertheless, some researchers do not appreciate fully the difference between neat cement paste, mortar, and concrete. They perform their tests on paste or on mortar, and then conclude what will be the behavior of concrete. I have commented on this in Section 1.2. There is, in addition, the broader issue of the extent to which the behavior of neat cement paste is replicated in concrete. This issue is discussed in Section 4.1.

Two sections, 4.2 and 4.3, deal with high-performance concrete. It could be said that this is a relatively new type of concrete; it could also be argued that high-performance concrete is an ordinary good concrete, tailored to fit the purpose for which it is intended. Section 4.2 aims to demystify that material and to show that it is not unusual or difficult. Nevertheless, careful selection of aggregate may be necessary and compatibility of cement and superplasticizer has to be ensured.

Section 4.3 gives an overview of high-performance concrete, and emphasizes that high-performance concrete is not synonymous with high-strength concrete. For some applications, high early-strength may be required; for others a high long-term strength may be needed; in yet other cases, it is not strength but the modulus of elasticity of concrete that has to be tailored to specific needs. For some applications, a particularly low permeability is required.

High-performance concrete generally has a very low water content (in liters per cubic meter or in gallons per cubic yard). This results in low drying shrinkage but in high autogenous shrinkage. This feature of high-performance concrete is discussed in Section 4.3 and developed further in Section 4.5, which considers also the curing needs of high-performance concrete.

Shrinkage of concrete has been known for a very long time. The problems arising from shrinkage are ubiquitous and, in my opinion, shrinkage cracking is the most common problem in construction. The use of so-called shrinkage reinforcement, as well as of temperature reinforcement, is universal. Nevertheless, the occurrence of shrinkage cracking in service is not unusual. Indeed, the ACI Guide for Concrete Floor and Slab Construction, ACI 302.1R-96, says: "Application of present technology permits only a reduction in cracking and curling, not their elimination....it is completely normal to expect some amount curling and cracking on every project...."

Drying shrinkage is not the only type of shrinkage. Four other types of shrinkage may occur: plastic shrinkage, autogenous shrinkage (sometimes known as self-desiccation shrinkage), thermal shrinkage, and carbonation shrinkage. These various types of shrinkage are explained in Section 4.5.

The inter-relation of strength, heat, and reduction in volume of concrete, presented in Section 4.5, explains the phenomena involved. Finally, Section 4.5 discusses the magnitude and consequences of the various types of shrinkage in columns, beams, and slabs.

Section 4.4 is fundamental in understanding the behavior of concrete with respect to its elastic deformation and the role of bond between aggregate and the surrounding hydrated cement paste, that is, the interface zone. Although the microstructure of this zone is a laboratory topic, its influence on the behavior of concrete is large. It is therefore useful to be aware of the nature of the interface zone, as well as of bond of aggregate to the surrounding hydrated cement paste. Section 4.4 pays particular attention to high-strength concrete and to lightweight aggregate concrete.

Section 4.1: Cement and Concrete: Their Interaction in Practice

The title of this paper may seem to state a truism because concrete must be made with cement and it would seem to follow that the properties of cement are an important factor influencing the properties of concrete. It would also seem that desired properties of concrete require the use of certain cements. It can therefore be inferred that there is an interrelation between the properties of cement and the properties of concrete. Thus far, the argument has considerable validity but the argument has, on some occasions, been extended so as to make a further, though tacit, assumption that the properties and behavior of cement paste control and determine the properties and behavior of concrete. With this assumption in mind, some researchers have made tests on cement paste and have drawn conclusions in terms of concrete.

It is the purpose of this paper to examine the validity of the transfer of test data on cement paste to the behavior of concrete. It is also intended to examine critically various cases of interrelation between cement and concrete in practice; the emphasis is on the word practice. The justification of this emphasis is that an important purpose of research in the field of concrete is to improve the quality of concrete in practice. This is not to deny the vital importance of scientific research that is necessary for an understanding of materials and their behavior, and that must continue if progress is to be made. The problem, which exists also in other fields of engineering, is that application of the research findings to actual practice is not simple and straightforward, and on occasion can lead to difficulties.

It may be useful to say at this preliminary stage that, in some cases, there is a clear interrelation between cement and concrete; in others, the interrelation is vitiated by important features of concrete which are absent in cement paste; yet again in others, an interrelation is present under ideal conditions but the conditions. In practice reduce the validity of that relationship.

What is meant by cement?

Originally, concrete was made from only three materials: cement, aggregate, and water; there was actually a fourth, hidden, element necessary to produce satisfactory concrete, namely know-how. In more recent years, the concrete mixture contains also other inorganic materials: ground-granulated blast-furnace slag, pozzolans of various kinds, including fly ash, and silica fume. In American Concrete Institute parlance, they are all referred to as mineral admixtures. Such terminology is outdated and gives the impression that portland cement is *the* cement, the pure product, other powders being supplementary or *ersatz* materials.

Moreover, the term mineral admixtures runs in parallel with the term chemical admixtures. Such classification is hardly sound, given that the mass of chemical admixtures represents a fraction of one percent of the mass of portland cement while the mass of mineral admixtures represents anything from a few percent to much more than the mass of portland cement. A majority "shareholder" cannot be described as an admixture. All these materials form part of what ASTM Standard C 1157-92 calls blended hydraulic cements and, as suggested elsewhere,[1] the various materials, whether hydraulic, latent hydraulic, or pozzolanic, will be referred to as cementitious materials or, for short, as "cement," recognizing that the presence of a certain amount of portland cement is essential for hydration to take place (See Section 1.1).

Do studies on cement tell us about concrete?

The extensive use of a variety of cementitious materials has necessitated a considerable amount of laboratory testing. Such testing has been highly successful and has illuminated our understanding of the properties of these materials and of their behavior. The problem, as already mentioned, is the transfer of the observations to concrete as used in practice.

In order to obtain a clear picture over a limited front, some of these investigations seek to establish the influence of a single variable, or occasionally of two or three variables; it is therefore highly tempting to some researchers to eliminate all the other parameters of the mixture that introduce variability. This desire instantly leads to the elimination of aggregate, and certainly of coarse aggregate. Not only do the properties of any specific aggregate influence workability, but the very presence of aggregate makes for greater variability and requires the use of relatively large specimens, leading to thermal strains and inhomogeneities. So it is convenient to conduct tests on very small specimens of neat cement paste or, at most, of sand-cement mortar, and this is occasionally followed by a second temptation: to generalize the results by putting the word "concrete" in the title of the resulting paper.

Other researchers are aware of the difference between the behavior of cement and the behavior of concrete but have an underlying belief that a better study of cement will somehow reveal all that happens in concrete. An honest and explicit example of this can be found in a paper,[2] published as recently as 1992, which says: "Unfortunately, there are other aspects of concrete performance whose links to cement behavior are not understood. Still, it is difficult to imagine that any aspect of concrete performance is not in some way related to cement behavior." It is further stated:[2] "For a given aggregate and aggregate content, one would expect a correlation between the consistency of cement paste at the same water-to-cement ratio and including mineral or chemical admixtures with the workability of concrete using that paste. If there were such a relationship between cement paste and concrete workability, then cement workability could be tested, controlled, and perhaps specified in order to provide predictable and uniform concrete workability. However, such a correlation has eluded researchers."

My view is that the correlation has "eluded researchers" because it is, in many cases, too weak to be of practical value. Specifically, as far as workability is concerned, a correlation "for a given aggregate and aggregate content" would be enormously limited in application as it is the various properties of aggregate that govern workability. We must remind ourselves again and again that concrete is not just a diluted cement paste but it is a different material of which the cement paste is only one component. Moreover, the treatment of concrete in practice—some would say maltreatment—can significantly affect its properties compared with the situation in the laboratory.

The preceding discussion is not meant to be a criticism of the paper quoted because the views in it are expressed with clarity, and they should prove helpful to researchers who have not asked themselves whether their work on cement can validly be presented as work "on concrete."

It may be useful to spell out two principal reasons why observations on cement paste cannot, in some cases, be automatically considered to be valid in concrete. First, in hydrated cement paste, there is absent the inhomogeneity arising from differential properties of the cement paste and of the aggregate with respect to: the coefficient of thermal expansion, modulus of elasticity, and Poisson's ratio. This inhomogeneity affects the stress field not only under load, but also in a drying situation. The inhomogeneity can be harmful, for example when shrinkage cracking develops at the surface of large aggregate particles, but it can also be beneficial, for example, when aggregate particles act as crack arresters. Thus, the interaction of aggregate and cement paste affects the behavior of concrete with respect to microcracking, or even the development of larger cracks, and also with respect to permeability.

The second reason why it may not valid to test neat cement paste and draw conclusions about concrete stems from the fact that concrete consists not only of hydrated cement paste and aggregate but contains also the interface between these two materials. This is forgotten by some. And yet, the interface creates a "wall effect" with a consequent differential local packing of dry cement and a differential microstructure of the hydrated cement paste in the interface zone. This zone has considerable consequences for permeability and durability. The importance of the interface zone is now recognized, especially insofar as the action of silica fume is concerned, but it is not so long since the pozzolanic activity of silica fume was considered to be of paramount interest and the physical effects of packing at the aggregate surface were ignored.

The two facets of the presence of aggregate discussed in this section explain why, in some cases, observations on the behavior of cement paste cannot be necessarily related to the behavior of concrete.

Strength of cement and strength of concrete

In the past, it was intuitively felt that the stronger the cement the stronger the concrete made with that cement. Although the two strengths need not be directly proportional to one another, the argument appears to be rational. However, when tests on the strength of cement were performed in the cement plant, and the relevant test certificate accompanied the shipment of cement, on site there did not seem always to be a corresponding test result on the strength of concrete.

Complaints arising from this lack of correlation between the strength of cement and the strength of concrete led to quite a remarkable "defense" publication[3] by the cement manufacturers to the effect that there is no such correlation; they "proved" their point by tests and concluded as follows: "No correlations of strength from any of the mortar tests with the 28-day strength of concrete were found in the study, even when the concrete was made to standard mix proportions."

My explanation of the apparent absence of a relationship between the strength of mortar tested at a cement plant and the strength of concrete on site is as follows.

Clearly, there is a variability in tests on the compressive strength of concrete; also, the mortar cube strength test used to determine the strength of cement, prescribed by ASTM C 109-92, has relatively poor precision. However, the crucial fact is that the cement used in the manufacturers' tests consisted of a *composite sample,* made up of sub-samples obtained over a 24-hour period. The sample was thus representative of thousands of tons of cement produced during that period so that the manufacturers' test results gave the *average* strength of the cement produced in a period of 24 hours. Inevitably, there are variations within that bulk of cement, only a small part of which is the particular shipment from which a concrete test specimen was made. No information about the cement in that particular shipment was available and even the manufacturer had no test data relating to that shipment. It follows that the result of tests on the given concrete specimen is affected by the properties of cement in that shipment of cement only; that is, not by the average properties of the cement produced in the 24-hour period. The same applies to all other shipments and corresponding concrete test specimens. It is therefore no surprise that no correlation between the strength of concrete and the strength of cement was established but the procedures used in the cement plant were not fully understood by those who tested concrete.

I will come back to this subject later in this section but, at this juncture, I should add that ASTM Standard C 917-91a provides a test method which makes it at least possible to appreciate the variability of the cement being produced. This method requires taking *grab samples* which are used in making mortar test specimens, a moving average of five grab sample strength test results being reported. To establish a correlation between the strength of concrete and the strength of cement we should use cement from the same shipment for both tests, and not from a composite sample over a 24-hour period. It is my view that, under such circumstances, a correlation exists. In fact, until the introduction of European Standards, there were two British standard methods of determining the strength of cement: one using mortar cubes, and the other using concrete test cubes. Although they yielded different numerical values, there was a definite implication of a correlation between the two strengths.

As an aside, I would like to make a comment about the use of cement manufacturers' test certificates in research. Often, the properties of cement, such as chemical composition or fineness, which are given in the certificate are used as a test parameter. If the test certificate refers to the average properties of a 24-hour production, the properties as listed cannot be considered as necessarily applying to the actual cement used by the researcher. If, nevertheless, they are so considered, spurious correlations with the property investigated may be found; alternatively, the experimental work may fail to show a real correlation through no fault of the researcher. Such a situation can be compared to deciding the treatment of an individual patient on the basis of average population data.

I have to admit to a worry that some of the otherwise well-performed tests may suffer from the "average-properties" flaw and may therefore be misleading. Indeed, many years ago I hypothesized that creep of mortar made with standard sand at a fixed water-cement ratio should be a function of the compound composition of cement. I tested mortars made with a number of cements and found few significant correlations with the compound composition calculated from the manufacturers' test data on cement. At that time, I was not aware that these were average values of the composition and, who knows, maybe such a correlation does truly exist. Luckily, I decided to look for correlations with parameters determined by me, and indeed I established a relation between creep and the strength of mortar at the time of application of load.[4]

From the discussion in this section we can conclude that there is a correlation between the strength of cement and the strength of concrete made with the same cement, provided the mixture ingredients and mixture proportions of concrete are constant.

Various cementitious materials and strength of concrete

The various cementitious materials—fly ash, silica fume, ground-granulated blast-furnace slag—are here to stay. In their various ways, and under various circumstances, they all have an important role to play in concrete; the days of "pure" or "unadulterated" portland cement, once upon a time lauded by cement manufacturers, have gone forever. Portland cement is a vital ingredient but the other cementitious materials are highly valuable.

This change has led to an enormous number of investigations of these materials and to a veritable flow of papers describing the influence of these materials on the properties of concrete made and tested in the laboratory. Various physical, and less frequently, chemical properties of these cementitious materials were studied, usually one at a time. And then, every year, we read, or perhaps just saw the title of, a review paper informing us that the presence of, say, x percent of carbon in fly ash results in a y percent decrease in the 28-day strength of concrete test cylinders cured in water at 20°C. This may be a perfectly true observation, and the accompanying statement about the cause (carbon) and effect (strength decrease) may be valid under the particular set of laboratory circumstances. But what is the effect in concrete as made and used in the field? *That* concrete is rarely kept at 20°C and even more rarely kept moist for such a length of time. The question to consider is: to what extent can the results of such laboratory tests be expected to apply to concrete in practice?

To the laboratory researcher, the preceding questions may be of little interest. Tests have to be conducted under strictly prescribed and controlled conditions and this is how the researcher performed them; these conditions are described in the paper under the heading "test details." It is, the researcher may well feel, up to the person on site or in the design office to control the conditions and, if this is not done, it is that person's look-out. Of course, only an unkind researcher would think so, but to consider site conditions is not perceived to be within the researcher's remit.

But life is not like that. Construction conditions are what they are, and construction practices are what they are, even if they are not what they should be. The sun shines on the concrete, or it does not; the temperature varies during the day, and even more so during the night; and as for curing, which is discussed in a later section, satisfactory wet curing is alas rare. These variations in temperature and, even more so, in humidity affect concretes made with the various cementitious materials to a different extent: some materials require extended wet curing. This results in an injunction to the effect that "curing is of particular importance." In practice, little is done on a run-of-the-mill site.

The true picture is not as gloomy as I have painted it because the engineer on site knows, for instance, that some cements lead to a slower development of strength than others. However, as the strength of site concrete is determined on standard test cylinders cured under ideal conditions of temperature and humidity, the engineer may not be fully aware of the actual strength of the concrete in the structure. In other words, the interrelation between the curing requirements of the cement paste and of site concrete may be largely vitiated by poor, but not uncommon, practice. This reality should not be forgotten when translating laboratory results involving various cementitious materials to concrete in practice.

Strength of cement and durability of concrete

In an earlier section, I discussed the interrelation between the strength of cement and the strength of concrete. This interrelation has had some unexpected results insofar as properties of concrete other than its strength are concerned. I am referring to the systematic changes in the strength of cement which occurred in the United Kingdom in the 1960s and which occurred in other countries as well, although not necessarily at the same time.

Although there has been such a continuing change over more than sixty years in consequence of the improvement in the manufacture of cement, it is the change around the 1960s that merits particular attention because it has had far-reaching consequences for the concrete production practice. I am referring to the increase in the 28-day strength, and also in the 7-day strength, of mortars made with a fixed water-cement ratio. The main reason for this was a large increases in the average content of C_3S: from about 47% in 1960 to about 54% in the 1970s. There was a corresponding decrease in the content of C_2S so that the total content of the two calcium silicates remained constant at 70 to 71%. This change was made possible by improvements in the methods of manufacture of cement but it was also driven by the benefits of a "stronger" cement as perceived by the users, namely: reduction in cement content for a given specified strength, earlier removal of formwork, and faster construction. Such benefits were, unfortunately, associated with disadvantages.

The rate of increase in the strength of cement up to 7 days (which is now higher) and the rate of increase between 7 and 28 days have changed in consequence of the change in the ratio of C_3S to C_2S and also because of the higher alkali content in modern cements. The ratio of the strength at 28 days to that at 7 days has decreased substantially. For concrete with a water-cement ratio of 0.6, a decrease in the 28 to 7-day strength ratio from about 1.6 prior to 1950 to about 1.3 in the 1980s was reported.[6] At lower water-cement ratios, the ratio of the 28-day strength to the 7-day strength is nowadays lower. Likewise, the increase in strength beyond the age of 28 days is much reduced so that the increase should no longer be relied upon in the design of structures that will be subjected to full load only at an advanced age.

An example of the change in the 28-day strength of cement between 1979 and 1984, reported by the Concrete Society Working Party[7] is as follows. Concrete with a cube strength of 32.5 MPa (4700 psi), which in 1970 required a water-cement ratio of 0.50, could be produced in 1984 at a water-cement ratio of 0.57. Assuming that, for the workability to remain constant, the same water content of, say, 175 L/m^3 of concrete was maintained, it was possible to reduce the cement content from 350 kg/m^3 to 307 kg/m^3. More generally,[6] over the longer period between the 1950s and the 1980s, for a concrete of a given strength and workability, it was possible to reduce the cement content by 60 to 100 kg/m^3 of concrete and concomitantly to increase the water-cement ratio by between 0.09 and 0.13.

Although the above data refer to some British cements, the changes have occurred worldwide as a concomitant of modernization of cement production. French figures reported by Divet[8] may be of interest: between the mid-1960s and 1989, the average C_3S content of portland cement increased from 42 to 58.4%, with C_2S decreasing concurrently from 28 to 13%.

The increase in the average 28-day strength appears to continue. In the United States, between 1977 and 1991, the strength of mortar made according to ASTM C 109-92 increased from 37.8 MPa (5470 psi) to 41.5 MPa (6020 psi) as reported by ASTM Committee C-1 in 1993.

While a higher 28-day strength of concrete at a given water-cement ratio can be economically exploited, there are consequential disadvantages. Concrete having the same 28-day strength as before (when the "old" cements were used) can nowadays be made using a higher water-cement ratio and a lower cement content, as shown earlier. Both these concomitant changes result in concrete with a higher permeability and which is therefore more liable to carbonation and to penetration by aggressive agents, and generally of lower durability.

Moreover, the absence of a significant increase in strength beyond the age of 28 days[6,7] removes a long-term improvement in concrete that reassured users in the past (even if such improvement was not taken into account in design).

The rapid early gain in strength also means that strengths adequate for removal of formwork are achieved earlier than was the case with the "old" cements so that effective curing ceases at an early age.[9] This has adverse consequences upon durability.

These various consequences of producing a stronger cement were not foreseen in the United Kingdom, partly because many concrete users were preoccupied with exploiting the high early strength properties of cement, and partly because the British concrete specifications were couched predominantly in terms of a 28-day strength, which remained the same as it had been when the "old" cements were used.

The preceding discussion shows a very definite, albeit complex, interrelation between cement and concrete that, despite certain economic attractiveness, has proved to be harmful in practice.

Hydration of cement and curing of concrete

In the preceding section, I referred to the effect of the high-early strength of modern cements upon a reduction in the length of the curing time necessary for the formwork to remain in place. I would now like say a little more on the subject of curing.

The strength development of various cements is almost invariably described on the basis of mortar or concrete specimens cured under specified and ideal conditions of humidity and temperature. Those conditions do not exist on a worldwide basis: much of the world is hot, much is dry, and much is both hot and dry. Moreover, it is in those parts of the world that a great deal of construction nowadays takes place.

Under hot and dry conditions, effective prevention of the loss of water from the concrete is essential and provision of water that can ingress into the concrete is desirable. But it is precisely under those conditions that achieving such controls is difficult. The difficulties are of various kinds: the rate of evaporation is very high; water is frequently scarce; and the climatic environment is not conducive to meticulous workmanship. In consequence, curing is often, or indeed usually, inadequate both in terms of performance and of duration. It follows that the behavior of cement as determined under ideal conditions is, in such a case, not replicated in concrete in practice.

An example of such interrelation between a cement which is "good" and the concrete which does not benefit from such cement is presented by the behavior of cements containing fly ash when the concrete is inadequately cured. In other words, cement that has desirable properties, say in terms of heat development, reduction in calcium hydroxide and durability, does not bestow those properties on concrete that is not adequately cured.

Composition of cement and durability

The action of aggressive agents on concrete has led to the search for cements that are resistant to those agents. If the attack is simply a chemical reaction at the surface of the concrete, then the behavior of the cement can be directly translated into the behavior of concrete. However, most durability problems in practice are much more complex. For example, the search for a cement that is particularly resistant to seawater has not led to the use of any particular cement; rather, the quality of the concrete in terms of its density and permeability seems to be the controlling factor. There is thus little interrelation between cement and concrete in practice in regard to durability in seawater.

The issue of the effect of the tricalcium aluminate content of the cement upon the attack of reinforcement by chlorides is still being debated. Laboratory tests show that tricalcium aluminate binds chlorides, chloroaluminates being formed. This is so when the chlorides are present in the mixture. It is less certain what happens when chlorides ingress into mature concrete from outside. In addition, under some circumstances, both chlorides and sulfates are present, but not in the same place, so that the concrete has to be capable both of minimizing the chloride attack of the reinforcement and of minimizing the sulfate attack of the matrix. For resistance to sulfate attack, it is considered that cement with a low content of tricalcium aluminate is preferable while for resistance to chloride attack a high content of tricalcium aluminate is believed to be preferable.

The problem of optimizing the resistance of concrete both to chloride attack and to sulfate attack is complex and requires a fuller discussion than is possible in this paper. My point is only that the behavior of the cement paste as such, when exposed to a single aggressive agent, is not an adequate indicator of the behavior of concrete in practice when more than one aggressive agent can be present.

Even with respect to the action of sulfates when no other attack is

Chapter 4: Special Aspects

likely, the relation between the behavior of cement and the behavior of concrete remains unresolved.[10]

Various cementitious materials and thermal problems in concrete

One argument, among many, in favor of using cements containing fly ash and ground-granulated blast-furnace slag is that such cements develop heat more slowly and, provided the heat can be dissipated, the peak temperature is therefore lower and occurs later. This situation is exploited in various types of structures including pavements.

It is, however, possible for concrete made with such a cement to lead to thermal problems when it is placed against fresh concrete containing a different cement. I have investigated such a situation in the construction of a highway pavement that was placed in two layers in order that dowels at locations destined to be contraction joints could be inserted half-way through placement. In this particular case, for reasons of economy and to facilitate the sawing of contraction joints, as well as to reduce the rise in temperature, the top layer was made using cement containing 40% of ground-granulated blast-furnace slag; the bottom layer contained portland cement only.

When the top layer was a few hours old, it was chiefly the portland cement component that had hydrated so that the peak temperature was lower than in the bottom layer. Also, because of the time lag in placing the second layer, and because it was exposed to cold air, the peak temperature in the top layer was lower than in the bottom layer. The cooling in the bottom layer started earlier than in the top layer, possibly even before the peak temperature in the top layer was reached. In consequence, over a certain length of time, the bottom layer was undergoing a restrained contraction, which put it into tension, while the top layer was under a low compressive stress.

There was also a second important difference between the two layers: at very early ages, the strength of the top layer was very much lower than the strength of the bottom layer. The effect of this was that, when the bottom layer was cooling and put the top layer into low tension, vertical cracking developed.

The above is a case of a complex interaction of the heat generation of the two layers, presented here to show that such a situation can lead to severe thermal stresses which, when combined with untoward ambient temperature changes, from hot weather at the time of placing the concrete to cold weather some 12 hours later, can lead to cracking. The description given above is an example of an interrelation between desirable properties of cement and undesirable thermal properties of concrete.

Of course, having read the preceding, one can say: this is obvious; you should calculate in advance the thermal stresses in concrete under all combinations of time and temperature. The principle of perfect hindsight cannot be faulted but the above example has shown how the characteristics of cement as generally described in the literature can, under a new set of circumstances, have an unexpected deleterious effects in concrete.

Concluding remarks

Casual reading of this paper might give the impression that I am denigrating tests on cement paste and advocating only tests on concrete. This is not so: tests on neat cement are necessary to give basic insights into the chemical and physical phenomena involved. However, some physical and mechanical aspects of behavior may be absent in neat cement paste, and the physical phenomena in cement paste may give an incomplete picture of the behavior of concrete.

I am sure that we should continue to study and test cement and cement paste. I recognize that improved tests on cement will give better insights into what happens not only in neat cement paste but also in concrete. However, I am far from persuaded that we can reach an adequate, let alone full, understanding of concrete from a study of cement only. I am saying this because, on the one hand, in so complex a material as concrete there is probably always some interrelation between cement and concrete. On the other hand, however, as shown in this paper, in many cases, the interrelation is not definite enough to explain the behavior of concrete. In consequence, prediction of its behavior from the behavior of cement paste cannot be made with adequate reliability and with adequate validity over the full range of concrete mixtures and procedures used in practice.

References

1. Neville, A. M., "Cementitious Materials—A Different Viewpoint," *Concrete International*, V. 16, No. 7, 1994. [Section 1.1]

2. Struble, L., "The Performance of Portland Cement," *ASTM Standardization News*, Jan. 1992, pp. 38-45.

3. Weaver, W. S.; Isabelle, H. L.; and Williamson, F., "A Study of Cement and Concrete Correlation," *Journal of Testing and Evaluation*, V. 2, No. 4, 1974, pp. 260-303.

4. Neville, A. M., "Tests on the Influence of the Properties of Cement on the Creep of Mortar," *RILEM Bulletin*, No. 4, 1959, pp. 5-17.

5. Corish, A. T., and Jackson, P. J., "Portland Cement Properties," *Concrete*, V. 16, No. 7, 1982, pp. 16-18.

6. Nixon, P. J., "Changes in Portland Cement Properties and Their Effects on Concrete," *Building Research Establishment Information Paper*, Mar. 1986, 3 pp.

7. Concrete Society Working Party, "Report on Changes in Cement Properties and Their Effects on Concrete," *Technical Report* No. 29, 1987, 15 pp.

8. Divet, L., "Evolution de la Composition des Ciments Portland Artificiels de 1964 à 1989: Exemple d'Utilisation de la Banque de Données

du LCPC sur les Ciments," *Bulletin Liaison Laboratoires Ponts et Chaussées,* No. 176, Nov.-Dec. 1991, pp. 73-80.

9. Neville, A. M., "Why We Have Concrete Durability Problems," *Concrete Durability: Katharine and Bryant Mather International Conference,* SP-100, V. 1, American Concrete Institute, Farmington Hills, Mich., 1987, pp. 21-30.

10. Cohen, M. D., and Mather, B., "Sulfate Attack on Concrete—Research Needs," *ACI Materials Journal,* V. 88, No. 1, Jan.-Feb. 1991, pp. 62-69.

Section 4.2: High-Performance Concrete Demystified*

For almost a century, structural concrete has been routinely produced with a 28-day strength in the range of 20 to 30 MPa (2900 to 4400 psi), or even up to 35 MPa (5100 psi). Occasionally, when special circumstances required it, higher strengths were obtained and such concrete was described as high-strength concrete. Thirty years ago the high-strength label was applied to concrete with a strength above 40 MPa (5800 psi) or thereabouts; more recently the threshold rose to 50 or 60 MPa (7300 to 8700 psi). Such concretes were not routinely produced but they were thought to be unremarkable. However, in the last 15 years, concretes of very much higher strength have entered the field of construction of high-rise buildings and bridges: 90, 100, 110 MPa (13,000, 14,500, 16,000 psi), with an occasional 120 MPa (17,000 psi). All of these strengths have been achieved consistently and on a routine basis.

This type of concrete is perceived by *some* engineers, owners, and specifiers as a material fundamentally different from the concrete we have all grown up with. They suspect that there is some mystery about the manufacture of such concrete, and possibly even a secret recipe. It is the purpose of this section to destroy this myth and to show that high-strength concrete is just concrete.

What is high-performance concrete?

First of all, we have to explain the apparent inconsistency between the term high-strength concrete used in the above paragraphs and the term high-performance concrete used in the title. It was natural, and correct, to refer to concrete that had a higher strength than usual as high-strength concrete. However, in practical applications of this type of concrete, the emphasis has in many cases gradually shifted from the compressive strength to other properties of the material, such as a high modulus of elasticity, high density, low permeability, and resistance to some forms of attack. It is therefore logical to describe such concrete by the more widely embracing term high-performance concrete (HPC). We propose, therefore, to use the term HPC even though the debate on terminology is still continuing.

The benefits of using HPC are many, but they are of value only when they can he exploited. To date, exploitation has been mainly in high-rise buildings, in bridges, and in structures under severe exposure conditions. There are many potential applications of HPC, but there is no doubt that in many locations the use of HPC is precluded by its unavailability. At the same time, the unavailability is said to be a reflection of the lack of demand. It is just possible that both the designer and the supplier are

---

*This section was co-authored by P.-C. Aïtcin.

reluctant to depart from the familiar, and who can blame them.

Conversely, nothing succeeds like success: in those areas where there are designers keen to exploit HPC, there is always a supplier or two who deliver HPC to the given specification. This availability of HPC, in turn, encourages other designers to use it, and so its use is spreading. The earlier reference to the reluctance on the part of some people to use HPC is not a criticism but rather recognition of the fact that HPC is considered to be something special or unusual, and certainly different.

What is different about HPC

I would like to start by quoting what I said, as far back as 1963, about the difference between bad concrete and good concrete:[1] "Bad concrete—often a substance of the consistence of soup, hardening into a honeycombed, non-homogeneous mass—is made simply by mixing cement, aggregate, and water. Surprisingly, the ingredients of good concrete are exactly the same, and it is only the know-how, often without additional cost or labour, that is responsible for the difference."

Ignoring the outdated reference to the consistency of concrete, we can make a parallel statement about the difference between normal strength concrete (NSC) and HPC: the know-how is the vital factor. The ingredients of both types of concrete are the same, namely, portland cement, aggregate, water, and admixture. To be more scrupulous about the ingredients, we should say that HPC invariably contains a high-range water-reducing admixture (or superplasticizer), while NSC does so only sometimes. As far as other ingredients are concerned, such as retarders, fly ash, blast-furnace slag and silica fume, they may or may not be present in either type of concrete.

The know-how necessary to produce HPC consists of a specific knowledge of the properties of the ingredients and of their interaction. The crucial outcome of the know-how is an extremely low water-cement ratio (w/c) coupled with a satisfactory workability at the time of compaction.

The necessity of using a superplasticizer in the manufacture of HPC merits brief explanation. Without a superplasticizer, even with ordinary water-reducing admixtures, the water content of the mixture cannot be reduced very far as this would result in an unworkable mixture. At the same time, the cement content cannot be raised excessively, not only because of cost, but also because a high cement content may lead to thermal problems. The combination of an upper limit on the content of cement and a lower limit on the content of water means that, without a superplasticizer, the w/c cannot be reduced below a value of about 0.4.

The constraint of workability of the mixture arises, in broad terms, from the tendency of cement grains to flocculate and thus to hold water and offer shear resistance during compaction. Without going into the details of electrical surface charges and the like, we can say that the superplasticizer deflocculates the cement particles and thus fluidifies the mixture so that a very low water content is sufficient for an adequate workability. In

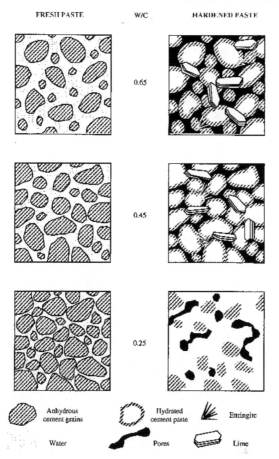

Fig. 4.2.1—Composition of fresh and hardened cement paste at maximum hydration at various *w/c* values

consequence, it is possible to obtain a mixture with a slump of 180 to 200 mm (7 to 8 in.) at a *w/c* in the range of 0.2 to 0.3 (based on the free water in the mixture). These values correspond to a water content of 125 to 135 L/m^3 (25 to 27 gal./yd^3) of concrete, as against 180 to 200 L/m^3 (36 to 40 gal./yd^3) in non-air-entrained NSC with a slump of 100 to 120 mm (4 to 5 in.). *This is the apparent secret of making HPC.*

The discussion so far has made it plain that, while the ingredients of HPC are substantially the same as those of NSC, their proportions are different. This is especially true of the water content of the mixture coupled with a large dosage of superplasticizer in HPC. Typically, 5 to 15 L/m^3 (1 to 3 gal./yd^3) of superplasticizer can effectively replace 45 to 75 L/m^3 (9 to 15 gal./yd^3) of water.

This drastic reduction in mixing water results in a reduced distance between the cement particles. In consequence, a much denser cement matrix than in NSC is achieved, the products of hydration of cement coalescing rapidly. By virtue of this high-density matrix, in addition to the chemical bonds created by the hydrates (which exist also in NSC), a very high compressive strength can be achieved. Since mixing water combines chemically or physically with cement and is lost by self-desiccation, the resulting hydrated cement paste in HPC has a very low porosity.

This very low porosity of the hydrated cement paste in HPC contrasts with NSC, in which uncombined mixing water creates an open network of pores that reduces the density of the matrix and thus leads to a lower compressive strength than is the case with HPC. The situation is illustrated in Fig. 4.2.1, which also shows that in HPC a significant proportion of the original cement particles remains permanently unhydrated. We may add that, in fundamental terms, the strength of concrete is a function of the total void content of the material, provided the bonds of cohesion and adhesion are strong enough; the degree of hydration of cement as such is not important, the unhydrated part of cement being simply a well-bonded, albeit expensive, aggregate.[2]

Optimum water-cement ratio

The preceding subsection leads to a number of questions:
- Is there a minimum w/c?
- Why use more cement than can possibly become hydrated?
- Is there an optimum w/c for maximizing the compressive strength?

It has been shown in the preceding subsection that the compressive strength of HPC depends not only on the w/c but also on the density of the resulting matrix. Furthermore, the strength of the concrete may be limited by the strength of the aggregate, as will be shown later. In other words, at first sight there appear to exist two factors influencing the strength of HPC: the w/c and the reduction in the distance between the cement particles at the outset of hydration.

It is well known that compressive strength is inversely related to w/c. It may also be useful to recall that during hydration portland cement binds water whose mass is about 22% of the mass of the unhydrated cement. For reasons that need not be discussed here, a mixture of 100 parts of cement and 22 parts of water cannot in practice achieve full hydration.

As already mentioned, in the case of HPC, an increase in the density of the matrix also increases the strength. The combination of the two effects—w/c and density of the matrix—provides a situation in which the w/c influences strength only above a certain minimum value of this ratio. For the currently available portland cements and superplasticizers, the usual mixing and placing methods, and the present-day curing practices, it has been found that the optimum value of w/c is about 0.22. At values higher than 0.22, the influence of w/c is operative; lower values of w/c are harmful because an adequately high density of the matrix cannot be achieved.

It may be appropriate to add that, despite the use of very low w/c, HPC may require air entrainment for adequate protection from repeated cycles of freezing and thawing.

Silica fume and HPC

Contrary to commonly held opinion, silica fume is not an essential ingredient of all HPC mixtures. In some of the projects described later in this section, strengths in the range of 60 to 80 MPa (8700 to 11,600 psi) were obtained without the use of silica fume.[3] Even higher strengths (up to 100 MPa [14,500 psi]) have been achieved, but only rarely. In our opinion there is no virtue in avoiding silica fume if it is available and economical, as its use simplifies the production of HPC and makes it easier to achieve compressive strengths in the range of 60 to about 90 MPa (8700 to 13,000 psi).

For higher strengths, the use of silica fume is essential. This inevitably affects the cost of the concrete. We can mention a case where an increase in the specified compressive strength from 90 to 100 MPa (13,000 to 14,500 psi) required the addition of silica fume in a quantity representing 10% of the portland cement used (500 kg/m^3 [843 lb/yd^3]). Taking the typical situation of silica fume costing ten times as much as portland cement, the use of the silica fume in this case represented a doubling of the cost of the cementitious material.

Controlling the rheological properties of fresh HPC

In the preceding subsections of this section we have discussed the strength of HPC and suggested that high and very high strengths can easily be achieved. Such success is clearly conditional on full compaction of the concrete, and this requires appropriate rheological properties—primarily workability—at the time of compaction. In the past, some concrete users have had experience of superplasticizers giving a very high workability at a very low water content of the mixture but only for a short time; subsequently, the concrete became so stiff as to make full compaction impossible. Indeed, achieving satisfactory rheological properties for a sufficiently long time presents the greatest difficulty in producing HPC.[4]

Experience has shown that it is not always possible to produce HPC that is still workable one hour after mixing, using 135 L/m^3 (27 gal./yd^3) of water, by simply choosing a random combination of portland cement and superplasticizer.[5] The reason for this is the interaction between the cement and the superplasticizer, which is rather complex and which can lead to a rapid slump loss. Indeed, in selecting materials for HPC, cements are more often eliminated because of their rheological behavior than because of their inability to achieve the required compressive strength.

The slump loss problem is essentially that of cement-superplasticizer compatibility.

Cement-superplasticizer compatibility

The problem, in essence, is that not every portland cement that complies with the appropriate national standard has the same rheological behavior when used with a given superplasticizer at a very low w/c. Similarly, not every superplasticizer that complies with the appropriate national standard reacts in the same way with a given portland cement. While we have always known that not all cements are compatible with all admixtures, in the case of HPC the problem of cement-superplasticizer compatibility is much more acute.

Studies on the interaction between cement and superplasticizer conducted at the Université de Sherbrooke have identified the important factors in their compatibility.[6] For the cement, these are the content of C_3A and C_4AF, the reactivity of C_3A, which depends on its morphological form and on the degree of sulfurization of the clinker, the content of calcium sulfate, and the final form of the calcium sulfate in the ground cement, namely gypsum, hemihydrate, or anhydrite.

For the superplasticizer, the important factors are the molecular chain length, the position of the sulfonate group in the chain, the counter-ion type, and the presence of residual sulfates, which affect the cement deflocculation properties.

On the basis of these factors, we can postulate an ideal cement for HPC from the rheological point of view: not too fine, with a very low C_3A content, and with an interstitial phase whose reactivity is easily controlled by the sulfate ions derived from the solution of the sulfates present in the cement.[5] An ideal superplasticizer should consist of rather long molecular chains in which, for example, the sulfonate groups occupy the ß position in a sodium salt condensate of formaldehyde and naphthalene sulfonates. As far as the content of residual sulfates in the superplasticizer is concerned, this depends on the content and solubility of the sulfates in the cement with which the superplasticizer is to be used: what is necessary is an adequate amount of soluble sulfates in the mixture.

These requirements (probably not applicable when superplasticizers are used in NSC) merit an explanation. In other words, why is there so much more of a problem with the cement-superplasticizer compatibility in HPC than in NSC?

The answer lies in the fact that virtually all national standards for portland cement specify tests on mortar at a w/c of about 0.5 and with no admixture present. In such mixtures, the relatively large amount of mixing water present plays a dominant role in determining the rheological properties of the mortar by keeping the cement particles separated. In addition, the relatively large amount of mixing water allows a larger number of ions to enter the solution before saturation is reached.

On the other hand, when a superplasticizer is used and the w/c is 0.3 or less, the distance between the cement particles is smaller and the number of ions that can enter the solution is smaller, too. Under these conditions, the kinetics of the initial hydration of cement are very different from those that exist at a w/c of 0.5. It follows, and has also been proven by experience, that compliance with existing standards for portland cement is no guarantee that a given cement is suitable for use in HPC. Indeed, tests prescribed by the standards give no indication as to which cement is most suitable for use in HPC.

To a large extent, the situation is the same for superplasticizers. The standards, which require the use of a standardized reference cement or a blended cement, were established at a time when the action of superplasticizer on cement was poorly understood. The standards were therefore based on those for water reducers, which had also been developed using a *w/c* of 0.5.

In view of the existing situation, and pending the development of new standards relevant to the mixtures used in HSC, how do we go about selecting cement and superplasticizer for HSC?

First of all, we can use the guidelines given in this section to eliminate inappropriate cements and superplasticizers. The next step is laboratory testing on a trial-and-error basis of a number of cement pastes containing combinations of different cements and superplasticizers, to establish the best combination from a rheological point of view. Tests such as the mini slump or grout fluidity[7] make it possible to narrow the choice to a few cements compatible with one or two commercially available superplasticizers. For the final selection of the cement and the superplasticizer, it is necessary to make tests on a trial concrete mixture, as only such tests give truly reliable data on the slump loss and strength gain.

Supplementary cementitious materials

In recent years, the use of supplementary cementitious materials, mainly fly ash and blast-furnace slag, has become increasingly common in NSC, partly for reasons of economy and partly because of technical benefits imparted by these materials. In the case of HPC, the reasons for their use are even stronger.

Given that the high strength of HPC is due, in part, to the presence of a dense matrix, replacing a portion of the portland cement with one or more supplementary cementitious materials would not unduly depress the early strength of the concrete. Also, the lower chemical reactivity of supplementary cementitious materials means that a partial replacement of cement is beneficial from the standpoint of controlling the rheological properties of HSC.

In most cases there is also the economic benefit of the price differential between cement and the supplementary cementitious material. In addition, partial replacement of cement nearly always allows a significant reduction in the dosage of the superplasticizer, which is a particularly expensive ingredient.[8]

It is worth remembering that fly ash and blast-furnace slag are not chemically inert in concrete and indeed contribute to strength development. In addition to supplementary cementitious materials, the modern trend in some commercial cements is to include an inert filler such as ground

Chapter 4: Special Aspects

Table 4.2.1—Composition of experimental concretes produced in a ready-mix plant

Concrete type		Reference	Silica fume	Fly ash	Slag + Silica Fume	
w/(c+m)		0.3	0.3	0.3	0.3	0.25
Water	kg/m³	127	128	129	131	128
Cement ASTM Type II	kg/m³	450	425	365	228	168
Silica fume	kg/m³	—	45	—	45	54
Fly ash	kg/m³	—	—	95	—	—
Slag	kg/m³	—	—	—	183	320
Dolomitic limestone coarse aggregate	kg/m³	1100	1110	1115	1110	1100
Fine aggregate	kg/m³	815	810	810	800	730
Superplasticizer*	L/m³	15.3	14	13	12	13
Slump after 45 min	mm	110	180	170	220	210
f_c at 28 days	MPa	99	110	90	105	114
f_c at 91 days	MPa	109	118	111	121	126
f_c at 1 year	MPa	119	127	125	127	137

* Sodium salt of naphthalene sulfonate

limestone. The use of fillers in HPC has not as yet been adequately investigated. In this connection, it may be noted that silica fume acts as a filler during the early stages of hardening of concrete and thus helps to increase the density of the matrix.[9,10] However, silica fume plays also other important roles in concrete since it is a supplementary cementitious material.

The race to reach ever-higher compressive strength has led to the use of combinations of silica fume with one or other of the two most common supplementary cementitious materials. For example, in Toronto, concrete containing 315 kg/m³ (530 lb/yd³) of portland cement, 137 kg/m³ (230 lb/yd³) of blast-furnace slag, and 36 kg/m³ (60 lb/yd³) of silica fume exhibited good rheological properties, led to compressive strengths of 70 to 85 MPa (10,100 to 12,300 psi) and proved to be economical.[11] In Montreal, a ready-mix concrete plant produced concrete with a one-year compressive strength of 130 MPa (18,900 psi) (measured on 100 × 200 mm [4 × 8 in.] cylinders) using a water-cementitious materials ratio (w/cm) of 0.25. The total content of cementitious material was 542 kg/m³ (913 lb/yd³), of which 60% was blast-furnace slag, 10% was silica fume, and only 30% was portland cement. Properties of this concrete and also of four other experimental concretes are given in Table 4.2.1.[12]

Which water-cement ratio?

In the preceding paragraph, we referred to w/cm rather than to w/c. With the growing use of supplementary cementitious materials in HPC, we should consider which of the two ratios is more appropriate in any given circumstance. Indeed, we might also give some thought to the ratio of water to the mass of all the finely divided particles present in the mixture.

There is certainly no simple or unique answer to the question "which water-cement ratio?" In ternary systems (cement-slag-silica fume or cement-fly ash-silica fume) the various particles participate in different ways and at different rates in the hydration process and in creating the bonds that result in the final strength of the concrete. Water also enters into combination with the different materials at different rates and at different times. Generally speaking, we can say that cement hydrates more rapidly than supplementary cementitious materials and therefore takes up most of the mixing water. Because of this last-mentioned fact, it is tempting to conclude that the density of the matrix at the early stage of hardening is governed by w/c, regardless of the presence of the other finely divided particles.

However, recent experimental work, as well as field observations of HPC, indicate that this is too simplistic a point of view.[13,14] In fact, silica fume, the finer particles of blast-furnace slag, and fly ash all enter into the hydration reaction before some of the coarser cement particles have done so.

We are obliged therefore to conclude that, in the present state of knowledge, it is preferable to quote both w/c and w/cm; it is also useful to state the water content of the mixture. The usefulness of both ratios, each for an appropriate purpose, has already been accepted.[15]

Aggregate strength

In NSC, the strength of the aggregate *per se* plays a minor role. While the properties of the aggregate influence the modulus of elasticity of the concrete as well as its thermal properties, they are critical mainly in cases when the concrete undergoes a large and repeated temperature variation.

We must not fall into the trap of belittling the importance of the quality of aggregate in NSC. Aggregate grading is still of paramount importance, and its shape and surface texture, as well as the presence of deleterious materials also influence the properties of the resulting concrete. However, the strength of the aggregate is not a factor limiting the strength of the resulting concrete, as witnessed for example by the fact that strength of lightweight concrete can be higher than the strength of the aggregate used. What controls the strength of NSC is the strength of the hydrated cement paste; this is why w/c is the major factor controlling the compressive strength of NSC.

The situation is rather different with HPC, where the bond between aggregate and hydrated cement paste is so strong that it results in a significant transfer of stress across the paste-aggregate interface. At the same time, the strength of the cement paste phase, for reasons given in the early part of this paper, is very high, and sometimes higher than the strength of the aggregate particles. Observation of fracture surfaces in HPC has shown that they pass through the coarse aggregate particles as often as, if not more often than, through the cement paste itself.[16,17] Indeed, in some instances the strength of the aggregate particles has been found to be the factor that limits the compressive strength of the concrete.[18]

In other words, HPC behaves very much like an ideal composite soft material, with the stress shared by the aggregate and the hydrated cement paste. The participation of aggregate in the load-carrying process can be discerned from the shapes of the hysteresis loops obtained in laboratory tests. The shape of the hysteresis loop is strongly influenced by the properties of the aggregate.[18,19]

From the above, it follows that the properties of aggregate, especially its coarse fraction, have a considerable influence on the properties of the resulting HPC. While we are not yet in a position to give theoretically based guidance (which is also the case with NSC), we have evolved some guidelines.[20]

One of these concerns the maximum size of crushed aggregate (MSA). Because the crushing process preferentially takes place along any potential zones of weakness within the parent rock, and thus removes them, smaller particles of the coarse aggregate fraction are likely to be stronger than the very large ones. Consequently, we have found that for HPC with a compressive strength in the 60 to 100 MPa (8700 to 14,500 psi) range, MSA $\geq$ 20 mm (¾ in.) can be used. However, when strength in excess of 100 MPa (14,500 psi) is required, MSA should not exceed 10 or 12 mm (˜½ in.) unless laboratory tests demonstrate that a larger MSA can be used.

Concerning the shape of the coarse aggregate, we have found glacial and fluvioglacial gravels to be excellent. Crushed aggregate can be equally good, provided the crushed particles are mostly cubic in shape, with a minimum of flaky or elongated particles. The latter would have an adverse effect on workability.

From the petrographic standpoint, we have found fine grained rocks such as some types of limestone, dolomitic limestone, and granite all to be equally good; there may, of course, be some other rocks which produce excellent aggregate for HPC but they would have to be proved by tests.

Examples of use of HPC

Rather than give details of proportioning HPC mixtures, which have been described in several publications,[20-23] we shall describe briefly several structures in which HPC was used. The structures have been chosen so as to illustrate each of the points discussed earlier in the paper; details of the concretes used are given in Table 4.2.2.

- **Water Tower Place (Chicago, 1975)** typifies the composition of HPC prior to the use of superplasticizers.[24]
- **Joigny Bridge (France, 1989)** was constructed using HPC without silica fume; this was done to demonstrate that HPC can be produced and used in construction even when silica fume is not economically available.[3]
- **La Laurentienne Building (Montreal, 1984)** contains two experimental columns in which the HPC contained a superplasticizer and a retarding agent. This was necessary because only one cement was available, and it had somewhat problematic

Table 4.2.2—Typical HPC mixtures

Mixture Number		1	2	3	4	5
Water	kg/m³	195	165	135	145	130
Cement	kg/m³	505	451	500	315	513
Fly ash	kg/m³	60	—	—	—	—
Slag	kg/m³	—	—	—	137	—
Silica fume	kg/m³	—	—	30	36	43
Coarse aggregate	kg/m³	1030	1030	1100	1130	1080
Fine aggregate	kg/m³	630	745	700	745	685
Water reducer	mL/m³	975	—	—	900	—
Retarder	L/m³	—	4.5	1.8	—	—
Superplasticizer	L/m³	—	11.25	14	5.9	15.7
w/(c+m)		—	0.35	0.37	0.27	0.31	0.25
f'$_c$ at 28 days	MPa	65	80	93	83	119
f'$_c$ at 91 days	MPa	79	87	107	93	145

1—Water Tower Place, Chicago (1975)
2—Joigny Bridge, France (1989)
3—La Laurentienne Building, Montréal (1984)
4—Scotia Plaza, Toronto (1987)
5—Two Union Square, Seattle (1988)

rheological properties, and the delivery time of concrete was long.[25]
- **Scotia Plaza (Toronto, 1987)** was built using HPC containing both silica fume and blast-furnace slag.[11]
- **Two Union Square (Seattle, 1988)** is an outstanding example of use of HPC in a major construction with clear economic benefits of the use of the material. The strength of the concrete was the highest used so far in actual construction.[26]

There exist numerous other examples of use of HPC in structures. In some cases, the mixture used was similar to one or other of those cited previously, in others it was radically different. Indeed, the objective of citing the preceding five examples of HPC was to show that there is no single recipe for HPC, any more than there is such a recipe for NSC. The selection of concrete mixtures is a flexible process that can be adapted to local conditions and often to local materials. As far as HPC is concerned, success depends on the use of a very low w/c coupled with the best combination of ease of placing, and cost. There is no single measure of success: depending on the availability and characteristics of local materials, it may be difficult to produce a 60 MPa (8700 psi) concrete or relatively easy to make a 120 MPa (17,400 psi) one.

Concluding remarks

HPC is not an unusual or difficult material. It is concrete. As with any concrete, the compressive strength of HPC depends on w/c but the strength of the aggregate itself can be a limiting factor.

For the first time in the history of concrete technology (as distinct from paste compacts) workable concrete can be made with a water content in the

4-26 Chapter 4: Special Aspects

mixture no greater than the amount of water theoretically required to hydrate all the cement present. This is possible because of the outstanding dispersing properties of modern superplasticizers.

The superplasticizer and the cement have to be compatible with one another so that the concrete has suitable rheological properties, and there exist means of establishing compatible pairs of these materials.

Using a compatible combination of portland cement and superplasticizer, suitable aggregates, and a very strict and consistent quality control, it is possible to produce, in a ready-mix plant, HPC with a compressive strength of up to 150 MPa (21,800 psi) and to maintain the production of concrete with this strength day in and day out.

HPC is there to serve engineers and clients and to offer economic technical solutions. It would be a pity if the benefits of this material were not exploited simply through lack of widespread knowledge of its properties and means of production.

References

1. Neville, A. M., *Properties of Concrete,* Fourth Edition, Longman London, 1995, and John Wiley, N.Y., 1996, 844 pp.

2. Neville, A. M., "Essentials of Strength and Durability of Various Types of Concrete with Special Reference to Sulfur," *ACI Journal,* V. 76, No. 9, Sept. 1979, pp. 973-996.

3. Malier, Y.; Brazillier, D.; and Roi, S., "The Bridge of Joigny," *Concrete International,* V. 13. No. 5, May 1991, pp. 40-42.

4. Aïtcin, P-C.; Bédard, C.; Laplante, P.; and Haddad, G., "Very High Strength Cement for Very High Strength Concrete," *Symposium of the Material Research Society,* Boston, Mass., Nov. 1984, pp. 201-210.

5. Aïtcin, P.-C.; Sarkar, S. L.; Ranc, R.; and Lévy, C., "A High Silica Modulus Cement for High-Performance Concrete," *Advances in Cementitious Materials, Ceramic Transaction,* V. 16, pp. 103-121.

6. Hanna, E.; Luke, K.; Perraton, D.; and Aïtcin, P.-C., "Rheological Behavior of Portland Cement in the Presence of Superplasticizer," *Superplasticizers and Other Chemical Admixtures in Concrete,* SP-119, American Concrete Institute, Farmington Hills, Mich., 1989, pp. 171-188.

7. de Larrard, F., and Puch, C., "Formulation des Bétons à Haute Performance: la Méthode des Coulis," *Bulletin de Liaison du Laboratoire Central des Ponts et Chaussées,* No. 161, May/June 1989, pp. 75-83.

8. Djellouli, H.; Aïtcin, P.-C.; and Chaallal, O., "The Use of Ground Granulated Slag in High-Performance Concrete," *Utilization of High-Strength Concrete—Second International Symposium,* SP- 121, American Concrete Institute, Farmington Hills, Mich., 1990, pp. 351-368.

9. Bache, M. M., "Densified Cement Ultra-Fine Particle-Based Materials," *Second International Conference on Superplasticizers in*

Concrete, June 1981, Ottawa, pp. 1-35.

10. de Larrard, F., "Modèle Linéaire de Compacité des Mélanges Granulaires," *De la Science des Matériaux au Génie des Matériaux,* V. 1, Chapman and Hall, 1987, pp. 325-332.

11. Ryell, J., and Bickley, J. A., "Scotia Plaza: High-Strength Concrete for Tall Buildings," *Proceedings on the Utilization of High-Strength Concrete,* Stavanger, Norway, 1987, pp. 641-653.

12. Baalbaki, M.; Sarkar, S. L.; Aïtcin, P-C.; and Isabelle, H., "Properties and Microstructure of High-Performance Concretes Containing Silica Fume, Slag and Fly Ash," *Fly Ash, Silica Fume, Slag and Natural Pozzolans in Concrete,* SP-132, American Concrete Institute, Farmington Hills, Mich., 1992, pp. 121-142.

13. Sarkar, S. L.; Aïtcin, P.-C.; and Djellouli, H., "Synergistic Roles of Slag and Silica Fume in Very High-Strength Concrete," *Cement, Concrete and Aggregates,* V. 12, No. 1, Summer 1990, pp. 32-37.

14. Sarkar, S. L.; Baalbaki, M.; and Aïtcin, P.-C., "Microstructural Development in High-Strength Concrete Containing a Ternary Cementitious System," *Cement, Concrete and Aggregates,* V. 13, No. 2, Winter 1991, pp. 81-87.

15. Neville, A. M., and Brooks, J. J., *Concrete Technology,* Longman, London, Revised edition, 2001, 438 pp.

16. Sarkar, S. L., and Aïtcin, P.-C., "The Importance of Petrological and Mineralogical Characteristics of Aggregates in Very High-Strength Concrete," ASTM STP 1061, 1990, pp. 129-144.

17. Sarkar, S. L.; Lessard, M.; and Aïtcin, P.-C., "Correlation of Petrographic Characteristics of Aggregates with Type of Failure in High-Strength Concretes," *Canadian Aggregates,* V. 4, No. 6, 1990, pp. 21-25.

18. Aïtcin, P.-C., and Mehta, P. K., "Effect of Coarse Aggregates Characteristics on Mechanical Properties of High-Strength Concrete," *ACI Materials Journal,* V. 87, No. 2, Mar.-Apr. 1990, pp. 103-107.

19. Baalbaki, W.; Aïtcin, P.-C.; Chaallal, O.; and Benmokrane, B., "Influence of Coarse Aggregate on Elastic Properties of High-Performance Concrete," *ACI Materials Journal,* V. 88, No. 5, Sept.-Oct. 1991, pp. 499-503.

20. Mehta, P. K., and Aïtcin, P.-C., "Microstructural Basis of Selection of Materials and Mix Proportions for High-Strength Concrete," *Utilization of High-Strength Concrete—Second International Symposium,* SP- 121, American Concrete Institute, Farmington Hills, Mich., 1990, pp. 265-286.

21. ACI Committee 363, "State-of-the-Art Report on High-Strength Concrete (ACI 363R-84)," American Concrete Institute, Farmington Hills, Mich., 48 pp.

22. Albinger, J.-M., "Le Béton à Haute Résistance aux U.S.A.," *Annales de l'Institut Technique du Bâtiment et des Travaux Publics, Les Bétons à Hautes Performances: Expériences Nord-Américaines et Françaises,* No. 473, Mar./Apr. 1989, pp. 170-181.

23. de Larrard, F., "A Method for Proportioning High-Strength

Concrete Mixtures," *Cement, Concrete and Aggregates,* V. 12, No. 1, Summer 1990, pp. 47-52.

24. "Water Tower Place, High-Strength Concrete," *Concrete Construction,* V. 21, No. 3, March 1976, pp. 102-104.

25. Aïtcin, P.-C.; Laplante, P.; and Bédard, C., "Development and Experimental Use of a 90-MPa (13,000 psi) Field Concrete," *High-Strength Concrete,* SP-87, American Concrete Institute, Farmington Hills, Mich., 1986, pp. 14-17.

26. Godfrey, K. A., Jr., "Concrete Strength Record Jumps 36%," *Civil Engineering,* V. 57, No. 10, Oct. 1987, pp. 84-86.

Section 4.3: High-Performance Concrete—An Overview*

In the last few years, the expression *high-performance concrete* and the acronym HPC have become very fashionable. But what *exactly* is meant by this? Is high-performance concrete a material really different from "just concrete"? Or is it concrete that is appropriate for a particular situation? It is the purpose of this paper to explore the broad issue of high-performance concrete and to put this material in what we perceive to be a proper perspective.

However much those involved in concrete and concrete structures enjoy this material and deal with it satisfactorily, it is a fact that the last half century has not ushered in any revolutionary changes. It is true that we now use a variety of admixtures, as well as air entrainment, and that we have broadened the range of cementitious materials in the mixture; nevertheless, the changes are nowhere near as revolutionary as the changes in telecommunications or even in motor cars. But this paper is not meant to look back or to offer a historical review, except in so far as high-performance concrete has appeared on the concrete scene. This is to recognize that, in the last 15 years or so, some new concepts in the field of concrete have appeared and it can be postulated that they have led to the advent of high-performance concrete.

What is high-performance concrete?

Let us start by defining high-performance concrete. It is arguable that the expression "high-performance concrete" is not felicitous. It sounds like advertising a new product but, in most respects, high-performance concrete is not fundamentally different from the concrete that we have been using all along, because it does not contain any new ingredients and does not involve new practices on site. Actually, high-performance concrete evolved gradually over the last 15 years or so, mainly by the production of concrete with higher and higher strengths: 80, 90, 100, 120 MPa, and sometimes even higher. Nowadays, in some parts of the world, 140 MPa can be routinely produced. But high-performance concrete is not the same as high strength concrete. The emphasis has moved from very high strength to other properties desirable under some circumstances. These are: a high modulus of elasticity, high density, low permeability, and resistance to some forms of attack.

What then is the difference between high-performance concrete and the usual concrete? We have stated that the ingredients are the same in both cases, but this is not entirely correct. First, high-performance concrete very often contains silica fume whereas ordinary concrete usually does not. Secondly, high-performance concrete usually, although not always,

---

*This section was co-authored by P.-C. Aïtcin.

contains fly ash, or ground-granulated blast-furnace slag (or slag, in short), or both these materials. The aggregate has to be very carefully chosen and has a smaller maximum size than is the case with ordinary concrete: in high-performance concrete, the maximum size is usually 10 to 14 mm. There are two reasons for this. First, with a smaller maximum size, the differential stresses at the aggregate-cement paste interface, which could lead to microcracking, are smaller. Second, smaller aggregate particles are stronger than larger ones; this is due to the fact that comminution of rock removes the largest flaws, which control strength.

Another point about ingredients is this: the low water-cement ratio and the inclusion of silica fume in the mixture necessitate the use of a superplasticizer. It is not good enough to use any superplasticizer with any portland cement; the superplasticizer must be compatible with the actual cement that is used. The problem of compatibility is discussed later in the paper, but at this stage some general comments, alas critical, about marketing of portland cement may be in order.

There is no doubt that cement manufacturers have an excellent knowledge of cement, but some of them are less well informed about the needs of the concrete producer and the concreting contractor. They are inclined to say that portland cement is an excellent product which always conforms to national standards and that it is not the user's business to look into the detailed physical and chemical properties of cement. The only "allowed" exceptions are the ASTM Type classification or the European strength classification, and also some other broad features such as an early high strength, low heat of hydration, or a low alkali content. For example, in the United Kingdom, in the past, the cement prices were fixed and the cement manufacturers strongly discouraged the purchaser from selecting supply from a particular cement plant. The marketing attitude was: all our ordinary portland cements are equally excellent. Even today, in most countries, it is not routinely possible to ensure that successive deliveries of cement are from the same plant, let alone the same production batch. The only exception is in the case of large bulk deliveries for major projects where negotiation is possible. Of course, from the commercial standpoint this attitude is understandable.

However, with changes in the economic climate, attitudes have altered; nevertheless, many cement manufacturers still display a sublime disinterest in the concrete user's specific needs. Admittedly, the mention of admixtures no longer provokes the reply: the best admixture is more portland cement! Nevertheless, the usual disinterest on the part of many cement manufacturers is unhelpful when the concrete producer wishes to make high-performance concrete: he buys portland cement from one source and the superplasticizer from another; the marriage of the two may be very unhappy.

Mixture proportions of high-performance concrete

It may be useful to give an idea of the typical mixture proportions of

Table 4.3.1—Mixture proportions of some high-performance concretes[5]

Ingredient (kg/m³)	Mixture								
	A	B	C	D	E	F	G	H	I
Portland cement	534	500	315	513	163	228	425	450	460
Silica fume	40	30	36	43	54	46	40	45	—
Fly ash	59	—	—	—	—	—	—	—	—
Slag	—	—	137	—	325	182	—	—	—
Fine aggregate	623	700	745	685	730	800	755	736	780
Coarse aggregate	1069	1100	1130	1080	1100	1110	1045	1118	1080
Total water	139	143	150	139	136	138	175†	143	138
Water-cementitious materials ratio	0.22	0.27	0.31	0.25	0.25	0.30	0.38	0.29	0.30
Slump, mm	255	—	—	—	200	220	230	230	110
Cylinder strength (MPa) at age (days)									
1	—	—	—	—	13	19	—	35	36
2	—	—	—	65	—	—	—	—	—
7	—	—	67	91	72	62	—	68	—
28	—	93	83	119	114	105	95	111	83
56	124	—	—	—	—	—	—	—	—
91	—	107	93	145	126	121	105	—	89
365	—	—	—	—	136	126	—	—	—

* Further information about the mixtures: (A) United States; (B) Canada; (C) Canada; (D) United States; (E) Canada; (F) Canada; (G) Morocco; (H) France; (I) Canada
† It is suspected that the high water content was occasioned by a high ambient temperature in Morocco.

high-performance concrete. Usually, Type I cement (in the ASTM classification) is used but, if a high early strength is required, Type III cement can be used. We have already said that other cementitious materials are also included in the mixture. The total content of the cementitious materials is very high: 400 to 550 kg/m³. The mass of silica fume, when used, represents 5 to 15% of the total mass of the cementitious material, the value of 10% being typical. The dosage of superplasticizer is 5 to 15 liters per cubic meter of concrete. The actual dosage required depends on the content of active solids in the liquid superplasticizer and on the "reactivity" of the cement, which, in turn, is a function of the C_3A content and its polymorphic form and of the amount of alkali sulfates, as well as some other factors. This dosage allows a reduction in the water content of 45 to 75 liters per cubic meter of concrete. The value of the ratio of the mass of water to the total mass of cementitious material is usually between 0.35 and 0.25, but even a value as low as 0.22 has been used.

In practice, the mixture proportions vary, depending on the properties of the individual ingredients and on the desired properties of the concrete in service; details of some actual mixtures used in the past are shown in Table 4.3.1. The properties of the ingredients of the mixture are discussed below.

As already said, the aggregate must not have too large a maximum size. The coarse aggregate must have a number of characteristics: it must be strong; absolutely clean, that is free from adhering clay or dust; it must not contain reactive silica; and it must be equidimensional in shape, that is neither flaky nor elongated. With very few exceptions, crushed aggregate is used. As for fine aggregate, it has to be coarsely graded, preferably with a fineness modulus of 2.7 to 3.0. It is worth remembering that, in order to achieve good packing of the fine particles in the mixture, as the cement content increases, the fine aggregate has to become more coarsely graded.

Let us now consider the inclusion of fly ash and slag in the mixture. First of all, these materials are generally cheaper than portland cement. Secondly, these materials hydrate or react chemically somewhat later than portland cement. In consequence, they develop the heat of hydration more slowly. This means that the very early temperature rise of the concrete is a little lower. Even a small reduction in the maximum temperature is important because, with the very high cement contents used in high-performance concrete, the temperature *rise* at the center of a massive section can be 50°C or even more. What matters, of course, is not the maximum temperature as such but the temperature gradient between the center and the surface of the concrete element, which is usually at a moderate temperature. It has been suggested[1] that if the temperature gradient does not exceed 20°C per meter, then thermal cracking due to differential cooling will not occur.

There is another reason why the use of fly ash or slag is beneficial, and this has to do with the slump loss of fresh concrete. Because these materials react only very little during the first few hours, they do not contribute to the slump loss, so that less superplasticizer needs to be used. In other words, the amount of superplasticizer necessary to ensure an adequate workability is governed only by the content of the portland cement. On the other hand, mixtures that have more fly ash or more slag develop a lower strength at, say, 12 to 24 hours than when all cementitious material is portland cement, but this can be compensated by lowering the ratio of the mass of water to the total mass of cementitious materials. It follows that, before deciding on the use of these materials, the structural designer must establish the age at which a given strength is necessary; he should also be familiar with concrete technology[2] (See Section 6.1).

The whole question of the use of superplasticizers is important because they are an expensive ingredient in the mixture, and yet they must be included in the mixture. Silica fume is also very expensive. In many countries, 1 kg of silica fume costs as much as 10 kg of portland cement. So, let us look at the question of whether silica fume is necessary.

In deciding whether or not to use silica fume, we are guided by experience:[3] mixtures without silica fume were found to achieve a 28-day compressive strength of up to about 90 MPa, albeit with some difficulty. To obtain a higher strength, silica fume must be included in the mixture. As already stated, the optimum content of silica fume is about 10% by mass of cement.

It follows from the above that if, instead of making a 90 MPa concrete, we make a 100 MPa concrete, we have to use silica fume. Because of the 10-times higher price of silica fume, the addition of 10% of silica fume to the mixture doubles the cost of the cementitious material. This represents a very large increase in the price of the concrete.

Nevertheless, we should remember the benefits of using silica fume. It is not only a highly pozzolanic material, but it is also an extremely fine powder whose particles are about 100 times smaller than cement. The particles of silica fume pack tightly against the surface of the aggregate and fit in-between the cement particles, thus greatly improving packing. It follows that, if there is too little silica fume, say less than about 5%, it is not very effective. If there is too much, say more than 15%, there is no room between the cement particles to accommodate all the silica fume, and some of it is wasted. Wasting a very expensive material is not good engineering practice.

Because the extremely fine particles of silica fume reduce the size and volume of voids near the surface of the aggregate, the so-called interface zone (also known as transition zone) has improved properties with respect to microcracking and permeability. The bond between the aggregate and the cement paste is improved, allowing the aggregate to participate better in stress transfer[4] (See Section 4.4).

We should not leave the topic of mixture proportions of high-performance concrete without considering the water-cement ratio. There are, in fact, two meanings of this word. We have known for about 80 years that the water-cement ratio is the controlling factor in strength because the relative volume of the space originally occupied by water determines the total volume of solid matter in the hardened concrete. In very general terms, the higher the volume of the solid matter, the higher the compressive strength.[5] It follows that, with high-performance concrete, just the same as with ordinary concrete, strength at, say 28 days, is a function of the water-cement ratio. By this time, slag will have reacted to a significant extent and fly ash somewhat less so; however, the extent of the reaction of fly ash is very sensitive to the effectiveness of curing.

But what is the situation at 24 hours or at two or three days? Fly ash and, to a lesser extent, slag occupy only the volume represented by their original, that is dry powder, form. It follows that the water-cement ratio relevant to the very early strength is approximately the mass ratio of water to the portland cement only. This ratio is much higher than the ratio of the mass of water to the mass of all the cementitious materials taken together. This is why the inclusion of fly ash or slag, at a given water content in the mixture, leads to a lower *very* early strength. The preceding statement is true not only for high-performance concrete but for all concrete mixtures.

Compatibility of portland cement and superplasticizer

At the beginning of the section, we said that high-performance concrete does not require a special cement and that usually a Type I portland cement is used. The problem is that the cement and the superplasticizer must be

Chapter 4: Special Aspects

suited to one another, that is, there must be no incompatibility between the two materials. This needs to be explained at some length.

First of all, we should re-state how a superplasticizer acts so as to give the concrete a high workability. Superplasticizers are long and heavy molecules, which wrap themselves around the cement particles and give them a highly negative electrical charge so that they repel each other. This results in a deflocculation and dispersion of the cement particles and, therefore, in a higher workability of the mixture. The fundamental structure of the hydrated cement paste is not affected, but superplasticizers interact with tricalcium aluminate (C_3A) in the portland cement. We should remember that C_3A is the first component of cement to hydrate, and this reaction is controlled by the gypsum added in the manufacture of portland cement.

We have thus a situation such that both the superplasticizer and the gypsum can react with C_3A. Although a certain amount of superplasticizer is necessary during mixing in order to achieve an adequate workability, it is essential that all the superplasticizer does not become fixed by C_3A. Such fixing would occur if the gypsum does not release the sulfate ions fast enough to react with C_3A. When the sulfate ions are released too slowly, the portland cement and the superplasticizer are said to be incompatible.

Thus, in practice, the controlling factor is the solubility of gypsum in the given portland cement. The term "gypsum" is used to describe the calcium sulfate in portland cement, but this calcium sulfate may exist in several forms, depending on the raw materials used in the manufacture of the cement. These can be: gypsum, that is calcium sulfate di-hydrate, hemi-hydrate, and anhydrite. Each of these has a different rate of solubility. Moreover, the solubility of anhydrite depends on its structure and origin. In practice, the situation is even more complicated. National standards for portland cement usually specify a maximum SO_3 content in the finished cement powder; for example, for Type I cement, ASTM Standard C150-00 specifies the maximum content of SO_3 of 3.0 or 3.5%, depending on the content of C_3A.

The crucial point is that it is not the gypsum added during the grinding of the clinker, but the total SO_3 content in the cement that is limited by the ASTM Standard. Now, there is very often another source of SO_3 in the cement, and that is the sulfur in the coal, or possibly oil, used in the cement kiln. Such sulfur is often present because it is the cheaper coal, petroleum coke or oil, all of which have a high sulfur content, that is used in cement making. What happens is that the sulfur in the fuel reacts with the volatile alkali oxides in the kiln to form alkali sulfates. These sulfates are highly soluble.

It follows from the above that two portland cements can have the same total sulfate content but, depending on the origin of the sulfates, more or less sulfate will be available for reaction with C_3A in the early stages. If there is too little soluble sulfate available, the sulfate ends of the superplasticizer become fixed, and the superplasticizer is not available for improving the workability of the mixture. This is why some of the superplasticizer is often added after the initial mixing operation, but this may not be enough, apart from complicating the mixing process.

The issue of compatibility can be readily resolved because it has been established that, for each portland cement, there exists an optimum amount of soluble alkalies (that is, those existing as alkali sulfates) which ensures compatibility with a given superplasticizer. It is to be hoped that the approach of buying portland cement as one item and the superplasticizer as an independent second item will soon come to an end; matching pairs of cement and superplasticizer will become available, thus reducing laborious testing prior to their use.

The problem of incompatibility, just described, can also exist in ordinary concrete, but it is much more acute in high-performance concrete. There are two reasons for this. First, in high-performance concrete, the water-cement ratio is very low so that there is less water available to accept the sulfate ions. Secondly, because the content of portland cement per cubic meter of concrete is very high, there is present more C_3A whose reaction must be controlled to ensure the desired workability.

There is an important general point that should be made: the behavior of cement at a high water-cement ratio does not tell us enough about its behavior at a very low water-cement ratio. The majority of standard tests are, therefore, not good enough for ensuring a satisfactory performance of cement in high-performance concrete. This is why special compatibility tests are required. Special tests may also be required for assessing the behavior of concretes with water-cement ratios of 0.35 or even lower, and these are the water-cement ratios used in high-performance concrete.

Shrinkage of high-performance concrete and curing

Concrete can undergo several types of contraction, generally referred to as various types of shrinkage,[6] but here we are concerned specifically with high-performance concrete (See Section 4.5). First, there is a contraction of concrete while it is still in a plastic state. The magnitude of this so-called plastic shrinkage is affected by the amount of the water lost from the exposed surface of the concrete. If the amount lost per unit area exceeds the amount of water brought to the surface by bleeding, plastic shrinkage cracking can occur. High-performance concrete has a very low water content (expressed in liters per cubic meter of concrete) and the developing capillary pores are consequently very small. There is, therefore, virtually no bleeding, and this would lead to plastic shrinkage cracking unless the loss of water from the surface of the concrete is prevented. Hence the need for wet curing from the earliest possible moment.

The second, and best known, type of shrinkage is drying shrinkage of hardened concrete; it is this that is referred to simply as "shrinkage." The cause of drying shrinkage is the loss of water by evaporation to the outside of the concrete. In high-performance concrete, there is very little drying shrinkage, partly because the capillaries are very small. But there are other reasons, too. The chief of these is that much of the water has already left the capillaries due to self-desiccation. This can cause autogenous shrinkage; this then is the third type of shrinkage. Autogenous shrinkage is the consequence of the continuing hydration of cement throughout its mass, and not just near the surface. This shrinkage is encouraged by the low water-cement ratio and, therefore, a smaller number and size of capillary pores. Silica fume, which reacts very early, rapidly uses up the water and also contributes to self-desiccation.

An important consequence of autogenous shrinkage in high-performance concrete is the development of internal microcracking throughout the concrete mass, and this can and must be prevented by wet curing. We should know, even if we often do not practice the rule, that curing all concrete is of great importance, the more so the lower the water-cement ratio. In the case of high-performance concrete, wet curing from the earliest possible moment is absolutely essential and must be continued until the tensile strength of the hydrating cement paste is high enough to resist internal microcracking.

We should state very clearly that membrane curing is not good enough in the case of high-performance concrete. All that membrane curing does is to prevent the loss of water from the concrete. This is good enough when the water-cement ratio is greater than about 0.42 because the amount of water in the mixture is adequate for full hydration. However, at the very low values of the water-cement ratio in high-performance concrete, it is essential for additional water from the outside to ingress into the concrete. Admittedly, in some cases, immediate fog misting or covering the surface of the concrete with water by means of ponding may be impracticable. What should be done then is to apply a membrane temporarily, for a few hours at the most, or to use a new type of admixture called "waterhold" so as to prevent the development of plastic shrinkage cracking. But as soon as significant hydration has occurred, it is essential for external water to ingress into the concrete. This can be done by placing pre-wetted burlap (hessian) or pre-wetted geotextile, covered by plastic sheets, with a perforated hose underneath the plastic keeping the burlap permanently wet.

Hydration of cement in high-performance concrete is very rapid and, if it is allowed to proceed uninterrupted by a continuous supply of curing water, no menisci will form in the capillary pores and there will be no autogenous shrinkage, at least close to the external source of water, that is, the exposed surface zone of the concrete member. Thus, with really good curing, there will be virtually no autogenous shrinkage and no drying shrinkage. Admittedly, if at a later date the concrete surface is allowed to dry out, there will be drying shrinkage but, by then, the tensile strength of concrete will be high enough for no shrinkage cracking to occur. It is worth remembering that it is not the shrinkage itself that matters but only shrinkage cracking.

Much space in this section has been devoted to the topic of curing, but we believe this to be well worthwhile: a successful use of high-performance concrete is conditional upon observing the simple curing requirements. There is absolutely no point in using excellent materials and being unsuccessful in the end product.

When to use high-performance concrete

At the beginning of this paper, we expressed the view that high-performance concrete is not a fundamentally different material from ordinary concrete, but a concrete fit for a given purpose. The specific needs are varied.

The earliest need was for a high-strength concrete. This strength can be required at a very early age in order to put the structure into service. More often, however, high strength is required at the age of 28 days or later. A relatively common requirement for high strength is in compression members. Here, high strength allows the use of smaller columns and, therefore, a reduction in weight and, hence, a lower load on the foundations. Also, a smaller part of the horizontal area is occupied by columns so that there is more of the economically valuable floor space. In flexural members, the benefits of high strength are more difficult to exploit. One reason for this is the problem of cracking in the tension zone of a beam because the tensile strength does not increase in proportion to the compressive strength. Another difficulty lies in the limitations imposed by the existing design codes, but these are likely to disappear in the near future.

High strength of concrete may also be required, not for itself, but because high strength concrete has a high modulus of elasticity. This is of importance with respect to deformation of structural members.

A particularly important use of high-performance concrete is in ensuring a very low permeability of concrete. This is essential in severe exposure conditions where there is a danger of ingress into the concrete of chlorides or sulfates, or of other aggressive agents. These conditions exist in many parts of the world where rapid deterioration of concrete is common.

We would now like to make a general comment about deterioration of concrete. The two conditions that lead to most damage are water movement and temperature change. We are referring to "water movement" because, if a concrete member is totally immersed in air-free water, even seawater, very little damage will occur. On the other hand, alternating periods of rapid wetting and prolonged drying are particularly harmful.

So is a cyclic temperature change. There is a synergy between the two: a combination of repeated wetting and drying with a frequently alternating temperature is likely to lead to considerable damage. High-performance concrete, with a very low permeability, ensures long life of a structure exposed to such conditions.

We should emphasize the fact that durability is not just a problem under extreme conditions of exposure. Carbon dioxide is always present in the air, and significantly so in cities. The resulting carbonation of concrete in the cover zone can destroy the passivation of the reinforcement and lead to corrosion. Aggressive salts are sometimes present in the soil. The surface of the concrete may be subjected to abrasion.

Concrete can sometimes be exposed to repeated cycles of freezing and thawing. The early self-desiccation means that there is little free water in the interior of the concrete, so that generally there is no disruptive formation of ice.[7]

We have to admit, however, that there are some situations where the very low permeability of high-performance concrete is a disadvantage. This is so in the case of fire that results in a rapid increase in temperature of the concrete. Because of the very low permeability, any water present inside the concrete cannot escape rapidly enough; the pressure of the resulting water vapor can cause bursting of the cement paste and spalling of the concrete.

A very important point is that we should not be concerned solely with the strength of concrete but *also* with its durability: concrete that is properly durable can be said to be high-performance concrete, or just "good quality concrete."

General lessons from high-performance concrete

In accordance with our thesis that high-performance concrete is not a "species" distinct from what we would like to call "good ordinary concrete," some comments on the latter are in order. Much of the concrete produced on a worldwide basis is not as good as it should be or even as it could be. We do not lack the knowledge to be successful in concrete-making and we do not need any new research to help us make good concrete. Bad concrete can usually be traced to bad workmanship and to bad construction practices. We believe that the use of high-performance concrete can teach us to make good-quality ordinary concrete. We will give two examples.

Earlier in the paper, we laid great emphasis on the importance of good curing. We stated very clearly that high-performance concrete that has not been very well cured will be of poor quality. Any contractor using high-performance concrete will, therefore, have to learn to follow proper curing procedures. When, on some other construction site, that contractor is placing ordinary concrete, he is likely to continue to use the good practice that he has learnt. In other words, he will apply good curing. In the case of ordinary concrete, which is concrete with a water-cement ratio of, say, 0.45 or more, good curing may be of somewhat lesser importance, but it will nevertheless improve the performance and durability of the concrete structure. By performance, we mean the absence of shrinkage cracking. By durability, we mean a low permeability of the concrete in the cover zone and, therefore, the protection of the reinforcing steel from corrosion. Actually, shrinkage cracking is also detrimental to the protection of the steel.

In the case of ordinary concrete, the importance of good curing is limited mainly to the concrete in the cover zone, but the consequences of inadequate curing are important. It is no use having very good concrete in the interior of a structural member when the concrete cover to the reinforcement is of poor quality, so that carbonation of concrete rapidly extends through the thickness of the cover zone, or chlorides or sulfates from the outside penetrate into the concrete and destructive reactions take place, leading to cracking, spalling or delamination.

The second lesson to be learnt from high-performance concrete is much more general. To make high-performance concrete, it is essential to have a very high quality control of materials and procedures. For example, batching must be very accurate, with the total, and not just added, amount of water in each batch being exactly the same. The final grading of aggregate must not vary. It is not acceptable to please the man who places the concrete by adding another bucket of water to the mixture! Thus the concrete producer must develop a first-class system of production. Our contention is that he should, and is likely to, use the same system when making ordinary concrete.

The first reaction to this proposal may well be that such a high level of quality control costs money, but we believe that it is not money wasted. There are two reasons for our opinion. One is that a higher level of quality control results in a lower variability of the properties of the concrete; for example, the difference between the mean strength of the concrete as produced and the minimum specified strength is smaller. Now, the price is based on the minimum strength, but the cost is related the mean strength. For a given minimum, having a lower value of the mean results in a cost reduction; this is money saved in terms of cement content.

The second reason is that a concrete supplier who can produce high-performance concrete has a reputation, and justifiably so, for being a good concrete supplier. Consequently, he may get more business from those who care about the quality of construction, even if his price per cubic meter is a penny or two higher than that of his competitors, with whom there are endless disputes about the quality of their work; resolution of these disputes is expensive.

The future of high-performance concrete

It is easy for older people to "predict" the future. They can be optimistic and write about progress and about wonderful future developments without running any risk. If, in the fullness of time, they turn out to be wrong, they will be no longer alive to hear the opprobrium. If they turn out to be right, everyone will regret the death of the wise old men. So, we will give our views.

The separate class of high-performance concrete will disappear and high-performance concrete and ordinary concrete will merge into simply "concrete," that is good quality concrete. At the same time, the present poor quality concrete, made with a water-cement ratio of 0.6 and more, will cease to be used structurally, although of course it will still be useful for backfill, for leveling, or in garden work by the do-it-yourself person. We expect that a maximum ratio of water to the total cementitious material in the mixture will be 0.45 at the most, and much lower when durability is an important criterion. This does not mean that the content of portland cement in the concrete will be very high; indeed, it will probably be lower than nowadays, because there will be hardly any mixtures that do not contain one or two other cementitious materials, as well as silica fume or metakaolin and various kinds of fillers. I hope that the cement manufacturers will come to realize that they have a better future with less portland cement per cubic meter, but with more cubic meters of concrete being used. Their rear-guard action of opposing the use of other cementitious materials, of fillers, and even of admixtures, which they are still pursuing in many countries, is not in their best interest, and it certainly is not in the interest of concrete users, and that means everyone.

Some of the materials needed to make good quality concrete are expensive. This is, for example, true of silica fume. And yet, silica fume is a waste product in the manufacture of silicon and ferrosilicon alloys from high purity quartz and coal in a submerged-arc electric furnace. There is very little further processing required to make the silica fume usable in concrete. Thus the price of silica fume is governed by demand and not by cost of production. Such a situation is absolutely appropriate in our system of economy, but this gives us scope for alternative materials.

One such material is rice husks, appropriately processed so as to produce a very fine powder with a high silica content. Rice husks are plentiful in many countries, but in the majority of those countries, the transportation and processing facilities are poor. On the other hand, the demand for a high quality, very fine pozzolanic material is mainly in the industrialized countries. There is thus a mismatch. However, this is likely to be resolved by technology transfer, and one large cement manufacturer is already involved in developing appropriate facilities for rice husk processing in Asia.

Superplasticizers are also expensive because of the high cost of producing long and heavy molecules, but there, too, ways of reducing cost are possible. For example, superplasticizers have been diluted with lignosulfonate water reducers, which are cheaper. This situation is, however, changing because modern paper mills use a technology that does not yield usable lignosulfonates as a by-product or produces lignosulfonates with so many impurities that they cannot be used as water reducers in concrete. No doubt, human ingenuity will find a way of producing reasonably-priced, highly effective water reducers.

Nevertheless, the cost of what we have called "good concrete" is likely to be higher than the cost of present-day concrete with a high water-cement ratio. This concrete, however, is not durable, and consequently often requires repairs, then repairs to the repairs, and possibly replacement. All this costs money, although the initial cost and the cost of repairs and maintenance may come from different purses. It is logical to look at the life-cycle cost, and we are beginning to use this approach. Once the value of life-cycle costing is fully appreciated, we shall, in the main, be producing good concrete.

Concrete, good, bad, or indifferent, has been the leading construction material for nearly a century. At the present time, in the world as a whole, it is used at an annual rate of 1 cubic meter or 2½ tonne per capita. There is no economically viable competitor in sight. We can therefore expect that high-performance concrete, or just good concrete, will serve the world for a long time to come.

References

1. Bamforth, P. B., "In-Situ Measurement of Effect of Partial Portland Cement Replacement Using Either Fly Ash or Ground Granulated Blast Furnace Slag on the Performance of Mass Concrete," *Proc. Inst. Civ. Engineers, Part 2,* Sept. 1980, pp. 777-800.

2. Neville, A. M., "Concrete Technology—An Essential Element of Structural Design," *Concrete International,* V. 20, No. 7, July 1998, pp. 39-41. [Section 6.1]

3. Aïtcin, P-C., "High Performance Concrete," First Edition, E & FN Spon, 1998.

4. Neville, A. M., "Aggregate Bond and Modulus of Elasticity," *ACI Materials Journal,* V. 4, No. 1, 1997, pp. 71-74. [Section 4.4]

5. Neville, A. M., "Properties of Concrete," Fourth Edition, Longman, London, 1995, and John Wiley, N.Y., 1996, 844 pp.

6. Aïtcin, P.-C.; Neville, A. M.; and Acker, P., "Integrated View of Shrinkage Deformation," *Concrete International,* V. 19, No. 9, 1997, pp. 35-41. [Section 4.5]

7. Pigeon, M., and Pleau, R., "Durability of Concrete in Cold Climates," First Edition, E & FN Spon, 1995.

Section 4.4: Aggregate Bond and Modulus of Elasticity

In discussing the properties of concrete, the topics of the bond between the aggregate and the matrix and of the modulus of elasticity of concrete are usually considered as separate items, even though they are not independent parameters. It is the contention of this paper that the two parameters are intimately linked and, in particular, that the difference between the moduli of elasticity of the aggregate particles and of the matrix (which influences bond stresses) strongly influences the mechanical behavior of concrete.

Research significance

Understanding the influence on the monolithic behavior of concrete of the differences between the moduli of elasticity of the aggregate particles and of the hardened cement paste is shown to make it possible to select a concrete mixture with desirable relevant properties.

Concrete as composite material

It is well established that concrete is a composite material. As far back as 1958, Hansen[1] suggested two fundamentally different models. The first of these, an ideal composite hard material, has a continuous matrix of an elastic phase with a high modulus of elasticity in which are embedded particles with a lower modulus of elasticity. The second model, an ideal composite soft material, has a continuous matrix with a low modulus of elasticity in which are embedded particles with a higher modulus of elasticity.

Methods of calculating the modulus of elasticity of composite materials corresponding to the two models have been discussed in Reference 2 and others. The two models represent the boundaries of behavior, neither of which can be achieved in practice because the requirements of equilibrium and compatibility cannot be simultaneously satisfied by either of the models. For this reason, a number of more complex models have been suggested.[2]

In concrete made with normal weight aggregate, where the modulus of elasticity of aggregate is higher than the modulus of elasticity of the surrounding matrix, the composite soft model gives a reasonable approximation to the actual behavior of concrete. A question not considered in the past is: what is the effect of the difference between the two moduli, not just on the modulus of elasticity of concrete, but on the entire stress-strain relation? The relevant factor is the bond beween the aggregate and the matrix, which is usually taken as mortar but in this paper as hardened cement paste.

Modulus of elasticity of concrete

The broad relation between the modulus of elasticity of concrete and its

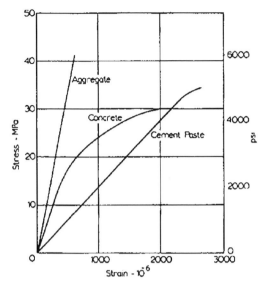

Fig. 4.4.1—Stress-strain relation for cement paste, aggregate, and concrete

compressive strength is well known but there is no agreement on the precise form of the relation. Indeed, there can be no unique relation because the modulus of elasticity of concrete is affected both by the modulus of elasticity of aggregate and by the volumetric content of aggregate in the concrete. While the latter is easily calculated, the modulus of elasticity of aggregate is rarely known. This is probably the reason why some expressions for modulus of elasticity of concrete, for example that of ACI 318-89 (Revised 1992),[3] allow for the modulus of elasticity of aggregate by a coefficient that is a function of density of concrete, usually unit mass raised to the power 1.5. Other expressions[4] give a power index of 2. Whatever the value of the power index, at a constant aggregate content by volume, the density of concrete increases with an increase in the density of aggregate. The consideration of density is equally applicable in the case of normal weight and lightweight aggregates, so that the ACI expression[3] can be used for concretes made with either type of aggregate. However, in the case of lightweight aggregate, its modulus of elasticity differs little from the modulus of elasticity of hardened cement paste, so that the volumetric content of aggregate in the concrete does not significantly affect the modulus of elasticity of lightweight aggregate concrete.[5]

None of the expressions for the modulus of elasticity of concrete considers the bond between the aggregate and the surrounding hardened cement paste. The bond depends on the interface zone, which is known to have a different microstructure from the bulk of the hardened cement paste and which is the locus of early microcracking, known as bond microcracking. Microcracking is relevant to the shape of the stress-strain relation of concrete, and it is this relation that influences the behavior of concrete rather than the modulus of elasticity, which is an idealized value of the ratio of stress to strain under fixed conditions.

Returning to the composite nature of concrete, we can observe that both the aggregate and the hardened cement paste, when separately subjected to load, have a sensibly linear stress-strain relation (Fig. 4.4.1). However, some suggestions about nonlinearity of the stress-strain relation of hardened cement paste have been made,[6] this arising from extremely fine cracking within the hardened cement paste at high strains. Nevertheless, the main explanation of the largely curvilinear stress-strain relation of concrete lies in the presence of interfaces between the aggregate and the hardened cement paste in which microcracks develop even under modest loading.[7] The progressive development of these microcracks was confirmed by neutron radiography.[8]

Aggregate-cement paste interface

At this stage, it may be appropriate to modify the concept of the composite nature of concrete, discussed earlier, by recognizing the interface as an element in the structural model of concrete.

Why is the interface zone different from the bulk of the hardened cement paste? The microstructure of the interface zone is greatly influenced by the situation that exists at the end of placing and compaction of concrete. At that stage, the particles of cement are unable to become closely packed against the relatively large particles of aggregate. This "wall effect" means that there is less cement present to hydrate and fill the original voids in the fresh mixture. In consequence, the hardened cement paste in the interface zone has a much higher porosity than the hardened cement paste further away from the particles of aggregate.[9] It is known that the higher the porosity the lower the strength.[10]

A brief description of the microstructure of the interface zone may be useful. The surface of the aggregate is covered with a layer of oriented crystalline $Ca(OH)_2$, about 0.5 µm thick, behind which there is a layer of calcium silicate hydrate, also about 0.5 µm thick. These layers are known as the duplex film. Further away from the aggregate, there is the main interface zone, some 50 µm thick, containing products of hydration of cement with larger crystals of $Ca(OH)_2$, but without any unhydrated cement.[11] The significance of this structure is twofold. First, the complete hydration of cement indicates that the water-cement ratio at the interface is higher than in the bulk of the hardened cement paste. Second, the presence of large crystals of $Ca(OH)_2$ indicates that the porosity at the interface is higher than in the bulk of the hardened cement paste; this confirms the "wall effect" referred to earlier.[10]

If pozzolans are included in the mixture, the strength of the interface zone increases with time. Silica fume, which has an extremely high fineness, is particularly effective and also insures superior packing.

The interface zone exists not only at the surface of coarse aggregate

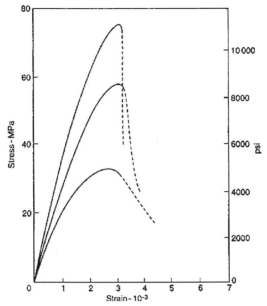

Fig. 4.4.2—Typical stress-strain curves for concretes of different strengths[15]

particles but also around fine aggregate particles.[12] Here, the thickness of the interface zone is smaller, but the various zones add up to a very considerable volume:[12] the total volume affected is between one-third and one-half of the total volume of the hardened cement paste.

Effects in high-strength concrete

It has been found that the use of large maximum aggregate size, even though it reduces the water requirement, is counterproductive when a high strength of concrete is sought.[13] The reason for this is the incompatibility between the aggregate and the hardened cement paste in terms of their moduli of elasticity, their Poisson's ratios, and other properties.[10] In consequence, more bond microcracking develops than when the maximum aggregate size is about 10 mm. A smaller maximum aggregate size also offers the benefit of a larger total surface area and therefore a lower bond stress, so that bond failure does not occur. Consequently, fracture surfaces pass through the coarse aggregate particles as well as through the hardened cement paste, both under compressive and under tensile loading.[14]

What is also important is that in high-strength concrete the strength of the hardened cement paste is, by definition, high because a very low water-cement ratio is used and the porosity of the hardened cement paste is very low. For this reason, the modulus of elasticity is high. There is therefore a smaller difference between the modulus of elasticity of the hardened cement paste and the modulus of elasticity of aggregate. This situation results in a better bond stress and a more monolithic behavior of concrete. It is arguable that the use of aggregate with a low modulus of elasticity would be beneficial with respect to bond stress and therefore to the strength of concrete, other factors remaining constant.

It has also been observed that the linear part of the stress-strain curve in high-strength concrete extends to a stress that may be as high as 85% of the ultimate strength, or even higher (Fig. 4.4.2). Subsequent fracture of concrete takes place through the coarse aggregate particles, as already mentioned; these particles cannot therefore act as crack arresters (as is the case in normal concrete) and failure is sudden[15] or even explosive.

It is useful to observe one further consequence of the good bond between aggregate and the hardened cement paste in high-strength concrete: the modulus of elasticity of the aggregate has a greater influence on the modulus of elasticity of concrete than is the case in concretes with higher water-cement ratios.[16] Consequently, there is no simple relation between the modulus of elasticity of high-strength concrete and its strength[16] so that, for structural design purposes, the modulus of elasticity of high-strength concrete should not be assumed to have a fixed relation to its compressive strength.

Effects in lightweight aggregate concrete

Another type of concrete in which the bond between the aggregate and the hardened cement paste is particularly good is lightweight aggregate concrete. One reason for this is the rough surface texture of many lightweight aggregates so that there is mechanical interlocking between the two materials. Indeed, often there is some penetration of the fresh cement paste into the open surface pores in the coarse lightweight aggregate particles. More importantly, lightweight aggregate particles have a low modulus of elasticity so that the difference between the moduli of the aggregate particles and of the surrounding hardened cement paste is smaller than when normal weight aggregate is used. Consequently, the differential stresses between the aggregate and the hardened cement paste are low, and the concrete behaves monolithically.

There is one other beneficial aspect with respect to bond of using lightweight aggregate. Lightweight aggregate usually absorbs some mixing water; with time, this water becomes available for the hydration of the hitherto unhydrated remnants of the cement particles. Such hydration takes place in the interface zone (because this is in the vicinity of the absorbed water) so that the bond between the aggregate and the hardened cement paste improves still further.

In this connection, Bremner and Holm[17] observed that air entrainment lowers the modulus of elasticity of mortar, thus bringing the value of its modulus nearer to the modulus of elasticity of lightweight aggregate. This reduction in the difference between the moduli is conducive to a better stress transfer between the aggregate particles and the matrix.

One consequence of the very good bond between lightweight aggregate

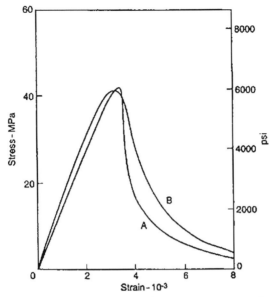

Fig. 4.4.3—Stress-strain curves for lightweight aggregate concretes; A = all lightweight, B = normal weight fine aggregate[19]

and the hardened cement paste is the absence of the development of early bond microcracking. As a result of this, the stress-strain relation is linear up to a stress as high as 90% of the ultimate strength[18] (Fig. 4.4.3). This behavior is similar to that of high-strength concrete. There is also little microcracking induced by temperature cycling.[10]

Another similarity in behavior between lightweight aggregate concrete and high-strength concrete is in the effect of good bond on the modulus of elasticity of concrete: the modulus of elasticity of the aggregate has a strong influence on the modulus of elasticity of concrete. However, unlike high-strength concrete, because the elastic properties of lightweight aggregate are affected by its voids content, which also influences the unit mass of concrete, the modulus of elasticity of lightweight aggregate concrete can be expressed as a function of density of the concrete as well as of its compressive strength.[10]

Concluding remarks

It may be instructive to consider now, in a more general manner, the bond between the aggregate and the surrounding hardened cement paste in three categories of concrete— ordinary normal weight concrete, high-strength concrete, and lightweight aggregate concrete—in so far as the bond is affected by the moduli of elasticity of the aggregate and of the hardened cement paste. In ordinary normal weight concrete, the modulus of elasticity of the typical hardened cement paste is generally much lower than the modulus of elasticity of the aggregate particles. In high-strength concrete, the hardened cement paste has a very much higher modulus of elasticity so that the difference between it and the modulus of elasticity of the aggregate is much smaller. In lightweight aggregate concrete, the modulus of elasticity of the aggregate is much lower than the modulus of elasticity of the normal weight aggregate; consequently, the difference between the moduli of elasticity of the lightweight aggregate and of the hardened cement paste is small. In consequence, both in high-strength concrete and in lightweight aggregate concrete, the stress concentration at the aggregate-matrix interface is reduced.

It should be added, however, that when the moduli of elasticity of the aggregate and the hardened cement paste approach one another, the resulting concrete has a more linear stress-strain relation; that is, the concrete may exhibit an increased brittleness.

A practical conclusion from the previous comments on the moduli of elasticity of the components of concrete is that when it is desirable to produce concrete with a strong monolithic action (for example, when temperature cycling is expected), means of reducing the difference between the moduli of elasticity of the aggregate and the hardened cement paste should be sought; in other words, lowering the former and raising the latter are desirable. Achieving this is feasible by using an appropriate aggregate and by incorporating silica fume and superplasticizer in a mixture with a very low water-cement ratio.

References

1. Hansen, T. C., "Creep of Concrete," *Bulletin* No. 33, Swedish Cement and Concrete Research Institute, Stockholm, 1958, 48 pp.

2. Neville, A. M.; Dilger, W. H.; and Brooks, J. J., *Creep of Plain and Structural Concrete*, Construction Press, Longman Group, London, 1983, 361 pp.

3. ACI Committee 318, "Building Code Requirements for Reinforced Concrete and Commentary (ACI 318-89/ACI 318R-89) (Revised 1992)," American Concrete Institute, Farmington Hills, Mich., 1995, 345 pp.

4. Lydon, F. D., and Balendran, R. V., "Some Observations on Elastic Properties of Plain Concrete," *Cement and Concrete Research*, V. 16, No. 3, 1986, pp. 314-324.

5. Klieger, P., "Early High-Strength Concrete for Prestressing," *Proceedings*, World Conference on Prestressed Concrete, San Francisco, Calif., 1957, pp. A5-1 to 14.

6. Attiogbe, E. K., and Darwin, D., "Submicrocracking in Cement Paste and Mortar," *ACI Materials Journal*, V. 84, No. 6, Nov.-Dec. 1987, pp. 491-500.

7. Shah, S. P., and Winter, G., "Inelastic Behavior and Fracture of Concrete," *Symposium on Causes, Mechanisms, and Control of Cracking in Concrete*, SP-20, American Concrete Institute, Farmington Hills, Mich., 1968, pp. 5-28.

8. Najjar, W. S., and Hover, K. C., "Neutron Radiography for Microcrack Studies of Concrete Cylinders Subjected to Concentric and Eccentric Compressive Loads," *ACI Materials Journal*, V. 86, No. 4, July-Aug. 1989, pp. 354-359.

9. Scrivener, K. L., and Gariner, E. M., "Microstructural Gradients in Cement Paste around Aggregate Particles," *Materials Research Symposium Proceedings*, V. 114, 1988, pp. 77-85.

10. Neville, A. M., *Properties of Concrete*, Fourth Edition, Longman, London, 1995, and John Wiley, N.Y., 1996, 844 pp.

11. Larbi, L. A., "Microstructure of the Interfacial Zone around Aggregate Particles in Concrete," *Heron*, V. 38, No. 1, 1993, 69 pp.

12. Monteiro, P. J. M.; Maso, J. C.; and Ollivier, J. P., "The Aggregate-Mortar Interface," *Cement and Concrete Research*, V. 15, No. 6, 1985, pp. 953-958.

13. Aïtcin, P.-C., and Neville, A. M., "High-Performance Concrete Demystified," *Concrete International*, V. 15, No. 1, Jan. 1993, pp. 21-26. [Section 4.2]

14. Remmel, G., "Study of Tensile Fracture Behavior by Means of Bending Tests on High-Strength Concrete," *Darmstadt Concrete*, No. 5, 1991, pp. 95-115.

15. Slate, F. O., and Hover, K. C., "Microcracking in Concrete," *Fracture Mechanics of Concrete: Material Characterization and Testing*, A. Carpinteri and A. R. Ingraffen, eds., Martinus Nijhoff, The Hague, 1994, pp. 137-159.

16. Baalbaki, W.; Aïtcin, P.-C.; and Ballivy, G., "On Predicting Modulus of Elasticity in High-Strength Concrete," *ACI Materials Journal*, V. 89, No. 5, Sept.-Oct. 1992, pp. 517-520.

17. Bremner, T. W., and Holm, T. A., "Elasticity, Compatibility, and the Behavior of Concrete," ACI Journal, *Proceedings* V. 83, No. 2, Feb. 1986, pp. 244-250.

18. Zhang, M.-H., and Gjorv, O. E., "Mechanical Properties of High-Strength Lightweight Concrete," *ACI Materials Journal*, V. 88, No. 3, May-June 1991, pp. 240-247.

19. Siebel, E., "Ductility of Normal and Lightweight Concrete," *Darmstadt Concrete*, No. 3, 1988, pp. 179-187.

Letter to the Editor

The key theme of the paper, as stated by the author, is that "the main explanation of the largely curvilinear stress-strain relation of concrete lies in the presence of interfaces between the aggregate and the hardened cement paste in which microcracks develop even under modest loading." That interfaces play an important role, there is no doubt. That they represent the "main explanation" is far from certain. This writer would like to raise three points: (1) unlike the example given in Fig. 4.4.1, the stress-strain curve of cement paste is highly nonlinear, even at low stresses; (2) bond per

4-50   Chapter 4: Special Aspects

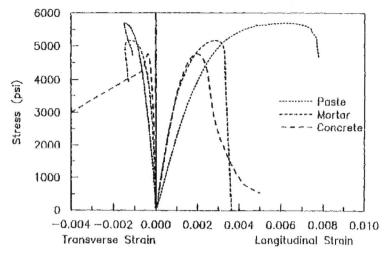

Fig. 4.4.4—Stress versus longitudinal and transverse strain for w/c = 0.5 cement paste, mortar, and concrete under monotonic loading (A) (1 psi = 6.895 kPa)

se has no effect on the initial modulus of elasticity, but will affect the modulus at higher stresses; and (3) it is the heterogeneous nature of concrete, not the presence of interfaces alone, that dominates the nonlinear stress-strain behavior of the composite material.

The stress-strain curve for cement paste illustrated in Fig. 4.4.1 of the paper, in which the linear portion reaches 90 percent of the compressive strength, bears little resemblance to stress-strain curves of bulk cement paste. Such a curve is illustrated in Fig. 4.4.4 (A), along with curves for mortar and concrete cast with the same cement and water/cement ratio. Other examples of the nonlinear stress-strain behavior of hydrated cement paste can be seen in References 6, B, and C. Thus, the argument can be made that the nonlinear stress-strain behavior of concrete is closely tied to the nonlinear behavior of the matrix portion of the composite material. As discussed in Section 2.2 of ACI 224R-90 (D) and in References E and F, analytical attempts to obtain a curvilinear stress-strain response for finite element models of concrete have succeeded only when the mortar constituent is modeled as a nonlinear material. Modeling microcracks as the only nonlinear behavior results in a nearly linear stress-strain response. Clearly, microcracking at the surface of aggregates contributes to the nonlinearity of concrete, but the contribution has not been shown to constitute the main explanation.

While changes in bond strength affect the overall stress-strain behavior of concrete, they do not apparently affect the initial modulus of elasticity. This is illustrated in Fig. 4.4.5. The figure shows stress-strain curves for two concretes—one with normal coarse aggregate and one with coarse aggregate for which the bond to the mortar has been

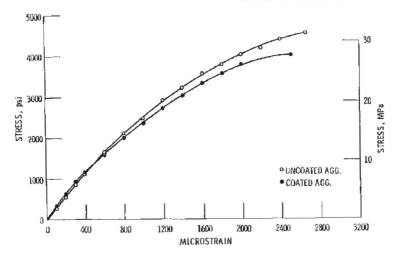

Fig. 4.4.5—Stress-strain curves for concrete comparing the effects of uncoated and coated coarse aggregate (G)

reduced by a thin coating of polystyrene. The curves in Fig. 4.4.5 demonstrate that a major reduction in the bond strength between mortar and coarse aggregate has no effect on the initial modulus of elasticity and a small but measurable effect on the overall response of concrete. Thus, changes in bond strength affect the modulus of elasticity to the extent that full stress-strain response is altered. At low stress, however, no effect is discernible. References E and H describe analytical studies that further support this point (also discussed in Refs. D and F).

The writer would like to suggest that the effects of the heterogeneous nature of concrete go beyond the aggregate-cement interface. The writer agrees with the author that differences in the moduli of elasticity between cement paste and aggregate play a crucial role. The closer the moduli, the lower the bond stresses. But also, the closer the moduli, the lower the stress concentrations due to differences in the properties of the materials. For concrete containing normal weight aggregate, the higher relative modulus of the aggregate causes some portions of the paste to be stressed more highly than others. The regions of cement paste that are stressed more highly are subjected to higher strains. As a result, these regions exhibit reduced incremental stiffness (Fig. 4.4.4) and, thus, contribute more to the nonlinear stress-strain behavior of the composite material. The greater the differences in the moduli, the greater the nonlinear behavior of the composite. For both lightweight and high-strength concretes, the moduli are closer, which results in more uniform stresses within the cement paste constituent, as well as the lower bond stresses described by the author. This delays the highly nonlinear stress-strain behavior of the cement paste constituent and, along with the reduction in bond microcracking, results in a longer linear region, as illustrated in Fig. 4.4.2 and 4.4.3 of the paper.

David Darwin
Ackers Professor of Civil Engineering and Director
Structural Engineering and Materials Laboratory
University of Kansas
Lawrence, Kans.

References

A. Martin, J. L.; Darwin, D.; and Terry, R. E., "Cement Paste, Mortar and Concrete under Monotonic, Sustained and Cyclic Loading," *SM Report* No. 31, University of Kansas Center for Research, Lawrence, Kans., Oct. 1991, 161 pp.

B. Attiogbe, E. K., and Darwin, D., "Strain Due to Submicrocracking in Cement Paste and Mortar," *ACI Materials Journal,* V. 85, No. 1, Jan.-Feb. 1988, pp. 3-11.

C. Harsh S.; Shen, Z.; and Darwin, D., "Strain-Rate Sensitive Behavior of Cement Paste and Mortar in Compression," *ACI Materials Journal,* V. 87, No. 5, Sept.-Oct. 1990, pp. 508-516.

D. ACI Committee 224, "Control of Cracking in Concrete Structures, (ACI 224 R-90)," American Concrete Institute, Farmington Hills, Mich., 43 pp.

E. Maher, A., and Darwin, D., "Microscopic Finite Element Model of Concrete," *Proceedings,* First International Conference on Mathematical Modeling, St. Louis, Mo., Aug. 29-Sept. 1, 1977, V. III, pp. 1705-1714.

F. Darwin, D., "The Interfacial Transition Zone: 'Direct' Evidence on Compressive Strength," *Microstructure of Cement-Based Systems/Bonding and Interfaces in Cementitious Materials,* S. Diamond et al., eds., Materials Research Society Symposium Proceedings, V. 370, 1995, pp. 419-427.

G. Darwin, D., and Slate, F. O., "Effect of Paste-Aggregate Bond Strength on Behavior of Concrete," *Journal of Materials,* ASTM, V. 5, No. 1, Mar. 1970, pp. 86-98.

H. Stankowski, T; Runesson, K.; Sture, S.; and Willam, K. J., "Simulation of Progressive Failure of Particle Composites," *Micromechanics of Failure of Quasi-Brittle Materials,* S. P. Shah et al., eds., Elsevier Applied Science, London and N.Y., 1990, pp. 285-294.

Neville's Response

I am grateful to David Darwin for saying that he "agrees with the author that differences in the moduli of elasticity between cement paste and aggregate play a crucial role" in the monolithic behavior of concrete and also for agreeing that the closer the moduli the lower the stress concentrations due to differences in the properties of the materials. The introduction to my paper says that bond between the aggregate and the matrix and the modulus of elasticity of concrete are intimately linked. The research significance of the paper, as stated, was to make it possible to select a concrete mixture with desirable properties to achieve monolithic behavior. I tried to show that this situation exists both in lightweight aggregate and in high-strength concretes; with this Darwin agrees.

However, Darwin challenges my views on the linearity of the stress-strain relation of hardened cement paste. His views on non-linearity are acknowledged in the paper (Ref. 6). Of the references in his discussion, six are by Darwin, but there is no additional experimental evidence. If such evidence becomes available, it might show that the presence of interfaces between the aggregate and the cement paste is not the main explanation of the largely curvilinear stress-strain relation (as I wrote) but only that "the interfaces play an important role" (as stated by Darwin). Future experiments alone will tell. The failure of the finite element models of mortar to support my point of view is not a valid argument because mortar contains interfaces, too. As for hardened cement paste, Darwin refers to his tests on cement paste with a water-cement ratio or 0.5; at such a consistency, the paste does not simulate the behavior of cement paste in concrete.

The purpose of my paper was to help users of concrete to select a satisfactory mixture for practical situations in which a strong monolithic action is required, by considering bond between the aggregate and the hardened cement paste and also the stress concentrations arising from the difference in the moduli of elasticity of true two components of concrete. My reference to "the differential stresses between the aggregate and the hardened cement paste" is echoed by Darwin when he says "the closer the moduli the lower the stress concentrations due to differences in the properties of the materials." His support is satisfying.

In summary, I feel that my attempt, not previously made, to show a link between the behavior of high-strength concrete and lightweight aggregate concrete with a view to practical application has been supported by Darwin. For this I thank him.

Adam Neville

Section 4.5: Integrated View of Shrinkage Deformation*

Shrinkage is a seemingly simple phenomenon of contraction of concrete upon loss of water. Strictly speaking, shrinkage is a three-dimensional deformation but it is usually expressed as a linear strain because in the majority of exposed concrete elements one or two dimensions are much smaller than the third dimension, and the effects of shrinkage are greatest in the largest dimension. In common usage, the term shrinkage is a shorthand expression for drying shrinkage of hardened concrete exposed to air with a relative humidity of less than just under 100%. However, there exist several other types of shrinkage deformation in concrete which, depending on circumstances, may or may not occur simultaneously or be independent of one another.

This section offers some practical suggestions for minimizing the often very harmful consequences of shrinkage of concrete, and also addresses high-performance concrete (HPC).

The five types of shrinkage

Drying shrinkage, which occurs in hardened concrete, is best known. Chronologically, drying shrinkage is preceded by plastic shrinkage, which occurs when water is lost from concrete still in a plastic state. Commonly, the water is lost by evaporation to the atmosphere, but the loss can also occur by suction of underlying dry concrete or soil.

Shrinkage deformation in hardening concrete occurs also as a consequence of ongoing hydration of cement. Because this type of shrinkage occurs within a concrete mass, that is, without contact with the ambient medium, it is often called self-desiccation shrinkage. An equally good description of contraction under such circumstances is given by the term autogenous shrinkage, which will be used here to achieve a symmetry of terminology with respect to all the types of shrinkage deformation. Occasionally, the term chemical shrinkage is used, but this has little merit.

Shrinkage deformation also occurs as a result of a decrease in temperature of concrete from the temperature at the time of setting, or soon thereafter, when the overall dimensions of a concrete element or mass become fixed. Strictly speaking, this deformation should be called thermal contraction but, once again for reasons of symmetry, the term thermal shrinkage will be used in this paper.

In addition, the reaction of hydrated cement paste with carbon dioxide in the air, in the presence of moisture, causes carbonation shrinkage. When all or some of these types of shrinkage occur, their sum is referred to as total shrinkage.

---

*This section was co-authored by P.-C. Aïtcin and P. Acker.

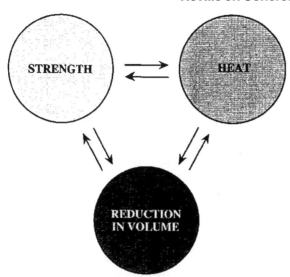

Fig. 4.5.1—The "eternal triangle" of hydration: strength, heat, and reduction in volume of the hydrating cement paste system

To appreciate fully the various mechanisms of shrinkage it is necessary to understand the hydration of cement and its physical, mechanical, and thermodynamic consequences. It is on the basis of such an understanding that it is possible to take appropriate measures to reduce the various types of shrinkage or to attenuate their consequences.

Hydration of cement and autogenous shrinkage

The term hydration of cement is a global description of several phenomena the root of which is the chemical reaction of portland cement with water. This reaction results in the formation of a cohesive and adhesive solid—hydrated cement paste, the essential element of strength of concrete. The reactions of hydration entail the generation of heat and a reduction in the volume of the hydrating cement paste system. Thus the two concomitants of the development of strength are heat and a reduction in the volume of the cement past system.

This "eternal triangle" is represented schematically in Fig. 4.5.1 in order to emphasize the concomitant nature of the three phenomena, two of which—the development of heat and the reduction in the volume of the hydrating cement paste system—may be considered to be harmful. They are, nevertheless, tolerable in that engineers have learned to manage their effects in concrete structures. Were concrete a material which swells upon hydration and whose temperature simultaneously decreases, the engineer's task would undoubtedly be much more difficult.

While the global terms "hydration of cement" and "heat of hydration" of cement are very useful, they could be misleading without a recognition that portland cement is a multiphase material whose composition can vary over a wide range. Currently, when portland cement is often used in conjunction with other cementitious materials, the situation is even more complex. Nonetheless, all the chemical reactions involved are exothermic in nature so that they result in an increase in the temperature of concrete unless externally induced heat extraction produces isothermal conditions or even a reduction in the temperature.

The main compounds of portland cement are tricalcium silicate (C_3S), dicalcium silicate (C_2S), tricalcium aluminate (C_3A), and tetracalcium aluminoferrite (C_4AF); also, calcium sulfate has to be added to the cement clinker in order to prevent too rapid a hydration of C_3A. Portland cement also contains so-called minor compounds and impurities such as alkali sulfates, free lime, unreacted silica, and magnesia; their influence is not always negligible.

In so far as hydration is concerned, the two calcium silicates result most probably in the same products: calcium silicate hydrate (with an uncertain and variable stoichiometric composition, described in abbreviated form as C-S-H) and calcium hydroxide [$Ca(OH)_2$], known also as portlandite. The C_3A, in the presence of calcium sulfate and water, reacts to form ettringite ($C_3A \cdot 3CaSO_4 \cdot 32H_2O$); when no more calcium sulfate is available, a metastable calcium monosulfoaluminate ($C_3A \cdot CaSO_4 \cdot 12H_2O$) is formed, and finally a stable calcium aluminate hydrate (C_3AH_6). As for C_4AF, it also reacts with calcium sulfate, but more slowly than C_3A, forming calcium sulfoaluminate and calcium sulfoferrite; the final products of hydration are C_3AH_6 and C_3FH_6.

The amount of heat generated and the development of concrete strength are influenced by several factors. The main ones are: the respective proportions of the four main compounds in portland cement, the specific surface of the cement, the initial temperature of the concrete, the ambient temperature during the progress of hydration, and the mass and shape of the concrete element (which control the heat flow to outside). Because the development of strength and of temperature are affected by the same parameters, it follows that the development of strength of concrete can be assessed from the development of its temperature. This is the principle of maturity meters. From the "eternal triangle" of Fig. 4.5.1, it follows that, in the reverse direction, temperature development can be assessed from strength, but this is of little practical importance. A note of warning should be added about the effect that temperature at the time of setting has on strength because it is then that the overall dimensions of the concrete element become fixed.

Because the calcium silicates represent the bulk of portland cement, the volumetric change that occurs during their hydration is of paramount importance; nevertheless, what follows refers in a general manner to the reduction in the volume of the whole hydrating cement paste system. If volume C of dry cement powder reacts with volume W of water, this being the nonevaporable water (that is, more or less chemically combined), the resultant volume P of the products of hydration is always such that $P < C + W$. There is some uncertainty about the precise magnitude of the reduction in the volume of the hydrating cement paste system, that is, the sum of the volume of the solid products of hydration and of the water-filled gel space, but exclusive of the volume of the capillary pores.

More than 100 years ago, Le Chatelier[1] estimated the volume reduction to be between 8 and 12% of the original space occupied by unhydrated cement and by water which was destined to become part of the hydrated cement paste system. Nearly 50 years ago, Powers[2] found the volume reduction to be 0.254 of the volume of nonevaporable water. He also found that the nonevaporable water represents about 23% of the mass of anhydrous cement (which has a specific gravity of about 3.15) and that the hydrated cement paste has a characteristic porosity of 28%. The volume reduction (considering 100 g of dry cement powder which hydrates fully) can be calculated as follows:[3]

Mass of dry cement = 100 g
Absolute volume of dry cement = 100/3.15 = 31.8 mL
Mass of nonevaporable water = 23.0 g
Volume of solid products of hydration = $31.8 + 0.23 \times 100 (1 - 0.254) = 48.9$ mL
Volume of gel water, w, is given by $w/(48.9 + w) = 0.28$
whence $w = 19.0$ mL.

Thus the volume of the hydrated cement paste system is: $48.9 + 19.0 = 67.9$ mL.

This value should be compared with the original volume of unhydrated cement (31.8 mL), nonevaporable water (23.0 mL), and gel water (19.0 mL), all these totaling 73.8 mL. Thus, the reduction in the volume of the hydrated cement paste system is: $(73.8 - 67.9)/73.8 = 8.0\%$. This autogenous shrinkage is the same as Le Chatelier's lower value for the volume reduction. This volume reduction is physically present as empty capillary pores distributed throughout the hydrated cement paste.

It may be worth noting that, because the products of hydration can form only in water-filled space, only a part of the water in the capillary system can be used up in hydration. For hydration to take place, there must be enough water present both for the chemical reactions and for filling of gel pores. Thus, if the water-cement ratio (w/c) is 0.42 (as in the example given) or greater, full hydration of cement is possible. If, however, the w/c is lower, at some stage of hydration there will not be enough water left to saturate the solid surfaces of the capillary pores; hydration comes to a halt when the vapor pressure in the capillary pores falls below about 0.8 of the saturation pressure.[4] This can be referred to as self-desiccation. However, this term is correct only if the concrete is isolated from an external moisture source or from a moisture sink. Such isolation exists when a concrete element is perfectly sealed; in practical terms the situation exists also in the

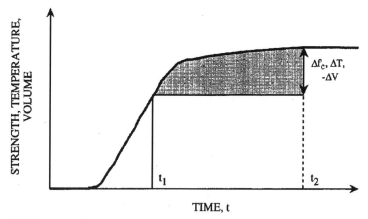

Fig 4.5.2—Schematic representation of changes in compressive strength (f_c'), temperature (ΔT), and volume of the hydrating cement paste system ($-\Delta V$) with time under adiabatic conditions

interior of a large concrete mass. In those cases, Le Chatelier's or Powers' reduction in the volume of the hydrated cement paste system takes place.

On the other hand, if water can enter into the hydrating cement paste from outside, hydration will continue until there is not enough space left to accommodate the products of hydration (remembering that $P > C$). This occurs when the w/c is smaller than about 0.38. It can be added that the presence of remnants of unhydrated cement is not disadvantageous; indeed, unhydrated cement is an excellent, albeit expensive, "aggregate."

Thus, when concrete is continuously wet-cured, the capillary system is always full of water so that hydration proceeds uninterrupted. From the "eternal triangle" of Fig. 4.5.1, it follows that strength increases, heat is generated, and the solid volume of the hydrating cement paste system decreases.

Hydration, strength, and heat

We shall now consider the other two concomitant effects of the hydration of cement. A mechanical effect is the development of strength. From the thermodynamic standpoint, heat is generated and this is relevant to thermal shrinkage.

Strength

Although the increase in the strength of concrete with the passage of time is well known, the fundamental causes of this increase are less well understood. Even though progress in scanning electron microscopy has made it possible to observe phase changes during the hardening of concrete, the detailed structure of the main strength-giving hydrate, C-S-H, for a long time known to be layered, is even now much less well-established than in the case of two closely related silicates: kaolinite (hydrated aluminum

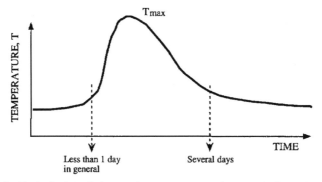

Fig. 4.5.3—Typical plot of change in temperature of concrete in a structural element

silicate) and chrysotile asbestos (hydrated magnesium silicate). Moreover, it is not well understood how the individual C-S-H crystallites are joined together to give strength to the cement paste and finally to concrete.

What is important and can be stated unequivocally is that, if there is a strength gain during a certain interval of time, there is a concomitant generation of heat and a concomitant reduction in volume of the hydrating cement paste system. The generation of heat may or may not result in a temperature increase, depending on the thermodynamic conditions of the concrete. Likewise, the reduction in the volume of the hydrating cement paste system may or may not result in overall shrinkage, depending on the curing conditions. If the concrete specimen is sealed and is cured under adiabatic conditions, then, during the time interval considered, the changes in strength (an increase), in temperature (an increase), and in the volume of the hydrating cement paste system (a decrease) all follow qualitatively the same curve, as shown in Fig. 4.5.2.

Heat

The hydration of cement is always accompanied by a generation of heat which results in an increase in the temperature of concrete. However, the magnitude of this increase depends on a number of factors. These include the cement content in the mixture, the type of cement, the thermal properties of the aggregate, the temperature of the concrete at the time of placing, the ambient temperature, the thermodynamic conditions of curing, and the shape and size of the concrete element. From the thermal standpoint there are two extreme conditions: isothermal curing (that is, a constant temperature) and adiabatic curing (that is, no heat exchange with the exterior). In practice, the concrete follows conditions somewhere between these two extremes.

Usually, there is a short period during which the temperature rise is negligible. This is followed by a period of an increase in temperature, more or less rapid and more or less intense. Finally, there is a prolonged period

4-60 Chapter 4: Special Aspects

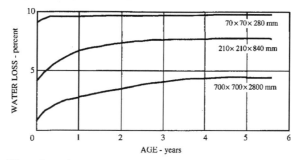

Fig. 4.5.4—Water loss (expressed as a percentage of total volume of mixing water) from concrete prisms of various sizes exposed to air with a relative humidity of 55%[5]

during which the temperature of the concrete gradually reduces to the ambient temperature. These changes in temperature are shown in Fig. 4.5.3, which will be useful in the discussion of appropriate ways of curing concrete so as to attenuate the effects of the various types of shrinkage.

Drying and autogenous shrinkage

For clarity, we should define the apparent volume of a concrete element as that contained within its external surfaces, regardless of the internal structure or porosity. The solid volume is that part of the apparent volume which is occupied by solid matter.

As far as hydrated or hydrating cement paste is concerned, the solid matter is that consisting of unhydrated cement and of the products of hydration inclusive of gel pores (full of water) but exclusive of capillary pores (whether full of water or empty). In the case of concrete, the solid matter also includes aggregate (assumed, for the present purposes, to have a zero porosity).

While the mass of solid matter may remain constant, the volume occupied by it can change. This occurs when the hydration of cement continues. On the other hand, when concrete is drying, hydration will slow down, but the loss of water results in a decrease in the apparent volume in consequence of drying shrinkage. A change in apparent volume is also caused by a variation in temperature: concrete has a positive coefficient of thermal expansion.

In very young concrete, it is possible for a decrease in the volume occupied by solid matter to be accompanied by an increase in the apparent volume: this occurs when concrete is permanently submerged in water as autogenous shrinkage is then accompanied by swelling (the opposite of drying shrinkage) through ingress of water into capillary pores. However, this expansion is small and stabilizes rapidly. A typical value for concrete is 150×10^{-6}, and much higher in cement paste.

The combination of autogenous shrinkage and an absence of drying shrinkage is of practical interest, as discussed later in this section. However, as soon as concrete is allowed to dry, even at an advanced age, drying shrinkage occurs.

Because autogenous shrinkage may or may not be accompanied by drying shrinkage, it is important to consider the phenomena involved. The driving force for drying shrinkage is evaporation of water from capillary pores in hydrated cement paste at their ends that are exposed to air with a relative humidity lower than that within the capillary pores. The water in the capillary pores, called free water, is held by forces which are stronger the smaller the diameter of the capillary pore. Therefore, the loss of water is progressive and proceeds at a decreasing rate as shown in Fig. 4.5.4.

From Fig. 4.5.4 it can be seen that the loss of water, expressed as a percentage of the apparent volume of concrete, is smaller the lower the surface/volume ratio of the concrete element. Other factors influencing the magnitude of the loss of water are the porosity of the concrete and the characteristics of the capillary pore system in the hydrated cement paste such as the size and shape of the pores and their continuity. As already mentioned, the relative humidity of the ambient air is also a factor.

From the practical standpoint, it is not the presence of drying shrinkage that matters but the occurrence of cracking caused by drying shrinkage. It is only when the tensile stress induced in the hydrated cement paste by the capillary forces exceeds the local tensile strength of the concrete that cracking occurs. It is possible also for autogenous shrinkage to induce cracking in a similar manner. There is, however, a difference between the two cases. Autogenous shrinkage develops isotropically within the concrete mass, provided that the distribution of the original cement grains is uniform in space. On the other hand, drying shrinkage always starts at the surface of the concrete, and indeed only at a surface exposed to unsaturated air; this could be a single surface or all the surfaces of a concrete element. The tensile forces in the concrete near the surface are balanced by compressive forces in the interior, which are relieved as the exterior part of the concrete undergoes cracking or as creep takes place.[6]

Brief reference should be made to carbonation shrinkage which takes place in a very thin surface layer of concrete exposed to air at a relative humidity of 30 to 70%. Under conditions of alternating drying and wetting, both carbonation shrinkage and drying shrinkage can occur and can cause shallow cracking, known as crazing.[3]

The various types of shrinkage discussed in this section have their locus in hydrated cement paste, but concrete, of course, also contains aggregate particles, which indeed occupy the major part of the volume of concrete. Aggregate does not undergo drying shrinkage, autogenous shrinkage, or plastic shrinkage, and its action is to resist the contraction induced by the hydrating cement paste. However, aggregate does undergo thermal shrinkage, although the coefficient of thermal expansion of aggregate is lower than that of hydrated cement paste.

The restraining action of aggregate with respect to shrinkage of hydrated cement paste is of great importance: neat cement paste would undergo shrinkage of such magnitude that this material could not be used structurally. To put it another way, if cement were free and aggregate expensive, we would still need to make concrete with the usual mixture proportions. It should be added, however, that at the interfaces between coarse aggregate particles and hydrated cement paste there is a certain incompatibility of strain, and hence bond microcracking may develop.

Although in well-mixed concrete the aggregate particles are distributed uniformly, the thin layer of mortar at a formed or finished surface contains few coarse aggregate particles. Thus, the drying shrinkage of hydrated cement paste near the surface is less restrained than deeper within the concrete element; hence, more drying shrinkage can actually develop at the surface and more cracking can take place than would be the case at, say, a sawn surface. A particular problem occurs in HPC, which has a higher cement content and a smaller maximum size of aggregate than ordinary concrete.[7] Hence, wet curing of HPC is particularly important; this is discussed in a later subsection.

Total shrinkage of concrete and curing

We should now consider together the various types of shrinkage which can occur in concrete. We can define total shrinkage as the sum of drying shrinkage, autogenous shrinkage, and thermal shrinkage, including cracks caused by shrinkage. These types of shrinkage are additive but there can be a certain interaction between them.

Plastic shrinkage, by definition, occurs prior to setting of concrete. The magnitude of plastic shrinkage is affected by the amount of water lost from the surface of the concrete, which is influenced by temperature, ambient relative humidity and wind velocity. However, the rate of loss of water does not by itself predict the magnitude of the plastic shrinkage, which also depends on the rigidity of the fresh concrete.

Essentially, if the amount of water lost per unit area exceeds the amount of water brought to the surface by bleeding, plastic shrinkage cracking will occur. This is a particular problem in HPC, which exhibits minimal bleeding, and will be considered in a later section. Using proper procedures, plastic shrinkage cracking can be avoided and reworking the concrete surface at an appropriate time can close the plastic shrinkage cracks, should these develop. Nevertheless, upon subsequent drying, these cracks can reopen as drying shrinkage cracks and can contribute to total shrinkage.

The development of autogenous and drying shrinkage can be delayed by wet curing, and prolonged curing can help avoid the development of cracking. Wet curing should be started as soon as hydration begins to take place. This is particularly important with rapid-hardening cements.

Wet curing should continue long enough to defer the onset of excessive autogenous shrinkage until the tensile strength of concrete is high enough to resist cracking. There is no conflict there because the increase in strength is

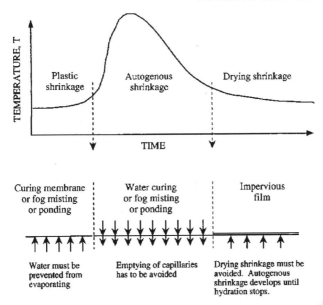

Fig. 4.5.5—Schematic representation of a desirable curing pattern

a direct function of the duration of wet curing.

To eliminate drying shrinkage it is not necessary to provide a source of water to be imbibed by concrete; it is only necessary to prevent evaporation of water from the surface of the concrete. Therefore, an application of an impervious film is sufficient. This, of course, represents an additional cost that may or may not be justified by the consequences of allowing shrinkage cracking to develop. The justification is likely to be in terms of durability and possibly also on aesthetic grounds.

It is worth noting that water curing has also a cooling effect so that the period of water curing should be adjusted to match the temperature–time curve. A schematic indication of the desirable curing pattern is shown in Fig. 4.5.5. It can be seen that to prevent plastic shrinkage a curing membrane should be used; alternatively, fog misting or ponding can be applied. During the period of temperature rise and early cooling, water curing is desirable, although fog misting or ponding can also be used. Following this period, an impervious film should be applied.

To apply this guidance in practice is difficult as curing is notoriously poorly executed, if not omitted altogether, despite the inclusion in all specifications of clear clauses prescribing curing. The remedy lies in making curing an item in the contract that is paid for separately and specifically. While this represents cost to the owner, this is likely to be outweighed by the benefits of improved durability and enhanced appearance. The importance of curing cannot be overemphasized.

A question that sometimes arises is whether wet curing is really necessary or whether the application of a curing membrane is sufficient.

4-64   Chapter 4: Special Aspects

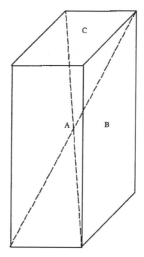

Fig. 4.5.6—Isometric view of a large concrete column

Briefly, if the *w/c* is greater than about 0.42, the amount of water in the mixture is sufficient for full hydration to take place so that membrane curing is adequate. Although there will be a decrease in the volume of the hydrated cement paste system (as stated earlier) autogenous shrinkage will not be high because it is mainly the larger capillary pores that will be emptied. The higher the *w/c* the better the concrete is able to withstand a lack of curing. Once the curing membrane has lost its effectiveness, drying shrinkage will develop.

If the *w/c* is smaller than about 0.42, autogenous shrinkage will develop rapidly despite the presence of a curing membrane. When the membrane has ceased to be effective, drying shrinkage will also develop. It is only at *w/c* in excess of about 0.5 that membrane curing is fully satisfactory. As HPC with very low *w/c* becomes more widely used, the importance of wet curing will be even greater.

Shrinkage in high-performance concrete

To ensure durability of a concrete structure it is necessary not only to select appropriate ingredients and their proportions, especially a low *w/c*, but also to ensure proper curing. The use of HPC provides the first of these requirements but this type of concrete is particularly demanding with respect to curing if problems arising from shrinkage are to be avoided. The reasons are as follows.

First, HPC exhibits very low bleeding and a very low bleeding rate. As discussed earlier, this is conducive to the development of plastic shrinkage. Second, HPC, having a high cement content, generates a large amount of heat of hydration, which induces steep temperature gradients, particularly harmful during the cooling stage. Third, because HPC has a high cement content and undergoes rapid hydration, the reduction in the volume of the hydrating cement paste system is large. However, this reduction and the associated autogenous shrinkage do not induce internal cracking because the capillary pores that are developed are small and the partial vapor pressure in them is rapidly reduced.

Because the w/c in HPC is very low, prevention of the loss of water by evaporation from the surface by means of the application of a curing membrane is inadequate: supply of water by wet curing is essential. The requirements with respect to the length of wet curing depend on the particular circumstances as illustrated by the examples of structures made with HPC.

Large columns

Consider a massive column with, say, dimensions of 1 × 1 × 2.5 m (3 × 3 × 8 ft) shown in Fig. 4.5.6. We shall look at shrinkage at three locations: A, at the center of the column; B, at its side surface; and C, at the top surface.

At A, there can be no plastic shrinkage or drying shrinkage (which would reach the center of the column only after many decades). On the other hand, thermal shrinkage will be very high because it is here that the highest temperature develops and the temperature gradient is steepest. The magnitude of thermal shrinkage can be readily calculated from the physical parameters of the concrete, the insulating characteristics of the formwork, and the ambient temperature. The autogenous shrinkage is not easy to predict but a typical value found in a column of the size given here was found to be 250×10^{-6} at the age of 4 days, and increased by 30×10^{-6} in the succeeding 4 years.[8] However, as the rapid hydration is accompanied by a rapid development of a high tensile strength of concrete, the risk of cracking is negligible.

At B, there can be no plastic shrinkage but, in the absence of curing following the removal of formwork, there will develop drying shrinkage; this was measured to be 330×10^{-6} on an actual large column.[8] In addition, there occurs autogenous shrinkage and thermal shrinkage, although the latter is much smaller than at A. The steepness of the temperature gradients depends on the insulating properties of the formwork: the better the insulation the lower the gradients, so that plywood forms or insulated steel forms are preferable to ordinary steel forms.

Still, in order to limit the effects of autogenous shrinkage at B, the forms should be eased at an early age and wet curing should start. If wet curing continues for 7 days, the concrete will have reached about three quarters of its long-term tensile strength so that, when wet curing ceases and drying shrinkage takes place, the concrete will be strong enough to resist the effects of drying shrinkage. If no wet curing is applied and curing is limited to leaving the formwork in place, drying shrinkage will be successfully deferred, but autogenous shrinkage will develop fully.

At C, plastic, drying, and autogenous shrinkage can all take place at an early age. However, at this location, it is easiest to ensure proper curing. Immediately following finishing, a curing compound or fog misting should be applied. Once the concrete has set, wet curing should start.

Large beams

These are taken to mean beams in which the smallest dimension is at least 500 mm (20 in.), so that significant temperature effects will develop. We shall look at shrinkage at three locations: A, at the center of the beam; B, at its bottom surface; and C, at the top surface.

At A, the situation is similar to that at A in the large column but the thermal shrinkage will be lower.

At B, the situation may be critical because, in addition to the shrinkage effects at that location, in service there is a tensile strain arising from the beam flexure under load. Moreover, the formwork at B remains in position for a long time (to prevent excessive deflection under self-weight) so that large autogenous shrinkage can develop. Wet curing should start as soon as structural considerations allow the easing of lateral formwork.

At C, the situation is similar to that at C in a large column, and cracking can be avoided by proper curing procedures.

Small beams

By this we mean beams in which the smallest dimension is less than 500 mm (20 in.). The locations A, B, and C are taken to be the same as those in large beams.

The thermal effects will be generally negligible so that the nature of the formwork is of small importance. However, at B, autogenous shrinkage can be very harmful so that the same curing procedures as in a large beam should be used.

The most serious problem is drying shrinkage because the surface/volume ratio of a small beam is large so that evaporation is large and continues for a long time. Applying an impervious film to the surface of the concrete is very helpful.

Thin slab

By this we mean a slab less than 300 mm (12 in.) thick. The locations considered are: A, at the center of the slab; B, at its bottom surface; and C, at the top surface.

There are no thermal problems because the heat of hydration is steadily dissipated and the temperatures at A, B, and C do not differ much from one another. At C, plastic shrinkage can be very serious, and drying shrinkage and autogenous shrinkage can also be large. However, proper curing procedures, similar to those recommended for the top of large columns can be easily applied.

It is worth emphasizing that, whereas proper curing of thin slabs is particularly easy, absence of adequate curing has particularly harmful effects.

Slabs-on-grade

In the case of thin slabs, if they are placed on wet soil, the lower part of the slab hardens under ideal curing conditions and does not develop drying shrinkage. However, the top part of the slab undergoes drying shrinkage as well as autogenous shrinkage. The differential in horizontal strain between the top and the bottom of the slab leads to curling (warping) with edges and corners rising. This situation can be prevented by providing a draining substrate or an underlying impermeable membrane so that water cannot be imbibed from the material underneath the slab.

In a thick slab, that is one at least 500 mm (20 in.) thick, high temperature gradients develop at the top of the slab and also at its underside. The higher temperature developed at the underside results in more rapid hydration near the underside than higher up. Consequently, a higher strength is developed in the lower part of the slab. This reduces the creep coefficient of the concrete[5] so that thermal stresses on cooling are not well relieved and cracking can occur.

For these reasons, the rate of development of the heat of hydration should be reduced by using low-heat portland cement and other appropriate cementitious materials. Cooling the fresh concrete by the use of chilled water, ice, or pre-cooled aggregate is particularly effective. In hot climates, placing the concrete at a time when the ambient temperature is lowest is also helpful. Sometimes, cooling the substrate is also feasible.

The cases considered illustrate the nature of the shrinkage problems in HPC and suggest appropriate procedures. These can be adapted for use in any situation.

Conclusion

It is hoped that an understanding of the various types of shrinkage and of the development of shrinkage under different circumstances will be of value in planning procedures to minimize the harmful consequences of shrinkage. In the past, when concrete with a high w/c was commonly used, the consequences of failure to apply wet curing were very serious and catastrophic only under a combination of adverse humidity and temperature conditions.

However, with the increasing use of HPC, which undergoes a large autogenous shrinkage, the need to control total shrinkage is becoming of considerable importance.

References

1. Le Chatelier, H., *Recherches Expérimentales sur la Construction des Mortiers Hydrauliques*, Dunod, Paris, 1904, pp. 163-167.

2. Powers, T. C., "Structure and Physical Properties of Hardened Portland Cement Paste," *J. American Ceramic Soc.*, V. 41, Jan. 1958, pp. 1-6.

3. Neville, A. M., *Properties of Concrete*, Fourth Edition, Longman,

London, 1995, and John Wiley, N.Y., 1996, 844 pp.

4. Powers, T. C., "A Discussion of Cement Hydration in Relation to the Curing of Concrete," *Proceedings*, Highway Research Board, V. 27, 1947, pp. 178-188.

5. L'Hermite, R. G., "Quelques Problèmes mal Connus de la Technologie du Béton," *Il Cemento*, V. 75, 1978, pp. 231-246.

6. Neville, A. M.; Dilger, W.; and Brooks, J. J., *Creep of Plain and Structural Concrete*, Longman, England, 1983, 361 pp.

7. Aïtcin, P.-C., and Neville, A. M., "High-Performance Concrete Demystified," *Concrete International*, V. 15, No. 1, Jan.-Feb. 1993, pp. 21-26. [Section 4.2]

8. Dallaire, E.; Lessard, M.; and Aïtcin, P.-C., "Ten Year Performance of High-Performance Concrete Used to Build Two Experimental Columns," ASCE Structures Congress—High-Performance Concrete Columns, Chicago, Ill., Apr. 1996, 10 pp.

Chapter 5

Construction

As emphasized in the Preface, this book deals with concrete insofar as it leads to producing better concrete structures. While achieving this aim has various facets, the actual construction is particularly important.

The vast majority of concrete structures contain reinforcing steel. The position of the steel is important in that it controls the flexural strength of members and also influences their deflection. Also, correct positioning of reinforcing bars ensures a satisfactory distribution of stresses within any structural member. Reinforcing bars are also placed close to the surface of concrete for the purpose of controlling shrinkage cracking. And yet, these and other steel bars need adequate cover to ensure protection from corrosion. It is the concrete cover to reinforcement that is the subject matter of Section 5.1.

Unfortunately, the thickness of cover is not always satisfactory in practice. Often, this is not realized because the steel is "buried" in concrete, and it is only when the consequences of corrosion of steel have become apparent that the actual depth of cover is investigated.

Strangely enough, the methods of specifying cover have not been adequately standardized. It is easy for the designer to say: I want cover of 50 mm (or 2 in.) but in practice, here and there, the cover will be too large or too small. Section 5.1 discusses the tolerance, both positive and negative. It reviews also the background to problems, which are not limited to unsatisfactory placing of the steel bars. In some cases, cover may be incorrectly specified, or the specification may be incorrectly formulated. Resolving these problems is of great importance; some suggestions are made in the article and amplified by a letter to the Editor, included after Section 5.1.

This book does not deal with construction techniques, but some techniques require particular properties of the concrete mixture. One of these techniques is slipforming. When I was involved in a dispute involving slipformed construction, I found myself in need of documented information about properties of concrete appropriate for slipforming. To my surprise, there was none other than the practical knowledge of slipforming contractors and specialists. There is available literature on formwork for slipforming and on the operation of slipforming, but when it comes to the properties of the mixture to be used, the designer of the structure has no sources of information.

And yet, when slipforming construction is to used, the specification for concrete must provide for the use of a mixture that will satisfy the slipforming contractor with respect to workability and rate of stiffening of concrete and, at the same time, also satisfy the various requirements of durability and strength. It is this combination of properties that is the subject matter of Section 5.2.

Sometimes, changes observed on the surface of concrete are perceived to be problematic or even an indication of unsatisfactory construction. Such changes may be due to weathering or to external attack of concrete surfaces; important as they are, I have not dealt with them.

On the other hand, occasionally there appear white deposits on the surface of the concrete. These deposits come and go with rainfall, and with changes in humidity and in temperature. Such deposits bear the charming name of efflorescence.

The question is: is efflorescence harmful and an indication of noxious changes within the concrete or is it just a surface blemish, and usually temporary at that? I should point out that brickwork is even more liable to efflorescence than concrete, and efflorescence in brickwork has been studied more than in the case of concrete. With respect to the latter, it is about 20 years since a serious review of the problem was published.

I have therefore believed that efflorescence is not of serious concern in construction but, literally a couple of years ago, the presence of efflorescence gave rise to allegations of faulty concrete in some court cases in which I was involved. Accordingly, I decided to study the literature on various aspects of efflorescence. Two sections represent the outcome. Section 5.3 reports the available knowledge of efflorescence, including its chemical composition and the possible mechanisms involved.

Section 5.4 deals with efflorescence observed in actual structures, discusses its significance, and describes methods of minimizing efflorescence.

Autogenous healing is a peculiar property of concrete. In essence, it means the ability of narrow cracks in concrete to close (akin to the healing of a cut in the skin) under suitable conditions. We could adopt the attitude: this is very nice, but so what? My involvement in autogenous healing stems from an important type of construction, in which drying shrinkage induced cracks; these cracks could give rise to corrosion of metal. The question posed was: will the cracks heal when the concrete is immersed in water, so that corrosion need not be feared? While my answer was case-specific, the review of the available information on autogenous healing under a variety of circumstances is of broader interest and is included in this book as Section 5.5.

Section 5.1: Concrete Cover to Reinforcement or Cover-Up?

Cover to reinforcement is the shortest distance between the surface of a concrete member and the nearest surface of the reinforcing steel. My earliest contact with the problem of cover to reinforcement was when I was a young engineer in charge of construction of drilling pits in which well-heads were to be located. The four reinforced concrete sides of the pit were 150 mm (6 in.) thick; the main reinforcement consisted of vertical bars 25 mm (1 in.) in diameter. When I came to inspect the reinforcement prior to concreting, I noticed that it was located about 75 mm (3 in.) from the inner face of the finished pit. From the structural point of view, this seemed a strange position, given that the earth on the outside would exert pressure on the pit walls and put their inner face in tension.

I spoke to the foreman, who produced a drawing, that clearly read: *minimum cover 25 mm (1 in.)*. "If the minimum is to be 1 inch," he said, "I thought I would do better than that and make it 3 inches." This taught me an important lesson: the designer or the detailer must not assume that the steel fixer (ironworker) or the operative necessarily understands the rationale of the instructions on the drawing or in the specification, or that an operative interprets these instructions on the basis of personal knowledge of structural behavior. Indeed, all instructions must be self-standing and self-explanatory.

Purpose of cover to reinforcement

It may be appropriate to remind ourselves of the reasons for providing the cover. There are several reasons, and at times we become so preoccupied with one of them that we tend to forget the importance of the others. For example, in the last decade, prevention of corrosion of reinforcement has reached such importance that we sometimes specify a large cover, without considering all the consequences of doing so. Let me, therefore, briefly list the purposes of cover.

Tensile force: Historically, the primary purpose is to put concrete around the reinforcing steel in a beam so that the strain in concrete in flexure is transferred to the steel which can then develop a tensile force. This is how reinforced concrete works and, if tension does not develop in the steel, it does not work! In other words, this purpose of providing cover to the reinforcing steel is essential and all-important. For this purpose, however, the cover can be very small; but what happens if the cover is excessive? The further the reinforcement from the tensile face of the beam the smaller its contribution to the carrying moment. In the extreme, if the cover is grossly excessive, the steel may develop no tension. This would be the case in a cantilever in which the cover from the top is so large that the steel is no longer on the tension side in flexure. This, too, would have been the situation in the walls of "my" drilling pit in service.

Cracking: Cover is also important from the standpoint of shrinkage cracking. Unreinforced concrete, if restrained (and restraint can be avoided only in some situations) will allow concentrations of tensile strain to develop. If this strain exceeds the strain capacity of the concrete, shrinkage cracking will develop. To prevent the development of such concentrations of tensile strain, it is necessary to provide reinforcement, spaced fairly closely, and located sufficiently near the exposed drying surface of the concrete member. Otherwise, cracking may occur. This is objectionable on aesthetic grounds if the concrete surface is apparent, and on durability grounds if an attacking medium can penetrate through the cracks toward the reinforcement.

The corollary of the above is that the thickness of cover must not be excessive; otherwise, the outer part of the concrete member would be, in fact, unreinforced and liable to shrinkage cracking. I shall come back to the issue of the maximum thickness of cover that we should use despite the occasional clamor for thicker and thicker cover by those seeking to provide protection of the reinforcement from corrosion.

Corrosion: This brings us to the need for cover for the purpose of protection of steel. Bare steel undergoes corrosion, that is, it rusts. When embedded in concrete, however, the surface of the steel is passivated and protected by the alkaline environment of the pore liquid in the hydrated cement paste. Continuation of this protection over the life of the structure requires that the alkalinity of the cement paste is not reduced. A common cause of such reduction is by carbonation, mainly of calcium hydroxide in the hydrated cement paste in the cover right up to the vicinity of the surface of the steel. Carbonation is progressive from the outer surface of the concrete: the progress is more rapid the greater the penetrability of the concrete, and this process effectively decreases the protective cover. Hence, the need for adequate cover.

Embedded steel becomes liable to corrosion, also in the absence of carbonation, if aggressive ions reach the surface of the steel. The most common ion is chloride, either from seawater (splashed or airborne) or from chloride salts used as deicing agents. The penetration of cover concrete by these agents is governed by the same factors as the progress of carbonation.

Fire: There is one more important reason for the provision of adequate cover to reinforcement—fire protection of the steel. Fire endurance of reinforced concrete elements is a complicated topic because it involves structural action, which may be impaired by flame penetration and heat transmission. In essence, however, design codes specify the minimum cover of various types of structural elements (beams, floors, ribs, and columns) necessary to ensure a fire resistance over a certain number of hours; this specification is sometimes known as fire rating.

Three types of problems with cover

A problem with cover means that it is unsatisfactory. There can be

5-6   Chapter 5: Construction

three reasons for it. First, the cover may be incorrectly specified. Secondly, the specification can be incorrectly formulated. Thirdly, the actual cover "as built" can be different from what was specified.

Cover incorrectly specified

As discussed in the preceding section, the requirements for cover include the development of the appropriate tensile force in the reinforcement, considerations of durability, fire resistance, and shrinkage distribution or restraint. As I have already intimated, the fire resistance requirements are too complex, as well as specialized, for inclusion in this section.

Let us now look at the consequences of incorrectly specified cover on the behavior of concrete elements.

To develop the calculated tensile force, the reinforcement must be in the position assumed in the calculation of the moment of resistance of the beam or its ultimate strength. If the cover is larger than assumed in the calculations, large cracks will open under a lower applied load than should be the case. Under an overload, failure may occur prematurely. If the position of the reinforcement is structurally correct but an excessive cover was achieved by an additional depth of concrete, the self-weight of the beam is greater than assumed in the design calculations. This additional weight has adverse structural consequences, too.

From the standpoint of durability, the protection of the reinforcement is a function of the thickness of cover and of the quality of concrete in it. It is believed that these two factors can be offset one against the other, and some codes of practice provide tables of alternative combinations of thickness of cover together with the quality of concrete to ensure durability under given conditions. The quality is described by minimum compressive strength or water-cement ratio or cement content. This is not the place to discuss which of these parameters is most appropriate. However, it is important to point out that a combination of a very large cover and a very poor concrete is entirely unsatisfactory: no matter how large the cover, if the concrete is porous and permeable, aggressive agents will rapidly penetrate through it to the surface of the steel reinforcement.

In other words, in my opinion, when concrete is to be exposed to conditions generally called severe or very severe, let alone extremely severe, the quality of concrete must be high or very high. Indeed, it is only the concrete in the cover zone that matters as far as the durability problems discussed here are concerned; the quality of the concrete in the interior of the member is almost irrelevant. The quality of concrete that is necessary is discussed fully in *Properties of Concrete*.[1] What is not discussed in that book, but must be considered here, is that the thickness of the cover must be adequate but it must not be excessive.

What is considered adequate cover is prescribed in design codes. Many codes in more-or-less temperate climates underestimate the fact that in

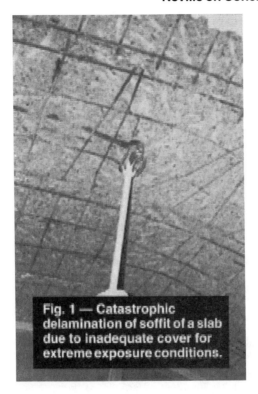

Fig. 1 — Catastrophic delamination of soffit of a slab due to inadequate cover for extreme exposure conditions.

Fig. 5.1.1

some other climates the severity of exposure can be much more acute than what is labeled "very severe" in Great Britain or in many parts of the United States. The Middle East, and especially the Gulf area, is a prime example of truly extreme conditions. Here, the temperature of the concrete is high and the insolation is severe, so that a considerable depth of concrete becomes very dry and "thirsty." At some later stage, the seawater, carried by wind in the form of droplets or aerosol, or salt-laden dust, wetted by dew, is deposited on the surface of the concrete and imbibed to a considerable depth. The process is cumulative and chloride ions reach the surface of the reinforcement (Fig. 5.1.1). The irregularly placed reinforcing is now "uncovered."

This section is not concerned with what happens next, but it is clear that an adequate thickness of cover is necessary. Codes of practice give advice on what is adequate but, as I have already pointed out, under extreme conditions the advice may be too optimistic. This has led some engineers to recommend a greater thickness of cover: 100 mm (4 in.) or even 120 mm (5 in.). In my view, this is wrong because such a large cover means that a considerable thickness of concrete is unreinforced; consequently, shrinkage cracks can open or flexural cracks can develop under load. Such cracking would allow ingress of aggressive agents so that the alleged remedy of a very thick cover is, in fact, detrimental. To quantify my opinion, I would say that cover should not exceed 80 mm (3¼ in.), or perhaps 90 mm (3½ in.). If this is still inadequate for the desired durability, a better quality of concrete, possibly containing some special ingredients, must be used. It is also possible that reinforced concrete is inappropriate for the given conditions. We sometimes forget that, at least in a particularly exposed part of a structure, unreinforced concrete masonry could be used. When there is no reinforcement, carbonation is not harmful and chlorides do not represent serious danger.

Specification incorrectly formulated

Let me now turn to problems arising from an unsatisfactory formulation of the specification. First, all embedded steel, not just structural reinforcement, is subject to corrosion. It follows that the specified minimum cover must apply to links (stirrups) and, indeed, to other embedded steel. This is sometimes forgotten; more often, the drawing states: "cover to steel, so much." The person on site cannot be blamed for interpreting this to mean "cover to main steel"—never mind bits and pieces.

A more serious problem arises with the precise meaning of the term "cover." To say "cover to be 40 mm (1½ in.)" and expect the cover everywhere to be exactly *that much* is entirely unrealistic. In reality, cover, here and there, must vary from the specified value. The issue is then how to interpret that value. For example, the British approach[2] is to operate in terms of what is called "nominal cover," that is, the value of cover used in the structural design calculations and indicated on the drawings. To allow for the variability in the thickness of cover in reality, the British code[2] says: "The actual cover to all reinforcement should never be less than the nominal cover minus 5 mm (¼ in.)." The code is silent on how much more than specified is tolerated. So, our aforementioned foreman was not wrong to exceed the cover by 50 mm (2 in).

The ACI Building Code,[3] Section 7.5, uses a similar approach: the tolerance on minimum cover is -10 mm (-3/8 in.) for members up to 200 mm (8 in.) deep, and for deeper members, -13 mm (-½ in.). The important point here is that the American tolerance is -10 or -13 mm (depending on the depth of the member) while the British tolerance is only -5 mm (-¼ in.) in all cases.

I am not arguing which code is right, but in my view they are both inadequate in that they do not lay down a positive tolerance. Some other codes do give both a negative and a positive tolerance, and many job-specific specifications do so likewise. I shall return to the issue of specifying the cover in an unambiguous and adequate manner later in this section. At this stage, I should consider two potential questions: How does cover vary in reality? Is the cover in actual structures too large as well as too small?

Actual cover not as specified

The underlying reason for writing this section is the fact that, in a

Fig. 5.1.2

number of actual structures, the cover varies, often considerably, from the specified value. This fact is not widely known and, even when the existence of improper cover is known, this is not considered to be a problem; certainly, nothing is done about it. It is only when a given structure has shown signs of serious deterioration, involving the corrosion of the reinforcement, that detailed inspection reveals the fact that the cover as executed was not as specified. It is this belated discovery, or uncovering, that has prompted me to refer in the title of this section to a "cover-up."

My first personal observation of improper cover was when external signs of large-scale corrosion of reinforcement (rust, cracking along the position of the reinforcement, spalling, and delamination) led to a detailed investigation of the position of the reinforcement. This was in several major structures in the Middle East, but the problem is certainly not limited to that part of the world. For example, in a high-rise building in Australia, described as prestigious (which translates into high rent), I observed what I call "negative cover" (Fig. 5.1.2). This term should be introduced into the ACI vocabulary to describe a situation where the reinforcement can be actually seen by the naked eye and felt by a finger. It is only fair to add that inadequate cover was not the sole cause of corrosion in all those cases, but it is a vital element in the deterioration of many structures.

In passing, I could add that the members, some of which had a negative cover, were cladding units without interior lining, 75 mm (3 in.) thick. The reinforcing bars were 25 mm (1 in.) in diameter, with a specified cover, front and back of 20 mm (¾ in.). Can this be achieved in real life? What happened in the event is that chloride ingress occurred from outside and carbonation from inside the building. This must be the classical case of reinforcement suffering from the worst of both worlds.

Those familiar with construction may wonder how woefully improper cover can exist, given that, on most well-regulated sites, the position and fixing of reinforcement is checked by the engineer, or his representative, prior to authorization of the actual concreting. There is a delightful story, I am assured not apocryphal, about the construction of a multi-story building. When the reinforcement for a given floor was in position, the engineer

Fig. 3 — Leaning on the shovel is not enough; the operative must stand on something, and reinforcing steel must be supported.

Fig. 5.1.3

verified it, ordered concreting to proceed, and departed. On the occasion of concreting the tenth floor, he happened accidentally to leave his briefcase behind but did not discover its loss until several hours later. When he returned to retrieve his briefcase, the concreting had been finished. To his astonishment, he observed a large number of reinforcing bars stacked to one side. The explanation was simple: after the engineer's inspection, the building owner ordered the removal of half the reinforcement at every floor level. He was going to sell the building as soon as it had been completed, so economy took precedence over safety.

In reality, fraud or malpractice is not the cause of improper cover, but there are several reasons for it. One of the main ones is a lack of appreciation of the importance of achieving the specified cover. In this respect, as in many others, some of the operatives are inadequately trained and knowledgeable about reinforced concrete. It has to be admitted that they often work under physically demanding conditions and under considerable pressure. Tying up the bars may be skimped, fixing may be inadequate, chairs, spacers and other supports may become damaged or displaced. Standing on the reinforcement by the vibrator operatives (Fig. 5.1.3) can temporarily displace it, but the weight of concrete makes the displacement permanent. These problems can become aggravated when the shrinkage reinforcement is relatively light.

The use of well-made reinforcement cages should minimize the above problems, but then whole cages have been known to be wrongly placed or to shift bodily when their support is too weak. The real trouble is that all the means of fixing the reinforcement are minor, small or flimsy, so that they do not attract major attention of those involved in concreting.

But the causes of improper cover are not limited to site operations. The design and detailing of reinforcement sometimes makes for serious practical difficulties on site. Occasionally, there is more steel than can actually be fitted into the space available, especially when lapping or cranking is necessary. The drawing often does not show how to achieve this, there being a simple instruction of the type: "laps to be 460 mm (18 in.)." In some types of structures, the amount of reinforcement required for structural reasons is so large that there are great difficulties in fitting it in, without shoving and pushing. On one occasion, the amount of steel was so large that the foreman was moved to ask the engineer: Do you want me to concrete it or to paint it?

The reason for the practical difficulties of placing the reinforcement as specified may lie in the fact that some designers (of course, not all) lack site experience and are simply not aware of how hard, if not impossible, it is to execute their designs and sketchy detailing. The use of computers should help in this respect but, sometimes, detailing is taken care of too far down the line.

Another source of difficulties is improper or incorrect bending of bars or cutting them to length. Relatively small errors can have serious consequences, given that bar lengths are handled in feet or tenths of a meter, whereas cover is measured in quarters of an inch or in increments of 5 mm.

Observed problems with cover

At this stage, I may well be called upon to demonstrate that the various problems with improper cover are real problems, frequently occurring in actual structures, and not just an imaginary or rare occurrence. There are several publications demonstrating the reality of my concern. I propose to refer to four of them, dealing with several types of structures in three countries.

In Canada, Mirza and MacGregor[4] studied the actual cover in a number of slabs, both cast on site and precast. In cast-on-site slabs, they found that the top reinforcement was more often affected than bottom reinforcement. The mean deviation of cover from the specified value was –20 mm (–0.8 in.) for the top reinforcement, and –8 mm (–0.3 in.) for the bottom reinforcement. In precast slabs, with only bottom reinforcement, the cover was virtually exactly as specified. This is not surprising, given that factory-style mass production can achieve perfection; moreover, in precasting operations, the reinforcement is generally not used to support the operative.

An Australian study[5] is particularly interesting because it combined the determination of actual cover with observations of corrosion; there was a good correlation between the two. This is not surprising because, at 227 "fault locations" in 95 buildings, the average cover had a shockingly low value of 6 mm (¼ in.). This decisive role of inadequate cover with respect to corrosion of reinforcement does not invalidate my earlier argument about the paramount importance of the quality of concrete because, in the elements studied, the quality of concrete was the same and only the thickness of cover varied. The highest occurrence of problem areas in building façades was in beam and slab end faces.[5] This is interesting because it suggests that it is the longitudinal displacement of reinforcement, or its excessive length, in a horizontal member that leads to inadequate cover.

Some other figures from the same study[5] are of interest. In as many as 18% of locations in buildings, the cover was less than 60% of the specified value. In bridges, the situation was very much better, with only 4% of locations having a cover of less than 60% of the specified value. Nevertheless, cover that is too small by a factor of 0.4 is a serious matter.

The data on excessive cover are of particular interest in view of my earlier comments on the occurrence of this phenomenon, rarely considered in design codes. It was found that at 62% of locations in buildings and 51% of locations in bridges, the actual cover exceeded the specified value.[5]

A major investigation of the state of 200 bridges in Great Britain provided a considerable amount of information about cover to reinforcement.[6] Of these bridges, 77 exhibited rust or spalling associated with low cover. The cover survey was carried out on 500 mm (20 in.) square test areas, and the minimum cover in each area was determined.

The value of the minimum cover in the bridge elements varied widely. For example, in deck soffits of bridges constructed between 1970 and 1980, the lowest value of the minimum cover was 10 mm (3/8 in.) and the highest value of minimum cover was 130 mm (5 in.). For bridges built between 1980 and 1985, the corresponding values were 20 mm (¾ in.) and 75 mm (3 in.) respectively. For abutments and piers, the spread was even larger: in the earlier period, 0 to 115 mm (0 to 4½ in.), and for the later period, 10 mm to 100 mm (½ to 4 in.). By way of comparison, I should add that the nominal cover specified for the deck soffits was generally 30 to 45 mm (1¼ to 1¾ in.), and for abutments and piers, 40 to 55 mm (1½ to 2¼ in.); a negative tolerance of 5 mm (¼ in.) was usually allowed.

For each period of construction, the mean minimum value was calculated, as well as the modal minimum value (that is, the minimum value most frequently encountered). In all cases, the modal value of minimum cover was lower than the mean value. This means that the low values of cover were more frequent, but the large values of minimum cover were very large. In other words, the distribution of minimum values of cover was skewed to the right.

Data on prestressed concrete decks are difficult to interpret because much depends on the specific method of prestressing. Generally, however, the control of cover in prestressed concrete members seems to be better.[6]

The final study to be considered in this paper is also British; it is very recent and deals with structures actually under construction on 25 sites.[7] The study is limited to vertical members: columns and walls; there was no difference in the pattern of cover between these two types of members. Many of the measurements make dismal reading. For example, in one major bridge with a specified cover of 40 mm (1½ in.), in two columns, the measured values of cover were all higher than specified, in some cases with values up to 93 mm (3¾ in.). In another bridge, also major, with a specified cover of 50 mm (2 in.) all but three measured values of cover were too small, down to 37 mm (1½ in.).

One more set of measurements should be reported because it serves a useful purpose, if only as a horrible example. On a more than a billion dollar building project, with a specified cover of 50 mm (2 in.), the actual cover in one wall ranged from 12 to 75 mm (½ to 3 in.). The incidence of too-small cover and of too-large cover was about equal; what is remarkable is that there were almost no measurements of cover between 43 and 53 mm (1-5/8 and 2-1/8 in.), that is, near the specified value.

These four studies, among them, show that the problem of improper cover is not limited to just a few structures or to particular types of structures or only to some countries. I can add from my personal experience the existence of inadequate cover in many major structures in the Middle East.

To demonstrate that I am not a lone crusader against improper cover, I would like to quote from Reference 7: "It is evident...that the required cover values and their allowable tolerances [negative and positive] have not been met, by wide margins, on most sites. Hence, it is confirmed that lack of cover is an extensive problem which is of a chronic rather than a sporadic nature."

Conclusions

The various data in References 4 to 7 can be analyzed further, but this would be of value only to an historian of concrete problems. For my part, I wish to look to the future, and I am sure that ACI is primarily concerned with doing better in the years to come. So what are the lessons?

I believe that there is an endemic problem of improper cover. Does it matter? I believe it does.

Even if, from the structural standpoint, having half the reinforcing bars with an inadequate cover and half with an excessive cover is not critical, this is not true for durability considerations. When one-half of the bars have corroded, major repairs will be necessary, and then even the strength of the structure can become impaired.

We must remember that for the protection of the reinforcing steel, having the appropriate thickness of cover is not enough: the concrete must be of appropriate quality, but it is only the quality of the cover concrete that matters.

Should we increase the specified minimum cover in the knowledge that, even where the actual cover is much less than specified, the actual value will be adequate for durability purposes? If we do so, we shall increase the weight of the structural members (with cost implications in materials, labor, and foundations), as well as in the size of the cross-section of the member required to carry the heavier loads. We shall also produce an unreinforced concrete tension zone, with cracks of considerable width; this will promote ingress of aggressive agents, and vitiate our attempts to minimize corrosion.

Should we shrug our shoulders on the grounds that, as stated in Reference 7, we live in a world of "contractual terms and conditions and a harsh economic climate which do not foster collaboration"? This may well be true, but it is a hard life for car makers, and aircraft manufacturers, too. I need not ask whether parallel consequences for safety and durability would be acceptable.

So what can we do? I do not presume to offer a recipe, but only to present a few ideas.

In the design office, we should pay much more attention to detailing the reinforcement; this is not a trivial task left to somebody down the line. The designer who has not got the requisite experience under his or her belt would be well advised to get thoroughly acquainted with site operations and the attendant difficulties of working under inclement conditions. The designer must also make sure that the structure is buildable in so far as fitting in the reinforcement is readily possible.

The chairs, spacers, and supports of the reinforcement are an integral part of the finished structure. Their quality should be assured, and the task of providing them should not be left to an indeterminate operative.

The output of the steel-bending shop should be more carefully verified than is sometimes the case. The approach of "adjusting" the reinforcement on site by a sledge hammer will not do.

The site operatives should be better trained and better aware of what the reinforcing steel does and why its cover matters very much. This theme of the need for training is recurrent whatever aspect of concreting is considered.

At the same time, there should be better cooperation and communication between the supervisory staff and the operatives. I am convinced that exacting supervision is helpful; so is frequent verification prior to concreting and also after. Modern covermeters are highly reliable and can deliver a printed output.

I know that it is easy to say that the pressure of time and the need to proceed with the job should not interfere with the quality. In my view, the quality must take precedence, if we are to continue to build concrete structures in an economic way. Poor quality is very expensive, even though the expense is incurred at a later date. This does not make economic sense.

The required cover should be very carefully specified both on the drawings and in the specification. The meaning of "minimum" should be defined. It could be an absolute minimum or a characteristic (say, 5%) value (which, personally, I do not favor because of the difficulty of defining the population to be tested). The tolerances should be defined, both positive and negative, but they should not be unrealistically small. The need for cover to the ends of reinforcing bars should not be ignored.

Many more suggestions can be made. But what is really required is recognition that cover does matter. Cover-up will eventually be exposed.

References

1. Neville, A. M., *Properties of Concrete,* Fourth Edition, Longman, London, 1995, and John Wiley, New York, 1996, 844 pp.

2. British Standard BS 8110: Part 1: *Structural Use of Concrete*, London, 1985.

3. ACI Committee 318, "Building Code Requirements for Structural Concrete (ACI 318-95) and Commentary (ACI 318R-95)," American Concrete Institute, Farmington Hills, Mich., 1995.

4. Mirza, S. A., and MacGregor, J. G., "Variations in Dimensions of Reinforced Concrete Members," *Proceedings* ASCE, 105, No. ST4, Apr. 1979, pp. 751-766.

5. Marosszeky, M., and Chew, M., "Site Investigation of Reinforced Placement on Buildings and Bridges," *Concrete International,* V. 12, No. 4, Apr. 1990, pp. 59-70.

6. Wallbank, E. J., "The Performance of Concrete in Bridges,*"* Report for the Department of Transport, U.K., Apr. 1989.

7. Clark, L. A., et al., "How Can We Get the Cover We Need?," *The Structural Engineer,* V. 75, No. 17, 1997, pp. 289-296.

Letter to the Editor
Concrete cover

It gave me great pleasure to read the article by Adam Neville...("Concrete Cover to Reinforcement—or Cover-up?" November 1998 issue, pp. 25-29)...and to find such an illustrious concrete technologist airing in very practical terms one of my hobby horses.

Neville suggests that the maximum cover should be no more than 3 to 3.5 in. (76 to 89 mm). I would even go further. Two inches (50 mm) would be nearer the mark. As a matter of fact, there was a time as long as 30 years ago or more when a steel mesh was specified in the unreinforced area for a large cover for concrete columns under fire situations. This is an improvement from having no steel reinforcement in thick covers but there was still the danger of the mesh being displaced, resulting either in exposed steel and/or the mesh touching the main reinforcing bars. During this period, the erroneous idea of equating cover to fire resistance of building elements unfortunately still existed.

In an ideal world with a perfect quality assurance regime in place, I would have preferred to see no more than 1 in. (25 mm) specified for the most severe exposure condition. If the high-performance concrete specified cannot resist these exposures, then appropriate admixture added to the concrete to reduce permeability in conjunction with impermeable layers and/or coatings appropriately bonded to the substrate would be a better bet than adding to the thickness of unreinforced concrete cover. This is for all the justifiable reasons with practical examples that Adam so eloquently cited in his paper.

Patrick J. E. Sullivan
City University
London, England

Neville's response

I am grateful to Patrick Sullivan for emphasizing the importance of cover in the behavior of concrete in service.

The message is clear—cover is not just a steel fixer's concern; achieving the right cover in reality affects the performance and durability of concrete structures.

Adam Neville

Section 5.2: Specifying Concrete for Slipforming

Slipforming is a well-established method of construction of placing concrete in moving formwork for vertical elements, usually of considerable height, such as towers, chimneys, bridge piers, shear walls, silos, oil platforms, water tanks, shaft linings, and nuclear reactor containment vessels (Fig. 5.2.1). Horizontal elements can also be slipformed but that method, akin to extrusion, is not considered in this section.

The main advantages of slipforming are the absence of construction joints (which mar the appearance and contribute to leakage of liquids or gases), a high speed of construction, and a good adaptability for construction in bad weather. Slipforming is a specialized operation that requires skill and experience, and it is usually performed by specialist contractors who have the requisite knowledge. Others can learn about slipforming operations from handbooks of slipform contractors or from publications on formwork in general, such as *ACI Special Publication No. 4,*[1] and the *ACI Guide to Formwork,*[2] or an equivalent guide published by the British Concrete Society.[3] However, readily available sources of information about the concrete technology aspects of slipforming do not seem to exist. And yet, designers need to know about the limitations that slipforming imposes on the concrete mixture. Likewise, concrete suppliers should be knowledgeable about the kind of mixture that may be required.

A particular problem that sometimes occurs is writing a specification for concrete that will be, or may be, slipformed. The method of construction is prescribed in one part of the specification, but it is in various other parts that the properties of concrete are laid down in terms of ingredients to be used, limits on mixture proportions, possibly workability, and durability (as influenced by the water-cement ratio and specific cementitious materials), as well as thermal properties in terms of a maximum temperature or thermal gradient.

If the latter parts of the specification are written without due consideration to compatibility with what is needed to ensure a smooth slipforming operation, then problems may arise. These problems can be avoided if the specification writer understands the relevance of concrete technology to slipforming. It is the purpose of this section to highlight this relevance.

Stiffening of concrete

Strictly speaking, slipforming is not a continuous operation in that the upward movement of the formwork occurs in steps of 25 or 50 mm (1 or 2 in.). The overall rate of climb of the formwork is variable, generally between 0.15 and 0.4 m (about 6 and 16 in.) per hour. Important factors affecting this rate are the need to place the reinforcement (which may be

Chapter 5: Construction

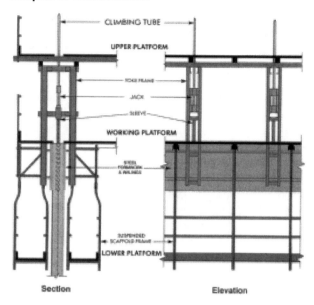

Fig. 5.2.1—An example of slipform[5]

more or less complex or plentiful) and the need to form box-outs for doors or passages. The required variability in the rate of climb imposes demands on the rate of stiffening of the concrete.

The concrete is placed in the formwork, which is commonly 1.2 m (4 ft) deep, in thin layers, about 200 mm (8 in.) deep. Once a layer has been completed over the entire operational surface, placing concrete in the next layer begins. It is essential that the underlying layer remains in a plastic state long enough for there not to be a "cold joint" or a plane of weakness between the layers. On the other hand, the concrete lower down in the formwork must stiffen sufficiently and achieve an adequate strength when the form is removed by its upward movement so that the concrete does not slough-off or bulge. In simple terms, the concrete must remain plastic "long enough" but must then stiffen and harden "soon enough."

The stiffening of concrete is not measured by any standardized method, such as a Proctor needle, which would be inappropriate in the congested and busy space available and which requires removal of coarse aggregate. This is why the term "stiffening time" is preferable to "setting time." The stiffening time is determined in a practical manner: a 16 mm (¾ in.) diameter bar is pushed vertically down into the concrete until its movement becomes halted by the resistance of the concrete. This should occur at a depth of about 0.6 m (2 ft), that is, half-way down the form. Clearly, the "test result" is greatly affected by the operator. Not surprisingly, such a requirement for the stiffening time cannot be readily translated into a specification clause, but then slipforming is an art, and not just a technique.

Given that the stiffening is required at approximately the mid-depth of the form, the time interval between placing of concrete and stiffening in a form 1.2 m (4 ft) deep is as follows: when the rate of climb is 0.15 m/h, the interval is 0.6/0.15 = 4 h; when the rate of climb is 0.4 m/h, the interval is 0.6/0.4 = 1.5 h.

Clearly, the faster the rate of climb of the slipform, the shorter the stiffening time required. For practical purposes, the stiffening time has to be reckoned from the time when the concrete is discharged from the mixer. It follows that, to the time interval calculated above, we have to add the time required to transport the concrete from the mixer to the slipform plus the time necessary to place and compact the concrete. Thus, much depends on the actual circumstances, but an additional period of 2 hours is likely. In such a case, the stiffening time required would be between 3.5 and 6 hours. The concrete emerging from the formwork would be between 5 and 10 hours old.

From the above values, it is clear that the concrete needs to retain a high workability for a long time; it follows that a superplasticizer needs to be used, especially at low rates of climb.

Naturally, it is desirable to slipform as fast as possible but an occasional slowing down is inevitable in order to place the reinforcement or box-out apertures, as mentioned earlier. If the density of reinforcement decreases with height, speeding up may be possible. Also, if the cross section being slipformed changes, adjustments to the formwork become necessary, and this takes time. There may also be occasional problems with equipment. With proper planning and good communication between the slipform team and the batcher, the delivery of concrete may be deferred, but the ability to vary the stiffening time is essential. Also, it should also be remembered that the stiffening time of a given mixture is affected by changes in temperature. Given that slipforming is a round-the-clock operation, temperature changes are bound to occur, even if the weather does not change.

Because of the multitude of factors involved, ensuring a suitable stiffening time cannot be achieved by a desk study alone. Trials must be carried out at the planning stage to establish reliable values of stiffening times for the likely range of rates of slipforming.

Workability and mixture composition

Concrete suitable for slipforming generally requires a slump of about 150 to 200 mm (6 to 8 in.). Much depends on the density of reinforcement, but this is true also in the case of nonslipformed concrete. It may be worth noting, however, that in some slipformed oil platforms the density of reinforcement has exceeded 1000 kg/m^3 (1700 lb/yd^3) of concrete.

To be suitable for slipforming, the fresh concrete has to possess, in addition to workability or "mobility," an adequate cohesion and also a low frictional resistance to the movement of the form; otherwise, despite the outward taper (batter) at the bottom of the form, a streaked or uneven surface may result. Methods of ensuring good cohesion are well established.

The workability and mobility of concrete are affected by the specific cementitious materials in the mixture, especially ground-granulated blast-furnace slag and silica fume. The use of these on the grounds, for example, of improved durability may be considered, and the specification may prescribe them, permit them, or forbid them. In taking an appropriate decision, the needs of the slipforming operation must be borne in mind.

The preceding discussion illustrates an important need for mixture selection such that an appropriate retention of workability is ensured, and that the period of retention can be modified by varying the dosage of the superplasticizer, possibly by a re-dosage at a later stage after the initial mixing, or by the use of a retarder. All this should be reflected in the part of the specification dealing with mixture ingredients and permitted admixtures.

The need for a high slump means that the mixture must have a high water content unless a suitable superplasticizer, and possibly also a compatible water-reducing admixture, is used. A high water content may have implications for the total content of cementitious material when a low water-cement ratio is necessary from strength or durability considerations. Now, a high content of cementitious material may have adverse effects on the maximum temperature in the interior of the concrete element and on temperature gradients in the concrete.

Thus, there is a potential incompatibility between the choice of mixture proportions to satisfy the thermal requirements and also to satisfy the requirements of fresh concrete suitable for slipforming. Such an incompatibility can be avoided if advance thought is given to the mixture desiderata. It is only when some parts of the specification are written without consideration of the consequences of slipforming that difficulties may arise.

Thermal problems in concrete can be alleviated by lowering the temperature of fresh concrete, for which standard techniques are available. In the case of slipforming, such lowering of temperature has the additional advantage of improving the workability and delaying the loss of slump.[4]

Aggregate requirements

For most concrete construction, the specification does not impose particularly onerous conditions with respect to aggregate: after all, it would not be economic to specify aggregate that cannot be obtained locally. For concrete that will be slipformed, it should be borne in mind that good workability is of particular importance. Consequently, the aggregate should have better grading than the minimum laid down in ASTM C 33. Furthermore, the proportion of flaky or elongated particles in the coarse aggregate should be fairly severely limited.

A fairly high proportion of particles passing the 5 mm (No. 4 ASTM) sieve, say 45%, contributes to the cohesion of the mixture, reduces friction at the surface of the formwork, and leads to a satisfactory finish.

If possible, the fine aggregate should not consist wholly of crushed material: rounded fine aggregate helps water retention by the concrete and reduces bleeding. Of the fine aggregate, almost 50% should be smaller than 0.6 mm (No. 30 ASTM) sieve. However, the optimum value would be affected by the content of cementitious material in the mixture. Generally, it is desirable to have a lower content of particles smaller than 0.15 mm (No. 100 ASTM) sieve than permitted in national standards. It may be desirable to make up the fine aggregate out of separated fractions. Minor changes in the proportion of smallest particles of aggregate have a large effect on the behavior of the mixture.[4]

Curing

Wet curing of slipformed surfaces is not easy, but it is highly desirable in hot weather, especially when accompanied by wind, or when aesthetic considerations preclude the use of membrane curing. One slipforming contractor[5] recommends a mist spraying circuit suspended from the scaffold frame hanging below the actual formwork, as well as protective plastic sheeting. Such a system requires a continuous water supply and hence pumps, and may appear to complicate the construction process. But then slipforming has many exigencies, including the provision of portable toilets on the platform!

Conclusions

I hope that this section has not given the impression that selecting a mixture suitable for slipforming is very difficult. In reality, a mixture that can be easily pumped needs only to be somewhat modified to make it appropriate for slipforming.

The rationale of this section is not to highlight the problems in slipforming in so far as mixture selection is concerned. Rather, I have tried to show that, when slipforming is to be used as a method of construction, the specification for concrete must take into account the required properties of the concrete. There is no inherent difficulty in this—all that is needed is knowledge and forethought.

Slipforming is an excellent method of construction. It is also "good" for concrete in that concrete suitable for slipforming is a "good" concrete with respect to mixture composition coupled with a low variability in mixture properties.

References

1. Hurd, M. K., *Formwork for Concrete,* SP-4, Sixth Edition, American Concrete Institute, Farmington Hills, Mich., 1995, 500 pp.

2. ACI Committee 347, "Guide to Formwork for Concrete (ACI 347R-94)," American Concrete Institute, Farmington Hills, Mich., 34 pp.

3. Concrete Society, *Formwork: A Guide to Good Practice,* 1995, 292 pp.

4. Haug, A. K., and Sandvik, M., "Mix Design and Strength Data for Concrete Platforms in the North Sea," *Second International Conference on*

Performance of Concrete in Marine Environment, CANMET, 1988, 33 pp.

    5. Rapid Metal Development, *Slipform Construction,* Walsall, England, 1997, 21 pp.

Section 5.3: Efflorescence—Surface Blemish or Internal Problem? Part 1: The Knowledge

There exist few publications devoted to efflorescence in concrete, and hardly any recent ones. A practical booklet on the topic was published 20 years ago.[1] It could be inferred from this situation that efflorescence is generally not a significant problem meriting a major study. This has been my own view for a long time, and indeed in *Properties of Concrete* I say: "Apart from the leaching aspect, efflorescence is of importance only in so far as it mars the appearance of concrete."[2] This view considers efflorescence as a *surface blemish*. Why then am I writing this article?

The answer is that, in reports on suspected or alleged sulfate attack of concrete, I have found statements to the effect that any efflorescence is a manifestation of attack on concrete, and indeed that the appearance of efflorescence is an indication of serious and terminal damage; in other words, a sign of an *internal problem*. Which view is correct?

To answer the question we need first to review such knowledge as is available; this is done in this section. Section 5.4 will deal with the situation in practice.

What does efflorescence look like?

The term efflorescence is used to describe white deposits on the surface of concrete. Efflorescence can be found both on horizontal and on vertical surfaces, usually exposed to outdoor air, and also on basement walls and garage floors. The deposits often have an irregular shape, perhaps redolent, if you have a vivid imagination, of flower petals. The New Shorter Oxford Dictionary, published in 1993, states that the word comes from Latin, partly through French, meaning "the process, or period, of flowering." A particular chemical meaning is a crystalline deposit of powder resulting from the loss of water or exposure to air. Given this origin of the term, it should not come as a surprise that the term "bloom" or "lime bloom" is sometimes used. We may recall Chairman Mao's reference to "letting a hundred flowers blossom," although he was not a concretor.

The term "lime weeping" is sometimes also used. It is applied to thicker deposits of efflorescence originating at cracks and joints, usually in more mature structures.[3]

According to the ACI terminology (ACI 116R-90), efflorescence is "a deposit of salts, usually white, formed on a surface, the substance having emerged in solution from within either concrete or masonry and subsequently been precipitated by evaporation." The British Standard Glossary, BS 6110: Subsection 1.3.7:1991, defines efflorescence as a "crystalline deposit of soluble salts on a surface that results from the migration and evaporation of water."

Categorization of efflorescence

The descriptions in the preceding subsection give a simple, usable working definition of efflorescence, but engineers love more elaborate classification. This is sometimes helpful but, at other times, introduces what lawyers call "a distinction without a difference." However, because a number of terms are used in the literature, I consider it useful to mention them here.

I should add that the bulk of the literature on efflorescence deals with brickwork and building stone, but sometimes broadly with building materials. Efflorescence in these various materials should be considered, not only because there is some similarity in the transport of water-borne salts in concrete and in brickwork and stone masonry, but also because much of the efflorescence in the latter two materials emanates from mortar containing portland cement.

In addition to salt crystallization on the surface of concrete, crystallization may also occur within the pores of the building material; one term for it is "cryptoflorescence."[4] I do not view this phenomenon as being within the scope of this article.

Another categorization is made by Aïtcin[5] who distinguishes efflorescence according to the age of concrete when it appears. Primary efflorescence appears during setting and at early ages, and involves chiefly water *in* the concrete; on the other hand, secondary efflorescence is caused mainly by water that travels through the concrete to the evaporative surface. This differentiation may be of significance in the case of paving slabs and in colored concretes, which is what Aïtcin is concerned with, but this is not the case with "grey" concrete. Consequently, I shall make no distinction on the basis of age.

Nevertheless, it is useful to note the temporary and weather-dependent character of efflorescence. Efflorescence forms most readily when concrete is wet but becomes apparent with the onset of dry weather.[6] Much depends on local changes in relative humidity, temperature, and insolation. Often, efflorescence appears in foggy periods in the fall.[7] Generalizations are difficult because words like "warm" or "humid" are relative and do not have absolute meaning. Generally, when the rate of evaporation is high, salts are left below the surface of concrete so that efflorescence is less apparent. On the other hand, when the rate of evaporation is low, efflorescence becomes more pronounced. What matters is that efflorescence often comes and goes.

Chemical composition of efflorescence

As I see it, efflorescence is a crystalline deposit of soluble salts from the cement that have migrated to the surface of concrete, precipitated there, and reacted with carbon dioxide. The main soluble salt is calcium hydroxide, which is one of the products of hydration of di- and tri-calcium silicate in clinker. The removal of calcium hydroxide has led some investigators to refer to it as "water soluble;"[8] in turn, this has been interpreted in some lawsuits as decalcification of hydrated cement past, leading to a loss of cohesion and of strength, with inevitable damage to follow.

In my opinion, these views are erroneous. First of all, portland cement is a hydraulic cement, that is, a material that develops strength by reaction with water, with the resulting solid mass remaining strong in contact with water. Indeed, these features are the *forte* of portland cement. The formal statement in ACI 116R-90 is: "cement, hydraulic—a cement that sets and hardens by chemical interaction with water and is capable of doing so under water." Secondly, the solubility of calcium hydroxide is low: typically 1.2 grams of CaO per liter. Interestingly, the solubility in water decreases with an increase in temperature, ranging (in grams of CaO per liter) from 1.30 at 0 °C to 1.13 at 25 °C.

There is also circumstantial evidence of the lasting nature of structures submerged in water as well as of structures in contact with water over prolonged periods. For example, many gravity dams have a permeability such that there is some transport of water through the concrete to galleries that collect and remove it, and yet I am not aware of any record of mass removal of calcium hydroxide and decalcification damage in dams.

These facts are very important and they should demolish the proposition that, because efflorescence is the consequence of leaching of calcium hydroxide, efflorescence is a proof of decomposition of hydrated cement paste and therefore of weakening of concrete.

The reaction of calcium hydroxide with carbon dioxide may be written as:
$$Ca(OH)_2 + CO_2 \Rightarrow CaCO_3 + H_2O$$
Gaseous CO_2 is not reactive so that presence of water is necessary, but it is water that brings calcium hydroxide to the surface of concrete. Subsequently, wetting by rain or other sources provides the necessary conditions for the formation of calcium carbonate.

Calcium hydroxide is not the only compound in hydrated portland cement paste that is carried by water to the surface of concrete. Alkali hydroxides can also be transported, and they become deposited as sodium and potassium carbonates; this is much less so with cements having a low content of water-soluble alkalies. The alkalies are conducive to efflorescence because they greatly increase the solubility of carbon dioxide. It is likely that the relatively insoluble calcium carbonate is precipitated and more calcium hydroxide is leached.[9]

Efflorescence may also contain salts from unwashed, or inadequately washed, seashore aggregate. I shall consider this in Section 5.4 in connection with the presence of chlorides in the mixture.

Other salt deposits on concrete

In addition to salts originating from the hydrated cement paste or from other ingredients of the concrete mixture, there may be, deposited on the surface of concrete, salts whose origin is *outside* the given concrete element but which have traveled through it. Often, these are sodium or calcium sulfates. Their origin is in the groundwater in contact with the concrete element. Their transport through the concrete may involve chemical reactions with the hydrated cement. Such reactions may be deleterious.

Specifically, when water is continuously transported through concrete to a drying surface, extensive reactions can take place: sodium sulfate attacks calcium hydroxide and calcium aluminate hydrate; calcium sulfate attacks calcium aluminate hydrate; and magnesium sulfate is particularly destructive in that it attacks not only the two above-mentioned compounds but also calcium silicate hydrates.

These reactions take place principally inside the concrete, and their products remain within the concrete. In the absence of a drying surface, for example, when concrete is immersed in a static sulfate solution, the extent of the reaction is limited and, of course, there is no dry deposit on the surface. It is also possible for these reactions to take place at the surface of a concrete element that is in contact with sulfate-laden ground or sulfate-bearing water.

All these reactions come under the heading of sulfate attack, and their avoidance lies in minimizing the reactive compounds in the cement and in preventing contact between sulfate-laden groundwater and concrete. The quality of concrete is relevant in so far as it influences the ease with which the sulfates are transported.

What is relevant to the present article is that surface deposits may consist of, or contain, salts other than calcium carbonate. The chemical composition of efflorescence is the key to deciding what is the origin of the salt deposit and, therefore, and what is the significance to the health of the concrete. The presence of *any* white deposit in not proof of damage *per se*.

It follows that the determination of the chemical composition of the deposits is very important and should be performed using accepted and reliable test methods, such as X-ray diffraction or wet chemistry if the sample is large enough. This statement may be thought to be so obvious as to be superfluous. And yet, I have encountered a chemist who used taste as a tell-tale of chemical composition of efflorescence, and insisted that his taste buds were reliable. As someone who has recently argued in a *CI* article[10] for the proper use of standard test methods, I deplore the use of a "suck-it-and-see" approach.

It is known that sulfate attack may result in damage to concrete consequent upon removal of a significant amount of hydrated cement and the formation of expansive compounds. On the other hand, I have come across surface deposits of sodium sulfate that were not the consequence of a reaction with hydrated cement paste. My understanding of the situation is that sodium sulfate in groundwater traveled through a small thickness of concrete to an evaporative surface, where it emerged as sodium sulfate; no damage had occurred. Brown and Badger,[11] in a study of 4-in.-thick (100 mm) slabs on grade, stated: "Depending on the rates of transport and the presence of evaporative surfaces, a portion of the alkali sulfates will pass through the concrete pore structure and produce efflorescence at these evaporative surfaces."[11] It is only fair to add that their views on sulfate attack do not agree with mine.

In several cases, these deposits were found at the leading edge of garage floors; the leading edge was adjacent to a drive slab but separated from it by a gap. Why the deposit formed there and not elsewhere was baffling, and it is only when reading a paper by Pickel, et al.[12] that I found a plausible explanation. A brief description of what happened is necessary.

Pickel, et al. describe how textured precast concrete paving slabs were laid on a bedding mortar placed on top of an in-place concrete slab.[12] To achieve a particular appearance, the gaps between the slabs were filled with sand; there was no adequate provision for drainage below the paving slabs. Some time later a white deposit was observed on the surface of the paving slabs in the vicinity of the joints. The explanation given by Pickel, et al.[12] is as follows. Rainwater penetrated through the sand and, not being able to drain, leached some calcium hydroxide from the bedding mortar, which was more permeable than the dense precast slabs. During a subsequent dry period, the leachate traveled upwards along the vertical faces of the paving slabs and, when water evaporated, efflorescence formed on the paving slabs near their edges. So it was not the concrete in the paving slabs that was the source of the salt deposit.

I find the above situation to be very similar to that in the garages except that, in the latter case, the salt being transported upwards was sodium sulfate from the soil at the bottom of the gap between the garage slab and the drive. Thus there was no significant chemical damage to concrete because the sodium sulfate in solution did not react chemically with the hydrated cement paste near the edge of the concrete garage floor, and was finally deposited as sodium sulfate.

Salt crystallization

Nevertheless, we should recognize that the formation of some salt crystals in the pores in concrete can be damaging. This is usually referred to as salt crystallization; when the crystals are sulfates, the term physical salt attack is sometimes used. A fairly common case is the conversion of thenardite (anhydrous sodium sulfate) to mirabilite (sodium sulfate decahydrate) and vice versa in consequence of change in the ambient relative humidity or in temperature, or in both of these. According to Hime, et al. the conversion involves a threefold expansion.[13]

The arguments about volumetric expansion and consequent disruptive pressure are complex and by no means universally agreed. These topics are outside the scope of the present section. Nonetheless, some remarks about salt crystallization are necessary because the conversion of an anhydrous salt into a hydrated form is not universally accepted as salt crystallization. Specifically, Hime, et al.[13] limit the term salt crystallization to salt crystallizing from a supersaturated solution and causing damage. According to them, the consequences of conversion and reconversion of a salt should be termed salt hydration distress.

While conscious of not being a cement chemist, and looking at the situation from the standpoint of concrete technology, I view the distinction as being somewhat pedantic. Moreover, the term distress is emotive rather than cooly descriptive. Given my age and appearance, I would not like my numerous wrinkles to be described as skin distress; I can think of other, kinder, terms such as skin folds, loss of skin tension, if not simply wrinkles.

Conclusions

This article has shown that our knowledge of the crystal deposits of efflorescence is not crystal clear (no pun intended). Nevertheless, I hope that my review has made it possible to assess the situation in practice in a way such that we shall avoid unnecessary problems with efflorescence and also avoid condemning concrete just because it has a surface blemish. What is on the surface may not be of paramount importance. As Thomas Gray says in the "Elegy Written in a Country Churchyard":

> *Full many a flower is born to blush unseen,*
> *And waste its sweetness on the desert air.*

References

1. Russell, P., *Efflorescence and the Discoloration of Concrete*, Viewpoint Publications, Eyre & Spottiswoode Ltd, Surrey, England, 1983, 41 pp.

2. Neville, A. M., *Properties of Concrete*, Fourth Edition, Longman, London, 1995, and John Wiley, New York, 1996, 844 pp.

3. Bensted, J., "Efflorescence—Prevention is Better Than Cure," *Concrete*, Sept. 2000, pp. 40-41.

4. Schaffer, R. J., "The Weathering of Natural Building Stone," *Special Report*, No. 18, Building Research, Dept. of Scientific and Industrial Research, London, 1932, pp. 56-72.

5. Aïtcin, P.-C., *Lime Binders*, Spon, 2002.

6. Higgins, D. D., "Efflorescence on Concrete," *Appearance Matters*, No. 4, Cement and Concrete Association, 1982, 8 pp.

7. Kresse, P., "Coloured Concrete and its Enemy: Efflorescence," *Chemistry and Industry*, Feb. 1989, pp. 93-95.

8. Chin, I. R., and Petry, L., "Design and Testing to Reduce Efflorescence Potential in New Brick Masonry Walls," *Masonry: Design and Construction, Problems, and Repair*, ASTM 1180, Philadelphia, PA, 1993, pp. 3-17.

9. Dow, C., "Efflorescence on Concrete Products," *PhD Dissertation*, University of Aberdeen, 1998.

10. Neville, A., "Standard Test Methods: Avoid the Free-For-All," *Concrete International*, V. 23, No. 5, May 2001, pp. 60-64. [Section 7.1]

11. Brown, P. W., and Badger, S., "The Distributions of Bound Sulfates and Chlorides in Concrete Subjected to Mixed NaCl, $MgSO_4$, Na_2SO_4," *Cement and Concrete Research*, V. 30, 2000, pp. 1535-1542.

12. Pickel, P.; Permesang, C.; and Hofmann, O., "Efflorescence on Terrace Paving Slabs," *Betonwerk und Fertigteil-Technik*, V. 55, No. 7, 1989, pp. 43-48.

13. Hime, W. G.; Martinek, R. A.; Backus, L. A.; and Marusin, S. L., "Salt Hydration Distress," *Concrete International*, V. 23, No. 10, Oct. 2001, pp. 43-50.

14. Neville, A., "Seawater in the Mixture," *Concrete International*, V. 23, No. 1, Jan. 2001, pp. 48-51. [Section 2.4]

15. Griffin, D. F., and Henry, R. L., "Integral Sodium Chloride Effect on Strength, Water Vapor Transmission, and Efflorescence of Concrete," *Journal of the American Concrete Institute*, Dec. 1961, pp. 751-770.

16. Butterworth, B., "Contributions to the Study of Efflorescence, Part VIII. The Camerman Theory," *Transactions of the British Ceramic Society*, V. 53, No. 9, 1954, pp. 563-607.

Section 5.4: Efflorescence—Surface Blemish or Internal Problem? Part 2: Situation in Practice

In Section 5.3, I reviewed our knowledge of efflorescence. This knowledge is not extensive, but we must answer, if only provisionally, the question posed in the title of Sections 5.3 and 5.4. To do this we need to review the observations of efflorescence, as well as their significance, and finally, to look at ways of minimizing the occurrence of efflorescence.

Reported observations of efflorescence

Traditionally, technical papers, especially those academic in character, start by reviewing the literature and end this by conclusions. In Section 5.3, I expressed my views and gave supporting evidence. It is only at this stage that I propose to discuss the literature on efflorescence, much of it dealing with masonry.

My reason for this sequence is that in most of the existing literature efflorescence is mentioned only peripherally. Usually, when discussing damage to a concrete structure, there is a remark on the lines that, in addition to spalling and cracking, efflorescence was observed; no further information on efflorescence, either in terms of its chemical composition or its significance, is presented and there is no discussion of the role of efflorescence in the damage to the structure concerned. In my opinion, bandying around expressions of the type that there was efflorescence in the damaged structure can be misleading, and is of no help to future investigators of damaged structures.

Here are some examples. In a description of the rehabilitation of a "massive concrete arch structure," published in 1994, Nagaraja and Khan[1] say, "reports of efflorescence from the viaduct and general awareness of its (mostly esthetic) impact go back to the late 1960s"; the parentheses are in the original paper. No further comment on efflorescence is made, presumably because it is only of aesthetic interest.

In a case history of a dam spillway, Coghlan and Vanderpoel note "progressive concrete deterioration ... evidenced by progressive cracking, spalling, raveling rock pockets, and efflorescence."[2] No further reference to efflorescence is made.

Similarly, Ojha describes the rehabilitation of a major bridge, stating in the synopsis that the "bare concrete decks exhibit extensive deterioration in the form of spalls, cracks, leakage, and efflorescence."[3] The word "efflorescence" is not even mentioned in the body of the paper.

By contrast, Hadden reports some useful facts.[4] He emphasizes the role of rain in the formation of efflorescence in the Baha'i House of Worship in Illinois: "white sheets and stalactites ... almost pure calcium carbonate formed when rainwater leaked through the gutter on top of the cornice and trickled down into the concrete."[4] The action of the rainwater is ascribed to its being a dilute carbonic acid, which slowly dissolves calcium hydroxide.[4]

The reference to the absence of gutters is relevant to the occurrence of efflorescence in some housing construction in Southern California where gutters are absent. Of course, this is a simple example of the fact that water continuing to come into contact with concrete is an essential ingredient of the development of efflorescence.

The formation of stalactites was reported also by Fukuda et al. in numerous bridges close to the seashore in Japan; they were a consequence of serious leakage of water between precast units and the in-filling concrete.[5] No information on the chemical composition of the deposited salts is given. I suspect that they did not emanate from the concrete and are therefore not related to concrete as a material, but rather to poor construction practice.

Efflorescence in masonry construction

Most of the early studies on efflorescence were done on masonry, either of the stone or brickwork type, joined by mortar. Brick itself may lead to efflorescence, but this is not relevant to concrete. If the brick does not release salts, efflorescence in masonry can be traced to jointing mortar containing portland cement. This is an important point, particularly because in 1949, Camerman, as reported by Butterworth, suggested that "neither bricks nor mortars normally cause efflorescence, but that efflorescence is the result of chemical reaction between bricks and mortar that occurs when new brickwork is saturated at the time of erection."[6]

Camerman's theory was shown to be incorrect,[6] and we now know that it is mortar containing portland cement that can lead to efflorescence in brickwork. An indirect confirmation of this situation is provided by the fact that soft lime mortar (which does not contain portland cement) does not lead to efflorescence whereas hard lime mortar, containing portland cement, may result in efflorescence on the surface of the brickwork.[7]

Although the underlying concern with efflorescence is that it is aesthetically displeasing, this view is not shared by everybody. In Southern California, where some lawsuits include claims involving efflorescence on concrete, I have seen, in the very same locations, bricks with mock efflorescence. Presumably, some people think that such bricks look old and weathered, and give the house an "old-world" charm.

Concrete containing chlorides

It is nowadays accepted that concrete containing embedded steel should not contain chlorides.[8] This rule does not apply to plain concrete; moreover, there exist some older structures that were built with chloride in the mixture, often originating from seawater used as mixing water. In such structures, there develops a deposit, predominantly consisting of sodium chloride, in a shape that has lead to the term "salt crystal whiskers."[9] Personally, for reasons obvious from my photograph on page 9-27 of this book, I take great exception to that term.

Detailed information about efflorescence in concrete containing chlorides is given by Griffin and Henry,[9] but nowadays this is of little interest. The only useful observation is that, because salt absorbs moisture from the air, such concrete may exhibit almost permanent dampness at surfaces exposed to the air so that plaster or paint cannot be applied; clearly, appearance is also adversely affected.[10]

Test for efflorescence

There exists no standard test method for efflorescence on concrete. With respect to bricks, as far back as 1933, Butterworth said that efflorescence is essentially a qualitative phenomenon so that it is not possible to devise a test method that would give a numerical answer.[11] The reason for this is that the extent of efflorescence in any given structure depends on a number of factors that cannot always be established or quantified. There exists an ASTM Standard Test Method C 67-00, which provides for a visual comparison between bricks that had been in contact with water and those kept dry; looking at the bricks from a distance of 3 m (10 ft) an observer classifies the bricks as "effloresced" or "not effloresced." This method is not intended to be used on concrete, and I do not think that to do so would be of value.

Minimizing efflorescence

Prevention or minimization of efflorescence lies in structural details that do not allow water to flow unnecessarily through concrete or over its surface; the solutions lie more in geometry and exposure of the structural elements than in the concrete mixture. Water, especially when containing dissolved salts, should be drained away from concrete foundations.

Likewise, water should be kept away from the exterior of basement walls because the higher temperature and lower relative humidity indoors will encourage the formation of efflorescence.[12] According to Anderegg, problems in basement walls "may have resulted from improper grading around the house or failure to drain off downspout water. A little carelessness with these details often gives much trouble."[13] These wise words were written in 1952, but even in 2001, I saw similar problems due to the same causes.

Less permeable concrete will lead to less efflorescence. Formwork conducive to concrete with low porosity near the surface (especially controlled permeability formwork) reduces the incidence of efflorescence.[14]

Inclusion of pozzolans and silica fume reduces efflorescence because these materials react with calcium hydroxide so that it is unavailable for leaching. Absence of calcium hydroxide in autoclaved concrete means that such concrete exhibits very little, or no, efflorescence. High-alumina cement in concrete does not lead to efflorescence because its hydration does not result in the presence of calcium hydroxide,[15] but such cement is not used in structural concrete.[16] In concrete made with alkali-activated slag cement, a small amount of efflorescence may occur but has been found in most cases not to be serious.[17]

Significance of efflorescence

We can now attempt to answer the question in the title of Section 5.3 and 5.4. In the vast majority of cases, efflorescence consists of the product of carbonation of calcium hydroxide leached from the concrete and deposited on its surface. In such a situation, efflorescence mars the appearance of concrete (which may or may not matter) but does not represent damage or weakening of the structure. This is the view of Bensted who says: "Efflorescence is ugly but not normally damaging to the exposed concrete or brickwork."[18]

Bensted also says: "Even with lime weeping, the durability of the structure is not normally in question."[18] However, lime weeping may be an indication that water is flowing through the concrete, and this in itself is undesirable.[19] Lime weeping is sometimes seen on earth-retaining walls with weep holes that do not shed water clear off the face of the concrete; this can be readily avoided.[19] Badly-made joints in such walls can also result in unsightly appearance, but then joints should not be made badly!

The presence of efflorescence on surfaces of colored concrete, be it structural elements, or paving slab or block, is undesirable because the deposit is usually non-uniform so that the apparent color is variable and certainly not as intended.[20] White portland cement leads to less, or even no, efflorescence because of its very low alkali content; also, white efflorescence on a white background is not readily discernible.

It would be wrong to interpret the above to mean that, generally, we need not avoid situations conducive to efflorescence. Nevertheless, we must recognize that, under certain conditions of construction and exposure, efflorescence is inherent in concrete construction and unavoidable.

If efflorescence consists of, or contains, large amounts of salts other than calcium carbonate, then the significance of efflorescence depends on its exact chemical composition. The presence of sulfates *may* indicate that sulfate attack is in progress but does not offer conclusive proof.

To establish that damage has occurred, it is essential to demonstrate that deleterious reactions have occurred in the interior of the concrete or at its surfaces in contact with sulfates in the surrounding soil. Specifically, surface deposits of sodium sulfate may be harmless because they have not reacted chemically with hydrated cement paste. However, even if these surface deposits do not signify sulfate attack, under some situations of exposure involving cycling of temperature or relative humidity, sodium sulfate may change its crystal form between thenardite and mirabilite, with consequent surface scaling.[21] When this happens, we are dealing with a physical phenomenon that is, so far, not well understood, and one that is outside the scope of the present section.

In my view then, the appearance of efflorescence—this blooming and flowering on the surface of concrete—does not usually indicate an internal problem. And I would agree with the words in the comic opera (nowadays called a musical) *The Mikado* by W.S. Gilbert and Arthur Sullivan:

"The flowers that bloom in the spring,
Tra la,
Have nothing to do with the case."

And further:

"Oh, bother the flowers that bloom in the spring."

References

1. Nagaraja, M., and Khan, R.A., "Rehabilitation of Flushing Viaduct," *Concrete International*, V. 16, No. 8, Aug. 1994, pp. 44-47.

2. Coghlan, G. T., and Vanderpoel, A., "Vesuvius Dam Spillway," *Concrete International*, V. 8, No. 5, May 1980, pp. 42-44.

3. Ojha, S. K., "Rehabilitation of Verrazano Narrows Bridge Approach Roadway Decks in New York City," *Concrete Bridges in Aggressive Environments*, SP-151, American Concrete Institute, Farmington Hills, MI, 1997, pp. 245-260.

4. Hadden, D., "Cleaning Restoration of the Baha'i House of Worship," *Concrete International*, V. 14, No. 9, Sept. 1992, pp. 44-51.

5. Fukuda, S., et al., "Durability of Prestressed Concrete Bridges in the North-East District of Japan," *Fourth CANMET/ACI International Conference on Durability of Concrete*, SP-170, American Concrete Institute, Farmington Hills, MI, 1997, pp. 1447-1465.

6. Butterworth, B., "Contributions to the Study of Florescences, VI," *Transactions of the Ceramic Society*, V. 32, June 1933, pp. 270-283.

7. Schaffer, R. J., "Weathering of Natural Building Stone," *Special Report*, No. 18, Building Research, Dept. of Scientific and Industrial Research, London, 1932, pp. 56-72.

8. Neville, A., "Seawater in the Mixture," *Concrete International*, V. 23, No. 1, Jan. 2001, pp. 48-51. [Section 2.4]

9. Griffin, D. F., and Henry, R. L., "Integral Sodium Chloride Effect on Strength, Water Vapor Transmission, and Efflorescence of Concrete," *ACI Journal*, Dec. 1961, pp. 751-770.

10. Lea, F. M., *The Chemistry of Cement and Concrete*, Third Edition, Edward Arnold, 1970, 727 pp.

11. Butterworth, B., "Contributions to the Study of Efflorescence, Part VIII, The Camerman Theory," *Transactions of the British Ceramic Society*, V. 53, No. 9, 1954, pp. 563-607.

12. Haage, R., "Efflorescence on Bricks and Masonry—Origin, Removal, Prevention," *Ziegelindustrie International*, V. 44, No. 4, 1991, pp. 170-175.

13. Anderegg, F. O., "Efflorescence," *ASTM Bulletin*, Oct. 1952, pp. 39-45.

14. Christen, H. U., "Conditions Météorologiques et Efflorescences de Chaux," *Bulletin du Ciment*, V. 44, No. 6, 1976, 8 pp.

15. Bensted, J., "Efflorescence—A Visual Problem on Buildings," *Construction Repair*, Jan./Feb. 1994, pp. 47-49.

16. Neville, A., "A 'New' Look at High-Alumina Cement," *Concrete International*, V. 20, No. 8, Aug. 1998, pp. 51-55. [Section 1.5]

17. Wang, Shao-Dong, et al., "Alkali-Activated Slag Cement and Concrete: A Review of Properties and Problems," *Advances in Cement Research*, V. 7, No. 27, 1995, pp. 93-102.

18. Bensted, J., "Efflorescence—Prevention is Better Than Cure," *Concrete*, Sept. 2000, pp. 40-41.

19. Higgins, D. D., "Efflorescence on Concrete," *Appearance Matters*, No. 4, Cement and Concrete Association, 1982, 8 pp.

20. Aïtcin, P.-C., *Lime Binders*, Spon, 2002.

21. Hime, W. G., et al., "Salt Hydration Distress," *Concrete International*, V. 23, No. 10, Oct. 2001, pp. 43-50.

Section 5.5: Autogenous Healing—A Concrete Miracle?

Many people are aware of autogenous healing, but a fairly common attitude is: so what? We cannot design a structure on the assumption that autogenous healing will take place but, under certain circumstances, the occurrence of autogenous healing can be highly beneficial. It is, therefore, useful to know how autogenous healing works, when it works, how to promote it, and how to take advantage of it.

Genesis of this article

Recently, I was asked to express an opinion on autogenous healing under somewhat unusual circumstances. What was required was an assessment of the extent of autogenous healing that can be expected and hence a prognosis for the durability of the particular structure. As soon as I approached the problem, I realized that our knowledge of autogenous healing is scanty and it has not been coherently reviewed for a long time. Indeed, the last overview of autogenous healing was written by Clear in 1985.[1] Thus, I was not able to answer immediately the questions put to me and, in order to obtain background information, I undertook a literature search. This is the genesis of the present article, written in the hope that it may be of help to others in the future.

What is meant by "autogenous"?

The word "autogenous" entered the English language from Greek in mid-nineteenth century; it means "self-produced." According to the *New Shorter Oxford Dictionary*, an especial meaning with respect to welding is "formed by or involving the melting of the joined ends, without added filler."

The word is, therefore, entirely appropriate to what happens in concrete when healing takes place by restoring continuity between two sides of a crack without a deliberate external intervention of repair.

Practical significance of autogenous healing

Situations where autogenous healing may be beneficial were given by Turner, as far back as 1937, as: damaged precast concrete elements; piles damaged by handling or driving; cracked water pipes made of concrete or lined with cement mortar; tanks that were allowed to dry out excessively; and green concrete disturbed by vibration or shock.[2]

The occurrence of autogenous healing and the benefits therefrom are especially significant in a reduction in water transport through the cracks and in improving the protection of embedded steel from corrosion. The latter is important mainly when the water contains chlorides. In some situations, the recovery of strength or of the modulus of elasticity is also of interest.

How does the healing work?

The process of autogenous healing occurs between opposing surfaces of narrow cracks. In the vast majority of cases, the cracks were caused by shrinkage extensive enough to induce locally a strain larger than the tensile strain capacity of the concrete; this means that the tensile strength of concrete at the given location has been reached.

Healing can take place only in the presence of water because the healing consists of chemical reactions of compounds exposed at the cracked surfaces. These reactions produce new hydrates and other minerals. The accretion of these from the opposing surfaces of a crack eventually bridges the crack so that continuity is re-established.

The essential requirement is for the presence of compounds capable of further reaction. Thus, it is the cement, hydrated or unhydrated, that is the essential element in autogenous healing. Clearly, we are concerned with cement at or near the surface of the crack; this cement is the parent part of concrete or mortar.

The reactions of healing

To my knowledge, it has not been established conclusively what are the chemical reactions of healing. There are two possibilities: the formation of calcium hydroxide and of calcium carbonate. The former requires the presence of water only; the second requires, in addition, the presence of carbon dioxide.

A third mechanism that can contribute to healing, but cannot provide it by itself, is silting up of cracks or deposition of debris.

Whatever the chemical reactions that take place, the presence of water is essential; this will be discussed more fully later. Because no simple statement about the crack-filling material (or filler) can be made, a brief literature review may be useful.

Is it continued hydration?

Opening of cracks, regardless of whether due to shrinkage or to excessive tensile strain, exposes the interior of cement paste, including cement hydrates as well as the hitherto unhydrated remnants of cement powder. As long as they are exposed to air, no autogenous healing takes place. However, when the air becomes replaced by water, hydration re-starts and calcium hydroxide, as well as calcium silicate hydrate, are formed. The presence of carbon dioxide is not necessary, and carbon is not involved in the new products. There exists considerable evidence of the above phenomena.

Hearn recognized the role of further hydration of cement and the formation of calcium hydroxide.[3] However, her literature study led her to report the formation of calcium carbonate as well. In addition, she introduced the concept of a self-sealing effect, but I have a difficulty in understanding this classification, especially since she says, in one place, that the self-sealing effect encompasses both autogenous healing and continued hydration; and, in another place, she distinguishes self-sealing, autogenous healing, and continuing hydration as three separate phenomena.[3] She mentions also physical clogging of cracks.

Turner also recognized further hydration of cement at cracked surfaces, as well as continued hydration of already formed gel, and also inter-crystallization of fractured crystals.[2] He did not explain the latter two phenomena, and their exact nature is not obvious to me.

Lauer and Slate determined by petrographic analysis that the new material in a healed crack in a tension briquet consisted of calcium carbonate and calcium hydroxide.[4] They explained the presence of calcium carbonate by the reaction of carbon dioxide in ambient water or air with calcium hydroxide present at the crack surface.[4] When this calcium hydroxide has been consumed in this reaction, more of it migrates from the interior of the concrete.[4] Unfortunately, Lauer and Slate did not explain the driving force for this migration. The calcium carbonate crystals grow preferentially outwards from the crack surface because the space available within the hydrated cement paste is limited.[4] Wagner also found the crack filler to consist of calcium carbonate.[5]

Is it formation of calcium carbonate?

We can see thus that the formation of calcium carbonate, alone or together with calcium hydroxide, is the second possible mechanism of autogenous healing. A condition for this is that the water in the crack contains a large amount of dissolved carbon dioxide.[6]

Clear found the formation of calcium carbonate to be significant in later stages of exposure of cracks to water, but this mechanism is not predominant in the first few days.[1] In his experiments, early reduction in the flow of water through a crack (which indicates progress of autogenous healing) was caused by blocking with loose particles already present in the crack.[1]

On the other hand, Edvardsen found that blocking and swelling of hydrated cement paste had minimal influence.[7] According to her, in the initial phase, there is a reaction between calcium ions and carbon dioxide at the surface of the crack.[7] Once the calcium ions at the surface have been used up, further calcium ions are transported from the hydrated cement paste deeper in the mortar or concrete, the process being diffusion-controlled.[7]

Edvardsen found that calcium carbonate is "almost the sole cause" of autogenous healing.[7] She expressed the view that the availability of carbonate ions is not the controlling factor in the formation of calcium carbonate.[7]

In any case, calcium hydrogen carbonate, which is one source of carbonate ions, is present in many waters. Carbon dioxide is also present in solution in water. Calcium carbonate is, of course, almost insoluble. In this connection, we should note Edvardsen's finding that water hardness seems not to influence the process of autogenous healing; surprisingly, nor does the value of pH.[7]

As for chemical effects, it appears that the reactions involved in producing the crack filler involve portland cement only. Specifically, Gautefall and Vennesland found that silica fume in the mixture had no influence on autogenous healing.[8]

What is the filler material?

The preceding review does not lead to a clear and unequivocal answer to the question: what is the filler material that has resulted in autogenous healing?

In *Properties of Concrete*, Fourth Edition, I said that autogenous healing "is due primarily to the hydration of the hitherto unhydrated cement."[9] I now believe that this is true only in very young concrete, in which the fracture is jagged so that it exposes some unhydrated parts of cement. However, later on, the predominant product in the crack filler is calcium carbonate. A practical conclusion from this is that, when it is intended actively to promote autogenous healing, it is highly beneficial to ensure an ample supply of carbon dioxide in the water involved in the healing process.

Requirements for the presence of water

The filler in the cracks is either the product of hydration or it is formed in water. It follows that the presence of water in the cracks is essential.

If the cracking was caused by shrinkage, the relevant part of the concrete must have been exposed to drying. As long as the drying conditions exist, the crack will remain "as is" or even become wider. It is only on wetting that autogenous healing can take place. The water can be stationary or flowing.

The wetting has to be thorough, that is, the crack has to be inundated. It was found that, even when the relative humidity of the air was as high as 95%, the extent of healing was much lower than in water; moreover, the healing was erratic.[4]

My interpretation of the very small extent of autogenous healing in humid air, even almost saturated, is two-fold. Hydration of hitherto unhydrated cement is faster in water, and also water encourages the leaching of calcium hydroxide from the parts of the concrete somewhat remote from the crack surface. More importantly, very little carbonation can take place in air because only carbon dioxide dissolved in the surface films of water is available.[4] Carbon dioxide in gaseous form does not react with calcium hydroxide.

In some situations, uninterrupted wetting of crack surfaces is not practicable. Fortunately, periodic wetting, but without periods of low relative humidity in between, results in the healing process, but may not produce a full closure of cracks.[4] It follows that, when it is planned to benefit from the process of autogenous healing, full contact of the crack surfaces with water is essential.

In addition to promoting chemical reactions, the presence of water in the cracks and in their vicinity has some other beneficial effects. For example, in pipes, autogenous healing may be supplemented by the expansion of the mortar lining owing to the absorption of water into the previously dried mortar.[5] It is thus that the wetting part of the moisture movement, which is a reversible deformation, augments the process of autogenous healing.

In some situations, temperature changes contribute to the closing of cracks. I have seen some very large diameter concrete pipes that had been exposed for a long period to drying in an arid climate; not surprisingly, extensive shrinkage cracks have opened. Subsequently, the pipes were put into service to carry water at a much lower temperature than the previous air temperature. The resulting thermal contraction had a positive effect on the closing of cracks. There is nothing surprising in this but the presence of thermal effects illustrates the difficulty of predicting autogenous healing and the associated closing of cracks. An additional difficulty arises from the fact that some pipes are made of reinforced concrete, others are prestressed, and yet others are metal pipes with mortar lining.

Maximum width of cracks that heal

An important practical question is: what is the maximum width of cracks that will be closed by autogenous healing?

Reports on various crack widths that have undergone healing are of interest in pointing towards "safe" crack widths. However, various investigators report different maximum widths of cracks. This is not surprising because the test conditions have varied widely. In some cases, the cracks were caused by shrinkage, in others by the application of tension, usually flexural, but in some tests by direct tension. The age at the opening of cracks varied, too. The healing took place in static water or flowing water. There was a head of water or not. The water was fresh or seawater. The material undergoing the autogenous healing was concrete or mortar. The combinations of these conditions are numerous, so that generalizations about the maximum width of cracks that will heal are not possible. Nevertheless, a review of the published test results presents a useful background.

Loving reported that large, 1.5 to 2.4 m (5 to 8 ft) reinforced concrete pipes that developed shrinkage cracks up to 0.8 and 1.5 mm (0.03 and 0.06 in.) wide, and were subsequently put into service, were found to have the cracks completely closed by autogenous healing five years later.[10]

Wagner reported autogenous healing in mortar lining of metal pipes: a crack 0.33 mm (0.013 in.) wide was still open after 30 days' immersion in city water, but healing had taken place below the surface, and was complete in places.[5] In another case, autogenous healing in a concrete pipe resulted in sealing of cracks up to 0.76 mm (0.03 in.) wide after 5 years; one crack, 1.56 mm (0.06 in.) wide became sealed.[5]

Gautefall and Vennesland reported that, when immersed in seawater, concrete specimens with cracks more than 0.6 mm (0.024 in.) wide were "susceptible to corrosion attack" but this did not happen when the cracks were less than 0.4 mm (0.016 in.) wide.[8] These tests were conducted under conditions such that ample oxygen was available at a separate cathode, which was remote from the anode; such a situation is unlikely to be common in real-life structures.

The relevance of the presence of seawater is that it may be conducive to corrosion of embedded or underlying steel exposed by the crack. The opinion of Lea on the maximum crack width sealed by autogenous healing carries considerable weight.[11] He wrote: "Provided the width at the surface is not more than about 0.2 mm (0.008 in.) the presence of such cracks does not usually lead to any progressive corrosion of the steel, though the critical width depends on the thickness of the concrete cover and the exposure conditions."[11] Fuller consideration of the possibility of corrosion will be discussed in a later subsection.

Edvardsen found that one-quarter to one-half of cracks 0.20 mm (0.08 in.) wide healed completely after 7 weeks of water exposure; the proportion of cracks closed depended on the water pressure.[7] With a crack width of 0.30 mm (0.012 in.), the flow through the crack was reduced five-fold after 15 days in water under a head of 2.5 m (100 in.). The use of mean values of crack width has to be interpreted to signify that some cracks had a greater width, and the bulk of flow of water would occur through the wider cracks.[7]

Jacobsen et al.[12] reported extensive data on autogenous healing, but this took place in lime-saturated water. As such conditions are unlikely to exist in a real-life structure, the results of their tests are of very limited interest with respect to the subject matter of this article.[12]

The widths of cracks that have healed and the length of the period of healing, cited earlier, are related to one another. This is only to be expected, but above a certain width, adequate autogenous healing will not take place. Also, beyond a period of about 3 months, significant healing stops.

Regardless of autogenous healing, cracks up to a certain width in reinforced concrete are inevitable and acceptable. For example, the British structural design code CP 110:1972 (drafted by a committee of which I was a member) gives the following as a serviceability limit state requirement: "An assessment of the likely behaviour of a reinforced concrete structure should show that the surface width of crack would not, in general, exceed 0.3 mm."

Tolerable crack dimensions in pipes are prescribed by the British Standard B.S. 534:1990 Specification for steel pipe ... for water and sewage. These apply both to concrete pipe and to mortar-lined pipe. The specific statement is: "Cracks up to 0.25 mm in width in saturated linings and not over 300 mm in length shall not be a cause for rejection." First of all, we should note that the British Standard applies to the acceptance of a newly manufactured pipe. At that stage, excessive shrinkage should not have occurred; indeed, the extent of drying shrinkage is largely under the control of the pipe manufacturer.

Secondly, we should note the word "saturated." At the time of acceptance of the pipe by the purchaser, the pipe can be maintained in a saturated state by sprinkling with water or by other means. However, on site, the pipe is likely to be exposed to dry air, and in some cases also to wind, and it is then that the cracks open or widen. A dispute may arise as to whether the cracks will close by autogenous healing in service. The problem is that inspection of the pipe and the measurement of crack width take place, of necessity, in an open pipe, and therefore a dry, or greatly dried out, state.

Recovery of strength on healing

In the majority of published studies, the parameter investigated was the extent of autogenous healing as evidenced by the filling of cracks or by the reduction of flow of water through the cracks. The strength of the healed concrete is rarely of interest and has not often been determined.

It is arguable that full healing of a crack makes the concrete or mortar monolithic, and therefore "as good as new," or nearly so. Strictly speaking, the development of strength is a function of the extent of complete bridging of the crack and of the proportion of the volume of the crack that has become filled by the new compounds.[4] Laboratory tests have provided detailed quantitative data on the relation between the extent of filling of cracks by autogenous healing and recovery of the initial tensile strength.[4] However, such information is largely of academic interest because, in actual structures, little is known about crack sizes: unlike laboratory experiments, the cracks vary in their dimensions. Thus, the extent of healing is difficult to quantify. What matters is the maximum width of cracks that are expected to heal; this was discussed earlier.

A condition necessary for a successful recovery of strength through autogenous healing is that there be no longitudinal displacement of the concrete on opposite sides of a crack; in other words, the "fit" must not be disturbed. On the positive side, sustained compression across the plane of the crack enhances the process of healing;[13] this is not surprising.

Tests on cracked cubes of mortar, subsequently allowed to undergo autogenous healing, indicated a higher percentage recovery in richer mortars.[14] In some cases, the recovery was 100%. The modulus of elasticity followed the pattern of strength.[14]

Lauer and Slate[4] reported that more healing occurs in cement paste with a higher water-cement ratio when the cracks open in the first few days after setting, but at later ages there seems to be no influence of the water-cement ratio on the recovery of strength.[4] There is, however, no corroborating evidence.

Role of autogenous healing in corrosion protection

The main function of mortar or concrete lining of a metal pipe is to provide corrosion protection of the metal, especially when the pipe carries seawater or industrial liquids. If such a pipe has developed cracks owing to drying shrinkage and it is then to be put into service, there may arise the question whether corrosion of steel can occur or, alternatively, whether autogenous healing will seal the cracks and prevent all contact between seawater and the steel. When confronted by this problem, I was not able to find any directly relevant information, but an answer may be inferred from the information so far presented in this section.

There is, however, an additional point. It is sometimes thought that a crack right through the lining, exposing bare steel, will automatically lead to corrosion because both chlorides in the seawater and oxygen dissolved in the water have access to the steel. However, corrosion is the consequence of the formation of an electrolytic cell. The exposed steel at the bottom of the crack is a possible locus of the anode, where the actual corrosion occurs, and the portion of the steel pipe covered by the mortar lining, is the possible cathode.

What happens then is that the oxygen, if present in the water, has to penetrate through the mortar lining to the surface of the steel. There, oxygen and water react with the negatively charged free electrons that have passed through the steel pipe at the cathode, and negatively charged hydroxyl ions are formed. They travel through the mortar to the anode where they react with positively charged ferrous ions to form iron hydroxide, that is, rust.

In other words, chlorides are involved at the anode while oxygen is required at the cathode. The crack allows the development of the anode whereas the cathode is the large area of lined steel away from cracks. The transport of oxygen through the intact lining controls the amount of oxygen that reaches the cathode. Now, the lining is likely to have a very low air permeability because it is made of mortar with a very low water-cement ratio; for example, the British Standard B.S.534:1990 Specification for pipe ... for water and sewage, limits the water-cement ratio to 0.46. Also, the pores in the mortar are saturated, being in contact with water, and this lowers the air permeability. Moreover, the amount of oxygen in the water depends on the source of water and on the extent of air entrainment in the water caused by turbulent flow.

These parameters are difficult to establish, and some help may be found in papers by Vennesland and Gjørv[15] and by Gautefall and Vennesland.[8] In tests on reinforced concrete immersed in seawater, Vennesland and Gjørv[15] confirmed that the corrosion of the exposed steel in the crack was a function of the ratio of the area of the cathode to the area of the anode. The effect of autogenous healing was to slow down the progress of corrosion of the steel at the root of the crack.[15] They reported that "although corrosion was observed for all crack widths of 0.4 mm (0.016 in.) or more, corrosion damage never developed in the 0.5-mm (0.02 in.) crack in spite of the galvanic coupling" (that is, a large cathode-to-anode area ratio).[15]

While these data are interesting, it is not possible to apply the numerical values to actual situations in a lined steel pipe because the area of the cathode cannot be estimated. Nevertheless, it is worth quoting the conclusion drawn by Vennesland and Gjørv from their tests on reinforced concrete blocks: "For crack widths smaller than 0.4 to 0.5 mm (0.016 to 0.02 in.), however, precipitation of reaction products may effectively clog up the crack and thereby inhibit the corrosion before any damage to the steel has occurred."[15]

Somewhat relevant are the findings of Jacobsen et al. who measured the effect of autogenous healing on chloride ion transport in concrete.[12] Healing resulted in about one-third reduction in the ion migration but in only a very small improvement in the compressive strength of the specimens. No explanation of this apparent inconsistency was offered by the investigators.[12]

One more observation in connection with corrosion and autogenous healing should be made. Gautefall and Vennesland reported that products of corrosion of steel may contribute to blocking of cracks.[8]

Conclusions

To a large extent, this article is a review of published information about autogenous healing. This includes: the nature of the filler material in the cracks; the maximum width of cracks that will close fully in consequence of autogenous healing; some mechanical properties of the material in the filler; the requirements for the water that will effect the healing; and the role of autogenous healing in the corrosion of embedded or underlying metal.

Unfortunately, from all these data it is not possible to deduce "rules" about what to do in order to achieve autogenous healing of concrete or mortar in a specific situation. The reason for this is that the vast majority of published data were obtained on specimens that were cracked deliberately so that the crack properties were well known. Such an approach is perhaps unavoidable, but the information deduced from those experiments is difficult to translate into practical situations. At the same time, in real life, we have little knowledge of crack widths and sometimes also of other conditions in place.

This is not to deny that the published data are valuable and, I hope, their review is useful. When problems in the field are encountered, intelligent guesses can be made and probable assessments put forward. Nevertheless, inevitably, there remains a gap between laboratory-based knowledge and the behavior of concrete and mortar in the field. This, of course, is always the case but, with respect to autogenous healing, especially so.

In connection with transfer of findings in laboratory experiments to the behavior in actual structures, I cannot resist expressing my usual words of caution. For example, test results on concrete immersed in lime water, such as those reported by Jacobsen et al.,[12] may well serve the experimenters' objectives, but they should be ignored with respect to autogenous healing.

Finally, the following may be salutary. In searching for information about the relation between autogenous healing and fluid transport through concrete, I came across a paper by Sandberg and Tan.[16] Although the paper did not provide any direct information on autogenous healing, reporting mainly on chloride ion transport, it contained an illuminating statement which it is worth bearing in mind in a broad context. This reads: "Maximum chloride diffusivities calculated from the field profiles after 4 years of exposure were more than 10 times lower than those obtained from the same concrete in the laboratory."[16] It is worth pondering on an "error" of fully one order of magnitude.

So, autogenous healing of concrete is not a miracle, but it is an interesting and, under some circumstances, useful self-healing process; we can thus look to concrete to heal itself in the words of St. Luke in Chapter 4: "Physician, heal thyself."

References

1. Clear, C. A., "The Effects of Autogenous Healing upon the Leakage of Water Through Cracks in Concrete," Cement and Concrete Association, *Technical Report* No. 559, May 1985, 28 pp.

2. Turner, L., "The Autogenous Healing of Cement and Concrete: Its Relation to Vibrated Concrete and Cracked Concrete," International Association for Testing Materials, London Congress, Apr. 19-24, 1937, p. 344.

3. Hearn, N., "Self-Sealing, Autogenous Healing and Continued Hydration: What is the Difference?," *Materials and Structures*, V. 31, Oct. 1998, pp. 563-567.

4. Lauer, K. R., and Slate, F. O., "Autogenous Healing of Cement Paste," *Journal of the American Concrete Institute*, V. 52, June 1956, pp. 1083-1097.

5. Wagner, E. F., "Autogenous Healing of Cracks in Cement-Mortar Linings for Grey-Iron and Ductile-Iron Water Pipes," *Journal of the American Water Works Association*, V. 66, June 1974, pp. 358-360.

6. Hearn, N., and Morely, C. T., "Self-Sealing Property of Concrete – Experimental Evidence," *Materials and Structures*, V. 30, Aug.-Sept. 1997, pp. 404-411.

7. Edvardsen, C., "Water Permeability and Autogenous Healing of Cracks in Concrete," *ACI Materials Journal*, V. 96, No. 4, July-Aug. 1999, pp. 448-454.

8. Gautefall, O., and Vennesland, Ø., "Effect of Cracks in the Corrosion of Embedded Steel in Silica-Concrete Compared to Ordinary

Concrete," *Nordic Concrete Research*, No. 2, Dec. 1983, pp. 17-28.

9. Neville, A. M., *Properties of Concrete*, Fourth Edition, Longman, London, 1995, and John Wiley, New York, 1996, 844 pp.

10. Loving, M. W., "Autogenous Healing of Concrete," Bulletin 13, American Concrete Pipe Association, 1936, Revised 1938, 6 pp.

11. Lea, F. M., *The Chemistry of Cement and Concrete*, Third Edition, Edward Arnold, 1970, 727 pp.

12. Jacobsen, S.; Marchand, J.; and Boisvert, L., "Effect of Cracking and Healing on Chloride Transport in OPC Concrete," *Cement and Concrete Research*, V. 26, No. 6, 1996, pp. 869-881.

13. Ngab, A. S.; Nilson, A. H.; and Slate, F. O., "Shrinkage and Creep of High Strength Concrete," *ACI Journal*, July-Aug., 1981, pp. 255-261.

14. Dhir, R. K.; Sangha, C. M.; and Munday, J. G., "Strength and Deformation Properties of Autogenously Healed Mortars,"*ACI Journal*, V. 70, Mar. 1973, pp. 231-236.

15. Vennesland, Ø., and Gjørv, O., "Effect of Cracks in Submerged Concrete Sea Structures on Steel Corrosion," *Corrosion '81*, Toronto, Ontario, Aug. 1981, pp. 49-51.

16. Sandberg, P., and Tang, L., "A Field Study of the Penetration of Chlorides and Other Ions into a High Quality Concrete Marine Bridge Column," *Durability of Concrete—Third International Conference*, SP-145, American Concrete Institute, Farmington Hills, Mich., 1994, pp. 557-571.

Letter to the Editor

The article by Dr. Neville in the November 2002 issue of *Concrete International* about autogenous healing is of particular interest to me because I have through the years used it in interpreting data from petrographic examinations, which is somewhat different than its benefits for strength purposes and for plugging cracks. I have noted it in varieties of concretes and masonry mortars. For example, I have found calcium hydroxide in mortars as fillings in microscopic-sized bladed imprints of former ice crystals, and in concrete that has been disturbed when it was weak and young so that short microfractures resulted. Subsequently, these microfractures filled with calcium hydroxide $(Ca(OH)_2)$. Petrographic observations such as these are of great assistance in diagnosing the early life of portland cement-based products because that early life history may lead to explanations of performance.

Calcium hydroxide has a hardness of 2 based upon Mohs scale of hardness and thus is very soft (it can be scratched by a fingernail). When in platelet form it is also flexible (like mica). Because calcium hydroxide is so soft, and even weaker due to its flexibility when in thin platelets and to its excellent basal cleavage, I have concluded, rightfully or wrongly, that "late" autogenous healing does some, but little, good to help regain strength. As pointed out by Dr. Neville, lots of water is needed for the process of autogenous healing (perhaps we should call it faith healing when it comes to strength), and water is also needed to continue hydration of the portland cement. Perhaps along with the autogenous healing, and equally as influential relative to strength gain during the "early" life of cracked concrete, is the concurrent development of calcium silicate hydrates from hydration of the cement as Dr. Neville implies in his article.

During research studies at the PCA Research and Development Laboratories some years ago, Dave Hadley and I noted that during cement hydration calcium silicate hydrates were encapsulated by calcium hydroxide. I also captured that encapsulation using time-lapse photomicrography of cement hydration. By use of the petrographic microscope we found residues of amorphous silica (from calcium silicate hydrates) after calcium hydroxide crystals that grew in a paste environment were treated with acid. Using transmission electron microscopical studies we found that the intergrown calcium silicate hydrates were cigar-shaped (tectoids) and formed preferentially oriented normal to the basal faces of the calcium hydroxide crystals. Perhaps the intergrowth with calcium silicate hydrates is what "reinforces" the cleavage-weak calcium hydroxide along the basal faces similar to the effects of load-transfer dowels—and perhaps makes autogenously healed young concrete stronger than it might have been because at that time cement hydration is more vigorous than at later ages. To me, the greater role of late autogenous healing (where there are far less calcium silicates to hydrate) is plugging cracks, which reduces permeability.

Calcium carbonate ($CaCO_3$) has a hardness of 3 based upon Mohs scale of hardness and thus is very soft (it can be scratched by a copper penny) but not quite as soft as calcium hydroxide. It has one excellent cleavage and several not so excellent cleavages (partings) so that it is more cleavage-weak than calcium hydroxide, except when it consists of mosaics of very fine calcite crystals such as some dense, fine-grained limestones that have relatively high strengths. Typically carbonated portland paste consists of microsized crystals of calcite with residues of amorphous silica intergrowths from the calcium silicate hydrates from which it formed. That usually is not the case for calcite-filled cracks.

At later stages in the life of cracks, after walls of cracks and flanking paste are carbonated, because of the carbonation cement hydration does not continue in the immediate vicinity of the cracks and leaching of the paste is needed for cracks to be autogenously healed.

In addition to the references provided by Dr. Neville, Chapter 15, "Autogenous Healing," the 1959 American Concrete Pipe Association Handbook, provides a discussion of that subject and relates some experiences with concrete cylinders and concrete pipe. The chapter includes quotations from several published papers, some of which were referenced by Dr. Neville.

It is said in Chapter 15 that autogenous healing was reported in 1836 by the French Academy of Science—that is shortly after Joseph Aspdin patented portland cement in 1824. Hydraulic lime was recognized long before then and was used by Smeaton in construction of the historically famous Eddystone Lighthouse in 1756—hydraulic lime can also provide the calcium hydroxide needed for autogenous healing.

In Chapter 15, several incidences about the effects of autogenous healing are reported:

(1) In 1917, the average strength of concrete cylinders tested by Duff Abrams at 28 days was 2382 psi. The cylinders were discarded outdoors in the Chicago area and when tested again eight years later had an average strength of 5096 psi—a strength gain of 214%. Presumably, the load was pulled before the cylinders were significantly damaged so that cracks were tight.

(2) In 1928, a section of unreinforced pipe that cracked at a load of 2430 lb/ft was laid aside outdoors in the coastal area of Washington and when tested three years later sustained a load of 2570 lb/ft.

(3) In 1929, an extra-strength reinforced concrete pipe liner placed near the summit of the Cascade Mountains cracked in the crown and invert areas during installation. In 1936, the liner was inspected and the cracks were "completely sealed by autogenous healing."

Among items not reported are the characteristics of the cements, such as their fineness—cements at the time of the tests were ground coarser than "modern" cements. Cement fineness plays a major role in the amount and surface area of residual cement available for hydration and the generation of calcium hydroxide (and calcium silicate hydrates) needed for autogenous healing.

Compliments to Dr. Neville for his return to a subject that can play a major role in concrete serviceability and durability, and also aid in interpreting data detected during petrographic examinations.

Bernard Erlin, President, Petrographer
Fellow, ACI International
The Erlin Company
Latrobe, PA 15650

Neville's Reply

Bernie Erlin's interesting letter gives an example of how petrographic observations make it possible to draw engineering conclusions about the state of concrete in a structure. My article refers to engineering observations in actual structures, which may help petrographers to diagnose the state of the concrete. This is almost a reversal of roles that shows the complementary nature of the engineering and petrographic approaches.

Like most of my articles in *CI*, the article on autogenous healing stems from my consulting activity in an actual problem. Unfortunately, as always, clients do not want their problems to be "advertised" so that I cannot give any details of the project. However, there is a general point worth making; we need an input from several sources: engineers, petrographers, and scientists.

In the case to which I referred in my article, when commenting on the consequences of cracks, which may or may not heal, someone wrote: "The width of many of these cracks will potentially allow ... saline water to reach the steel ..." behind the mortar. He then continued to say that the "water contains oxygen and chlorides providing the necessary reagents for corrosion to occur." My view is that, whereas it is correct that for corrosion to proceed both chlorides and oxygen need to be present, their presence is required at two different locations. Specifically, chloride ions initiate and progress corrosion at the anode, while the oxygen has to be present at the cathode. The crack is the location of the anode, but the cathode in this particular case was a large area of steel away from the crack, the two being separated by very rich and dense mortar. All this is relevant to the interpretation of the autogenous healing process.

The references cited by Bernie Erlin are ancient, some even more ancient than I. This is why I did not cite all of them in my article, destined for practitioners and not for Ph.D. examiners who require a comprehensive review of literature. Nevertheless, the facts quoted by Bernie support the general theme of my article, and for this I am grateful to him. I also appreciate his petrographic description and interpretation of calcium hydroxide and calcium carbonate in hydrating and hydrated cement paste: Bernie is good at educating engineers!

Adam Neville

Chapter 6

Technology and Design

This chapter is headed Technology and Design, and its six sections have in common one theme: the knowledge of construction materials and of structural design are *both* needed to achieve successful, robust, durable, and serviceable structures. I do not apologize for this preoccupation with the "marriage" of technology and design because I firmly believe in its importance. It has been said that a building designed by an engineer without an architect is horrifying, and a building designed by an architect without and engineer is terrifying. A similar argument can be applied to materials technologists (terrifying?) and structural designers (horrifying?).

In the past, say 30 years ago and more, there were many engineers who combined a good working knowledge of materials properties and of structural analysis, and indeed this is how I was educated and trained. In fact, I conducted research both in the field of concrete as a material and in structural behavior, and I have published in both fields. My attitude has persisted into an era when some people undertake the design and others are involved in the selection of concrete mixtures to achieve the various chemical and mechanical properties of concrete. I am tempted to paraphrase Rudyard Kipling and say: "…and never the twain shall meet."

The various sections in this chapter make my point. If someone feels that I have been unduly persistent in my message, my defense is that the message is, in my view, so important that we have to ensure that it is heard loud and clear by everybody.

Very recently, I read about an interesting example of a mind receptive to new materials. This was in the obituary of Sir Alan Harris, a designer of early prestressed concrete structures. His consulting partner wrote that Harris had invited him to the famous British Air Show in Farnborough. Why an air show? The obituary says: "As it turned out, it was not so much for the aeroplanes, which were incidental, but for the new materials on the manufacturers' stands, which he thought might be useful for our structures, now or in the future."

Section 6.1 (and postscript by way of my letter to the Editor of *Concrete International* reproduced at the end of Section 6.1) present concrete technology as an element of structural design. Examples of this, in my view, essential link include durability and deformational characteristics of concrete. An important deformational characteristic is creep—a topic in which I undertook research over many years and which led to two books. Because of my close connection with creep, I wrote Sections 6.5 and 6.6. The first of these gives examples of problems induced in structures by creep; Section 6.6 reviews methods of dealing with creep problems.

To introduce the fairly extensive topic of the knowledge of creep of concrete as a material (and also of the other non-elastic deformations) I wrote Section 6.4. The title of Section 6.4 is neutral and cool, but I was tempted to use the following titles: *May a structural designer be ignorant about concrete?* and *If you don't know the material, can you design a structure?*

Section 6.4 includes consideration of the need for comparable levels of accuracy and precision in using quantitative properties of materials and in the analytical calculations. Examples of spurious precision with respect to the materials are cited; I commend these examples to the readers for their careful consideration.

Another topic mentioned in Section 6.4 is the investigation of problems in existing structures. I express rather strong views as to who is qualified to undertake such investigations and to proffer reliable and competent opinions.

Although Section 6.6 deals with creep of concrete, it introduces some ideas that may be of wider interest. This is particularly the case with what used to be called curve-fitting and is now a computer-based optimization of mathematical expressions for creep as a function of selected variables. This is an established technique, but the proponents of the various expressions base them on test data for concretes used in the past. However, modern concretes have changed substantially: cementitious materials used are more varied than portland cement alone (which has also changed in composition); much lower water-cement ratios are used; and strengths are higher.

I argue that a deductive approach to complex curve-fitting should be replaced, or complemented, by an inductive approach, starting with a single conceptual model. This kind of approach is used in other fields of engineering, for example, in flood forecasting. An "isolationist" attitude to creep is not serving the cause of concrete construction well.

My criticism of the almost purely mathematical and formal approach to the search for *the* creep expression is not just recent. In July 1991, at the invitation of ACI, together with two other member of ACI Committee 209 (Creep and Shrinkage) I wrote a note on that committee's activities. We said: "... the task for the Committee is to generate more data on the creep of concrete containing ingredients for which historical data are not available and to systematize these to obtain an understanding of the physical (and not just mathematical) factors involved." We also said: "no significant gain in accuracy of the prediction of creep of a 'new' concrete under 'new' circumstances has been achieved."

Perhaps the above, fairly extensive, comments on creep, explain why this chapter contains three sections dealing with creep.

Section 6.2 presents my view that there are some fundamental aspects of concrete with which everyone who is "in" concrete should be familiar. The approach is very much a broad-brush one, and I do not propose to summarize it in this introduction.

My views about the need to use a range of cementitious materials are neatly encapsulated in the comment by the Editor-in Chief of *Concrete International*, Bill Semioli on the front page of the April 1999 issue. He says:

> Speaking out again in *CI...* , ACI Honorary Member Adam Neville points to the days of "just ordinary concrete" being numbered except possibly—as he describes it—for "simple domestic applications." Neville, who is the author of the noted technical compendium "Properties of Concrete," now in its Fourth Edition, defines ordinary concrete as that containing just portland cement." While citing the growing use of a wide range of various admixtures, he colorfully describes purely portland cement, or PC, concrete as "everyman's concrete." And as such he points to the ever-growing requirements for "designer" or "custom-built" concretes—what are more commonly referred to today as high-performance concretes. Neville refers also to the paradox inherent in the trend of mixing more materials than just PC, water, and aggregates for concrete. The seeming ambivalence is that as more use is made of admixture materials, it will lead in turn to a greater use of PC. But how can this be so: that less will be more? No mystery here! Obviously, as blended cements increasingly improve the performance of concretes, even with their lesser proportions of PC, the greater will be the volume use of the latter material as more and more end users choose concrete as their preferred material for construction. As a great sleuth of old used to say: "It's elementary my dear Watson!"

In a way, Section 6.3 is a complement to Section 6.2. While recognizing the pre-eminence and ubiquity of concrete in construction, it warns that it is dangerous to assume that concrete is irreplaceable. Napoleon is reputed to have said that cemeteries are full of people who believed that they were irreplaceable.

I recognize the need for generalists, but specialists are, of course,

Chapter 6: Technology and Design

needed, too. There are probably few readers who may still remember the archetypal specialist in privies, Elmer Ridgeway. In a book, by Chick Sale, bearing the title *The Specialist*, first published in 1929, Elmer, who was first and foremost a carpenter and not a concretor, looking at a really superb privy, says: "I know I done right in Specilizin'; I'm sittin' on top of the world." (On top of what?) So, I recognize the real need for specialists and also the difficulty of being a fully competent generalist who knows enough about design, construction, and materials. But the designer should know *enough* in order to know what is possible and to know what advice is needed. Elmer improved only the superstructure and never modified the underlying can; a chemical process of course eventually superseded it.

Section 6.3 gives some examples of technologies that have become replaced. It may be salutary to add further examples where dramatic changes in design took place. For example, washing machines started by applying electric power to the old tub and hand wringer, and were therefore top-loaded with the mechanical wringer above the tub. The advent of a drum on a horizontal axis, with integrated washing and spinning, has led to a market of "one household, one washing machine." The very recent development, for once in England, of a dual drum, represents further progress. Likewise, domestic plumbing has been revolutionized, both in terms of materials used and of jointing systems. The same applies to fittings and fixtures in buildings.

Car bodies are no longer made of the same materials as before, nor are they welded as before. Even military tanks, traditionally made of steel, can now contain a composite material consisting of a woven glass fiber matrix impregnated with phenolic resin.

I am not listing the above examples of radical changes in technology, and above all in design-cum-materials, because I am a prophet of doom. On the contrary, I take the view that we have every right to be proud of concrete. But, they say, pride goes before a fall.

In any case, Section 6.3 is not a message of doom. Recent progress in concrete is reported. Overall, the emphasis in Section 6.3 is on symbiosis between the structural designer and the materials scientist. I also point out the need for much greater mechanization of concrete construction and for the use of robots, rather than an increased importation of unskilled labor. The theme of the need for more training and education—areas where ACI is active—recurs also in some other sections.

Section 6.1: Concrete Technology—An Essential Element of Structural Design

The title of this section spans two traditionally separate activities involved in achieving a structure or a building: structural design and concrete technology. The perception of the former, widespread even if not entirely correct, is that of analytical calculations, these days involving a computer. Concrete technology, on the other hand, conjures the image of a "practical" man who, if he does not wear gum boots, uses a laboratory mixer or tests numerous concrete cylinders. This is a deliberate and provocative distortion, but it is a fact that the structural designer and the concrete technologist are perceived as people involved in quite separate activities, clearly compartmentalized. It is my contention that concrete technology is an essential element of structural design. In this section, I hope to demonstrate that this is so.

Structural engineering and properties of materials

We live in an age of increased specialization so that design is usually effected by teams of engineers, often also including architects and other professionals. The structural engineer is responsible for ensuring the strength and robustness of the structure, as well as its serviceability, and also its durability. All these requirements of design inevitably involve the properties and behavior of the structural material used in the construction.

In structures not connected to the ground, such as automobiles or aircraft, there is a very close link between the choice of the structural form and shape, and the material used. Indeed, the choice of the structural aspects of the design and of the material of construction is made at the same time because the properties of the material determine what is achievable. If necessary and possible, material development is undertaken so as to suit structural requirements. The progress in this development in recent years has been phenomenal.

Not so in the vast majority of structures connected to the ground. Broadly speaking, the construction material is steel or concrete. I shall not consider steel, not only because it is outside the remit of ACI, but also because the material—steel—is factory-produced in a sophisticated manner, its properties are well-defined, described, and guaranteed; only site fabrication is job-specific.

In concrete structures the situation is substantially different. The structural designer does the analysis and design. He or she then specifies the concrete. This used to be done by prescribing the proportions of the various ingredients. As to the properties of these ingredients, the requirements are extremely wide. For example, the cement has to conform to an ASTM standard, but that standard lays down, quite rightly, wide limits. With respect to aggregate, it is sometimes specified that the aggregate must be

Fig. 6.1.1—Severe sag at midspan of a bridge caused by creep

clean and hard; this is hardly quantifiable. The grading requirements are such that they are satisfied by a number of practical gradings. This is right and proper because at this stage, when the contractor has not yet been chosen, it is not known which source of aggregate will be used. Thus the detailed properties of the materials to be used are unknown. The same applies to the plant or the method of transportation, compaction, curing, and other site activities.

With the move to performance-type specifications, the situation has improved somewhat in that at least certain behavioral characteristics of concrete are ensured. Nevertheless, many other properties of concrete are unknown and unpredictable at the structural design stage. Is it surprising then that, in many cases, problems arise during the life of the structure?

At this point, I would like to forestall the criticism that all this is inevitable because, unlike automobiles and aircraft, concrete structures are nearly always of a one-off variety and are built in-place in the open, in variable weather, often under difficult conditions, using a labor force that is not static and, let us be frank, less skilled than their factory-based counterpart. Yes, this is so, but need the consequences continue forever?

In what follows I propose, first, to show the close interdependence of structural design and the properties of concrete, and then to offer some pointers toward ameliorating the present situation.

Deformational characteristics of concrete

The structural engineer is well-educated and trained in structural analysis and design. He or she can derive the appropriate equations, or use existing analytical solutions. But these equations and solutions involve parameters describing material properties. In the case of concrete, compressive strength is the obvious one and it is generally most easily dealt with. But for the strength of a structure, we require also the deformational characteristics of the concrete: its modulus of elasticity, including the fact that the material is not truly elastic, and we have to account for creep of concrete.

In numerous design calculations, creep is dealt with as a coefficient with a single, or at best a two- or three-step value. However, in reality, creep is a complex function of concrete mixture proportions, age at loading, age at load removal, temperature, and exposure conditions (see Fig. 6.1.1). In addition to creep, concrete undergoes drying shrinkage, which is not stress-related, but is influenced by numerous factors. In the days of an increasing use of high-performance concrete, we require the knowledge of autogenous shrinkage in the concrete element; indeed, this is a topic just now being intensely studied.

Bridge piers under some conditions offer an excellent example of shrinkage-induced problems. In the case of continuous girders on tall supports, there is a substantial difference in the shrinkage of those over water and those over dry land where the shrinkage is much higher; this induces stresses in the girder.

As for creep, there can be a large differential vertical deformation in high-rise buildings. Some of this is avoidable by controlling variations in the sustained stress and in the reinforcement in the columns, and also their conditions of exposure (although this is not always possible). However, at the design stage, the creep characteristics of the concrete to be used, which are age-dependent, are unknown and cannot be controlled. The same applies to the effects of the creep of concrete columns on the cladding whose deformational characteristics are likely to differ from one another. Clearly, an allowance for creep of concrete is necessary, but how much creep will there be?

Most engineers are familiar with cases of unexpected shrinkage cracking which may well be due to the shrinkage characteristics of the aggregate or of the mixture used. At the design stage, these features are unpredictable, and specification limits are generally very broad.

My contention is that the influence of these various parameters is so great that the designer must be fully familiar with them. One cannot simply use a handbook of physical constants.

Structural design and durability

The same applies in the case of durability, which is by far the major problem with concrete structures in many parts of the world. Numerous cases of inadequate durability of concrete structures built in the 1960s and the 1970s were traced to the selection of concrete mixtures on the basis of strength alone. In fact, because of the changes in the cement properties, the same strength as previously specified could now be achieved at a higher water-cement ratio. In consequence, for a given 28-day strength, the 1970s concrete was more permeable than the concrete of the 1950s.

The factors affecting the durability of concrete are extrinsic as well as intrinsic, so that to take them properly into account the designer must have a good knowledge of chemical and physical phenomena of the interaction between concrete and the environment. There exist other examples of the relation between the behavior of concrete and the performance of concrete structures in service. One example is the influence of shape of the structure on durability; slab jetties in seawater are less liable to reinforcement corrosion than beam and deck construction. Such knowledge is especially important when the design is to ensure a specific service life of the structure—a requirement increasingly being invoked.

Structural engineer's knowledge of concrete

My contention is that most structural designers, with honorable exceptions, lack the necessary knowledge of the behavior of concrete. Universities do not teach the requisite knowledge, as demonstrated by the Portland Cement Association survey in 1995: only 22% of American civil engineering departments require a full semester hour course in concrete technology. On the job, acquisition of knowledge is too fragmentary, or dependent on chance, and without adequate scientific rigor.

It can be argued that one person cannot have all the necessary knowledge so that the designer should simply consult a materials specialist. The trouble is that, in the vast majority of cases, the materials specialist is a "pure" scientist who lacks knowledge of structural action or structural behavior. In consequence, he or she does not know what questions to answer, and the designer does not always know what questions to ask. It is almost as if the surgeon lacked an adequate knowledge of pathology and relied exclusively on the laboratory scientist for the decision on what to do.

Design without adequate knowledge of concrete

There is a quip to the effect that a structure designed by a structural engineer without an architect is horrifying, and a structure designed by an architect without an engineer is terrifying. It is arguable that design without an intimate understanding of concrete is inexpert: there can be a mismatch between structural design and material behavior. The major mistakes in the past include the use of an entirely unsuitable concrete, such as that made with high-alumina cement, or reinforced and, above all, prestressed concrete containing calcium chloride.

In 1995, in reviewing the general topic of structural safety, J.B. Menzies, formerly of the British Building Research Establishment, wrote that the use of both these materials had been "an error." Their use was withdrawn from British design codes in the 1970s. The entire question of structural use of high-alumina cement concrete will be considered in a sequel article

Fig. 6.1.2—Creep-induced sagging at midspan of a bridge in an American territory in the Pacific Ocean

published in *Concrete International* (See Section 1.5).

We no longer make these particular mistakes, but we must make sure we do not make other mistakes in the future. For structural designers as a group to learn from their own mistakes is not good enough, bearing in mind our social as well as legal responsibility for safety. We must also not shortchange our clients by giving them inadequately durable structures with too low a service life.

There are other examples of the importance of the knowledge of the behavior of concrete in structural design. The very structural form may be influenced by the thermal behavior of concrete. This behavior is different in the case of high-performance concrete than in many cases of the old-style concrete. In some large and massive concrete structures, the corrosion aspects of durability should lead designers to ask themselves at the very outset whether the use of reinforcement is essential, or whether a masonry-type construction is practicable; ship docks are an example of such a structure. The same question can be asked in the case of tunnel lining, where reinforcement is required only during handling and installation, and yet it is the corrosion of reinforcement that is often the limiting factor in the life of the tunnel lining.

Structural assessment and knowledge of concrete

I can offer two more, different examples of the importance of the knowledge of material behavior. One is the structural assessment of existing structures. The other, quite common, is the investigation of failures, luckily usually only local or partial. For example, is cracking or spalling solely stress-induced, or is it related to thermal causes, or to differential shrinkage, or inadequate curing? A "pure" materials scientist cannot answer these questions because he or she does not understand the structural behavior. A structural engineer can answer them, but only if he or she understands how concrete behaves under a whole range of conditions.

6-10 Chapter 6: Technology and Design

How to do better in the future

The corollary of my article is that a structural designer who is ignorant about concrete is not a truly competent designer. So, I am arguing for a broader education and training of structural engineers. This may be thought to be unpalatable, but the additional learning will pay off in achieving better designed and more durable structures.

This additional learning need not require the introduction of yet more courses to the already overcrowded undergraduate curriculum. In any case, it would be difficult to find enough university teachers who do actually combine the knowledge of structural analysis and design with a knowledge of the behavior of concrete. And if they know only the latter, they would not be able to relate the behavior of concrete to structural behavior. However, we live fortunately in the age of computer-based interactive learning methods: therein lies the solution.

For example, we could devise solutions on the assumption of certain properties of concrete, say a given creep coefficient. The value of the coefficient is then changed, up or down, and the effect on deformation and stress distribution is shown. Likewise, we could design a structure assuming a certain concrete mixture, and then see the consequences of using aggregate with a higher or lower modulus of elasticity. Simulating the effects on durability of varying the permeability of concrete would be equally instructive. These exercises would have to be prepared by the few who straddle the divide between structural design and materials engineering, but the computer programs could then be used in all engineering schools as a follow-up to learning the basic behavior of concrete.

The objective is to blur the divide between those who read the *ACI Structural Journal* and those who read the *ACI Materials Journal*. This way we shall assure better concrete structures, which is what ACI strives to do.

Letter to the Editor

There is a sad postscript to my article ("Concrete Technology — An Essential Element of Structural Design," July 1998 issue, pp. 39-41). The article includes a photograph (Fig. 6.1.2) captioned, "Creep-induced sagging at midspan of a bridge in an American territory in the Pacific Ocean." I took that photograph when on a holiday 10 years ago and I did not feel I should identify the bridge in my article. Alas, events overtook my comments about the excessive deformation even before my article was published.

The bridge, which can now be identified as the vital and sole inter-island bridge in the Republic of Palau, Caroline Islands, collapsed in September 1996, less than 20 years after construction. The total length of this balanced cantilever box girder was 1264 ft (385 m) and the main span was 790 ft (241 m). The cause of the collapse has not been established but it is known that, a few months earlier, repairs to the bridge were effected in order to reduce the creep-induced sag of about 3 ft (1 m) at the mid-point of the center span, shown in the photograph (Fig. 6.1.2). The repairs involved a recovery of deflection of about 12 in. (300 mm) by prestress.

A secondary concrete slab, 8 in. (200 mm) thick, was placed on top of polystyrene void formers 10 in. (254 mm) deep, over the central 260 ft (80 m) of the span. On top of the concrete, there was placed a 2 in. (50 mm) wearing course of asphalt, the original wearing course having been removed. I suspect that there was some increase in the dead load and, of course, it is dead load that would have induced future creep deformation. My comment is aimed at those who, without proper advice, remedy the sag of a flat roof by an additional layer of concrete or asphalt.

Whatever the causes of the collapse, there is no doubt that the need for the very major repairs arose from the excessive sag induced by creep. So, a good knowledge of the creep properties of concrete as a material is essential for the design of a concrete bridge. It is sad that this major accident confirmed the validity of the thesis of my article.

Adam Neville

Section 6.2: What Everyone Who is In Concrete Should Know About Concrete

The title of this section is long but this does not mean that it is a condensation of a long book on concrete. My intention is to mention those aspects of concrete which, in my opinion, everyone in the "concrete business" should know. By "know" I don't mean that he or she needs to have a detailed technical knowledge, but to realize what is important. If you know what is important, you look out for it and, as the occasion arises, you find the detailed information or, more likely, ask someone in your organization to obtain this information.

This does not mean that I will present a check list. Rather, I hope to explain why the various topics considered here are important. I should add that I am not going to discuss research or speculate about possible future developments; I will base myself on today's knowledge.

Problems with concrete

This section is addressed to people from different backgrounds and with different professional activities. Unfortunately, what we all probably share is that too often we find ourselves dealing with bad concrete. I hope the readers are not offended by this remark. I expect that everyone is doing a very good job but, somehow, the end product—concrete in the structure—is often not as good as it should be or could be, and problems with concrete are commonplace.

Some of the problems are real; for example, deterioration through corrosion of reinforcement, or poor appearance as a result of weathering and discoloration. Other problems are real only occasionally, but get blown up out of proportion by academics who spend days and years on a favorite problem: for example, delayed ettringite formation or even alkali-aggregate reaction. These researchers inevitably publish the findings of their laboratory work, and this publicity gives concrete an unjustifiably bad name. I don't mean to say that these problems are not real, but many of them are specific to some parts of the world or to some heat treatment method used in the production of precast concrete. These problems are not worldwide or endemic. Just because there is malaria in some parts of the world does not mean that you should not take a holiday abroad.

I want to make one thing clear: we should not try for the very best concrete but rather make concrete that is good for the specific purpose for which it is intended. To make my point, I shall occasionally use deliberate provocation, but my message is nevertheless serious.

What concrete we should make

My first statement is that the days of making what I would call "everyman's concrete" are gone or soon will be. What I mean by everyman's concrete is an idea that was actually promoted by some ready-mixed concrete people in Great Britain. They said that the customer should specify just two parameters: strength and workability. You can have high-strength concrete or low-strength concrete. And you can have high-workability concrete and low-workability concrete. Either strength can be combined with either workability, so that there are four possibilities. This is tremendously convenient for the ready-mix supplier but it does not serve the client well enough. Life is more complicated than that, so that the mixture must be suited to the specific purpose.

In fact, there are two sources of complication. The first one is that strength is not the sole criterion of satisfactory concrete in the hardened state. The second complication arises from the fact that nowadays the ingredients of concrete include a range of cementitious materials, and not just portland cement. Also, the appropriate workability can be achieved in more than one way.

Let me deal with strength first. The strength of concrete is nearly always specified as so many MPa or psi. This is necessary from the requirements of structural design, and the strength at, say 28 days, can be readily checked for compliance. In my experience, it is rare for concrete to have an inadequate strength, and cases of outright failure or collapse caused by too low a strength are extremely rare. However, there are numerous cases of inadequate performance of concrete through a lack of durability.

Durability depends on the specific conditions of exposure of the given concrete structure or even element. I am saying "element" because different parts of a given structure may be exposed to different conditions, sometimes referred to as a microclimate. For example, the foundations in sulfate-bearing ground may require a particular concrete mixture. The same applies to elements or parts of the structure exposed to periodic wetting and drying, especially by seawater. Parts of a structure may be subject to alternating freezing and thawing. The interior of a building is protected from freezing, from sulfate attack, and from seawater. On the other hand, carbonation in the interior of a building, especially when air-conditioned, may be more severe than the carbonation of a façade because a very high or a very low relative humidity is less conducive to carbonation, and also because rain washes down the exposed surface.

The concrete specification

It could be argued that all these factors should be considered by the designer of the structure who should specify the concrete that is needed. Very often, this is not done for two reasons. First, the designer is rarely well informed about the properties of various concrete mixtures, and is unable to tell exactly what kind of concrete to use. In addition, the designer is not aware of the sources of materials that the contractor may wish to use for economic reasons.

The second reason why the structural designer does not specify the concrete mixture is that a prescriptive specification is rarely used. The days of a specification that laid down in detail so many sacks of cement per cubic yard or kilograms of cement per cubic meter of concrete, or an exact value of the water-cement ratio *(w/c)* have gone. They have gone because such an approach is not conducive to economy and competition: all contractors would do the same thing, and a better contractor could not compete against other contractors because he had a better knowledge and a better technology. The contractor would know in advance which ingredients are likely to be suitable, and how to match the methods of transporting and placing (such as pumping or slipforming) with appropriate admixtures.

What is used nowadays, at least in many countries, is a performance-type specification. The performance may be specified in terms of satisfying a particular test. Examples are: resistance to so many cycles of freezing and thawing using an ASTM test; resistance to sulfate attack using another ASTM test; resistance to chloride penetration using an ASTM or some other specified test. Satisfying these test requirements means that the contractor, or the people on whom he relies, must be well familiar not only with the tests themselves but also with the parameters in the concrete mixture that influence the performance of the concrete under test conditions.

I want to elaborate this a little. To know how the test is performed is easy. What is less obvious is to appreciate how the various mixture ingredients and mixture proportions affect the performance under tests. Let me give an example. Chloride penetration of concrete is affected by many factors. The influence of the *w/c* is probably well known. But even at a given *w/c*, the penetrability of concrete depends on the pore structure, in fact, mainly on the pore size, of the hydrated cement paste. A finer pore structure is achieved by the inclusion of silica fume in the mixture because the particles of silica fume fit in-between the particles of cement, and especially because they fit into the spaces between the cement particles and the surface of the aggregate. Also, some of the chlorides become bound by the cement paste; it is useful to know that the extent of binding is affected by the content of C_3A (tricalcium aluminate) in the portland cement and that the binding by ground-granulated blast-furnace slag is greater than if portland cement only is present.

My point is this: If we know these effects of the various mixture ingredients, then when we read the specification we instantly realize that we must think about using silica fume or metakaolin, or maybe burnt rice husks, in selecting the mixture. If we think about all this, we will realize the necessary cost of the mixture ingredients, and we shall not make the mistake of using an inappropriate concrete mixture. In some cases, the specification is even more demanding and less obvious. For example, in large and important bridges, and also in tunnels, the specification may be couched in terms of an expected life: 120 years seems to be a popular number. When we see such a requirement, it is important to realize that we must think about a fairly complex mixture.

Mixture selection without specification

I would not like to convey the impression that durability aspects in the choice of the mixture arise only in large, prestigious, or unusual structures. There may be problems, real or suspected, of this kind in the construction of homes in an area where the soil is rich in sulfates. In such construction, there may well be no engineer involved in the design and no engineering specification. The contractor works to an architect's drawings, which include details of concrete footings and slabs on grade, possibly including wire mesh reinforcement. Notes on the drawing may simply say: to conform to the national code, or perhaps to the municipal code. Drawings may state that the soil or the groundwater contains sulfates; or else, they may say: consult a geotechnical specialist. This specialist may provide information about the severity of the sulfate conditions and may or may not design the concrete mixture for the contractor.

The important point is that the concrete must be fit for the purpose, that is, it must have an appropriate durability under the actual conditions of exposure. It may be left to the contractor to translate this indirect information into the selection of a concrete mixture. The contractor may well feel that the ready-mix suppliers are the parties to make the decision. But are they? A large and well-established ready-mix supplier may be knowledgeable. On the other hand, in many cases, the ready-mix supplier does not have the drawings in front of him or her, and probably lacks the expertise of interpreting drawings. It is also not his or her business to be fully familiar with the building code or the design code. And you have to be fully familiar with the code in order to know where to look for possible information relevant to the choice of the mixture.

By and large, house foundations are not viewed as a special kind of concrete. And yet, if the exposure conditions are severe, then someone in the concrete design-supply-place chain must be knowledgeable. Who is knowledgeable may be governed by national or local usage and practice or may be agreed upon between the parties. But if no one is knowledgeable, there may be litigation, probably involving all the parties, and the good name of concrete suffers.

I should add something more about exposure conditions because the mere fact of the presence of sulfates in the ground is not enough to determine to what conditions the concrete foundations or the slab are exposed. We need to know whether this is a slab on grade or a suspended slab; we also need to know whether any impermeable membranes are used and whether some drainage system is provided to isolate the concrete from the aggressive medium. We could consider these details as a kind of microclimate. The important point is that the person who makes a decision about the concrete mixture is aware of the potential for attack on the concrete and is familiar with an appropriate choice of ingredients and proportions of the mixture.

Concrete fit for the purpose

What general conclusions do we draw from all this? As I said earlier, we no longer live in a world of "everyman's concrete" or just ordinary concrete. What is often required is what we might call "designer concrete" or "custom-built concrete." I realize that this is not a particularly apt terminology but I am using it to show an analogy to *haute couture* or a custom-built house, as opposed to a mass-produced garment in a rack of identical clothes or a row of identical houses.

In fact, there is a term to describe a particular kind of that type of "designer" concrete: it is called high-performance concrete (HPC). High-performance concrete is no more than concrete fit for a particular purpose or performance. This performance may be in terms of compressive strength, or tensile strength, or permeability, or some other durability requirement, or modulus of elasticity. What I would like to suggest is that we can no longer consider HPC as something unusual or special, but rather as a part of a continuous spectrum of concretes. Accordingly, I will continue to refer to concrete as a generic term.

Let me now consider other aspects of hardened concrete with which we have to be familiar in 1999, but which barely figured in applied concrete technology even ten years ago. I would like to emphasize the word "applied" because the knowledge existed in laboratories, universities, and books. Why do we have to consider these aspects?

Designers are becoming, or even have become, more demanding in their structural design. This is right and proper because we should strive to produce better, more efficient, and more economical structures. Many designers are now more concerned about shrinkage, creep, and thermal expansion. Sometimes, their demands are expressed explicitly, sometimes implicitly.

As an example, I shall consider rigid cladding applied so as to connect with reinforced concrete columns. The columns will shorten under a sustained load in consequence of creep of concrete, but the cladding will not. Some details of fixing the cladding may be given, but whether or not the movement of the columns can be accommodated depends on how much creep, as well as shrinkage, will take place in the columns (see Fig. 6.5.1).

A similar problem arises when balconies in a high-rise building are connected by ornamental, non-load bearing vertical fins to provide shade or to enhance appearance. The shortening, due to creep, of load-bearing columns behind the balconies induces a compressive stress in the fins. I have seen buckling and spalling of the fins in countries as far apart as Colombia, Turkey, and Australia (see Fig. 6.5.2).

It is likely that the engineer's instructions are not explicit on the subject of creep in concrete. A wise contractor will be aware of the potential problem, even though he cannot be expected to be an expert on creep.

What does the contractor do? If aware of the potential problem, the concrete mixture will be chosen on the basis of past experience, avoiding, for example, aggregates known to exhibit shrinkage. Or the contractor may approach the designer and enquire about how the creep movement has been accommodated in the design. All I am trying to say is that awareness on the part of the contractor, or whoever selects the mixture, is essential.

The same applies in the case of a long building with widely spaced contraction joints. Here, thermal movement may be critical and, in the selection of the concrete mixture, it is important to avoid aggregates with a high coefficient of thermal expansion. Again, someone should be aware of the existence of aggregates with differing coefficients.

Shrinkage is another problem, and probably the most common problem in a wide range of types of construction. Strictly speaking, shrinkage as such does not matter but what does is cracking resulting from restrained shrinkage. A very simple example is flatwork. If you place a concrete slab, such as a garage slab or a patio or a drive, you have to provide contraction joints. These joints have to have a maximum spacing, perhaps 4 m (13 ft), and the parts of the slab surrounded by the joints must not be long and narrow, but have a length, say, not more than 1.5 times the width. Above all, if there are re-entrant angles in the slab, then cracking at the inner corner is virtually certain to develop. Such a configuration must be avoided, and those who provide customer care are well advised to educate all those concerned.

The joints in the slabs can be formed or sawn. If sawing is to be used, it is important to avoid siliceous aggregate because ravelling may result. So, in selecting the mixture, someone has to consider the choice of an appropriate aggregate. The choice of aggregate is also of importance with respect to skid resistance of the concrete surface.

Some readers may think that all this is well-known and obvious. And yet, we continue to see numerous problems caused by shrinkage. Some other readers may think that all this is the architectural or engineering designer's business. It is arguable that it is, but it is a fact of life that designers know less, and not more, about concrete than they did 20 years ago. This may be due to the fact that engineering education has moved more and more towards refined structural analysis and the concomitant preoccupation with computer-aided design, to the exclusion of good understanding, let alone a detailed knowledge, of concrete. The engineers' lack of knowledge may also be due to the fact that concrete has become a much more complex material than it was a quarter of a century ago, when all we needed was cement, sand, coarse aggregate, and water. (This of course is an example of my provocative exaggeration.)

Mixture ingredients of today's concrete

I might just as well start with another exaggeration: The days of concrete in which the only cementitious material is portland cement are numbered. Portland cement is a wonderful material—it has to be as it was invented in England! But portland cement causes a great many problems in concrete. Let me list them.

Hydration of portland cement generates heat which causes a rise in the temperature of concrete and the associated expansion, followed by contraction and possible development of cracks. It is the hydrated portland cement paste that causes shrinkage. The same paste is the seat of creep. It is through the hydrated paste that much of the ingress of water and aggressive agents occurs. Damage by freezing and thawing occurs within the hydrated cement paste. Chemical attack occurs either within the paste or at its interface with reactive aggregate particles. Also, the pore system in the hydrated cement paste is such that there is a practical limit on compressive strength and durability in general.

So, the solution lies in using other cementitious materials as well as portland cement. Portland cement is a necessary ingredient, but a choice of one, or often more than one, other material is greatly beneficial. I am referring to fly ash, ground-granulated blast-furnace slag, and silica fume. There exist other materials such as natural pozzolans, lime fly ash, rice husks, metakaolin, and non-reactive fillers. They all have their place and specific contributions to make. So have admixtures of various kinds, not forgetting superplasticizers, which can be highly beneficial.

This section cannot describe or discuss the use of these various materials. All I am trying to say is that all those concerned with concrete should be reasonably familiar with the action of these cementitious materials. I should add that not all of them are truly cementitious but I find this a convenient and logical terminology. I know that some people use the term replacement materials; in my view, these various cementitious materials do not replace cement, but are coequal with it in the mixture. Other people refer to supplementary materials; again, in my view, they are not just a supplement or addition but, as I said, they are coequal with portland cement.

I have found in the past that some producers of portland cement are opposed to, or even offended by, my views. I believe that it is better not to fight my notion, but rather to realize that if composite or blended cement results in better concrete, in the longer run their use is in the interest of the producers of portland cement. In commercial terms, it is better to put less portland cement into every cubic meter of concrete, but to produce more cubic meters of concrete. Every cubic meter of good concrete ensures future sales; every cubic meter of bad concrete turns people away and encourages them to look for new materials and new solutions.

Mixing and curing

So far, I have discussed only the composition of the concrete; after all, the ingredients of the mixture have a major influence on the behavior of the finished product. But you can have excellent ingredients and cook a bad meal. So I now want to turn to two treatments applied to the ingredients of the mixture.

The first of these is the process of batching and mixing. Despite all the computer controls, the quantity of water that goes into the mixture is not always as rigidly controlled as it should be. The usual culprit is the aggregate, especially fine aggregate. The mass of surface water on the aggregate is sometimes determined with poor accuracy. Even when the moisture content of the aggregate is measured accurately, the measurement is not repeated with sufficient frequency. If this is coupled with poor drainage in the aggregate stockpiles or with an inadequate drainage period, or else with interference by rain, then it is not surprising that a fixed weight of aggregate as batched contains a variable amount of water. Of course, the actual weight of aggregate in a saturated-and-surface-dry condition varies in a corresponding manner. In some climates and under some wind and sun conditions, the aggregate may contain less absorbed water than corresponds to a saturated-and-surface-dry condition. Such a situation again leads to a variable amount of free water in the mixer.

In some cases, there is a problem of inadequate mixing. This is especially true when several cementitious materials are used and, above all, when silica fume is included in the mixture. I have seen several cases of very good ingredients ending up as concrete with agglomerations of silica fume, and sometimes with parts of the concrete devoid of fly ash and other parts with an excess. In all cases, I tracked this down to too short a mixing time. There exist ASTM and ACI standards prescribing the mixing time as a function of the size of the mixer, but they are rarely mandatory. Moreover, mixing time is a matter for the contractor, and it always wants to increase its output per mixer by using a very short mixing time. In my opinion, this is an area deserving a close scrutiny.

One more comment about mixing is worth making. I have seen mixers, fortunately rarely, that were heavily worn and did not operate efficiently. I have also seen mixers that had not been properly cleaned for a long time so that they had a heavy encrustation of hard cement paste that interfered with proper mixing. When there are problems it may be worth inspecting the mixer.

The second treatment is that of curing, which those with a university degree often consider as too low tech to be bothered with. Curing is always prescribed in detail in the specification, but it is rarely done properly. Short of watching all the time, it is difficult to be sure that a given concrete has been properly cured. But a lack of proper wet curing shows. Unfortunately, it does not show immediately, but it does later when the poorly cured outer zone of concrete undergoes excessive carbonation or permits too rapid an ingress of chlorides. Curing affects almost exclusively the cover zone of the concrete, and it is precisely the cover zone that influences the durability of the structure in very many cases. When the mixture contains silica fume, good moist curing is even more important. So, my view is: pay thorough attention to curing.

Summary

Obviously, this section does not say *all* that those "in concrete" need to know; I do not want to injure the sales of my book *Properties of Concrete*. What I hope to have done is to highlight some of the aspects of concrete-making which are important and to which we should all pay attention.

The days of just ordinary concrete, containing portland cement only, are going and will soon be gone except for simple domestic applications. A whole range of cementitious materials must be considered for most purposes. Admixtures, especially superplasticizers, have an important role to play: more water is not the only, nor the best, way to improve workability. All this has to be borne in mind by the person interpreting a performance-type specification. Otherwise, it might turn out that the explicit requirements of the specification and the underlying desiderata are misunderstood, and become obvious only when trial mixtures are found to be non-compliant. At that stage, costs may rise above what was, incorrectly, expected.

The requirements of the specification are increasingly concerned with durability, in addition to the more traditional clauses. Sometimes, the durability is expressed as an expected life of the structure in years. To convert this figure into mixture ingredients requires a very good knowledge of the influence of the various mixture ingredients, and their proportions, on the properties of concrete. Not everyone needs to have such detailed knowledge, but there must be someone in every organization who can be consulted.

The situation is even more difficult when there is no separate specification written by an engineer, and the requirements have to be gleaned from notes on drawings and general references to national design codes, but without listing of specific clauses. In such cases, the person responsible for mixture design or selection of ingredients, whoever that may be, must be well aware of conditions of exposure and the appropriate mixture ingredients.

Speaking generally, the concrete mixture must be chosen so as to be fit for the particular purpose for which it is intended. In a way, this is what some people call HPC. In my view, the division between HPC and ordinary concrete will soon disappear, and we shall be providing custom-made concrete or tailor-made concrete for most of our clients.

If such an approach represents a higher cost, this is compensated for by improving the image of concrete and ensuring that it will continue to be used for all purposes for many years to come. The alternative is repairs, repairs to the repairs, and possibly demolition. One can laughingly say that this eventually provides recycled aggregate, but this encourages dissatisfied users to avoid concrete, if they can. It would be foolhardy to assume that concrete can never have serious competition.

I have concentrated on mixture composition, but this does not ensure good concrete in the finished structure. Batching is generally good but the control of water in the aggregate needs improvement and careful watching. The mixing time must be adequate to ensure a uniform distribution of all ingredients, especially when silica fume is used. The tendency to under-mix, so as to increase the output of a given mixer per hour, must not be tolerated. And finally, when the concrete has been placed and compacted, it must be properly cured. This is especially true when silica fume is present in the mixture but, with all concrete, good curing is essential to ensure that the concrete in the cover zone minimizes the ingress of deleterious agents and is fully capable of protecting the steel reinforcement.

All that I have said requires a knowledge and understanding of concrete. This is right and proper because concrete should no longer be regarded as a cheap and low-tech material.

Section 6.3: A Challenge to Concretors

My starting point is that, just because concrete is a highly successful and the most widely used construction material, there is no guarantee that it will continue indefinitely to keep its preeminent position. Life is competitive, and there needs to be continual progress and change. This section is a deliberate provocation to help foster an evolution in concrete practice.

Change has to be rooted in research. I recognize that much research in the field of concrete continues to be undertaken, but is it always being accomplished in the right place and by the right people? In the past, much research work was performed, at least in some countries, in government research organizations, and in trade association laboratories. For example, in the United Kingdom, in the half-century starting in the 1930s, considerable progress was achieved by bodies such as the Building Research Establishment, the Road Research Laboratory, and the Cement and Concrete Association. This is no longer the case.

In the United Kingdom, in consequence, the bulk of research on concrete is performed in universities. Now, universities, with their departmental organizations and with their emphasis on personal achievement and its related individual promotion, inevitably encourage an ever-narrowing specialization by individual researchers. I am aware of the fact that interdisciplinary centers are encouraged, but there have not been many success stories of materials specialists and structural engineers *jointly* developing better concrete structures *in practice*.

Specialists

The preceding statement is not a criticism, but rather a recognition of the fact that the complexity of structural engineering *and* of concrete as a material, with its numerous and varied ingredients, is beyond the capacity of a small research group, let alone an individual. The consequence is a greater specialization by individual professors. Though individual circumstances do differ, it is conceivable that a professor may spend a whole life at one and the same university and also has spent it working on a single subject, possibly a continuation of a Ph.D. topic. The professor may have moved through all the professorial ranks, even beyond a Full Professor to a Distinguished Professor. Apropos, I have heard of a University President (and I used to be one) who questioned the use of the term Full Professor and asked: full of what?

This professor, after what I consider to be prolonged drudgery, might have become an outstanding specialist in a narrow field, which, with luck, continues to be of importance. This work is useful and valuable. It is valuable to other engineers who deal with structures, their design and construction. They turn to the specialist for specialist advice, but the specialist is unlikely to advise radical changes in the design that would remove the underlying problem for which the specialist is needed.

Let me hasten to emphasize that specialists are, of course, needed. There is also the difficulty of there being fully competent generalists who know enough about design, construction, and materials. But, the designer should know enough in order to know what is possible and what advice is needed.

This situation is not limited to structural engineering; it can even be more severe in other disciplines, for instance medicine. A friend of mine recently had serious problems with his cardiovascular system—and at his somewhat advanced age, this is almost a *déformation professionelle*. Upon a specialist's advice, he had to undergo major heart surgery. All went well, because these days the heart surgeons are extremely competent and are supported by excellent technology devised by engineers who, in the eyes of the public, tend to be nameless and faceless. However, upon his return home, my friend became, in his words, "very anemic." He was rushed back to the hospital where a consultant in internal medicine discovered a very old and neglected duodenal ulcer. When this was cured, my friend felt "fighting fit" for the first time in years. Why this dreary story? We live in days of high specialization, with generalists hardly in existence. Even a general medical practitioner, as soon as he or she identifies a particular problem, refers the patient to a specialist who never looks outside of his or her field of knowledge and thus can miss much.

Let me tell an anecdote from a book by a well-known Scottish writer, Eric Linklater, himself a medical doctor. A medical student was taking a practical examination in respiratory medicine: he was told to examine the patient, identified some problem in the chest, and then stopped. The examiner prodded the student by asking whether he had observed anything else. The student could nor identify anything more untoward in the thorax, whereupon the examiner uncovered the full body of the patient: both his legs had been amputated!

Generalists

The extent of specialization is necessitated by the explosion of knowledge and the impossibility today of knowing much beyond the narrowest of fields. I was familiar with a number of people who were civil engineers by training, with a continuing interest in real-life construction: Eduardo Torroja in Spain, Hubert Rüsch in Germany, Bob Philleo in the U.S.; they are gone now. There probably are a number of such people still alive, but the only ones whose names spring rapidly to mind are Pierre-Claude Aïtcin and Mohan Malhotra in Canada, and Chester Siess in the U.S.

I think that a critical characteristic of these people is that they were trained as civil or structural engineers and had their roots in engineering. They may have not learned much about concrete at the university (I had a total of nine lectures in my degree course) but later they could acquire the necessary knowledge of materials. I believe that, with good grounding in engineering and a knowledge of chemistry and physics acquired as a part of their engineering courses, they knew the essentials and they knew enough to know what to seek from specialists. The converse is not true: a chemist, a geologist, or a materials science graduate is most unlikely to acquire *on the job* an adequate knowledge of structural behavior, let alone structural analysis.

In case the reference to structural analysis causes raised eyebrows, let me mention the consideration of creep in the design of reinforced and prestressed concrete structures. I have met some designers who can only use a few standard coefficients to allow for creep in their analyses, and who are not sufficiently aware of the influence of the composition of concrete upon creep. At the other extreme, there are creep specialists whose main preoccupation is with developing elaborate, or even extremely complex, formulae that describe creep in laboratory specimens. They, however, do not seem to concern themselves with the behavior of structures in service, in which the stresses are far from constant over time, and in which creep recovery as well as relaxation occur. Of course, people whose knowledge spans both creep of concrete as a material and of the variable and multiaxial stresses involved in a structure do exist. Two examples are Walter Dilger and Amin Ghali, both in Calgary.

What I have described so far are examples of what I see as root problems in concrete but, what is more important, they underlie the recurrent problems with concrete structures in service. It is extremely rare that such structures collapse, and this is a tribute to structural designers. There is, nevertheless, an inherent problem in design, and this is the economic consequence of an excessive variability of concrete. By the term "excessive" I mean that a reduction in variability would lead to economy, and I propose now to discuss the main reasons for variability of concrete as produced nowadays.

Safety factors

Modern structural design involves a probabilistic assessment of safety. With respect to concrete, the characteristic strength of standard test cylinders or cubes is, so to speak, converted into the service strength of concrete in the structure by a partial safety factor for material. Such factors vary between different design codes, and I shall limit myself to the example of the British code for structural concrete. First of all, I wish to consider the partial safety factor for strength: the design strength is the characteristic strength of concrete divided by this factor. According to the 1972 British code, the rationale of the partial safety factor for the strength of materials is "to take account of possible differences between the strength of the material in the actual structure and the strength derived from test specimens." The 1985 code explains that the partial safety factor for the strength of materials "takes account of differences between actual and laboratory values, local weaknesses, and inaccuracies in the assessment of the resistance of sections." As I was a member of the committees drafting both of these codes, I have to accept the above statements as being correct.

The value of the British factor for concrete is 1.5; by contrast, the value for steel is 1.15. This difference between the two materials is striking, and it reflects that steel is factory-produced, with actual coupons of some steel rods tested for yield stress and for extension. On the other hand, concrete is placed, compacted, cured (or not cured), and treated (or mistreated) by concretors on site. This is not all. What is often forgotten is that this partial safety factor of 1.5 is applied to the characteristic strength of concrete. The characteristic strength allows for 5% of the standard cylinder strengths being too low (but not too much so) so that the value used in design is well below the mean strength of the cylinders. In other words, 95% of the cylinder strengths are greater than the value of strength that is subjected to the partial safety factor for the material.

Variability of concrete strength

I presume it is not necessary to explain that the difference in the values of the mean strength and the characteristic strength is a reflection of the scatter of the strengths of the test cylinders. The scatter is expressed as the standard deviation. For a value of characteristic strength of 5%, the abscissa at that strength, reckoned from the mean strength in the normal distribution, is equal to 1.64 times the standard deviation. Now, the concrete mixture has to be designed on the basis of the *mean* strength. It follows that any reduction in the difference between the mean strength and the specified (and achieved) characteristic strength represents an economic gain because a smaller proportion of test cylinders has what might be termed an excessive strength. Because the difference is 1.64 times the standard deviation, a reduction in the magnitude of the standard deviation is clearly beneficial.

I strongly feel that a reduction in this "difference" when achieved in practice will represent the greatest economic benefit for concrete. I see this as far more important than obtaining a higher characteristic strength. And yet, as far as I know, apart from internal work by ready-mixed concrete producers, this "difference" is not being researched. Why? Because such work cannot be done in the laboratory, but only on site, dealing with "real materials" as they exist in place and with actual site procedures.

The point that I am trying to make is that there is a large difference between the strength of concrete called for to satisfy the structural design requirements and the strength that the ready-mixed concrete plant must produce to allow for the variability of the mixture as it leaves the mixer. Also, there is the fact that the processing on site is far from perfect. I am not saying that this difference is not necessary: it *is,* because safety is paramount. But, it is vital only because the procedures both of batching and of site operations are such as they are. To put it bluntly and generally, these could be much better. More to the point, if they do not improve, concrete is likely to lose ground to other construction materials.

Doing better, then, means reducing the scatter of strength of concrete coming out of a mixer and also improving site treatment so the difference between the *potential* strength of concrete and its *actual* strength in-place is reduced. Only then can we produce concrete more economically without sacrificing safety.

Possible improvements in concrete

The question then is: how can we do better? I am not competent to write a recipe, but I can identify a few areas in which improvement is possible. Let me start with mixture ingredients.

One source of variability of concrete at the point of discharge from the mixer is variability in aggregate grading. I know that aggregate is screened into several size fractions, which are combined in the batcher. But each size fraction covers a fairly large range of sizes, and *within* a specific range it is possible to have different grading. For example, in the fraction comprising particles 10 to 20 mm (3/8 to ¾ in.) it is possible to have more or fewer particles nearer the smaller size. If there are more small particles present, the water demand is likely to be greater or the workability lower. As workability *must* be as specified (otherwise the concrete cannot be consolidated with the means foreseen and provided on site), the ready-mix plant must provide for the worst-case scenario. The influence of precise grading of the finer fraction of fine aggregate is even greater, and sometimes there is large variation in the grading of the finest particles. This means that, not to compromise the characteristic strength, the average strength has to be higher than would be the case with a much more closely controlled grading. Aiming at a higher *average* strength represents, of course, a higher cost and makes concrete more expensive than could be achieved with a better control of aggregate and the moisture it contains.

The same argument applies to the presence of flaky or elongated particles of coarse aggregate. The usual controls are on the maximum proportion of such particles, but the actual proportion may be much lower. However, the design of the mixture must allow for the worst situation, which may be rare—but nobody knows when it will occur. This is especially so when the aggregate is supplied from more than one crusher. On an earlier occasion, I described a situation when one crusher was badly worn and consequently delivered aggregate with a significantly different shape from the other, but newer, crushers. The differences in the surface texture of the aggregate play a similar role with respect to the variability in the workability of the concrete coming out of the mixer. Physicists are able to measure the texture of particles, and we should endeavor to apply such techniques to aggregate as actually batched.

Still on the topic of variability of aggregate, there is the vexing problem of its moisture content. This varies notoriously, and usually the determination of moisture content is sporadic and infrequent. Determination of the *real* amount of water in *every* individual batch of aggregate as it enters a mixer is rare. In any case, I for one have not seen evidence of the reliability of such a determination coupled with an automatic adjustment of the amount of added water on a batch-by-batch basis. The point that I am trying to make is that if *precise* grading of aggregate, an assessment of its shape and texture, and measurement of moisture content were made on a batch-by-batch basis, then a considerable reduction in the variability of concrete coming out of the mixer could, indeed would, be achieved. This would lead to economy, but of course only after the cost of developing the new methods has been amortized.

As far as the development of a much better control of aggregate grading and shape is concerned, I think there is an additional factor at play, namely the use of recycled aggregate. Unless such aggregate is "good" and has well-controlled properties, there will be difficulties in its use. And yet we must use more and more recycled material. In many areas, sources of natural aggregate are running out or are conserved by legislation. Moreover, in a number of European countries, natural aggregate is subject to fiscal measures. Also, the use of a specified minimum proportion of recycled aggregate is mandatory. In the United Kingdom, during the last decade of the twentieth century, the sales of crushed rock, gravel, and sand, taken together, fell by 30%. In 1999, aggregate from other sources represented 18% of the total market. The other sources are: processed demolition material, used railway ballast, incinerator bottom ash, slag, slate waste, and other waste products. What is needed is the means of measuring precisely the properties of these aggregates.

The feeling of many people involved in the production of concrete is simply: this is how it is. They have little interest in achieving a closer control of aggregate characteristics, and even less interest if the source of aggregate belongs to the ready-mix concrete producer. A righteous, but perhaps naive, person might ask: why not? The answer is that such improvements would require a great deal of money and, in the short term, their costs would outweigh the economies gained.

A friend of mine, who is a prominent and highly successful concrete entrepreneur, once explained to me: "As long as everybody else does equally badly, there is no benefit in spending money to do better."

Of course, this is true, provided the customer *must* buy concrete because there is no alternative material that could be used. I shall come back to this later.

Ecological considerations

With a tight control of aggregate grading, there would also be a reduction in concrete wasted because it has been rejected for want of adequate workability or because of other deficiencies. Economy would lie in less cement being used, and this would please those preoccupied with ecological considerations and the popular concept of sustainability.

Obviously, we should reduce the emission of carbon dioxide into the atmosphere, but is concrete really a significant culprit? For instance, at the risk of offending many people, I cannot refrain from commenting on the use of motor cars. One example: on Californian freeways, for every car in the pool lane (obliged to carry at least two people) there are probably 10 cars that fit the description "one car, one person." Maybe the current car usage in California and elsewhere is essential for the comfort and well-being of people living there, but if we are serious about reducing noxious emissions, we should press on with cars and not with concrete. In comparison, to concentrate on concrete can be likened to reducing the price of caviar by using cheaper wrapping.

Possible improvements in site work

On site in some countries, concrete construction suffers grievously from the poor education and technical skills of the local operatives. There is, at present, little incentive to employ more highly skilled labor, even if they could be found. Such people would naturally and justifiably have to be paid more, but many contractors are shy of such expenditure. Again, it is a situation where there is no incentive to do better than your competitors. On other occasions, I have suggested that the solution to this problem lies in a mandatory provision in contracts to employ a proportion of trained and certificated operatives and trades people. To begin with, this could be done in construction for government agencies. Once contractors employ better-skilled people for these contracts, they would find such labor beneficial, albeit more highly paid, in construction projects for other clients as well.

It is also possible to reduce the overall number of people required for concrete construction. In the March 2001 issue of *Concrete International*, authors P. Kumar Mehta and Richard W. Burrows stated: "Globally, we do not have a labor shortage." But such laborers are not skilled concretors, and they would perpetuate the existing system. Moreover, to move these people around the world to the highly industrialized areas needing them would be a modern equivalent of the indentured labor once employed in sugar, tea, and coffee plantations in the bad old days. Rather, I believe that the solution lies in the use of robots and in a much greater mechanization of concreting. Some mechanization, but not enough, exists in highway construction. Much progress, especially in the use of robots, has been made in Japan, a country opposed to importing foreign labor. Moreover, with a very large proportion of the population of Japan educated to university level, there are few of its people keen to work as concretors.

As I have said earlier, this is not the occasion for suggesting specific remedies, and I am not qualified to develop them. Many technologies existing in other fields of endeavor could be applied fairly readily, for example, to classify aggregate and to determine its shape and texture. Of course, all this takes money but, in the long run, it would ensure that concrete will continue in its preeminent position as a construction material.

My frequent trips abroad had to do with the investigation of problems, real or alleged, with concrete and concrete structures. Much effort, time, and money are devoted to the resolution of such problems. However, initial conjectures on the causes of many of these "problems" were misdirected. For example, there was a case where it was alleged that the workmanship was poor where in fact the structure was under-designed. In another case, concrete was accused of not being water-resistant, yet the real culprit was excessive irrigation of the surrounding soil and a lack of adequate drainage.

Of course, there are cases where the concrete itself was not satisfactory for the given conditions in a particular structure. Would it not be better, as well as more economical, to make absolutely sure that the concrete is always "fit for the purpose," in the material itself *and* in the execution of the construction so that the concrete in the structure is accepted unequivocally as such by all involved? In other words, concrete should be tailored to the needs of the designer. But, the designer must know what can be demanded, and that this will truly exist in the finished structure.

Is concrete under threat?

A possible rebuttal of my point of view might be that there is no competitor to concrete in sight. A similar attitude was held by Swiss manufacturers of watches with a mechanical movement. But when Japanese quartz watches arrived, a large number of small Swiss watchmakers disappeared, and highly-skilled personnel became unemployed.

There are examples elsewhere, for instance, in aircraft propulsion. Manual typewriters have been relegated to a museum; for that matter, so have their successors, electric typewriters.

Some changes have occurred principally in design, with consequent alterations in materials. More importantly, there have been cases where evolution in design occurred only when new materials with very specific properties were developed. A striking example is the design of aircraft bodies. Changes there were effected only by using new materials having properties required by the designer. To satisfy this, there developed a kind of symbiosis between the aircraft structural designer and the materials scientist. This has occurred in other fields as well.

In concrete structures, the situation is often that a range of mixtures is available, and designers have to serve themselves from that "menu." Of course, there have been changes in the types of concrete available, but there has been no great progress or dramatic developments. It would be unrealistic to expect changes in concrete comparable to those in the aircraft industry; our industry's structural designers cannot develop new concrete-like materials, and concrete scientists are not in a position to envisage the design changes they could meet. But, the challenge to do better is there, and we need a "brains trust" to find a way forward.

Chapter 6: Technology and Design

Recent progress

What I have said in the preceding section may not be palatable to some—even many—and I admit that this is a deliberate provocation to galvanize all of us in concrete into action. Of course, there have been changes in concrete and concreting: slipforming, self-compacted concrete, and roller compacted concrete, to name a few. But, what significant overall progress has been made? Air entrainment is now 60 years old, and this was an accidental discovery. Superplasticizers had to wait 30 years until they came to be used in concrete, but undoubtedly they were a very significant development.

The use of concrete mixtures containing a high proportion of fly ash and ground-granulated blast-furnace slag also represents a significant development. There must be other changes that have improved concrete but, overall, concrete construction is, depressingly, much the same as in my younger days. In other words, the changes that have taken place are not as significant as they will need to be to keep out future competition. We should not forget that the consumption of steel has decreased dramatically. Cannot the same happen to concrete as we know it, even though the current use is so huge and widespread?

The future

I dare say it is my advanced age that leads to my pessimism. I may be exaggerating the situation, but my aim is to help jolt the concrete world out of its complacency. What we are doing, I believe, is largely tinkering at the edges. Even the holistic approach is really not much more than another look at concrete as a material, and not at the entirety of the structure, which requires input by the structural engineer.

While this goes on, someone somewhere will find better materials *and* better techniques to build better structures. These materials may include polymers of various kinds as well as all types of fiber components. Portland cement might still be used, but not as the main bulk component of concrete, which at present leads to the frequent problem of shrinkage cracking. It is quite likely that steel reinforcement will no longer be used, thus removing the vexing problem of corrosion.

These are not just flights of fancy. For example, the Department of Mechanical and Aerospace Engineering of the University of Alabama at Huntsville has developed a material composed of portland cement, glass microbeads, latex acrylic fortifier, and water; reinforced with a graphite fiber mesh that can be used for boat hulls, and just 7 mm (¼ in.) thick, providing strength, flexibility, and crack resistance. Its inventors are looking forward to the use of the new material in spacecraft. This could be an example of symbiosis of structural design and material development. I cannot help observing that, much to my chagrin, it was not in a department of civil engineering where this development occurred. The preceding is, of course, a highly specialized example, but history is full of specialized products being modified in due course into more mundane, widely-used materials.

Moreover, there has been some important development in large structural elements prefabricated to include load-bearing members, internally and externally finished, and connections. This approach provides better overall quality control than traditional construction methods. Specifically, monocoque construction, using composites reinforced with glass fiber, gives great freedom of shape and high durability.

These are not future possibilities: such structures already exist—for example, a five-story building called *Eyecatcher*, built in 1999 in Bâle, Switzerland (described by Thomas Keller in the *Bulletin of the Swiss Association of Engineers and Architects*, September 2001).

These large, innovative elements made of new materials will not be used as a substitute for concrete in structures of the type and form employed at present, which are largely rectilinear and angular. Rather, the direction of development will be to produce more rounded and softer shapes, made possible by the new materials and methods of fabrication and construction. Here, once again, development is a combined change in materials and in design.

My view is: It is the symbiosis of the material specialists and structural designers that is needed. The latter need to know better what is possible with respect to concrete as a material and its processing; the former need to know what is desirable. More mechanization of concreting operations and a greater (than zero) use of robots would undoubtedly improve the uniformity of concrete *in structures*, making them more reliable and, in the long run, more economical.

For us to wait until a "replacement" of "old-fashioned" concrete has arrived is to bury our heads in the sand. For concrete to have a good future, I feel we need a new approach concentrating on the problems I have mentioned. We still have time to find such an approach, but time is not on our side. We have to move more quickly if concrete is to maintain its preeminence as a construction material.

Section 6.4: Concrete Technology and Design—The Twin Supports of Structures

This is the first of three linked sections: the present one deals mainly with the role of concrete technology in the design of structures. The next two will give examples of the influence of the knowledge of technology upon design, including cases of design not soundly based on such knowledge.

For the purpose of this section, I would like to introduce myself as a structural engineer with a strong interest in concrete as a material. An elaboration of the two parts of the preceding sentence is necessary.

When I say "concrete as a material," I do not mean a study of the properties of concrete for its own sake, that is, just to find out what is the effect of factor A on parameter B. Such an investigation, often unrelated to practical issues, is of no interest in the context of the present article, and I leave it to pure scientists. For myself, I see no value in answering questions that practicing engineers do not ask, amusing though such an exercise may be.

Rationale of this section

I consider myself to be a structural engineer, not only because I have qualified as a chartered structural engineer, the term "chartered" corresponding to the American term "licensed," but also because, in the past, I have worked in a design office and later conducted research on reinforced and prestressed concrete, and even received an award from the Institution of Structural Engineers. The two books on creep of which I am the author or coauthor deal both with creep of concrete as a material and with the consequences of creep in various types of structures. Because I have been quite extensively involved in creep, I shall use that topic as the prime example of my concern in the next two sections.

Before readers point out that I have in the past already referred to the link between structural design and concrete technology, I had better confess that this topic is a bit of a hobbyhorse of mine; but then I have other hobbyhorses in my stable. My excuse is that, if a point of view is to be taken seriously, it needs to be expounded and stressed more then once.

The title of the present article is cool and almost anodyne. I have to confess that I was tempted to use one of the following titles:

May a Structural Designer be Ignorant About Concrete? and
If You Don't Know the Material, Can You Design a Structure?

Teaching concrete technology

The underlying cause of the problem—and I consider the present situation to be a serious problem—is the inadequate teaching of concrete as a material at the undergraduate level. My comments apply primarily to the United States and the United Kingdom, but I suspect that in other countries the situation is not all that different.

A survey by the Portland Cement Association in the United States in 1995 showed that only 22% of civil engineering departments teach concrete technology one hour per week during one semester. Bob Tobin has pointed out that this represents on average 18 hours of classroom work.[1] He also emphasized the fact that 78% of civil engineering undergraduates receive less instruction, or none at all, in what he calls "the greatest building material on earth."[1]

Of course, university courses are not the sole source of learning. However, a total absence of concrete from the curriculum is unlikely to whet a student's appetite for a study of the topic.

On the British scene, the situation used to be remedied by excellent courses on concrete technology for young engineers run by the Cement and Concrete Association, but all that ceased some time ago. Clearly, such courses cost money; seemingly, the cement manufacturers felt no need to promote their excellent product and assumed that engineers would naturally want to learn about it and its use. Ralph Waldo Emerson wrote: "If a man...make a better mousetrap than his neighbor, tho' he build his house in the woods, the world will make a beaten path to his door." This assertion is manifestly untrue as demonstrated by the enormous advertising industry. Moreover, Emerson's house was not made of concrete.

Returning to university teaching, we should consider not only the number of semester hours devoted to "concrete" but rather the fact that the topic of concrete is not considered exciting. If a few "hard" facts are presented without a connection to design, students can scarcely be expected to become interested or excited, and to wish to pursue the subject matter on their own. Of course, selfishly, I would like them to purchase a copy of *Properties of Concrete* for bedside reading!

In reality, if an undergraduate hears anything about concrete, it is "hard" facts presented often by a chemist. An example of this situation is a graph relating compressive strength to the water-cement ratio described as a "law" without a discussion of the underlying porosity, pore size distribution, load-carrying capacity of solids, water, and air. Also, it is not pointed out that the relation between strength and void ratio is not limited to concrete but applies equally to many other materials. Who can get excited by standard test methods for measuring slump or making cylinders (which are often the only practical laboratory experiments) or by rules about minimum curing times at different temperatures?

Concrete technologists in construction

So the situation is that graduates in civil engineering have been taught very little about concrete as a material; neither have they been made interested in the topic or appreciative of its importance. And yet, when it comes to construction, decisions involving knowledge of properties of concrete and concrete technology are necessary. Someone has to fill the gap.

In many cases, the person who undertakes the task of filling the gap is a graduate in geology, chemistry, physics, or materials science. Some of those people are well versed in laboratory studies but rather unaware of full-size and real-life behavior. This in itself does not matter.

What is important, however, is that these scientists, in most cases, totally lack knowledge of structural design and analysis, and of the behavior of structures in service. They may become knowledgeable about the response of the material to the environment but less so about response to different kinds of loading such as wind, earthquake, and fatigue-inducing stresses.

On the other hand, as I have argued earlier, the structural engineer is extremely well versed in analysis, especially in operations using a computer. This engineer has no problem including material properties in terms of symbols, but is not highly knowledgeable about substituting actual numerical values for those symbols: Which concrete will have a modulus of elasticity of a certain magnitude? What will be the effect of a particular local aggregate? What are the consequences of alternating humidity, or of the presence of particular salts in the ambient soil? And so on.

Of course, many of the relevant values are listed in codes and guides, but these can never be sufficiently case-specific. Moreover, without a detailed knowledge of the various important factors, the designer cannot choose the structural form to minimize any harmful effects. I propose to elaborate this topic in Section 6.5.

Investigation of problems

A particular difficulty arises in the investigation of problems such as the development of cracking. Cracks may form owing to numerous causes: some of these may be the consequence of external chemical attack, or of alkali-aggregate reaction, or of shrinkage. These the concrete technologist may well identify.

The situation is less clear when corrosion of reinforcement is suspected of being the culprit or if there is an overload or an inadequacy of reinforcement, such as a poor layout of reinforcement at a beam-column connection. To deal with these possible causes it is essential to understand structural action, load paths, and complex stresses. As these are not the provinces of the scientist, he or she may have difficulty ruling out the less likely causes and identifying the more probable ones.

More generally, there is a risk of a situation such that the concrete technologist does not know what are the underlying questions that should be *answered*. At the same time, the structural engineer does not know enough about concrete behavior to know what questions to *ask*. A parallel situation in the medical field would be if the surgeon did not know much pathology and had to rely on the laboratory technician for directions on what surgical action to take.

I am not alone in holding strong views on the qualifications of people who should investigate problems in structures. Zein al Abideen, Deputy Minister of Public Works and Housing in Saudi Arabia, wrote a specification for "evaluating experts," which included the following: "To be a civil engineer who actually practiced the job and in particular structural design calculations"; To "have a good knowledge and in depth experience in: statistics, properties of materials and specifications, destructive and nondestructive tests, requirements of design and execution...."[2]

Of course, there exist many structural engineers well knowledgeable about concrete properties; likewise, there exist concrete technologists who have studied structural behavior. In those cases, there are no problems but, alas, there are other situations, too, and these provoked Zein al Abideen to be outspoken.[2]

Accuracy and precision in structural analysis

The title of my article suggests that achieving successful structures requires the twin supports of structural design and concrete technology. Each of these involves quantitative data, and I would like to consider the accuracy and precision of such data; specifically, are the two inputs comparable?

Let me start with design. Nearly always, design involves structural analysis. There are two distinct aspects of accuracy in such an analysis. The first one arises from the fact that the analysis is performed on an idealized structure. This is rarely mentioned in textbooks on structural analysis; as an example, I can say that it is only in the preparation of the fifth edition of our book that Amin Ghali, the senior co-author, and I decided to discuss this topic.[3] Once stated, the situation is obvious, but not until then.

Idealized structures

The equations of equilibrium and compatibility, which are at the heart of structural analysis, are written for idealized structures. For example, beams are represented as being uni-dimensional; joints are considered to be points; concentrated loads are taken as acting at a point with no width; uniformly distributed loads are considered as having exactly the same magnitude over every inch or millimeter of the span; freely-supported beams are assumed to permit wholly unrestrained rotation; conversely, built-in (encastré) beams are assumed to permit no rotation whatever. None of these assumptions occurs in reality.

I realize that detailing of reinforcement takes care of some of these factors, but the calculation of bending moments, shearing stresses, and axial stresses is based on the idealized structure. There are, of course, other assumptions made, for example, about linearity of the stress-strain relation at service loads.

As for the magnitude of the loads themselves and of the dimensions of structural members, these are assumed to have certain values in the analysis. The dimensions are reasonably accurate but there are occasions where members are thicker or deeper or are made with a material of higher density than assumed. Nevertheless, dead loads are known with reasonable accuracy. The values of live loads are usually more arbitrary. Some loads, such as wind and earthquake, are derived from observed data on a probabilistic basis; from time to time, major changes are made in the values assumed to be appropriate in design. Despite this, it is a convenient and necessary assumption for the purposes of analysis that the loads are known accurately; however, to take care of the fact that in the real world no data are precise, various load factors and factors of safety are applied.

There is no problem with such an approach in design. However, when assessing the strength of existing structures, it is important to remember that stresses computed by analysis do not truly represent stresses in the actual structure. This is relevant to the accuracy and precision of properties of material, whose influence is considered in a later subsection.

Still on the topic of structural analysis, it should be remembered that, in the past, exact solutions of the various equations were obtained only when the number of simultaneous equations is small. Before the advent of computers, structures with a higher degree of indeterminacy led to a system of simultaneous equations that was too large for a solution using the then-existing mechanical calculators. Consequently, relaxation methods were utilized, pioneered in the United States by Hardy Cross and in the United Kingdom by Richard Southwell.

These methods gave an approximate solution, leaving a residual. With more cycles of relaxation, this residual could be made as small as you wished, but the final solution was always approximate. It is engineering judgment that determines when the residual is small enough to be acceptable, but judgment is by definition not exact. Southwell pointed out that, in mathematical terms, the residuals are in fact errors; in physical terms, the residuals are changes in data that, if actually made, would give an exact solution.[4]

Numerical methods used nowadays allow us to find exact solutions for complex structures, but they apply to idealized structures and linear materials. Specifically, the equations of equilibrium are based on the geometry of the structure as idealized, without recognizing the effects of deformation by loads, including self-weight. In other words, there is no geometric nonlinearity. In the case of the vast majority of concrete structures, the resulting inaccuracy is small. It is worth noting also that the finite element method determines stresses and deformations in a structure consisting of finite elements which, however well chosen and however fine, are only a substitute for a continuum in the actual structure.

In determining deflections, cracking of concrete under service loads has to be recognized, and the effects of creep need to be considered. For reasons to be discussed in Section 6.5, upper and lower bound values of creep are assumed. Iteration in the solution of the equations is necessary in order to deal with residuals.

The outcome of structural analysis serves the designers well but the accuracy and precision are not perfect.

Technologists' understanding of design

I must emphasize that there is nothing wrong with the situation described in the preceding section, but it should be borne in mind when we assess the properties of materials in place for the purpose of establishing structural adequacy. It is because I have found, on occasion, a preoccupation with spurious precision in tests on concrete that I feel that the topic should be aired.

Let me give a few examples of measurements of properties of concrete that exhibit a lack of understanding of structural design on the part of the tester. A tester in concrete, who was not a designer or a licensed engineer, opined on footings and slabs on grade. They conformed to the requirements of the Uniform Building Code, but he said: "The subject concrete slabs and footings contain some rebars, and are under-reinforced from the structural viewpoint. However, they cannot be considered as plain concrete slabs or footings since they do contain some rebars."

From the point of view of a structural engineer, the presence of "some rebars" does not make a concrete element into reinforced concrete. Specifically, ACI 318 defines plain concrete as "Concrete that is either unreinforced or contains less reinforcement than the minimum amount specified in ACI 318 for reinforced concrete." The Uniform Building Code gives a very similar definition. A mistake of this kind would not be made by a licensed structural engineer.

The same tester, when considering slabs on grade in single-family homes, criticized the presence of cracks and quoted ACI 360R, which says that a slab should "remain uncracked due to loads placed on its surface." In my opinion, the cracks were caused by shrinkage, and ACI 360R says: "Shrinkage cracking is controlled by a nominal or small amount of distributed reinforcement...." What I see as crucial is the ability to distinguish between load-induced cracks and shrinkage cracks, and also an appreciation that *controlling* cracking is not the same as preventing it. Perhaps the wording of ACI 302.1R is better because it refers to "a minimum amount of steel for the purpose of *limiting* the width of shrinkage-temperature cracks." The term "limiting" is more explicit than "controlling." As ACI 302.1R says, "Since drying shrinkage is an inherent characteristic of portland cement concrete, it is normal to experience some curling and cracking on every project." The preceding neatly makes my earlier point about understanding the origin of a given crack.

Moreover, ACI 318 clearly states: "Design and construction of soil-supported slabs, such as sidewalks and slabs on grade, shall not be governed by this code unless they transmit vertical loads from other parts of the structure to the soil." A structural engineer would know whether or not loads are transmitted onto the slab; such an engineer would also know that

6-38   Chapter 6: Technology and Design

ACI 318 is the mandatory document whereas ACI 360R is a Committee Report on whose front page it is stated that "ACI Committee Reports, Guides...are intended for guidance...."

Interpreting numerical values of properties of materials

Some people are fascinated by operations on numbers without realizing that the precision of derived quantities cannot be greater than the precision of input data. In some cases, their reports on tests can be highly misleading. Before discussing two actual cases in the field of concrete, I would like to consider a very simple example.

Let us consider a sack of cement whose nominal weight is 94 lb (418 N); let us assume that the weight was verified on scales with a resolution of 1 lb (4 N). We wish to place the sack on a three-legged stool, each leg having a cross-section 1 in. square (25 mm square) as determined by a ruler graduated to 1/16 in. (1 mm), the load being evenly distributed among the three legs. The area of the three legs is 3 in^2 (1936 mm^2). The resulting stress upon the floor on which the stool rests is 94/3 = 31.3 lb/in^2 (216 kPa), which should be reported as 31 lb/in^2 (220 kPa). It is not 31.33333 lb/in^2 (216.04333 kPa) that is given by a calculator because a difference of 1 lb (4 N) in the weight would alter the stress to 31.6666 lb/in^2 (218.3416 kPa), representing a difference in stress of 0.3 lb/in^2 (2.3 kPa). Likewise, if one of the legs has a cross section of 15/16 ×1 in. (24 × 25 mm), the other two legs being 1 in^2 (645 mm^2), then the stress would be calculated to be 94 divided by 2.9375, which comes to 32.0 lb/in^2 (221 kPa). Thus, the difference in stress would be 0.7 lb/ in^2 (5.0 kPa), so that, calculating stress to the nearest 0.1 lb/in^2 (0.7 kPa) is meaningless and misleading.

Readers might regard the above as insulting to their intelligence, and yet I have seen the use of spurious accuracy leading to conclusions about the safety of a structure.

Ultrasonic pulse velocity

I would now like to consider the interpretation of numerical values in ultrasonic pulse velocity (UPV) tests. These tests were performed by a professor on concrete cores for the purpose of establishing any possible variation in the deterioration of the concrete at different levels in the core. Consequently, UPV was determined at levels 15 mm (0.6 in.) apart. At each level, four measurements were made on diameters 45 degrees apart; the average of the four measurements represented the value of UPV at the given level.

An example of the test results is shown in Table 6.4.1. These results were interpreted by the tester to show that there was a real difference in the UPV at different levels, and therefore differential deterioration of the concrete had taken place.

My concern is with "finding" differences between the values of UPV at different levels, which are not statistically significant in view of the

Table 6.4.1—Values of ultrasonic pulse velocity (m/s) obtained by a professor

Level	Diameter				Mean	Standard deviation
	A	B	C	D		
1	3334.99	3444.77	3703.50	3339.72	3455.74	172.77
2	3456.15	3489.44	3310.37	3310.37	3391.58	94.76
3	3612.58	3311.53	3482.99	3280.39	3421.87	155.24
4	3345.65	3953.82	3471.43	3386.55	3539.36	281.23
5	3337.35	3732.86	3448.55	3837.82	3589.14	234.99
6	3576.91	3370.79	3627.89	3526.02	3525.40	111.15
7	3616.74	3510.25	3755.18	3488.15	3592.58	122.08
8	3278.11		3278.11	3393.87	3316.70	66.84

variation in UPV along the four diameters at a given level. Moreover, and equally importantly, I find reporting the value of UPV to the nearest 0.01 m/s (0.03 ft/s) quite unrealistic.

My criticism is as follows. Let us consider the values of UPV at Level 1: the standard deviation $s = 172.77$ m/s (566.8 ft/s). Hence, there is a 95% probability that the "true" value of UPV is within $\pm 1.96\ s$ of the mean value, that is, between 3455.74 + 338.63 and 3455.74 – 338.63. Thus the values of the upper and lower bound are, respectively, 3794.37 m/s (12,450 ft/s) and 3117.11 m/s (10,225 ft/s). If we now look at the mean values at *all* the levels, we find that they lie between the above extremes. In other words, the differences between the mean values of UPV at the different levels are not statistically significant and can be explained by chance variations.

None of this is surprising: ASTM C 597 gives a precision statement for UPV (admittedly for a distance between the transducers in excess of 0.3 m (1 ft) while the cylinder diameter was a little under 0.1 m (4 in.) to the effect that repeatability of test results is within 2%. Now, 2% of the average value of 3794.37 m/s (12,450 ft/s) is 76 m/s (250 ft/s). Looking still at Level 1, we find the difference between the highest and the lowest value of UPV to be 3703.50 – 3334.99, that is, 368.51 m/s (1210 ft/s).

It is worth noting also that ASTM C 597 says: "The accuracy of the measurement is dependent upon the ability of the equipment and the operator to determine precisely the distance between the transducers and the time of the arrival of the wave at the receiver." In the case of cores, the diameter varies locally owing to the actual coring operation so that, even with careful measurement, there may well be a variation of, say, 1 mm (0.04 in.). As the diameter is about 100 mm (4 in.), an error of 1 mm (0.04 in.) represents 1% of the distance measured. It follows that the *calculated* UPV is at best determined within 1%. With a velocity of 3300 m/s (10,825 ft/s), this represents 33 m/s (108 ft/s). It follows that the values shown in Table 6.4.1, which are given to the nearest 0.01 m/s (0.03 ft/s) bear no relation to reality. I come back to the issue of reporting a derived quantity with a greater precision than the input data in the next subsection.

It follows that no inferences can be made from the UPV tests as performed so that what the concrete tester tells us is positively unhelpful to the structural engineer charged with assessing the health of the concrete from which the core was taken. For the sake of completeness, I should add that ASTM C 597 says that in deteriorated concrete the variability is much greater but the calculated values of UPV are low. The values in Table 6.4.1 are not low but, of course, UPV depends on the aggregate and on its content in the concrete.

Strength of concrete

I would like to present one more example of confusion sown by manipulation of numbers arising from concrete tests with a spurious precision. This involves "reasoning" in concrete technology on the basis of spuriously precise numbers as well as a series of sequential corrections for various features of the test.

Tests were performed by a professor to determine the direct tensile strength of concrete cores; the test method was that of ASTM D 2936. That standard test method applies to rock core specimens, and in an earlier article I argued that there is no justification in using a standard method for a material for which it was not intended.[5] Ignoring this aspect, we can note that ASTM D 2936 prescribes the result to be expressed to the nearest 5 lb/in^2 (about 50 kPa).

In a particular case, the concrete core as tested was 2.82 in. (72 mm) high and 3.76 in. (96 mm) in diameter, and the direct tensile strength was reported as 203 lb/in^2 (1400 kPa). The calculations then ran as follows. First, it was said that "the size effect correction factor is about 0.8 to convert the test core to a 6 in. (152 mm) diameter core." (Incidentally, I am not aware of such a correction factor.) Hence, the corrected value was calculated to be 162 lb/in^2 (1120 kPa). Then, there was applied "the moisture effect - 5%, core effect + 5% and age effect - 10% after 10 years"; hence, the corrected direct tensile strength was found to be 146 lb/in^2 (1010 kPa). (Again, none of these values is known to represent an accepted "correction factor.")

The tester then said, "If we consider the height/diameter correction, the direct tensile strength would be even lower." As I showed in an earlier article,[6] ASTM does not allow for correction of core strengths with a height-diameter ratio smaller than 1.0.

The outcome of these calculations, all to the nearest 1 lb/in^2 (7 kPa), was then used to "prove" that the concrete had not conformed to the specified *compressive* strength of 2000 lb/in^2 (14 MPa) at the time of placing. None of this would matter were it not for the fact that such calculations may be highly misleading to those not well versed in structural engineering, and that their spurious precision is of a different order of magnitude from the precision of the structural analysis.

Conclusions

The above makes depressing reading, and some people might wonder whether it needs publishing. There is obviously great value in articles about good practice and good performance. But I have described here actual cases, and it is possible that some of those involved are not even aware that they are using poor and incorrect practices. It is for the benefit of others who may be unwittingly following these bad practices that some criticism should be voiced. I suppose it is cruel to say that everyone serves a useful purpose in life, even if only as a horrible example.

A good knowledge of concrete technology is just as important as a good knowledge of structural analysis and design in order to achieve good concrete structures. The former cannot be divorced from the latter; it is not possible to interpret properties of concrete without an understanding of structural function and behavior. I wonder whether the pendulum has not swung too far toward structural analysis at the expense of concrete technology. Maybe some structural engineers should take more interest in *their* material, concrete.

Let me finish by a true story about structural assessment. Some 30 years ago, a department store in a provincial English city decided to promote the sale of toys in the run-up to Christmas by hiring an elephant from a local zoo. The elephant was brought to the store and taken to the fifth floor in a freight elevator, which was not much larger than the animal. The elephant proved to be a great attraction and all went well until, late on Christmas Eve, it was time for the elephant to return to the zoo.

When the elephant was taken to the elevator, it remembered the cramped space and simply refused to get into the cage-like box. Persuasion proved fruitless, and it is not easy to shove and push a very large animal.

At this stage, the store manager approached my friend, a very well-known and successful structural engineer with the question: is the staircase strong enough to support the elephant? With Christmas around the corner, there was no time to indulge in load tests or to locate the reinforcement by a covermeter or to establish the strength of concrete: a decision, yes or no, was needed there and then. Fortunately, the engineer was very knowledgeable and was able, on the back of the proverbial envelope, to compare the load transmitted through the elephant's feet with the uniformly distributed load of people. He was also able to assess the concrete and the reinforcement, if only because of his experience in design and in construction and because of his sound judgment. The decision was: yes, go ahead.

Luckily, the elephant could be prodded onto the staircase and, once the descent started, it could not turn around. All was well and a merry Christmas was had by one and all, the elephant included.

References

1. Tobin, R. E., "Letters to the Editor," *Concrete International,* V. 20, No. 11, Nov. 1998, p. 8.

Chapter 6: Technology and Design

2. Zein Al Abideen, H. M., "Expensive Errors in Buildings Assessment and Repair," *Deterioration and Repair of Reinforced Concrete in the Arabian Gulf,* Fifth International Conference, V. 2, Bahrain, 1997, pp. 707-731.

3. Ghali, A., and Neville, A. M., *Structural Analysis: A Unified Classical and Matrix Approach,* Fourth Edition, E&FN Spon, 1997, 831 pp.

4. Southwell, R. V., "On the Philosophy of 'Relaxation'," *Applied Mechanics Review,* V. 10, No. 9, 1957, pp. 387-390.

5. Neville, A. M., "Standard Test Methods: Avoid the Free-For-All," *Concrete International,* V. 23, No. 5, May 2001, pp. 60-64. [Section 7.1]

6. Neville, A. M., "Core Tests: Easy to Perform, Not Easy to Interpret," *Concrete International,* V. 23, No. 11, Nov. 2001, pp. 59-68. [Section 7.2]

Letter to the Editor No. 1

I would like to congratulate Adam Neville on once again highlighting the limited teaching of concrete technology in most undergraduate engineering courses ("Concrete Technology and Design—The Twin Supports of Structures," *CI*, April 2002). Only where a faculty member is a particular enthusiast for the materials aspects of concrete is this situation likely to be remedied.

The engineering community as a whole, whether in the United Kingdom or the United States, should aim to support and encourage those isolated concrete specialists who continue to champion the study of concrete in universities.

However, the Advanced Concrete Technology Course, run in the United Kingdom by COMPACT (a consortium of U.K. universities), in South Africa, and in Ireland, under the auspices of the Institute of Concrete Technology (ICT), continues to provide an excellent means for young engineers to improve their knowledge of concrete.

The main thrust of Dr. Neville's argument is the need to link skills in structural design with those in concrete technology (as an aside, there is, however, much more to concrete technology than just "testing"). That this would be of benefit to specialists in both disciplines cannot be denied and the construction client or building owner would undoubtedly gain in terms of a more durable asset with reduced maintenance.

With the exception of water, concrete is the most widely used material on Earth. The study of the structural properties and long-term performance of this fascinating material deserves a far higher profile within the overall field of engineering.

Bill Price
The Institute of Concrete Technology
Crowthorne, UK

Letter to the Editor No. 2

Adam Neville's sequence of articles makes very compulsive reading and

CI is to be congratulated for publishing them. Most of Adam's observations make a lot of common sense and structural and forensic engineers should take note. However, there are certain points that I cannot agree with.

Scientists are a very important part of a building or civil engineering investigation. I have found in my material and structural investigations that material scientists form an extremely important part of a team. The samples that my technicians extracted from site were always tested in the laboratory by technical staff under the direct supervision of a chartered engineer. Chemists or physicists and, sometimes most importantly, petrographers would have carried out some of the tests. It is only by such a holistic approach that the actual condition of a structure can be appropriately evaluated.

Some examples were given of ultrasonic pulse velocity (UPV) test results with a conclusion that I cannot reconcile. It is suggested that a value of approximately 3.5 km/s is indicative of good concrete. For normal weight concrete, I would beg to differ, as I would consider anything below 4 km/s as poor quality structural concrete. I would go even further to say that although UPV tests give a very good indication of quality (for example, cracking, etc.), they are not very sensitive at estimating strength especially for higher strength concrete. I know that here, Adam is giving an example of undue precision in test results. With this, I would agree; hence, the usefulness of using appropriate units for velocity, such as km/s instead of m/s. UPV testing not only requires accurate distance measurements between the transducers, but also requires the transducers to have intimate contact with the concrete. So on a dusty site, it is imperative that the contact surfaces are perfectly flat and free of dust, as is the grease or jelly used to achieve perfect contact.

Adam gives an example of a loaded three-legged stool under the section on interpreting numerical values of properties of material. This section is meant to emphasize the unnecessary fascination that some people have for precision of derived quantities, to which one must agree. However, to make the example simple, I am afraid he has fallen into a trap quite a few structural engineers make. He states that the resulting stress on the floor is 31.3 psi (217 kPa), when in fact it is probably not. This value is not a stress but an applied load of 31.3 lb for every square inch or 0.87 kg for every mm of floor area. The stress (in kPa) upon the floor depends on the dimensions, properties of the slab upon which the stool stands and the subgrade reaction if the slab is on grade, or the beam spacing if the slab is suspended. I can understand the confusion in BS units, but it is much clearer in SI units if kPa or N/mm^2 is reserved for stress tensors and kg for force vectors or loading.[1]

I am convinced that this is an oversight of Dr. Neville's in his attempt to make the example he gave as simple as possible. My article in 1998 made a point of differentiating between stress tensors and load vectors and I could not resist the temptation of mentioning it again.

I am sure Adam has not exhausted his thoughts on concrete technology and structural engineering and I look forward to reading the next article in this interesting sequel of papers.

Patrick Sullivan
City University School of Engineering
London, UK

Reference

1. Sullivan, P. J. E., "Using the Language—The Terminology of Engineering," *Concrete International*, V. 20, No. 5, May 1998, pp. 55-56.

Neville's Response

The letter from Bill Price, who is President of the Institute of Concrete Technology, gives me particular pleasure because, for over 20 years, I chaired the Moderating Committee for the Diploma in Advanced Concrete Technology that governs admission to the Institute. Also, the Institute has been good to me in giving me its top award: the Silver Medal.

Bill Price's "aside comment" to the effect that there is "much more to concrete technology than just 'testing'" is important, but professional testers sometimes forget it. For my part, in my book *Properties of Concrete* I devote only one chapter out of 14 to testing concrete. As this book has been selling for 40 years, with the fourth edition now going into the eighth impression, I interpret this to mean that the "concrete world" seems, fortunately, to look beyond mere testing. I am grateful to Bill Price for his contribution.

Patrick Sullivan's comments on my example of spurious precision in the determination of the ultrasonic pulse velocity (UPV) should not be addressed to me: the values in Table 6.4.1 of my article are those of the original tester who is an academic and not a structural designer. Thus, it is actual test results that I was discussing, and the units of measurement (meters per second) were his, and not mine.

Be that as it may, the values of UPV should not be translated into strength because UPV is affected also by factors other than strength; in *Properties of Concrete*, p. 631, I say that "there is no physical relation" between UPV and strength. The modulus of elasticity of the aggregate and its content in the mixture, as well as the moisture condition of the concrete, affect the two parameters in different ways.

With respect to the stress applied by a stool *to* the floor surface, Prof. Sullivan extends my example of a misguided interpretation of the output from a calculator, given to several decimal places, to a consideration of stress distribution *within* the thickness of a slab. Load is a force expressed in pounds or Newtons; stress is the average value of the force per unit area, expressed in psi or N/m^2. ACI 116 defines stress as "intensity of internal force (i.e., force per unit area) exerted by either of two adjacent parts of a body...." On the other hand, kilograms are a measure of mass. Prof. Sullivan's excursion into tensors goes well beyond the realm of analysis and design of a concrete slab. Moreover, tensors and vectors are not mutually exclusive: rank 1 tensors are called vectors, and rank zero tensors are scalars.

As acknowledged by him, all this goes beyond my warning about spurious precision. For supporting this point of view, I am grateful to him.

Adam Neville

Section 6.5: Creep of Concrete and Behavior of Structures—Part I: Problems

In Section 6.1 titled, "Concrete Technology and Design—The Twin Supports of Structures," published in the April 2002 issue of *Concrete International*, I expressed the view that "a good knowledge of concrete technology is just as important as a good knowledge of structural analysis and design."[1] My intention and hope was to encourage more civil engineers to concentrate on concrete and thus apply the knowledge of concrete to construction. To give more credence to my view, I would like to cite some examples of design in which specific values of some properties of concrete—in particular, creep and shrinkage—are essential. I will do so in two parts. This section deals with problems caused by creep; in the second section, published in *Concrete International* in June 2002, I consider how to deal with creep in design and construction.

Why discuss creep?

My choice is the creep of concrete. The reason for this choice is that, some time ago, I undertook considerable research on creep and published numerous papers. Some of them dealt with laboratory tests and measurements of creep of concrete as a material, both under uniaxial and triaxial compressive stress and under tensile stress. Other papers dealt with the effects of creep in reinforced and prestressed concrete.

I also wrote, in 1970, what was probably the first major book on all aspects of creep titled, *Creep of Concrete: Plain, Reinforced, and Prestressed,* to which Walter Dilger made a contribution.[2] He and J. J. Brooks were my coauthors of the second book titled, *Creep of Plain and Structural Concrete,*[3] which was published in 1983.

I am mentioning all of the previous not to boast (especially as, in recent years, I have not studied the topic of creep) but to justify my choice of the topic of the present section. I should explain that this is not a scientific paper explaining and discussing creep, but only an attempt to persuade the readers of the importance of concrete technology in design and possibly to encourage designers to improve their knowledge of the properties of concrete. My contention is that a number of designers rely on standardized values of creep; they assume those values in their analysis and often cannot verify them in terms of the actual properties of concrete in the structure. If the discrepancy is large, the structure may behave in a way different from that assumed in the design.

What is creep?

Creep is a nonelastic deformation of concrete under sustained stress. It occurs in addition to stress-induced elastic deformation and also stress-independent strains known as shrinkage and thermal movement. Creep occurs under both a compressive stress and a tensile stress. Upon removal of a sustained stress, there occurs a reverse strain known as creep recovery; the two strains are not equal and opposite, so that creep is not a reversible phenomenon.

A further complication arises from the fact that the magnitude of creep is a function of the age when the stress begins to act and also of the age when it ceases to do so. Although creep is proportional to the applied stress, it is a function of numerous variables, which can be expressed in a variety of ways. Different researchers choose different groups of parameters, and different codes give standard values of creep characteristics in terms of different factors. To give an example, creep may be considered as a function of the water-cement ratio or of the strength of concrete at the time of application of the sustained stress. Other features of the concrete mixture that influence creep relate to the aggregate: its type and density, its volume per cubic meter of concrete, and its modulus of elasticity.

The age of the concrete when the sustained load is applied is also a factor. It is worth remembering that only sustained loads induce creep, and such loads may begin and cease to act at various ages. So overall, the situation is complicated.

In addition to intrinsic factors in the concrete itself, creep is affected by extrinsic factors such as the relative humidity of the ambient medium and the temperature of the concrete. Specifically, concrete under steady hygrometric conditions undergoes only what is known as basic creep. When drying takes place concurrently with sustained loading, there occurs in addition the so-called drying creep, and of course shrinkage. The larger the surface-volume ratio of a concrete member, the greater the proportion of concrete that undergoes drying; and therefore, the larger the drying creep and the shrinkage. It follows that, other things being equal, a smaller member with all its surfaces drying exhibits a larger creep and a larger shrinkage.

Creep is a dimensionless quantity, usually expressed in microstrain; that is, 10^{-6}. Some people write "inch per inch" or "mm per mm," but this is superfluous and even unhelpful. The magnitude of creep is small, but certainly not negligible. Creep is usually at least as large as the elastic strain, and can be two or three times larger. I should add that reinforcing steel modifies the strain actually developed.

I do not propose to discuss further the phenomenon of creep, but I shall give examples of several types of structures and load situations in which the effects of creep are significant. In addition to creep, I shall refer also, in some cases, to the effects of shrinkage. My treatment will be descriptive and qualitative. In Section 6.6, I shall comment on the current development of creep expressions and their relevance to the needs of designers.

Creep in supports of beams

Broadly speaking, we can distinguish between supports of beams and columns, although there is no fundamental difference between the behavior of the two and between their consequences.

Supports of a turbo-alternator

The first structure I came across was a very wide beam supporting a very large turbo-alternator. The beam rested on several supports, unevenly spaced and of varying height depending on the equipment and arrangements underneath. The supports were made of concrete, but they were not all of the same size because of their unequal spacing. They were also of unequal height. Under the sustained load of the turbo-alternator, the supports underwent creep and, consequently, they became shorter. The amount of shortening was variable depending on the characteristics of the supports. The main characteristics are height and volume-surface ratio, which controls the drying of the concrete that influences its drying creep and its shrinkage as well.

Now, the operation of the turbo-alternator requires that the rotating shaft remains perfectly straight and level at all times, so that unequal shortening of the supports cannot be tolerated. When unequal shortening occurs, it is, of course, possible periodically to stop the turbo-alternator and to insert shims on top of the supports in order to restore it to a level position. But such outage for a few hours interrupts the electricity supply and represents a loss of revenue.

The question to consider is: how to equalize the creep deformation and also the shrinkage of the various supports? I shall come to the answer after the next two examples, which show that the problems arising from creep are not very rare and are not limited to horizontal beams. The point that I want to make at this stage is that, unless the designer is sufficiently familiar with creep, he or she may not even be aware of the seriousness of the potential problem.

It is not good enough for the designer to look up the value of the creep coefficient in a design code or in a handbook: it is necessary also to have detailed information about the concrete mixture and the exposure conditions.

A bridge continuous over several supports

This brings me to my second example: a long bridge consisting of continuous beams over a number of supports. In some cases, the bridge runs not only over a river but also over low ground, which may be permanently dry or may be subject to periodic flooding. The aspect of this situation that is of interest is that some bridge piers were permanently in water or in humid air; other piers were in fairly dry air. Now, as I already mentioned, creep of concrete that is undergoing drying is much larger than creep of concrete at a steady humidity. In addition to creep under a sustained load, shrinkage takes place also when the concrete is drying; there

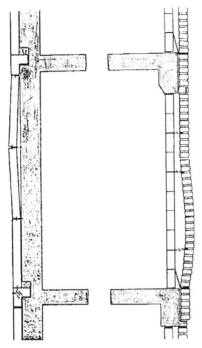

Fig. 6.5.1—Effects of creep in concrete columns

is no shrinkage under wet conditions. A further complication arose from the fact that the piers were of unequal height; and therefore, at a given magnitude of creep and shrinkage, the shortening over the whole height was variable.

Now, if the creep of the different piers varies in magnitude, the piers will undergo different amounts of shortening, even when all the piers are of equal height. If that happens, the supports undergo differential settlement, and the continuous beams resting over these supports will suffer parasitic bending moments and associated parasitic shearing forces. The structural design must provide for these induced stresses in the beams. To do so requires a calculation at the *design stage* of the differential shortening of the piers exposed to different conditions. To estimate the magnitude of that shortening, it is necessary to be knowledgeable about creep.

A suspension bridge

A long-span suspension bridge has high towers supporting the cables from which the bridge deck is suspended. There are various ways of arranging the cables at the top of the towers. What matters is that if a tower shortens after the initial installation of the cables, this geometric change affects the force in the cables. How significant this change is depends on the height of the towers and on the length of the bridge span.

I was involved in a case concerning a very large bridge with towers

206 m (676 ft) high and a suspended span of 1377 m (4197 ft). Four years after the installation of the cables, the towers had shortened an additional 150 mm (6 in.). As this was an important structure, the designer was fully cognizant of the phenomena involved. I am mentioning the relevant facts to illustrate the importance of creep; shrinkage of the massive concrete towers (with only a very small proportion of the cross section exposed to drying) was negligible.

Creep in buildings

Like bridge piers, columns in buildings are subject to shortening by creep. In buildings of no more than several stories, problems arise rarely, except when there is an incompatibility between the deformation of load-bearing concrete members and of cladding.

Problems with cladding

Strictly speaking, cladding problems may occur even in low-rise buildings, but they are more common in higher buildings and where the appearance of the façade is of greater importance.

From what I have already said, it is obvious that columns carrying the dead load of the building will shorten due to creep and possibly also shrinkage. This in itself need not create any problems. What happens, however, when cladding is attached to the columns or to load-bearing walls? Cladding does not carry a sustained load so that, even if it consists of concrete panels, creep does not enter the picture. Moreover, cladding is often made of a material that does not undergo creep: tiles, stone, or bricks. If the cladding is applied in such a way that there are gaps provided to accommodate the shortening of the load-bearing vertical concrete to which the cladding is attached, no problems arise.

A competent contractor knows that decorative precast concrete panels should be suspended, and not fixed rigidly at both the top and the bottom. Moreover, the contractor provides horizontal gaps in the cladding that can accommodate the shortening of the columns. It is up for discussion whether the responsibility is solely the contractor's or whether the designer should establish the magnitude of the movement to be accommodated and specify the necessary provisions. To do so, the designer must be well versed in the relevant properties of the materials used.

The fact is that I have seen on numerous occasions ornamental stone cladding or brickwork applied without any provision for accommodating the shortening of the columns or the walls behind. When this is done, something must "give." First, the cladding cracks, and the tiles, stone slabs, or bricks, or their parts, may drop onto the ground below; this is clearly a hazard for pedestrians.

Secondly, depending on the details of fixing, the cladding may bulge as shown in Fig. 6.5.1. This is a potential hazard and is, of course, unsightly. The remedy is expensive. The situation should not have been allowed to develop in

Left: General view of façade

Inset below: Detail seen from balcony

Fig. 6.5.2—The load-carrying columns are shortened by creep; the vertical fins, rigidly connected to balconies, are not loaded so that the imposed shortening causes spalling and buckling

the first place, and could have been readily avoided by a provision for the differential movement between the cladding and the load-bearing members.

A few years ago, I was involved in a related problem with cladding panels made of glass-reinforced cement. The panels were very thin and were of a fancy and complex shape. They were exposed to large temperature variations, but the resulting expansion and contraction of the panels could not be accommodated because each panel was connected to the building frame at four corners. In consequence, the highly restrained thermal expansion and contraction resulted in the tearing of some of the panels.

The situation was particularly acute because glass-reinforced cement (which is an erroneous British name) has a higher content of hydrated cement paste than ordinary concrete. Because the coefficient of thermal expansion of hydrated cement paste is about twice that of typical concrete, the coefficient of thermal expansion of the glass-reinforced cement was higher, maybe twice as high as in concrete. Thus, the *restrained* thermal

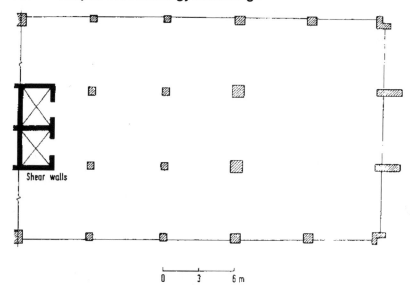

Fig. 6.5.3—Schematic plan of a tall building

movement was particularly high.

Should this have been foreseen by the designer? In other words, was the designer sufficiently knowledgeable about the influence of the cement content in the mixture upon thermal movement? Also, should someone— the designer or the contractor—have paid attention to the temperature, that is, the time of the day when the panels were being fixed? And, of course, should the fixing have allowed for thermal movement? I expect that some readers will immediately point out that they are fully aware of the factors involved and would have avoided the problem. I am sure this is so, but the fact is that there *were* serious problems in the case in which I was involved so that new cladding had to be provided, and this caused a disruption to the operation of what was a prestigious building commanding a high rent.

Problems with vertical service attachments

The shortening of reinforced concrete columns in a low-rise building, provided it is the same in all the columns, does not lead to problems caused by creep. However, in a tall, or high-rise, building, problems may arise both when there is a uniform shortening of all the columns and when there is differential shortening.

Dealing first with the overall shortening, the columns will contract due to creep and shrinkage. In a tall building, this shortening can be about 100 or 150 mm (4 or 6 in.). Fintel, Ghosh, and Iyengar quote figures of 180 to 230 mm (5 to 7 in.) in an 80-story building.[4] The actual figure that matters is the shortening *after* the installation of the vertical attachments. Those of interest are water and gas pipes, and guide rails for elevators.

One example that I heard about was a hotel in which the elevators ran along rails that were fixed to the lift shaft at quite a large spacing. When the building was more than a year old, some of the rails buckled. The explanation was that the creep of the concrete in the shaft walls put the slender rails into compression and induced buckling. That mishap occurred a number of years ago, and I expect that designers are now aware of the situation; perhaps it is more important that those installing the vertical attachments are also aware.

Problems with vertical architectural attachments

I observed an unusual problem when I stayed in a hotel in Rio de Janeiro. Figure 6.5.2 shows external vertical slats running on the outside of hotel balconies.[5] To the uninitiated, they look like columns, but they are not load-bearing and therefore not subject to creep. They are merely decorative slender elements connected horizontally at intervals. The inset in Fig. 6.5.2 illustrates my point; the slats do not transfer any load downwards.

Now, the columns behind the balconies, which are cantilevers, are clearly load-bearing, and as such they undergo creep, which causes their shortening. This shortening is also induced in the slats, so that they are put into compression. This compression can be so large that the slender slats buckle and spall.

A purely architectural feature, the slats must have become popular with architects because, in my later travels, I saw very similar damage in Turkey, Colombia, and Australia. The failures were not dramatic, but were aesthetically unacceptable. My point is that they indicate the designer's lack of appreciation of the effects of creep.

Differential creep in tall buildings

Let me now consider the differential effects of creep; that is, a situation where supports of beams or slabs that were originally level are no longer at the same level. Figure 6.5.3 shows a schematic arrangement of a tall building with columns spaced at 6 m on center, but with some columns omitted where a large uninterrupted area is required, as, for example, in a hotel ballroom or in a major auditorium. If the different columns undergo unequal shortening, any horizontal members spanning continuously over the columns will be subject to parasitic stresses of the type I mentioned previously in connection with bridges.

Strictly speaking, the shortening of the columns is caused by elastic deformation, as well as by shrinkage and creep but, for simplicity, I will limit myself to creep. Also, each floor, when placed, is in a horizontal position so that it is only affected by the strains induced *subsequent* to the time of placing. This is not as complicated as it may seem but, for the purpose of the present section, a qualitative description suffices.

Causes of differential creep

There are five possible causes of differential creep in columns in tall

6-54   Chapter 6: Technology and Design

buildings. First, as I have already mentioned, the relative humidity of the ambient medium affects creep. Thus, if a building has exposed external columns, they may be at a high relative humidity, while the adjacent columns in the air-conditioned interior may be exposed to quite dry air. In the simplest terms, we can say that the external columns undergo only basic creep, while the interior columns undergo both basic and drying creep. More accurately, we could establish the average relative humidity of the different columns and use appropriate values of creep.

The second cause of differential creep arises from the fact that, as discussed earlier, drying creep is affected by the ratio of the volume of concrete to the drying surface; the larger this ratio the less drying that takes place because water from the interior of the column cannot readily reach the surface and evaporate. Consequently, the concrete in the interior of a large column undergoes very little drying creep. Establishing the volume-to-surface ratio, which is the same as the ratio of the cross-sectional area to the perimeter of the columns, requires some care because the surfaces that are relevant are those that are allowed to dry. Column surfaces that are covered by an impermeable material are not considered in the volume-to-surface ratio. In practice, there may be differences between columns of the same size depending on where they are located: for example, whether they are exterior or interior. The largest difference is likely to be between the columns and the shear walls (Fig. 6.5.3), which have a much greater volume-surface ratio.

The drying factors mentioned so far apply as much to shrinkage as to creep.

Now, the third cause of differential shortening of columns is limited to creep. This cause is the elastic strain due to the *sustained* load on the concrete, excluding other loads. Creep is proportional to the applied stress, so that columns carrying unequal sustained loads will undergo differential shortening. However, from the design standpoint, the *total* load, not the sustained load, governs column size and spacing. Transient loads such as wind or earthquakes do not induce creep, but are relevant to the choice of column size. Corner columns are particularly critical because, other things being equal, they carry only one-half of the sustained load of other external columns; however, they may resist significant wind and earthquake loads, which are not sustained loads.

The fourth factor influencing the shortening of a column is the percentage of longitudinal reinforcement in the column. This reinforcement restrains the potential creep of plain concrete. The percentage of reinforcement may vary between columns because it is governed by the *total* design load and not only by the sustained load, which induces creep; this is similar to the situation discussed in the preceding paragraph.

The fifth and last important factor is the age at which the sustained load begins to act. Creep is significantly larger in concrete first loaded at an early age. The age of concrete at which dead loads begin to act varies from floor to floor, and of course, the application of loads is incremental. This has to be taken into account by proper procedures.

Effects of speed of construction

It follows from the influence of age that the speed of construction affects the magnitude of creep. Ghali, Dilger, and Neville have shown that a given load applied gradually over one year would result, in the long run, in one-half of the creep that would be caused by the same load applied all at once at the age of 14 days.[6]

This begs the question of what the designer should know about the likely speed of construction. Usually, the designer knows nothing. It is arguable that, when the system of design and construction makes it possible for the designer to be well informed, the process of design may benefit from it. Specifically, I would suggest that, in the case of important tall buildings, the designer might—or even should—lay down limits on construction speed differentials.

In particular, the designer should recognize that, whereas columns are constructed floor-by-floor, shearwalls are often constructed very rapidly using slipforming. Consequently, when a sustained load begins to act, the concrete in the shear wall may be much older than that in an adjacent column. On the other hand, the column may be more heavily reinforced and, very likely, it will have a much lower volume-surface ratio.

Structural consequences of differential creep

First of all, I should give an indication of the order of magnitude of the differential shortening of adjacent columns. According to Fintel, Ghosh, and Iyengar,[4] when all the factors influencing creep act in the same direction, the differential strain can be 200×10^{-6}. In a 50-story building, this represents a difference in level of nearly 40 mm (1½ in.). This differential shortening has the same effect on horizontal members—beams or slabs—that are continuous over several spans as the effect of differential settlement of supports. Bending moments and shearing forces are induced in the horizontal member, and they produce stresses in concrete and in the steel that are additional to those caused by design loads. The moments and shearing forces are greater the shorter the span is between the adjacent columns.

The situation is actually more complicated because there is a redistribution of loads between supports (from the more highly loaded to the less-loaded support) that, in turn, creates a new, modified level of stress that induces creep. Thus, creep has both harmful and beneficial effects, and this is of fundamental importance. It follows that what might be thought of as "erring on the safe side" and overestimating creep is conservative at one stage in the design but is not conservative at another stage. This is why the designer must consider both the upper and lower bounds of creep. Iteration may be necessary, as I mentioned in a previous article.[1] This approach is explicitly stated in the current British design code BS 8110: Part 2: 1985: "It may be advisable at the design stage to consider a range of values to bracket the problem, since an overestimate may be just as bad as an underestimate."

Problems with camber of prestressed units

In some countries, prestressed concrete floor slab units are often used. They are usually made with a camber so that, after the early loss of prestress, they are level. There have been cases of uneven deflection in service, and complaints have been leveled at the manufacturers about the uniformity of production. In one case, this accusation was unjustified, and the following explanation was established.

The loss of prestress is a function of the sustained stress in concrete at the level of the tendon. This stress induces creep. If the sustained stress begins to act at an early age, the magnitude of creep may be considerable; the resulting loss of prestress will be significant, and some camber will be lost. This loss of camber will be larger than when a similar load is applied at an advanced age. In the case in question, soon after installation, some units were subjected for several months to temporary loads of stored materials; the remainder of the floor was unloaded. It is not surprising, therefore, that, overall, the floor was not level. The manufacturer of the units was not at fault.

In another case, similar units were stored under varying conditions: some in the dry indoors, others outdoors in a humid atmosphere. When taken to be installed, the units were chosen in a random manner from the two stockpiles. Moreover, the age of the various units varied considerably, so that they had undergone different amounts of loss of camber. In this case, it is arguable that the manufacturer of the floor units should have been aware of creep effects in the units, and should have delivered them in a systematic manner. The moral of this story is that, when a very level floor is required, a careful control of various procedures is necessary.

An increase in camber of 50% or more in large prestressed concrete bridge girders, while in storage for two months, was reported by Yazdani, Mtenga, and Richardson.[7] Generally, with very large prestressed concrete units such as bridge beams, a high level of awareness of the importance of creep is essential and nearly always exists. Nevertheless, special circumstances may result in potential problems.

This was the case in Italy, when an industrial dispute stopped the construction of a prestressed concrete box girder bridge, launched in segments, for nearly two years. To allow for the differential in creep between young and old concretes when the construction restarted, it was necessary to make adjustments to the launching bearings and to apply temporary prestress.[8] In the process of the resumed construction, it was confirmed that, after two years under load, the ratio of creep to the elastic strain (known as creep coefficient) was twice as large as that after 6 months.[8]

In 1998, I described a balanced cantilever box girder, with the main span of 241 m (1264 ft), in which a creep-induced sag of about 1 m (3 ft) occurred at midspan; the consequences of the sag and of the remedial work were extremely serious.[9]

Dealing with problems

This section has described several examples of problems caused by creep of concrete, which may affect the behavior of structures. The next section will consider ways of dealing with these problems in design and construction.

References

1. Neville, A. M., "Concrete Technology and Design: Twin Supports of Structures," *Concrete International*, V. 24, No. 4, Apr. 2002, pp. 52-58. [Section 6.4]

2. Neville, A. M. (with Chapters 17 to 20 written in collaboration with W. Dilger), *Creep of Concrete: Plain, Reinforced, and Prestressed*, North-Holland Publishing Co., Amsterdam, 1970, 622 pp.

3. Neville, A. M., Dilger, W. H., and Brooks, J. J., *Creep of Plain and Structural Concrete*, Longman, Harlow, 1983, 361 pp.

4. Fintel, M.; Ghosh, S. K.; and Iyengar, H., *Column Shortening in Tall Structures—Prediction and Compensation*, Portland Cement Association, Skokie, Ill., 1987, 34 pp.

5. Neville, A. M., "Concrete Technology—An Essential Element of Structural Design," *Concrete International*, V. 20, No. 7, July 1998, pp. 39-41. [Section 6.1]

6. Ghali, A., Dilger, W., and Neville, A. M., "Time-Dependent Forces Induced by Settlement of Supports in Continuous Reinforced Concrete Beams," ACI Journal, *Proceedings* V. 66, No. 11, Nov. 1969, pp. 907-915.

7. Yazdani, N.; Mtenga, P.; and Richardson, N., "Camber Variation in Precast Girders," *Concrete International*, V. 21, No. 6, June 1999, pp. 45-49.

8. Rosignoli, M., "Creep Effects During Launch of the Serio River Bridge," *Concrete International*, V. 22, No. 3, Mar. 2000, pp. 53-58.

9. Neville, A. M., "Letters to the Editor," *Concrete International*, V. 20, No. 9, Sept. 1998, p. 8. [Section 6.1]

Letter to the Editor

I am hesitant to offer comments on Adam Neville's first of two articles on creep that appeared in the May 2002 issue of *Concrete International* ("Creep of Concrete and Behavior of Structures—Part 1: Problems"). My hesitance stems from the fact that I have commented on several of his recent series and some may think that we are in cahoots. In a sense, we are in cahoots in that I agree with what he says and in particular with some of the more important aspects of concrete as a building material.

On numerous occasions, opportunities to observe excessively deflected floor slabs have been made available. Invariably, the design engineer will comment that "it is all due to creep" in an effort to preserve and demonstrate an infinite knowledge of the material. The design engineer may even comment further that the deflection had been calculated using code formulas and had satisfied the code. In all probability, the contractor had built the forms "absolutely flat" without any allowances for camber ("the specs or the drawings didn't require any"). Thick slabs will produce creep due to long-term sustained dead loads not to mention the inevitable settlement of the forms themselves due to the immediate weight of the plastic concrete.

The importance of accurately calculating creep is particularly true in prestressed concrete. If the amount of creep is estimated too high, it could lead to excessive upward deflection at the midsection of a prestressed beam. Dr. Neville will undoubtedly include some comments on this and other very interesting subjects in his subsequent paper on creep.

Robert E. (Bob) Tobin
Structural Engineer
Arcadia, CA

Neville's Response

Bob Tobin's comments on deflection of floors and the need for providing camber in construction are highly pertinent and amplify my remarks on the variable deflection of prestressed concrete floor units. His reference to the use of code formulas for creep echoes my gently scathing remarks (in the Part 2 article, June 2002 issue of *CI*) about relying solely on codes for the calculation of the effects of creep in structures.

The need to avoid an "absolutely flat" (as Bob Tobin calls it) underside of a bridge antecedes modern creep studies. I remember learning in my (far and distant) youth that the human eye perceives a horizontal line of a bridge soffit as sagging, and provokes pedestrians under the bridge into alerting the relevant authority to the fact that "the bridge is falling down." Accordingly, I was taught to provide camber in such situations.

As Bob Tobin surmised, in Part 2 of my article, I refer to overestimating creep being just as harmful as underestimating it. This is an example of a case where "more is not better." The same applies to making concrete stronger than specified: this will reduce creep and hence reduce the redistribution of stresses in a continuous beam—a topic considered in my second article on creep.

Mr. Tobin's comments are well worth remembering by structural designers and contractors alike, and I am grateful to him for writing his letter.

Adam Neville

Section 6.6: Creep of Concrete and Behavior of Structures—Part II: Dealing with Problems

In the last section,[1] I discussed creep as a property of concrete. I also described several types of problems caused by creep in various types of structures. I now propose to consider, in a descriptive manner, ways of dealing with the consequences of creep in design and construction.

What should the designer do?

First of all, designers must be aware of the likelihood of differential settlement in structures with vertical members. Secondly, they must know enough about creep to assess the likely magnitude of differential settlement. The necessity of possessing this knowledge is the point that I am trying to make in this section. The designer can then minimize any possible differential creep by appropriate arrangement of the columns.

For example, referring to the tall buildings discussed in Part I,[1] the differences in their effective volume-surface ratios can be minimized by cladding of some columns. Column-to-column differences in cross-section size and in the amount of reinforcement can also be minimized, regardless of design requirements. Because there is an interplay between these two parameters,[2] it is possible to vary them so as to minimize the differences in creep of adjacent columns.

An alternative approach to mitigate differential creep effects is for the designer to accept the expected parasitic bending moments, shearing forces, and the resulting stresses, and to design the structure so that it is strong enough to resist them. In either case, a good knowledge of expected or possible creep is essential.

Relevance to the contractor

The contractor must know the properties of the concrete mixture that affect creep and, therefore, must not arbitrarily change the type of aggregate used or some other relevant parameter. In other words, the contractor must be aware of structural consequences of the composition of the concrete.

There is a further aspect of the contractor's work that affects differential creep: the application of temporary loads, such as those from bricks or concrete block, stored on some floors during construction. Although these are temporary loads, they may act for a month or two, and thus cause creep to take place. If these loads are applied at an early age, the magnitude of creep in one column, but not another, may be significant. The safest remedy is prevention: the contractor must be aware of the potential problem, and the designer must issue strict instructions about storage of materials during construction.

Are useful data on creep provided to designers?

The majority of reinforced concrete structures are not highly sensitive to creep and the use of a simple, standard coefficient to design for creep effects does suffice. But in many other structures, creep can have a significant influence on the performance of the structure. Some examples were discussed in the last section.[1]

The containment vessel for a nuclear reactor, a large structure with a complex prestressing system, illustrates an extreme case of the need for reliable and accurate knowledge about creep properties in a given structure. In this case, it is essential that no cracks develop and that the concrete in the vessel is always in compression. Several loading conditions need to be considered because the pressure and temperature in the vessel are not constant, such that creep, creep recovery, and relaxation need to be known.

I was involved with an experimental determination of such properties for one of the early British nuclear power stations whose construction started in 1963. The schedule of design and construction was such that I was able to perform tests on specific concrete mixtures, which were to be actually used in the structure. No recourse to "creep tables" was necessary.

I also did work later on nuclear power stations in the United States and in Scotland. In some of those stations, and in some others, a way of ensuring adequate time for creep tests was to designate the contractor ahead of completion of design, so that enough time was available for creep tests using the actual materials proposed by the contractor. Such an approach precludes, to some extent, competitive bidding, but "fairness" is achieved by nominating different contractors in turn. On the other hand, this militates against the advantages the contractor's experience offers during the bidding process.

For less critical structures that are nevertheless sensitive to creep, the question is: When a good knowledge of creep is required, is it available to the designer?

Use of codes

The first source of design information is usually the code. But codes often lag behind state-of-the-art knowledge; for example, the first British code for prestressed concrete was published in 1959, which is more than 10 years after prestressed concrete structures began to be built. This is not surprising because, in order to write the code, there must exist experienced members of the drafting committee. Incidentally, that code gave a simple expression for the loss of prestress in the steel due to creep of concrete per unit length: 48×10^{-6} per MPa (0.33×10^{-6} per psi).

The current British code, BS 8110: Part 2: 1985, has tables and nomograms providing data on creep and shrinkage. The parameters used are: age at loading, volume-surface ratio (expressed as effective section thickness), and ambient relative humidity. Creep recovery is given as a single coefficient, applied to the modulus of elasticity, at the time of unloading. Such an approach is grossly oversimplified but the code recognizes this, stating: "It is stressed that these statements provide only general guidance and are based primarily on laboratory data." As someone who was a member of the committee that drafted the code, I am aware that the code represents the "lowest common denominator" that could be agreed upon.

The difficulty with reliance on the code alone is that some of the concrete properties involved are unknown to the designer. For example, it is only when the contractor has been chosen that it is known exactly what kind of aggregate will be used, and yet aggregate has a considerable influence on creep; this was re-emphasized in Daye and Neville's publication.[3] Moreover, with the increasing use of performance specifications with respect to durability or temperature gradients, much leeway must be given to the contractor to choose the concrete mixture.

One could ask: Why not specify the performance with respect to creep? The problem is that, by their very nature, creep tests take a long time. Brooks and Neville tried to develop a system of relatively short-term tests coupled with extrapolation formulas.[4] This is helpful, but even short-term tests take several months, and the contractor can rarely wait that long before work starts. Furthermore, the contractor's bid is predicated on its freedom to choose the aggregate to suit the contract price. (In accordance with court usage, the pronoun "it" applies to a contractor, which is, in reality, a firm.) And in the interval between the call for tenders and their submission, there is no time to perform creep tests. I am mentioning this to show that technical considerations are not the only factor in selecting a mixture; economic considerations play a large role, and laboratory-based technologists sometimes forget this.

Creep committee reports

The mission of ACI Committee 209 includes the words "develop and maintain standards for creep and shrinkage of concrete and concrete structures." I chaired that committee some 35 years ago, and I am now an associate member. The latest committee report, "Prediction of Creep, Shrinkage, and Temperature Effects in Concrete Structures," was published in 1992 and reapproved in 1997. However, the 1992 report was based on the 1982 report and, as stated in a footnote, the revisions to the preceding 1982 report, reapproved in 1986, "consisted of minor editorial changes." In turn, the 1982 report was said to be "generally consistent with ACI 318-77."

I am giving this historical background because, in effect, there has been no change in the creep and shrinkage report in a quarter of a century whereas, during the same period, ACI 318, was modified and changed several times, including substantially new thinking.

ACI Committee 209 has produced several draft reports, but has not reached consensus on a single set of mathematical expressions for creep and shrinkage. I expressed my view of the situation in the Foreword to the *Proceedings of the Adam Neville Symposium* generously organized by ACI:[5] "I am somewhat bemused by these continuing efforts, each on an individual track, but the tracks are not convergent." It has been difficult on both sides of the Atlantic to develop a unique "creep formula." In 2001, Stuart Alexander discussed creep and the transfer of stress from concrete to steel, and he referred to "textbooks, which go to considerable lengths to model it [creep] mathematically, usually incomprehensibly and possibly incorrectly."[2]

There is a story that bears upon the search for a single creep prediction expression. Imagine a room painted black, with people in it all dressed in black, and a black cat in it. They are trying to catch the cat; that is philosophy. Imagine the same black room, people, and cat. Every now and again, someone calls out, "I've got it;" that is religion. Again, imagine a black room, with people dressed in black, trying to catch a cat, and every now and again someone calls out, "I've got it," but there is *no* cat. That is the search for *the* creep expression and, in my opinion, a universally valid expression cannot be found by curve-fitting.

What worries me is that the various expressions for creep are based on experiments performed in the past, some of them 50 years old. Such data are an inadequate base for the prediction of creep in present-day concretes. Nowadays, concrete often contains many cementitious materials and admixtures; it also has a much lower water-cement ratio and a much higher strength than in the past.

Engineers who work in a single field for a long time sometimes forget the benefits of cross-fertilization with other disciplines and do not look at other fields. Perhaps the "curve fitting" of creep could benefit from the work of Peter Young in flood forecasting.[6] Possibly, a deductive approach may be replaced, or complemented, by an inductive approach, starting with a simple conceptual model that has a physical interpretation, leading to a more general model.[6] Young points out that if the end-user understands the nature of the forecasting algorithm, this engenders confidence in its use.[6]

Sometimes, the expressions being derived contain terms not adequately related to the physical phenomena involved. For example, an ACI-RILEM workshop reported that "creep and shrinkage kinetics arise from the same mechanisms in the calcium silicate hydrates."[7] And yet, as far back as 1970, I showed that concrete made with high-alumina cement has a similar pattern and magnitude of creep—and also of creep recovery—as concrete made with portland cement.[8] Given that high-alumina cement does not produce calcium silicate hydrates, I suggested that creep is related to the grosser structure of the hydrated cement paste.[8]

Each of the various expressions canvassed within ACI 209 has wide confidence limits: to know that the "true" (that is, most probable) value of creep has a 95% probability of being within 25% or even more of the predicted value is not very useful. Moreover, the various expressions deal with the simple situation of creep under a single sustained stress. It is a long way from there to creep effects in prestressed concrete, where stresses and strains vary all the time. I am sorry to sound so strident and, if I am wrong, letters to the editor expressing contrary views will certainly pour in.

Nevertheless, I am not being negative or pessimistic, and I believe that we can live with the present situation. We should acknowledge that, for the present, no reliable and useable data in the form of mathematical expressions can be produced because we are dealing with a composite material, consisting both of manufactured and of natural ingredients and partly processed in a semi-industrial manner. So, we should accept that this is the reality of "concrete life." For most concrete structures this is tolerable. For complex, sensitive, and special structures, an experimental determination of creep is necessary.

What really worries me are offers of "quickie" solutions. I recently received one of these from someone whom I do not know. It came as an unsolicited fax offering "an evaluation of creep and shrinkage strains and creep coefficient according to the recently developed B3 model... the evaluation is instantaneous and requires no special qualification." The fax advertisement also said: "The task does not even require a qualified engineer. The possibility of error is excluded." It is a brave person, or a well-insured one, who dares to offer such an off-the-shelf design. For myself, I would not wish to enter an important and creep-sensitive structure, the design of which relied on someone other than "a qualified engineer."

Conclusions

My intention in writing this section, and the previous part, was to point out the interrelation between the structural design and the properties of concrete, using creep as an example. As I said at the outset, a good knowledge of concrete technology is just as important as a good knowledge of structural analysis and design. At one extreme, the designer must have a knowledge of properties of concrete; at the other, the materials specialist must understand the structural aspects of what is being built.

I chose creep as my example because of my interest in the properties of concrete developed when, during my final undergraduate year (1949-50), I made prestressed concrete beams and measured the loss of prestress due to creep. Prestressed concrete was still very new in England. But a similar argument could be woven for other properties of concrete. For example, corrosion of steel in reinforced concrete jetties in the sea is affected significantly by the geometry of the structure in that a beam-and-slab construction promotes the presence of wet and dry areas conducive to the development of electrochemical cells. On the other hand, flat slabs lead to better durability.

Those involved in designing and building structures can be justly proud of their works. But creating structures is an art as well as a scientific process. So perhaps it is appropriate to conclude by quoting E. H. Brown who, in his book *Structural Analysis* published in 1967, defined structural engineering as follows: "the art of moulding materials we do not really understand, into shapes we cannot really analyze, so as to withstand forces we cannot really assess, in such a way that the public does not really suspect." It is time to remove my tongue from my cheek and stop.

References

1. Neville, A., "Creep of Concrete and Behavior of Structures, Part I: Problems," *Concrete International*, V. 24, No. 5, May 2002, pp. 59-66. [Section 6.5]

2. Alexander, S., "Axial Shortening of Concrete Columns and Walls," *Concrete*, March 2001, pp. 36-38.

3. Daye, M., Neville, A., and Ghosh, S. K., "Concrete Creeps into the Future," *Concrete International*, V. 13, No. 7, July 1991, pp. 58-59.

4. Brooks, J. J., and Neville, A. M., "Predicting Long-Term Creep and Shrinkage from Short-Term Tests," *Magazine of Concrete Research*, V. 30, No. 103, 1978, pp. 51-61.

5. Neville, A., Foreword in *The Adam Neville Symposium: Creep and Shrinkage—Structural Design Effects*, SP-194, Akthem Al-Manaseer, ed., American Concrete Institute, Farmington Hills, Mich., 2000, pp. V-VI.

6. Young, P., "Advances in Real-Time Flood Forecasting," Flood Risk in a Changing Climate, *Philosophical Transactions: Mathematical, Physical, and Engineering Sciences*, The Royal Society, London, 2002, 17 pp.

7. Al-Manaseer, A.; Espion, B.; and Ulm, F. J., "ACI-RILEM Workshop on Creep and Shrinkage in Concrete Structures," *Concrete International*, V. 21, No. 3, Mar. 1999, pp. 25-27.

8. Neville, A. M., (with chapters 17 to 20 written in collaboration with W. Dilger), *Creep of Concrete: Plain, Reinforced, and Prestressed*, North-Holland Publishing Co., Amsterdam, 1970, 622 pp.

Chapter 7

Testing

Testing is an important part of achieving satisfactory concrete structures. Tests are routinely used for quality control and to confirm compliance, and of course, they are used extensively in research and development. In this chapter, there are only two sections on testing, and both are related to the investigation of failures or to resolving suggestions or suspicions of non-compliance.

Section 7.1 discusses the use of standard test methods for the purpose of ascertaining some specific property of concrete so that a quantified measure of this particular property can be compared with some other, usually specified, value. It might be thought that there is not much to be written on this topic, and yet, in a number of court cases in which I was engaged as an expert witness, I came across the use of standard ASTM test methods being used other than as specified.

In some cases, the specified test procedures, such as specimen size or duration of some treatment, were not properly followed. In other cases, a test intended for determining a specific property, such as resistance to chloride ion penetration, was actually used to measure permeability of concrete. In yet other cases, tests explicitly prescribed to be performed on a different type of material were used on concrete. One case was the use of a test intended for sheet-like materials on slices of concrete.

Another example is the use of a test intended for the determination of the direct tensile strength of rock cores on concrete cores. Of course, this approach would be perfectly acceptable in research or in the development of a test method for concrete. However, in litigation, there is an absolute need for using tests whose interpretation is objective and not dependent on the interest of any party to the proceedings.

A further comment with respect to the determination of the direct tensile strength of concrete may be appropriate. I understand that some work is in progress on the development of such a standard test. When, or perhaps if, such a standard is available, clearly its use will be proper and admissible for all purposes. However, until then, I stand by my view that concrete should not be subjected to tests not intended for it.

In any case, I do not foresee any test method for the direct tensile strength of concrete becoming a routine test for compliance. The application of axial force to a concrete specimen such that the tensile stress is uniform over the entire cross section and length of the specimen, is not likely to be easy and quick to perform. Moreover, the need to know the direct tensile strength of concrete, rather than to infer it from other properties of concrete, is bound to remain limited to very special cases.

The original publication of Section 7.1 led some readers of *Concrete International* to write letters to the Editor. They represent a valuable contribution, and are therefore included immediately after Section 7.1, and are followed by my reply.

Probably the most common uncertainty about properties of concrete arises in the determination of its compressive strength. The cylinder (or, in some countries, cube) test is one of the oldest tests on concrete and it is routinely performed, day in and day out, on, I believe, every construction project. In my opinion, it is not a good test but it is the best we have for some purposes.

When the 28-day strength of standard test cylinders is lower than specified, this represents *prima facie* non-compliance. But it may not be necessarily so. The original standard test cylinders may have been badly made or they may have been incorrectly treated or tested. How do we know? Well, generally we do not, or the structure is some years old and there are no further cylinders or cubes available. Recourse is then made to drilling cores in the actual structure, and then determining their compressive strength.

All that is normally quite easy, but it is not so easy to interpret the numerical values of the strength of cores in relation to the values of strength specified or necessary for safety. Section 7.2 discusses these matters in considerable detail.

Section 7.2 also discusses testing plans and the appropriate level of competence of people who decide on testing and on the interpretation of results. A correct approach before the commencement of taking the cores is vital if a dispute about the compressive strength of the concrete in the *structure* is to be resolved once-and-for-all. And if it is not, then I submit that undertaking the core tests was a waste of time and money.

The publication of the article that is now Section 7.2 produced considerable interest, and the resulting letters from readers of *Concrete*

International and my reply are included immediately following Section 7.2.

I have also studied a different kind of test, namely a petrographic estimate of the water-cement ratio of hardened concrete whose properties and history are unknown. I am conscious of the fact that I am not a petrographer, but the purpose of estimating the water-cement ratio is to provide information for an engineer who is concerned about the quality of the concrete in a structure, or about its safety, or alternatively is involved in a forensic investigation of alleged non-compliance in the past.

My evaluation, using a statistical approach, of the test results of the determination of the water-cement ratio by the optical fluorescence microscopy method, shows that highly optimistic claims of precision of the test results are unsubstantiated and may be misleading. Such a warning is salutary if we are not to be led into a false feeling of having "the right answer." This does not mean that the test does not serve a useful purpose, but it is applicable only as a tool in the quality control in the production of concrete, and of course with appropriate reference samples for comparison purposes.

Testing is an important and almost invariably an essential element in the production and evaluation of concrete. However, we should always make sure that our tests are appropriate for the intended purpose.

Section 7.1: Standard Test Methods: Avoid the Free-For-All

In the many years that I have used and relied upon standard tests, I always assumed that test methods had to be followed meticulously if other people—building officials, contractual parties, buyers, or sellers—were to rely on the test results. It never occurred to me that it could be otherwise, and that test methods could be varied, modified, or more or less developed on an *ad hoc* basis in order to achieve a seemingly desirable result. Was I naive? I was sure I was not. Indeed, it is arguable that what I have said thus far is obvious, so much so that there is no need to state it, and there is nothing further to add. If so, this section should end here.

However, recently I have come across several reports of what appear to me to be unusual tests. These developments in test methods make me wonder whether my views are, after all, too narrow and too literal: perhaps test methods are to serve us liberally rather than constrain us. If so, I expect that those who have been involved in the "developments" will tell us.

Purpose of standard test methods

Testing forms an inherent part of design and construction. There is a wide range of test types. At one extreme, before a novel structure is built, it may need to be tested on a model, physical or computer-based; in some cases, it may even be necessary to build a full-scale prototype and to subject it to service conditions. Tests of this type are on a one-off basis, and no general test methods are prescribed except for the need to ensure that the test procedures do not present danger to life or limb, or damage to property.

Testing an existing structure may be necessary when it has been damaged, say, by fire, or if a change of use is contemplated and the strength and serviceability of the structure are unknown or uncertain. A similar situation arises in the case of existing bridges that were never *designed* in the modern sense of the word but were built for the traffic of the day. For example, in the United Kingdom, there exist numerous masonry arch bridges built for horse-drawn traffic; they have happily carried motor cars, but with the arrival of heavy-axle loads their load-carrying capacity needs to be confirmed by full-scale load tests. The necessary load tests, which include deflection recovery, are prescribed, for instance, in British standards and in ACI 318,[1] and their execution is invariably in the hands of a structural engineer.

While such load tests are not usually prescribed by standard test methods of the ASTM type, their execution may include elements that are prescribed by ASTM. For example, it may be necessary to take cores from the structure and establish their strength. Taking cores, conditioning them, and testing them are all covered by ASTM standard test methods. It is the application and use of some of these methods that are the subject of the present section.

There is no problem in using the ASTM standard test methods, but it is important to know that, in the given case, these methods are applicable and valid. Validity of tests is a complex issue, and it is still a subject of debate and study in fields such as psychology and education. A simple approach is to consider a test to be valid if it measures the property that it is intended to establish. This, however, is easier said than done, even in the supposedly simple and most common case of the compressive strength of concrete.

The required average compressive strength of concrete is f''_{cr} determined on standard test cylinders (or, in some countries, on standard cubes) following the standard test method of ASTM C 39. The resulting value is used in structural design, but no one would pretend that f''_{cr} is the actual strength of concrete in place. Even if f''_{cr} represents the strength of concrete under uniaxial compression (which it does not because of platen constraints in the testing machine and because of radial and hoop stresses induced in the test cylinder) the concrete in place is subjected to a more complex system of stress than uniaxial compression. At best, the compressive stress is accompanied by a lateral constraint due to the surrounding concrete; more usually, there are other stresses involved so that the actual stress system is multiaxial. In service, perhaps the nearest to the standard test cylinder is a pedestal.

Nevertheless, we have successfully learned to live with the understanding that it is valid to *interpret* the specified strength f''_{cr} for the purpose of satisfying the stress conditions in a structure under a complex stress system and even under dynamic loading. All this works very well in the hands of a structural engineer.

It works well as long as the determination of f''_{cr} takes place following strictly the requirements of ASTM C 39. These requirements include not only the geometric properties of the cylinder, such as size—especially in relation to the maximum aggregate size—and the height-diameter ratio, but also age, curing, and the rate of application of load in the testing machine.

Requirements for satisfactory testing

If all the requirements of ASTM C 39 are not strictly adhered to, then the structural designer cannot rely on the reported value of the compressive strength as satisfying the value of f''_{cr} required by ACI 318.[1] For example, a higher rate of application of load in the testing machine would give a spuriously higher value of compressive strength; so would a height-diameter ratio smaller than the specified minimum of 1. Arbitrary departures from ASTM standard test methods are discussed in the latter part of this section.

At this stage, I wish to consider the possibility of using tests that are not valid in the sense that they do not measure the required property of concrete. Let me give a simple example. It is known that, under certain circumscribed circumstances, the density of structural lightweight concrete varies approximately in the same manner as its compressive strength.

However, in my opinion, it would be foolhardy to pretend that the value of f'_{cr} can be established by ASTM C 567 Standard Test Method for Density of Structural Lightweight Concrete.

A particular purpose of testing is to establish the uniformity of a product, conformity with the specification, control of production, or quality control. All these require standardized procedures, methods, and conditions. When we move to disputes and litigation, the exact conditions of testing acquire an even greater importance.

Thus, broadly speaking, there are two issues. The first of these is whether the tests should be applied exclusively for the purpose for which they were devised, and not applied to some other material or used to determine some property not included in the scope of the standard test method. The second issue is whether test methods, such as those of ASTM and their equivalent in other countries, should be, indeed must be, followed to the letter.

I am raising these issues because I have recently come across what, in my eyes, are infringements of my interpretation of the above two issues: using tests only for the purpose explicitly prescribed in the standards, and performing tests strictly in accordance with the standards. To test (no pun intended) my opinions and to clarify the situation and encourage others to express their views, I would like to describe a few examples of what I have encountered and what bothers me.

Testing in research

Clearly, all the above does not apply to research. In research, as in war and love (as the old saying goes) all is fair: to discover new knowledge, the experimenter is free to use any technique in a completely untrammeled manner and to use any test method, however unorthodox. The researcher's goals are to understand behavior and to establish cause and effect. This is done by testing under a chosen set of circumstances and then under another set; for example, by changing concentration, temperature, rate of loading, or pressure. Such freedom is not only acceptable, but even necessary to extend our knowledge, and by experimentation we develop new materials and processes, as well as improve standard test methods. All this comes under the heading of research and development.

Actual problematic testing

I would like to start with the first issue raised earlier, that is, the recognized or specified scope and application of the test method, because the procedure as laid down is not questioned in that case; only the field of application and the inferences are questioned.

Let me give some examples of what I have encountered so as to put my concerns in context.

Resistance to chloride ion penetration

The first example is ASTM C 1202 Standard Test Method for Electrical

Indication of Concrete's Ability to Resist Chloride Ion Penetration. The scope of this test method is given as "The determination of the electrical conductance of concrete to provide a rapid indication of its resistance to the penetration of chloride ions." In other words, the purpose of the test is to establish the resistance of a given concrete to penetration by chloride ions. This, of course, is of importance when concrete is exposed to water containing chlorides because if they penetrate the concrete cover and reach the vicinity of steel reinforcement (or other steel such as anchor bolts) corrosion of steel can occur. The consequences of this upon the integrity of the concrete member and upon its load-carrying capacity are well-known.

Because the test establishes the amount of electrical current passed through the concrete specimen 102 mm (4 in.) thick under an impressed potential of 60 volts dc from the side exposed to a sodium chloride solution, to the side exposed to a sodium hydroxide solution, the result is reported as charge of so many coulombs. The ASTM C 1202 Standard Test Method gives values of the charge that represent a low, medium, and high resistance to chloride ion penetration.

The problem with which I am concerned is the use of the ASTM C 1202 Test Method for the purpose of establishing the water permeability of concrete. This is what I have encountered: the charge in coulombs is interpreted as indicating that the concrete has a low, medium, or high water permeability. Now, ASTM C 1202 does not as much as mention the words "water permeability," and this surely is not the consequence of an unduly narrow outlook of the committee that wrote the document. Rather, the prescribed test conditions involve ionic movement of electrons. This is quite distinct from the flow of water through a continuous pore system in concrete.

There are published data supporting my view that there is no simple relationship between water permeability and chloride diffusion; Zhang and Gjørv concluded that "No direct relationship between water permeability and electrical conductivity (AASHTO T 277 test method) was observed, but a direct relationship between water permeability and accelerated chloride penetration was observed."[2] Despite this, in several instances, the results of tests according to ASTM C 1202 that give a high value in coulombs have been presented as support for the thesis that the given concrete has a high water permeability.

Moreover, I would like to express the view that water permeability of concrete should be determined primarily when this method of transport is relevant, that is, when flow under a head of water, or pressure, takes place. When transport of water is by sorption or by diffusion of water vapor, the mechanisms involved are different, and, therefore, different test methods are appropriate.

In the vast majority of actual structures, there is no significant head of water acting on the concrete. Perhaps this is why there is no ASTM test method for water permeability of concrete. The recognized tests are the

Chapter 7: Testing

German DIN 1048 test[3] and the U.S. Bureau of Reclamation Procedure for Water Permeability of Concrete 4913.[4] The latter, which uses a pressure of 2758 kPa (400 lb/in^2) corresponding to a head of water of 281.6 m (924 ft) is of course relevant to dams, in which a large head of water acts on one face of the concrete structure.

Overall, then, I view ASTM C 1202 Standard Test Method for Electrical Indication of Concrete's Ability to Resist Chloride Ion Penetration as not giving valid information about water permeability of concrete.

Vapor transmission

This test method is little known to those involved in concrete construction, and not surprisingly so: the title of ASTM E 96 is Standard Test Methods for Water Vapor Transmission of Materials. The scope is given as "The determination of water vapor transmission of materials through which the passage of water vapor may be of importance, such as paper, plastic films, other sheet materials, fiberboards, gypsum and plaster products, wood products, and plastics." In my opinion, it would require a considerable stretch of imagination to consider concrete as falling within the range of materials listed under the scope of ASTM E 96. Surely, had the committee that wrote this standard considered its use for concrete, it would have mentioned it by name; rather, concrete is excluded by implication in that it is not a "sheet-like" material.

Moreover, the maximum thickness of the specimens to be tested according to ASTM E 96 is given as 32 mm (1¼ in.). Now, it is possible, of course, to prepare a 25 mm (1 in.) thick sample of the concrete to be tested, but it would not be representative of the concrete in question if the maximum size of aggregate is 20 mm (¾ in.) as is very often the case in construction; this, indeed, was so in the structure for which ASTM E 96 was used to establish the vapor transmission of the concrete with which I was concerned. When the thickness of the specimen is not much larger than the maximum aggregate size, the interface zone between the aggregate and the hardened cement paste represents an unduly large proportion of the specimen being tested;[5] this may lead to a spuriously high value of vapor transmission. Indeed, it is common in various tests on concrete to specify a certain minimum ratio of specimen thickness to the maximum aggregate size.

So, in my opinion, ASTM E 96 is not an appropriate test method to establish the vapor transmission of concrete. It is arguable that research should be undertaken to develop an appropriate standard test method, but until this has happened, using an existing test established for another purpose is hardly justified.

Direct tension test

There exists no ASTM Standard Test Method for the determination of the strength of concrete in direct tension. At one time, there existed standard test methods for the direct tension strength of mortar. For example, the British briquette test was used to establish the strength of cement: the briquette was pulled apart, and tensile failure occurred in the narrow waist. Difficulties in avoiding eccentrically applied tension (which would give an unduly low value of tensile strength) led to abandoning that test some 30 years ago. Research involving very large cast specimens with a complex shape[6] has not led to the development of a standard test method; anyway, specimens cut from a concrete structure are inevitably of constant cross section.

Outside the United States, there exists a RILEM Recommendation CPC7 Direct Tension of Concrete Specimens. This is explicitly "essentially destined for research works," as stated in the document.[7] The restriction of the test method to research was stated when the test method was first published as a draft in 1973; presumably, since then, RILEM has not thought fit to broaden the scope of the test to concrete used in structures.

The only American test for the strength of concrete in direct tension is that found in Procedure 4914 of the U.S. Bureau of Reclamation.[8] This allows the use of cores and uses bonded end plates. I have to admit that I am not aware of the purpose and field of application of this test, and I have never heard of its use, but this may simply be my ignorance.

On the other hand, no ACI document requires, at least to my knowledge, the use of the value of direct tensile strength of concrete. If such a value is required for design purposes, then it is derived either from the splitting tensile strength or from the modulus of rupture, that is, the flexural strength. Usually, a relation between one of these and the standard compressive strength of concrete, f'_c, is established prior to the commencement of construction. Subsequently, the compressive strength is used for compliance purposes, it being implied that a satisfactory tensile strength is thereby ensured.

In the case when the concrete has already been placed but there are some doubts about its quality, ACI 318-99 is very specific about tests to be performed.[1] Section 5.6.5.2 states that: "Tests of cores drilled from the area in question in accordance with 'Method of Obtaining and Testing Drilled Cores and Sawed Beams of Concrete' (ASTM C 42) shall be permitted." The Commentary on the entire Section R5.6.5, titled Investigation of Low-Strength Test Results, makes no mention of testing concrete in tension.

Thus, when investigating problems, which may be real or alleged, with an existing concrete structure, ACI relies exclusively on compression testing of cores. And yet, I have come across the use of a direct tension test on concrete specimens cut out of an existing structure.

The rationale of this approach was ingenious and went as follows. There exists ASTM D 2936 Standard Test Method for Direct Tensile Strength of Intact Rock Core Specimens. The scope of this is given as "Determination of the direct tensile strength of intact cylindrical rock specimens." There is no hint that the test method may be applied to concrete. In the case with which I am familiar, the justification for extending to concrete the test method for rocks was that concrete is an artificial rock. On that basis, other test methods applicable to rocks could be indiscriminately used for concrete. By analogy, it could be argued that artificial flowers fall under the generic term of flowers, and that their manufacture should be within the purview of the Department of Agriculture.

My view is that neither ASTM D 2936, which is intended for rocks, nor the RILEM Recommendation CPC7, "essentially destined for research works," should be used to establish the quality of concrete in an existing structure.

There may, however, for some reason, be a particular concern about the tensile strength of concrete in place. For example, I have come across allegations that some deleterious chemical action on concrete affects its strength in tension but not in compression. I have considerable difficulty envisaging a physical set of circumstances supporting this proposition, but this does not mean that it is impossible. In my opinion, the solution to the problem is to determine the splitting tensile strength or the modulus of rupture of the suspect concrete. If the ratio of either of these to the compressive strength determined on cores is the same as for undamaged concrete, then the proposition that the strength in direct tension is the only means of establishing the deterioration of concrete in service is not justified. Extensive data on the usual relation between the splitting tensile strength and compressive strength have been collated by Oluokun[9] and are cited by Neville.[10]

Splitting tensile strength

In this case, I am concerned with a significant departure from an ASTM Standard Test Method, namely ASTM C 496 Splitting Tensile Strength of Cylindrical Concrete Specimens. This method is well established, and there should be no problem with its use. The standard permits the use of cores and refers to ASTM C 42, which requires that the length-diameter ratio of the test specimen be not less than one.

I have come across a case in which it was alleged that the deterioration of concrete had occurred in layers so that, in a core, different layers were affected to a varying degree, with the consequence that they would have a variable splitting tensile strength. To prove this, the core, about 125 mm (5 in.) in diameter, was sliced into discs, each about 20 mm (0.8 in.) thick. Each disc was then tested in splitting tension. In such a test, the length-diameter ratio is about 0.15 and thus violates ASTM C 42. We can note in passing that the effect of the specimen size of the resulting splitting tensile strength is still a subject of study: a recent paper concludes that, because of the influence of specimen size and of the width of bearing strip, "the splitting tensile strength cannot be considered a material property."[11]

I am not entirely surprised by this observation because the compressive strength of concrete also depends on the geometry of the specimen, its moisture condition, and rate of application of load. All of that reinforces the argument for testing strictly in accordance with prescribed test methods.

Clearly, tests on discs do not conform to ASTM C 496. But my concern is with the interpretation of test results on discs with a very low length-diameter ratio. Is the splitting strength as determined by a test on a disc representative of the value that would be obtained for a core with a length-diameter ratio between 1.0 and 2.0? Discs such as those described are slender, and I suspect that, therefore, the load as applied could be eccentric and could induce buckling; some popping out of coarse aggregate particles is possible. I cannot prove that my concern is justified, but I wonder whether such a significant departure from an ASTM Standard Test Method is appropriate. Perhaps some research is necessary before testing of discs is justified in resolving concerns about the quality of concrete in existing structures.

There have been other recent cases of using test methods prescribed by ASTM but with significant arbitrary departures from specified procedures. I am not describing them, as the above examples are adequate to identify the problems and to raise the issue of whether we should be concerned.

Conclusions

The purpose of this section was to point out the vital importance of strict adherence to standard test methods and to the use of tests deemed to be appropriate by the bodies that promulgate them. I used to think that these requirements were self-evident, but recently I have encountered "innovation" in test methods applied to actual structures (not in research) and arbitrary modifications of established methods. My concern is about what happens when these novel approaches are used to evaluate the state of existing structures when there is uncertainty about the safety of the approaches or when a change of use is proposed. Likewise, I worry about obtaining comparative data when some people use arbitrarily chosen, and not prescribed, test methods, or when they arbitrarily modify established test methods.

Because some people do so, I have to ask: is ASTM too cautious? Or is the temptation to suit one's aims too strong? My view is that now is the time to re-emphasize the importance of a meticulous adherence to the recognized test methods, and not to practice a free-for-all approach.

There is a form of wrestling, used in the Olympic Games, in which almost any hold, trip, or throw is allowed; this is known as catch-as-catch-can. Should we similarly accept test-as-test-can?

I believe that if test results are to be used for compliance purposes or to resolve disputes, then more, and not less, strict adherence to the use of standardized specimens and procedures is necessary. But am I old-fashioned and censorious in insisting that only proper test methods be used to obtain answers to questions about the state of concrete in a structure? I hope some readers will pick up, if not pen and paper, then computer and printer, and share their views with the concrete community.

References

1. ACI 318-99, "Building Code Requirements for Structural Concrete (318-99) and Commentary (318R-99)," American Concrete Institute, Farmington Hills, Mich., 2000, 391 pp.

2. Zhang, M.-H., and Gjørv, O. E., "Permeability of High-Strength Lightweight Concrete," *ACI Materials Journal*, V. 88, No. 5, Sept.-Oct. 1991, pp. 463-469.

3. DIN 1048, "Testing of Hardened Concrete Specimens Prepared in Moulds," *Deutsche Normen*, Part 5, 1991.

4. U.S. Bureau of Reclamation 4913-92, "Procedure for Water Permeability of Concrete," *Concrete Manual,* Part 2, 9th Edition, 1992, pp. 714-25.

5. Larbi, L. A., "Microstructure of the Interfacial Zone Around Aggregate Particles in Concrete," *Heron*, V. 38, No. 1, 1993, 69 pp.

6. Phillips, D. V., and Zhang, B., "Direct Tension Tests on Notched and Un-Notched Plain Concrete Specimens," *Magazine of Concrete Research*, V. 45, No. 162, 1993, pp. 25-35.

7. RILEM CPC7, "Direct Tension of Concrete Specimens," Nov. 1975, 2 pp.

8. U.S. Bureau of Reclamation 4914-92, "Procedure for Direct Tensile Strength, Static Modulus of Elasticity, and Poisson's Ratio of Cylindrical Concrete Specimens in Tension," *Concrete Manual,* Part 2, 9th Edition, 1992, pp. 726-731.

9. Oluokun, F. A., "Prediction of Concrete Tensile Strength from its Compressive Strength: Evaluation of Existing Relations for Normal Weight Concrete," *ACI Materials Journal*, V. 88, No. 3, May-June 1991, pp. 242-246.

10. Neville, A. M., *Properties of Concrete*, Fourth Edition, Longman, London, 1995 and John Wiley, New York, 1996, 844 pp.

11. Rocco, C., et al., "Size Effect and Boundary Conditions in the Brazilian Test: Theoretical Analysis," *Materials and Structures*, V. 32, July 1999, pp. 437-444.

Letter to the Editor No. 1
Standard Test Methods

Adam Neville has written a thought-provoking polemic ("Standard Test Methods: Avoid the Free-For-All!" *CI,* May 2001) on the necessity of following test methods "meticulously" if the results are to be relied upon by others for the purposes of quality control or conformity with specifications. It is difficult to argue with this proposition if the tests are indeed to be used only for these purposes. However, Dr. Neville considerably overstates his case in his discussion of the testing of concrete in an existing structure that is suspected of having undergone some damage. His remarks on tension testing in particular are misleading, at best.

It is true that ACI relies entirely on compression testing of cores to determine the quality of concrete in a structure. However, as Dr. Neville has pointed out in the 4th Edition of his *Properties of Concrete* (1996), the compressive strength test is the part of the *"baggage culturel"* of engineers. It is always carried out, in much the same way that a physician always begins an examination by taking temperature and blood pressure readings, regardless of the nature of the complaint. However, just as temperature and blood pressure can't be used as the sole indicators of the state of health, so the compressive strength shouldn't be seen as the only (or necessarily even the best) indicator of concrete quality. The assumption that if the concrete is strong enough in compression, then it must also be adequate in all other respects, is simply not true at all in all cases, particularly in those cases involving durability. One can certainly imagine instances in which the *tensile* strength, or even the permeability, would be the best indicator of concrete quality.

Direction-tension test

The fact that there is currently no ASTM Standard Test Method for determining the direct-tensile strength of concrete doesn't mean that this property is unimportant, or that it can't be measured. It simply means that the concrete committee within ASTM hasn't seen fit to make the development of such a standard test method a priority. As Dr. Neville points out, however, there are at least two other test methods that have been developed for this purpose, one by RILEM and the other by the U.S. Bureau of Reclamation. The real question to ask about a test method isn't whether it conforms exactly to some arbitrary standard, but whether it can, if carried out properly, provide the desired information about the property in question. Thus, in the case of direct tension, does the test setup provide essentially a pure tensile stress to the specimen? If so, it is a valid methodology. Both the RILEM and U.S. Bureau of Reclamation tests would appear to impose a pure tensile stress, and so either one should provide useful data. The fact that the RILEM test was, in 1973, "essentially destined for research works" doesn't imply that it can't now equally be used in engineering practice. Indeed, many (if not most) of our current test methods were first used as research tools before the became adopted as "standards." Similarly, if ASTM D 2936, Standard Test Method for Direct-Tensile Strength of Intact Rock Core Specimens, truly measures the tensile strength of rock, then there is no reason to believe that the same test methods wouldn't also measure the direct-tensile strength of concrete. The people who developed that test couldn't be expected to list all of the other materials for which the test methodology would be suitable.

Thus, contrary to the opinion expressed by Dr. Neville, any of the three test methods mentioned above could be used to establish the tensile properties of concrete in a structure. Further, in a particular situation, the tensile strength may well be a better indicator than the compressive strength of the quality of concrete in an existing structure.

Splitting-tension test

The splitting-tension test (ASTM C 496), like all other strength tests, depends not only upon the size and stiffness of the bearing strip, but also upon specimen geometry, moisture state, and so on. Thus, if the splitting tensile strength is for some reason specified in the contract documents, then it must certainly be determined in strict accordance with the procedures laid down in ASTM C 496. However, as with other strength standards, the procedures established in ASTM C 496 are to a large degree quite arbitrary. If the test is theoretically valid for the specimen geometry and test conditions specified by ASTM, then it is equally "valid" for larger or smaller specimens—one would derive the same equation for splitting-tensile strength of a cylindrical specimen regardless of specimen dimensions. Similarly, one could determine the compressive strength of concrete on cores with a diameter of, say, 600 mm (24 in.) and a length of 1200 mm (48 in.). While this would be most impractical for a whole host of reasons, the test would no more or less "valid" than the conventional tests on 100 or 150 mm (4 or 6 in.) specimens; it would, however, yield systematically lower strength values.

Thus, in the case cited by Dr. Neville, testing cylindrical disks only about 20 mm (0.8 in.) thick would certainly violate the provisions of ASTM C 496, and would result in *systematically* different values of splitting-tensile strength. However, the test would still be an indication of the splitting-tensile strength of the concrete in question. A comparison of the *relative* strengths of the disks cut from a single core would indeed be an indication of whether different sections of the core were damaged to a different degree.

In this regard, it could also be argued that the "forensic" examination of an existing structure that is suspected of having undergone damage is closer to research than it is to determining conformity with specification. Thus to quote Dr. Neville again, in this case "the experimenter is free to use any test method, however unorthodox. The researcher's goals are to establish cause and effect. This is done by testing under a chosen set of circumstances...."

Resistance to chloride-ion penetration

It is hard to disagree with Dr. Neville's comments about the scope of the ASTM C 1202 Standard Test Method. As mentioned by Dr. Neville, the determination of the electrical conductance provides a rapid indication of the resistance of (*saturated*) concrete to chloride penetration. In that respect, information derived from the test has no *direct* relationship to the water permeability of concrete. While the conductance of concrete (as measured according to ASTM C 1202) may vary from a few hundred to a few thousand coulombs, its permeability (either expressed in m/s or in m^2)

often varies over many orders of magnitude.

Dr. Neville is also right to emphasize the fact that the determination of water permeability is only relevant when the flow of water through concrete is initiated by a pressure gradient. For too many civil engineers, permeability is the sole property that controls the resistance of concrete to chemical degradation. In reality, the durability of concrete is influenced (at least in part) by its ability to impede the diffusion of ions.

It should, however, be emphasized that the various transport properties of concrete (permeation, ionic diffusion, etc.) aren't totally *unrelated* to one another. Research done over the past 50 years clearly indicates that they are all more less directly linked to the porosity of the material. For instance, a reduction of water-cement ratio not only reduces the ionic diffusion coefficients of concrete but also contributes to decrease its permeability (often to a point where it can't be measured by most methods). In this regard, concrete that is found by ASTM C 1202 to have a low resistance to chloride penetration is also *likely* to be highly permeable to water.

Vapor transmission

Over the past decade, civil engineers and those involved in concrete construction have been increasingly concerned by water intrusion problems. For instance, the transport of moisture through concrete has been extensively studied in Europe (particularly in Scandinavia). At the last ACI Spring Convention in Philadelphia, a special session was entirely devoted to the subject.

The reasons behind this interest are numerous. Moisture intrusion is often at the origin of various practical problems with floor-covering materials (loss of bond, carpet staining, etc.). The excessive penetration of moisture can favor the growth of mold and mildew and result in health problems. In many instances, these problems are directly related to the use of poor quality concrete in the construction of slabs and foundations.

As emphasized by Dr. Neville, there exists no ASTM Standard Test Method for determining the resistance of concrete to moisture intrusion. This is the reason why some have relied on a modified version of ASTM E 96 to evaluate the ability of concrete to transmit moisture. The method has the advantage of being relatively simple, well controlled, and quite reproducible. Its operating principles rest on a good understanding of the parameters that influence the transport of moisture through porous materials such as concrete. In this regard, the argument of Dr. Neville that the determination of materials properties should rest solely on the use of standard test methods is hard to follow. The investigation of the quality of existing concrete structures should rather be based on good understanding of the material. In any event, it would appear to be unwise ever to rely upon any *single* test to establish the quality of existing concrete structures. Thus, in a proper field investigation, ACI 318-99 notwithstanding, it may often be necessary to carry out a number of different tests (compressive strength, tensile strength [splitting or direct], permeability, microstructural characterization, and so on) to determine what is really happening to the concrete.

In conclusion, then, we would draw a distinction between testing for compliance purposes, and testing to "resolve disputes" (i.e., to evaluate the serviceability of the concrete in an existing structure). In the former case, we agree wholeheartedly that strict adherence to standard test methods is a requirement. In the latter case, however, there must be more latitude given in the choice of test methods, so that the true state of the concrete may be determined.

Sidney Mindess, FACI
Department of Civil Engineering
University of British Columbia
Vancouver, B.C., Canada

Jacques Marchand
CRIB-Department of Civil Engineering
Laval University
Sainte-Foy, Quebec, Canada

Letter to the Editor No. 2

An unreserved admiration for Adam Neville leads me to read all his articles; for this one I picked up my pen and paper, as suggested by the author. This article, apparently a reaction to some indiscriminate use of standardized tests, points to the "vital importance of strict adherence to standard test methods."

There are a number of reasons to agree that standard tests have to be followed literally — primarily, the discipline of their development calls for equal discipline in their use. Any comparison or compliance purpose is lost; any safety criterion becomes indefensible if it relates to tests subject to individual modifications. There are enough other uncertainties inherent in the transition from test results to the real structure; no need to add individual "improvements."

However, reading this article, I sensed the tenor of admonishing caution, of a "don't touch the ivory tower of established standards," of this unconditional call for "meticulous adherence to the recognized test methods." Yes, nobody would oppose this view as long as it relates to an officially approved oeuvre, as Dr. Neville's reasoning implies.

But how about more symbiotic togetherness between work on standards and research, and some loosening of censorship—not for existing standards but for newcomers? In this "other" aspect of judging standards, wouldn't less censorship here be beneficial?

An example is the wedge-splitting test. This promising test for fracture mechanics properties is, to my knowledge, not standardized. There are

RILEM recommendations to determine fracture mechanics parameters, but such tests are not used in common concrete testing. This is unfortunate because the wedge-splitting test (or any equivalent) provides a "two-dimensional" result. This is more informative than the single point value of traditional tests. Wedge-splitting results also enhance constitutive modelling FE-analysis (ACI 446.3R), for example, a strain softening curve provides a much better fingerprint of concrete than $f'_c = 42$ MPa (6000 psi).

Wedge-splitting testing uses a simple machine, and it can easily be extended to simulate a biaxial state of stress, the most common state of stress in many structures. Wouldn't it therefore also comply better to simulate the exposure in the real structure? I think a number of traditional tests are the answer to this question.

I know this is a subjective view and it might do injustice to the author's intentions, which seem to be towards discipline and not against openness for new ideas. The point I wish to make is that adherence to established standards should never contribute to widening the sometimes already large gap between the practicing and research community.

Harald Kreuzer
ACI Member
Brunegg, Switzerland

Letter to the Editor No. 3

I agree emphatically with Adam Neville's position that standardized tests can only be valid within given parameters, and the use of these tests should follow the testing protocols. This should be borne in mind especially where the tests may be subsequently used in legal proceedings, or where these results impeach the work of other engineering professionals. I share his concerns regarding misapplication of standardized tests and agree that whenever possible, engineers should use applicable standard tests.

However, that having been said, I don't believe that the sheaf of standard testing methods available are adequate to all situations. It may well be that a particular field engineering problem may require a non-standard testing procedure to be carried out. It is the nature of most construction work that such testing would be carried out under time pressure, and the luxury of waiting two or three years for a thorough academic examination of all the permutations and limits of the test simply is not feasible.

This explains, but doesn't excuse, the tendency to modify and incorrectly apply standardized testing. Engineers should be careful not to label a test with a standardized procedure number unless the test in fact conforms to all requirements. In those cases where a customized testing regimen is adopted, the engineer should carefully examine the effects of deviations from standard procedures. The engineer should be prepared to justify his testing approach on a technical basis, a financial basis, and a schedule basis.

Engineers are first and foremost problem solvers, not scientists. As long as new problems are encountered in the field, it will be futile to attempt to limit "non-standard" testing to scientific research and exclude it from the engineering field altogether. Nor does scientific research always yield effective engineering solutions—witness the quest for effective pavements and soil substrates carried out after World War I. That engineering problem was solved by the simple engineering expedient of destruction testing a number of possible paving solutions, and was completed decades before science was able to provide results.

This is not to say that engineers don't need science, or that science hasn't significantly advanced the engineer's capability to solve problems, but rather to say that one historically effective engineering technique has been, and should continue to be, problem-specific testing designed and carried out under the supervision of the engineer in charge. Careful review by competent professionals of customized tests are essential to understanding the limits of the data developed. Standardized tests, correctly applied, are an important part of that solution, but they aren't the whole solution, nor should they be.

Laurence M. Wright
Project Engineer
American Red Cross
Manassas, Va.

Letter to the Editor No. 4

I enjoy reading Adam Neville's instructive sequel of articles in your magazine and appreciate his emphasis on good concrete practice. I rarely contribute because I agree with so many of the points he makes.

Here again, I agree with most of Dr. Neville's observations in his last article May 2001. For example, there is no argument as to chloride ion diffusion being a different phenomenon to water or moisture flow in concrete, or that tests for diffusion through thin membranes aren't applicable to concrete. However, I would take issue with some of the statements made. I am sure these controversial statements were made to generate reactions from readers.

A standard test is never a realistic test and should only be used for comparison of physical or chemical properties of similar materials or for control purposes. So, a standard cube or cylinder test in compression cured in a standard condition and tested at a given rate bears very little relationship to the strength of concrete it is supposed to represent in the structure. Dr. Neville, of course, doesn't disagree. Structural engineers use a factor on the standard cube strength, say, to take account of the complex state of stressing in the cube to the assumed uniaxial state in a beam. They use another factor to transform the characteristic design strength to the design compressive strength. These factors, however, strictly apply to new construction as these factors have been calibrated to old codes, which have appeared to work safely over long periods of time.

In old construction where the actual strength can be found by testing standard cores, different "statistical" factors are used to allow structural engineers to revert to the control cube strength for the designer to "evaluate" the member ultimate strength. In the computer age, I find it difficult to comprehend why, having the actual strength, we don't use a more accurate way of assessing the member strength capacity rather than slavishly following the convoluted methods in Codes of Practice and Standards.

Standard fire tests are useful for comparison of similar materials under certain standard conditions at elevated temperature and are often used in Codes of Practice to design concrete protective cover for structural steel or concrete. This test, however, originating more than a century ago, bears no relationship to the material behavior in a fire. When beams using different materials such as steel and concrete are being compared, the standard test requires the furnace to follow a given time-temperature curve. The furnace has to supply more heat to the concrete, which has a higher thermal inertia than steel. An actual fire would never recognize the material it impinges on and wouldn't be able to comply with the so-called "fire test"! In addition, "fire tests" are normally performed on simply supported structural members rarely found in good construction. This exemplifies the errors that one can make if standard tests alone are relied upon to compare the behavior of materials under actual practical conditions.

I can understand the predisposition that arises in litigation cases, where certain materials, which have been specified to certain standards, must be shown to comply. In such circumstance, if the standard test specified as applicable to the condition that the material has to be, or was applied, then a standard test is the arbiter. However, the expert witness is in a position to recognize whether the specification is inappropriate, in which case the standard test may no longer be applicable. One can cite an extreme example of the use of seawater in a reinforced concrete structure exposed to a vicious environment and attempting to find the cause of deterioration by using standard tests alone. The article doesn't necessarily imply agreement with this notion, but the emphasis on standard testing may be easily misunderstood to accept it.

In assessing the material and structural properties of existing concrete structures, nondestructive testing is the most efficient and cost-effective method of testing. This would include both invasive tests such as cores and noninvasive tests as the latter generally require a limited number of cores for calibration or confirmation. Extracting of cores is a time-consuming and relatively expensive nondestructive test, and it would be a great waste of resources if these cores were only used to test for compressive strength as specified in most standards. If the cores are longer than required for compression, other parts may be used for other physical and chemical tests.

There are also nondestructive tests one can carry out on the cores prior to testing for strength, which can then subsequently be used on the structure to assess the performance of the remainder of the in-situ concrete.

Dr. Neville stresses that extracting and testing of cores should be carried out to a standard, but this isn't always possible. Although it is recommended that 100 or 150 mm (4 or 6 in.) cores should be extracted, there are cases where it is required to test the concrete in webs of beams, which may be no more than 75 mm (3 in.) thick. In this instance cores as small as 50 mm (2 in.) have been found to be very effective. Of course when such small cores are extracted, not only does extreme care have to be exercised on the extraction, but also on the preparation before testing. The surfaces prior to testing have to be as perfectly square to the axis as possible and optically flat. A simple test for flatness is for the core to be able to lift flat glass by pure suction. Even with these precautions, the variability of strength resulting with small diameter cores is higher than that with larger diameter. However, additional tests prior to testing in compression such as moisture absorption with time, density, and other NDTs improve the confidence to the strength results and will help pinpoint any outliers.

In conjunction with the author, I am conscious of the difficulty of interpretation of some results from nonstandard tests, but this also applies to quite a few standard tests, which may not perfectly recreate the conditions and behavior of the concrete in the structure. I am sure that there are many eminent engineers such as the author of the article who can discriminate between tests that may or may not be applicable for the assessment of safety of existing structures.

I am sure Adam Neville has not exhausted his thoughts on concrete technology, and I look forward to reading the next article.

Patrick Sullivan
Department of Civil Engineering
City University
London, England

Neville's reply

CI Editor-in-Chief Bill Semioli's exhortation to write letters on the topic of standard test methods has resulted in a varied discussion that should further stimulate the readers of *Concrete International*. I am grateful for such a response.

Professors Sidney Mindess and Jacques Marchand represent the academic view of those not involved in construction. As such, they deal with what I discussed in my article under the heading of research, namely studies of cause and effect. But even in that respect, the tests must measure the relevant property. For example, a splitting-tension test performed on disks 100 or 150 mm (4 or 6 in.) in diameter and 20-mm (0.8 in.) thick is likely to result in buckling of such slender specimens. Moreover, the presence of aggregate particles spanning the whole thickness of the disk could lead to debonding and pop-outs, and it is known that elongated particles can have one dimension greater than 20 mm. Also, it should be remembered that, for cylinders and cores, a minimum value of the ratio of diameter to the maximum aggregate size is prescribed by ASTM. In consequence, differences in the calculated tensile strength of thin disks may well be spurious, and not representative of the tensile strength of concrete as a continuum at different levels in the core. So the results of such tests are not valid. As Harald Kreuzer says: "There are a number of reasons to agree that standard tests have to be followed literally…" It is also satisfying that Laurence Wright says: "I agree emphatically with Adam Neville's position that standardized tests…should follow the testing protocols." Mindess and Marchand repeatedly refer to the ASTM procedures as "arbitrary." My experience of standard-writing committees is that, far from being plucked from the sky, the details of standard test methods are carefully selected and optimized. The first word in the title of my article was "Standard," and it is adherence to the *standard* methods that I am concerned with. So is Mindess[1] in his book. In a section headed "Need for 'standard' tests," he says: "To try to minimize the confusion that would result if everyone were to use different test procedures, various 'standard' test methods have been proposed."

Contrary to Mindess's book, Mindess and Marchand refer to modified and truly arbitrary methods. Of course, they can use such methods, but they must not pretend that they use the ASTM approach. As Wright says, "Engineers should be careful not to label a test with a standardized procedure number unless the test in fact conforms to all requirements." The words "all requirements" are, in my opinion, crucial. And yet, I have recently seen reports of tests on cores by a commercial laboratory, headed "tests according to ASTM C 42" in which the cores had a length-to-diameter ratio well under 1 (down to 0.6); worse than that, an unexplained correction factor for this ratio was applied. Now, not only does ASTM C 42 not provide correction factors for ratios smaller than 1.0, but it explicitly stipulates that the ratio must not be smaller than 0.95 before capping, or 1.0 after capping. The opinion that the length-to-diameter ratio of cores must not be less than 1.0 is supported by Patrick Sullivan who points out that, when the length of a core is, say, 75 mm (3 in.) then a diameter of 50 mm (2 in.) has to be used. Tests on small cores are discussed in *Properties of Concrete*.[2]

As for applying a test method destined for one material to another material, it is disingenuous to say that "the people who developed that test (ASTM D 2936) couldn't be expected to list all the other materials for which the test methodology would be suitable." Why not? It would be strange to omit such a major material as concrete while many less common materials are listed. And, it should be remembered that the scope of any test method is part and parcel of the ASTM standard. In this connection, Wright says: "I share his (Neville's) concern regarding misapplication of standardized tests...."

With respect to the Bureau of Reclamation test method for the tensile strength of concrete, and a similar method for the U.S. Army Corps of Engineers, I understand from Bryant Mather that most requests for the use of such a test came from people studying "the behavior of concrete under explosive loading or projectile penetration." RILEM is not a standard-writing body, but only prepares recommendations. Moreover, the comment from Mindess and Marchand that the RILEM Technical Recommendation TC 14-CPC Direct Tension of Concrete Specimens "was, in 1973, essentially destined for research works doesn't imply that it can't now equally be used in engineering practice" is not correct. The final version of the Recommendation was approved by RILEM in November 1975, with the reference to research works unaltered, and it still appears in the RILEM compendium.

Now, for international *standards*: ISO (International Organization for Standardization) is a worldwide federation of national standards bodies that prepares standards that may be adopted as national standards or used direct. The European Standards are produced under the aegis of CEN (Comité Européen de Normalisation) and are mandatory in the European Union and some other European countries.

I am pleased that Professors Mindess and Marchand agree with me that "information derived from the test (ASTM C 1202 Standard Test Method for Electrical Indication of Concrete's Ability to Resist Chloride Ion Penetration) has no direct relationship to the water permeability of concrete." And yet, I have read a description of tests by Marchand in which chloride ion movement is interpreted as a measure of permeability. Patrick Sullivan also agrees with me and says: "...there is no argument as to chloride ion diffusion being a different phenomenon to water or moisture flow in concrete, or that tests for diffusion through thin membranes are not applicable to concrete." Mindess and Marchand say that "...the various transport properties of concrete (permeation, ionic diffusion, ...) aren't totally unrelated to one another." This is true: likewise, height and weight of a person are not "totally unrelated" to one another, but we must not measure one of these for the purpose of establishing the other.

The statement that the various transport properties "are all more or less directly linked to the porosity of the material" may be useful in teaching undergraduates, but it is well known that the pore-size distribution is an important factor as well: at the same porosity, finer pores lead to a lower permeability of concrete than when the pores are few in number but large; indeed, a single large hole would allow a significant flow and thus result in high permeability. In *Properties of Concrete*, I say: "The pores relevant to permeability are those with a diameter of at least 120 or 160 nm. These pores have to be continuous;" this is followed by a list of "pores which are ineffective with respect to flow, that is to permeability."[2] More generally, in *Properties of Concrete*, I say : "in addition to total porosity, the effect of pore-size distribution on strength must be considered," and also: "The dependence of the strength of hydrated cement paste on its porosity and pore-size distribution is fundamental."[2]

Mindess and Marchand say, in defense of the determination of the transport of chloride ions as a measure of permeability, that "...the durability of concrete is often influenced (at least in part) by its ability to impede the diffusion of ions." But what if no chloride ions are involved or no ions of any kind? This is like saying, that because a deaf driver may fail to hear an approaching car and be involved in an accident, a determination of the number of deaf drivers is a measure of frequency of traffic accidents.

Mindess and Marchand refer to a medical examination of one of them (but which one?) that starts with taking temperature and blood pressure readings as being analogous to the determination of compressive strength of concrete. This is not a bad analogy because, according to the Merck Manual,[3] secondary hypertension can be caused by various diseases, and the Manual says: "Therefore every reasonable opportunity should be taken to measure blood pressure." The need for standardized measurement procedures is essential because leg pressure is usually 20 mm (0.8 in.) of mercury higher than arm pressure, and there can be up to 20 mm pressure difference between the right and left arm. The details of the cuffs are also prescribed . As for temperature, it must be taken at the same place because it can vary by 0.9°C (2°F) depending where on the body it is measured.[4] Thus, using standard procedures is vital (no pun intended).

The final remark from Mindess and Marchand about "more latitude...in the choice of test methods" in the resolution of disputes fails to take note of the rules of U.S. courts with regard to the use of new and novel test methods that are not accepted by the scientific community, the so-called Kelly-Frye legal test. Courts may rely only on tests that the various parties to the dispute can replicate, verify the results of, and interpret in an unequivocal manner. As a structural engineer, I find that, if there is a dispute involving a designer, contractor, supplier, or owner, then the issue is about the financial consequences of test results against the background of contractual relations. This is why standard test methods must not be arbitrarily modified to suit a given party. To allow "more latitude" in modifying standard test methods would simply result in a new dispute about the choice of test methods. Sullivan, also a structural engineer, seems to agree with me when he says that in litigation "a standard test is the arbiter."

Sullivan is of course right that a "standard test...should only be used for comparison of physical or chemical properties of similar materials or for control purposes," and he acknowledges that I do not disagree. For my part, I do not disagree that structural engineers translate the results of standard tests for compressive strength into stresses in a structure by the use of various factors.

Sullivan's comments on fire tests—and he is an acknowledged expert in that field—are of course valid. In real structures, there is an infinite variety of fire situations but, for compliance with codes and for assessment of safety of usual structures, it is inevitable that we use standard tests: This way, the fire rating of, say, partitions between apartments can be assessed in a standard manner and compliance can be verified. As Sullivan says, "...in litigation cases, where certain materials, which have been specified to certain standards, (they) must be shown to comply."

Sullivan's further comments about "attempting to find the cause of deterioration" come under the heading "Testing in research." In this connection, I fully accept Wright's views and also those of Kreuzer that engineers have to solve problems facing them by using various methods and various tests. This is what real-life engineering is about. Solving problems is likely to suggest new test methods, and in many cases, these will contribute to the development of new standard test methods. Hampering such development is clearly undesirable, and this is what I said when referring to freedom in testing in the section on "Testing in research." Although I ended that section with the words "research and development," I would have done better to call the section explicitly "Testing in research and development." For this I apologize.

Finally, it is highly gratifying that what Bill Semioli calls "Neville having a dialectic with himself" has provoked four letters from four countries; this is a testimony to *Concrete International*'s truly international character. The views of academics have to be heard because it is largely their work that deals with cause-and-effect and prepares the basis for standard test methods. Such a contribution of researchers is vital. In this connection, Kreuzer is absolutely right about the need for "more symbiotic togetherness between work on standards and research...." His specific recommendation with respect to the wedge-splitting test is a good example of what can, and should, be done by way of establishing new standard test methods. I would hope that his contribution will lead standard-writing organizations to do so.

Turning to engineering practice, as Kreuzer says: "Any comparison or compliance purpose is lost; any safety criterion becomes indefensible if it relates to tests subject to individual modifications...no need to add individual 'improvements'." It is the views of prominent practitioners that are particularly welcome to the author of an article in *Concrete International*; after all, those who write about concrete should have, as the ultimate aim, improvements in concrete structures. Hence, the great importance of contributions from practitioners—and that means engineers, architects, contractors, and suppliers—in the debate on concrete.

Adam Neville

References

1. Mindess, J., and Young, J. F., *Concrete*, Prentice-Hall, N.J., 1981, p. 408.

2. Neville, A., *Properties of Concrete*, Fourth Edition, Pearson

Education, U.K. and John Wiley, New York, 1995, 844 pp.
 3. *Merck Manual*, 16th Edition, 1992, p. 370.
 4. Murtagh, J., *General Practice*, McGraw-Hill Book Co., 1994, p. 427.

Section 7.2: Core Tests: Easy to Perform, Not Easy to Interpret

Many engineers have the experience of ordering the taking of cores. The operation is not difficult, usually undertaken by skilled specialist personnel. Once cut out of the concrete structure, the cores are sent to a testing laboratory for determination of compressive strength, and sometimes for other tests, generally petrographic in nature. The determination of strength is quite similar to the determination of compressive strength of standard test cylinders and yields in the first place the value of the load in Newtons (or pounds), under which failure by crushing occurs, which is then to be divided by the cross-sectional area of the core in square millimeters (or square inches). Dividing the first of these by the second gives a number in megapascals (or psi), but does this number represent the compressive strength of concrete in the structure from which the core was cut?

The answer is no. Not only must the number be processed, but the resulting value of strength also must be carefully interpreted. Because cores are generally taken when there is a problem, or suspected problem, with concrete, the situation usually involves two or more parties, and they may have different views of what is an appropriate interpretation of the core test results.

Why take cores?

The taking of cores most commonly occurs when the results of tests on standard test cylinders to determine the 28-day compressive strength indicate noncompliance with the specification. Such noncompliance may be due to the fact that the concrete that was placed in a given part of the structure, as well as in the test cylinders, is indeed noncompliant because its 28-day strength is lower than specified. But there may be other reasons: the cylinders may have been incorrectly consolidated (compacted); they may have been damaged in transit, subjected to freezing at a very early age, badly cured, or incorrectly tested; or the resulting compressive strength may have been incorrectly calculated or recorded.

The contractor has reasons to suggest that it is the cylinders that are unsatisfactory, while the concrete in the structure is as specified. On the other hand, the engineer has a professional responsibility to ensure the structural adequacy of the concrete, as well as a responsibility to the client (or owner) to ascertain that the quality of concrete corresponds to the price paid.

It may also happen that while the strength of test cylinders is satisfactory, there are suspicions that the concrete being placed in the actual structure has segregated or has been inadequately consolidated. This can be resolved by the inspection and testing of cores.

There are other situations where taking cores may be desirable, or even essential. For example, it may be required to subject an existing structure to heavier loads than hitherto; or a change of use may be proposed, and the load-carrying capacity of the structure needs to be verified; or it may be necessary to ascertain that the strength of concrete has not been impaired by overloading or by fatigue, fire, explosion, chemical attack, or some other deleterious agent.

Where to take the cores?

The decision on this depends on the question to which the engineer wishes to have an answer. Several questions are possible. The two most common are: does the concrete comply with the specification? and, what is the compressive strength of concrete in the structure in general? To answer these questions, cores should be taken at a number of locations spread over the whole structure. But there is a caveat: concrete in the uppermost part of any member or concrete lift is nearly always weakest and, therefore, nonrepresentative.

There may also be this question: is the strength of the particularly highly stressed parts of the structure adequate? In such a case, the locations for core-taking should be carefully chosen so that the load-carrying capacity of the structure is not impaired by removing some concrete from the place where it is most needed. It follows that only an engineer familiar with structural action can identify the appropriate locations.

An engineer would also be aware of where to test to answer the question: is the strength adequate for the *actual* loading system, even if the strength is not adequate for the *designed* loading system? In other words, if the structure is not adequate for the original design situation, is it strong enough for more modest loading? The owner may be willing to compromise in such a manner, perhaps by being recompensed by a lower price for the work done than stated in contract. Such a solution may well be acceptable to the contractor in preference to the wholesale removal of the concrete already placed and replacement by new concrete. There is a benefit to the owner, too: the considerable delay due to removal and replacement is avoided. It can be noted in passing that a contractual provision for penalty payment for concrete with a strength lower than specified, but accepted by the engineer, is likely to have a salutary effect on the contractor's efforts with respect to the quality of concrete.[1]

How to decide on a testing plan?

I have already referred to the need for a clear decision on the location of the cores to be taken. In my opinion, this has to be taken by a structural engineer and not left to the person on the job. This requirement is obvious, perhaps so obvious that it is not considered to be worthy of explicit instructions in codes of practice. The Commentary on ACI 318, Section R5.6.5 wisely says: "The building official should apply judgment as to the significance of low test results and whether they indicate need for concern."² The same section says: "Lower strength may, of course, be tolerated under many circumstances, but this again becomes a matter of judgment on the part of the building official and design engineer."²

This approach is right and proper, but ACI 318 is not normally consulted by a testing contractor who is more concerned with ASTM C 42-99, "Standard Test Method for Obtaining and Testing Drilled Cores and Sawed Beams of Concrete." Quite rightly, the ASTM standard does not concern itself with the testing location and plan or with the interpretation of test results. It is in these respects that I have seen problems arise.

I have encountered problems with testing plans in countries other than the United States, and, because *Concrete International* has a large worldwide readership (and, I hope, this book as well) I feel that they should be aired. Accordingly, I would like to refer to European Standard (which is also a British Standard) BS EN 12504-1:2000.³ Because this standard was first published in 2000, it is not yet well known. It is likely, however, that the standard will be used either as-is or in a modified form not only in Europe but also elsewhere, mainly in Commonwealth countries.

An important feature of BS EN 12504-1 is that it says explicitly what it does not cover. The Foreword states: "The standard ... does not consider a sampling plan." Notes to the section headed "Scope" say: "The standard does not give guidance on the decision to drill the cores or on the locations for drilling," and "This standard does not provide procedures for interpreting the core strength results." These deliberate omissions mean that it is the structural engineer involved in the core exercise who should decide on such matters. I believe this is important because I have encountered cases where the testing laboratory interprets the results of core tests vis-à-vis the specification for construction and vis-à-vis structural safety. The demarcation between testing and structural interpretation should be carefully maintained.

With respect to the choice of the location of the cores, ACI 318 says that cores may be drilled from the area suspected of having a low strength and expects the core tests to represent that area.² British Standard BS 6089:1981,⁴ while giving no guidance on the selection of locations of cores, proffers important advice in clause 4.2: "Before any program [of testing] is commenced, it is desirable that there is complete agreement between the interested parties on the validity of the proposed testing procedure, the criteria for acceptance...."

Following this wise advice obviates subsequent hassle over the interpretation of the core test results. The "agreement" referred to above should not be simply a compromise between the engineer and the contractor; it must take into account the questions that the engineer wishes to have answered. European Standard BS EN 12504-1 states explicitly in clause 5.1: "It is essential that full consideration is given to the aims of the testing and the interpretation of the data, before deciding to drill the cores."³ The word *before* should be noted.

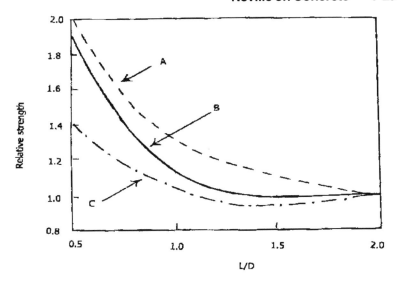

Fig. 7.2.1—Effect of *L/D* on the strength of test cylinders.[6] The 28-day strength at *L/D* = 2 in MPa (psi) is: A, 14 (2000); B, 26 (3700); C, 52 (7500).

Why should length-diameter ratio be 2?

The simple answer to this question is: because this is the value of the length-diameter ratio (L/D) of a standard cylinder. However, more should be said. In particular, it is worth giving the reasons for the choice of $L/D = 2$ as a standard. It is also worthwhile to discuss the situation in cores, because, unlike a test cylinder where the mold controls the value of L/D, the choice of the value of L/D lies in the hands of those involved in testing.

Test programs that led to the choice of $L/D = 2$ go back a long time. In 1925, Gonnerman wrote: "The 6 by 12 in. cylinder is now generally used for compression tests."[5] Supporting arguments for the choice of $L/D = 2$ were given by Murdock and Kesler in 1957.[6] In essence, they showed that the variations in the calculated compressive strength of a cylinder were small for L/D values in the vicinity of 2, and were more pronounced for values between about 1.0 and 1.6 (see Fig. 7.2.1). The reasons for this situation lie in the mode of failure of the test specimen: in a squat specimen, the restraining effect of the platens of the testing machine is much more significant than in a more slender specimen. In other words, the use of correction factors necessary to normalize the test results to the value of strength of a standard specimen is greater at smaller values of L/D; it is clearly preferable to minimize the need for correction factors.

The use of cores with the value of $L/D = 2$ is appropriate only when standard molded test cylinders have that value. In countries using cubes, the core should preferably have $L/D = 1$; this is recommended in European Standard BS EN 12504-1: 2000, which caters both to countries using cylinders and to those using cubes.[3] In the United Kingdom, which uses cubes, BS 6089:1981 requires the value of L/D to be not less than 0.95

Chapter 7: Testing

Table 7.2.1—Correction factors for L/D according to ASTM C 42.

Year of ASTM 42	Strength correction factor† Value of L/D				
	2.00	1.75	1.50	1.25	1.00
1927	1.00	0.98	0.95	0.94	0.85
1949	1.00	0.98	0.96	0.94	0.85
1961	1.00	0.98	0.96	0.94	0.89
1968	1.00	0.99	0.97	0.94	0.91
1977 1990 1999	1.00	0.98	0.96	0.93	0.87

* Based on ref. 7

† The strength of a core with L/D = 2 is equal to the strength determined on a core with the actual L/D multiplied by the correction factor

before capping, and not more than 1.3.[4]

Although it has not been unequivocally demonstrated to be the case, I am of the opinion that a cylinder (molded or a core) with the value of $L/D = 1$ has approximately the same strength as a cube whose edge is equal to the diameter of the cylinder.

What are the correction factors for L/D?

The use of correction factors to "convert" the strength determined on a test specimen with a given value L/D to the strength that would be obtained in a test on a specimen with $L/D = 2$ is well known. But how good are those factors?

Murdock and Kesler tell us that the factors are a function of the level of strength of the concrete.[6] Specifically, stronger concretes are less affected by the value of L/D than concretes of lower strength; this is shown in Fig. 7.2.1. According to ASTM C 42-99, concretes with strengths above 70 MPa (10,000 psi) are even less affected by the value of L/D. It follows that the correction factors tabulated in ASTM C 42 and by other standards are in reality typical values or average values. Using a single set of correction factors, such as that given by ASTM C 42, may overcorrect the test results on cores with a small value of L/D drilled from low-strength concrete; and yet, it is for this type of concrete that an estimate of strength may be particularly important. We can see thus that the use of correction factors increases the uncertainty about the strength of the concrete being tested, as compared with a situation where all the cores have $L/D = 2$. In other words, if at all possible, tests should be performed on cores with $L/D = 2$.

It is interesting to note that even the same standard, namely ASTM C 42, in its various editions, gives somewhat different sets of correction factors. These are shown in Table 7.2.1, based on a paper by Bartlett and MacGregor.[7] As they pointed out, the "true" value of the correction factor may be a function of the moisture condition of the core at the time of testing.[7] All this militates against the use of cores with varying values of L/D coupled with reliance on correction factors.

There is one other important observation in the paper by Murdock and Kesler, derived from earlier tests: "the data on the specimens whose L/D was less than 0.5 were erratic."[6] This situation is apparent in Fig. 7.2.1. Other tests showed that the "correction factor" for $L/D = 0.5$ was 1.98 for 21 MPa (3000 psi) concrete, and 1.68 for 31 MPa (4500 psi) concrete.[5] A value as high as 2.09 is quoted by Murdock and Kesler.[6] It is not surprising, therefore, that BS 6089: 1981 states: "Little reliance can be placed on results obtained on cores having length/diameter ratio of less than 0.5."[4]

My conclusion from the preceding discussion is that cores with L/D smaller than 1 should never be used, and yet I have seen the use of cores with $L/D = 0.5$; cores with a diameter of 6 in. (150 mm) were extracted from slabs 4 in. (100 mm) thick or even thinner. Testing such cores could be easily avoided either by taking cores with a smaller diameter or by subsequently subcoring the larger cores. Admittedly, there remains the issue of the minimum diameter of the core to be tested in compression.

Of course, there is no problem with cores that are too long: they can be cut so as to achieve the value of the ratio of the capped length to diameter of between 1.9 and 2.1, described as preferred by ASTM C 42. It is only when L/D is less than 1.8 that a correction factor needs to be applied.

Although it is not always stated explicitly, the various correction factors for L/D apply to normal weight concrete only. According to ASTM C 42-99, they apply also to lightweight concrete with a density between 100 and 120 lb/ft^3 (1600 and 2000 kg/m^3). However, there is some uncertainty about correction factors for concretes of lower density.[8]

What size cores?

It is considered preferable to use cores with a diameter of 6 in. (150 mm), with the alternative of 4 in. (100 mm) being acceptable. ASTM C 42-99 prescribes a minimum diameter of 3.75 in. (95 mm). It allows the use of smaller diameters when it is impossible to obtain a core with L/D of at least 1, but only for "cases other than load-bearing situations."

According to ASTM C 42, the minimum diameter of the cores is governed by the maximum aggregate size: it "should preferably be at least three times the nominal maximum size of the coarse aggregate and must be at least twice the nominal maximum size..." European Standard BS EN 12504-1: 2000 simply says that, when the core diameter approaches a value that is less than 3 times the maximum aggregate size, there is "a significant influence on the measured strength."[3] An Annex to this

Table 7.2.2—Effect of maximum aggregate size (M.A.S.) and core diameter on measured strength (based on ref. 3)

Core diameter	Relative strength for m.a.s. (mm)	
mm	20	40
100	1.00	1.00
50	0.94	0.86
25	0.78	0.72

Standard gives the relevant data, from which Table 7.2.2 has been derived.

The reason for the limitations on the core size is that, unlike a molded cylinder, in a core, some coarse aggregate particles are cut in the drilling process and are, therefore, not wholly bonded to the cement paste matrix. The adverse effect of incomplete bond is aggravated by the difference in the modulus of elasticity between the aggregate and the cement paste.[9] When a significant proportion of coarse particles is in that state, some of them may become partially loosened during the test and cease to carry their share of the applied load.[8] When this happens, a lower value of compressive strength is recorded. An indirect evidence of such behavior is offered by Malhotra, who found that cores with a smaller diameter exhibit a higher variability.[10] This situation is recognized in the assessment of the precision of tests on cores of various sizes.

Nonetheless, there are some proponents of the use of cores with a diameter as small as 1.2 in. (30 mm), even when the maximum aggregate size is ¾ in. (20 mm).[11] The confidence limits of the predicted strength of test cubes from the strength of such small cores are very wide, and I, for one, remain skeptical about the interpretation of the test results of compressive strength of such small cores. It has also been reported that the correction factor for the value of L/D is dependent on the diameter of the core.[12]

Which moisture condition?

Cores are usually obtained using water-cooled drills with diamond bits attached to a core barrel so that, when extracted from the parent concrete, the cores are superficially wet. The interior of the core may be wet or dry, depending on the situation of the concrete in service. The question to consider is: in which moisture condition should the cores be tested to ascertain their compressive strength? Different standards and guides offer differing views. ASTM C 42-99 states that the cores "shall be tested in a moisture condition representative of the in-place concrete" or alternatively "as directed by the specifying authority." There is a section in ASTM C 42 that allows this authority freely to choose any moisture condition. I would like to emphasize the provision for the decision by the "specifying authority," which usually is the project engineer.

In the absence of a specially chosen moisture condition, ASTM C 42 provides for two conditions. One of these is the "as-received condition," which requires drying for 12 to 24 hours at a temperature of 60 to 80°F (16 to 17°C) and a relative humidity below 50%. The second condition is the "dry condition," which requires drying at the same temperature but at a relative humidity below 60% for seven days.

The significance of the careful control of the moisture condition of the core at the time of testing lies in the fact that this condition influences the apparent strength of the core; broadly speaking, wet cores record a lower strength than dry cores. The problem is that the difference is variable and uncertain.

This section does not purport to present a literature review, and I shall limit myself to the work by Bartlett and MacGregor.[13] They found that, on average, the strength of cores dried in the air for seven days is 14% higher than the strength of cores soaked in water for at least 40 hours. The value of 14% appears to apply over the range of strengths from 15 to 92 MPa (2200 to 13,400 psi). The crucial words are "on average" because the scatter about the 14% difference is considerable.

Bartlett and MacGregor point out that the actual situation is more complex than "dry" and "wet": what influences the magnitude of the difference in strength of cores in the two conditions is the presence of a moisture gradient between the exterior and the interior of the core.[13] In the absence of such a gradient, the difference is not 14% but only 9%. They state "that the effect of seven days of air drying is to cause a moisture gradient that artificially increases the strength of the specimen by about 5% above the true in-situ strength."[13] Personally, I do not accept that *any* core test gives us the value of "the true in-situ strength." In a later paper, Bartlett and MacGregor reported that the apparent reduction in strength of wet cores is greater in cores with diameters of 2 in. (50 mm) than in larger cores.[14]

For these reasons, I am of the opinion that mimicking the moisture conditions in service is too elusive to be useful. In any case, the strength of the core as determined by the compression test does not represent the actual strength of the concrete in the structure in a meaningful way.

What the test result gives us is a value that has to be interpreted by the engineer who has to take into consideration a number of factors relevant to the core as well as to the history of the concrete in service. I shall discuss this topic more fully later.

Overall, then, I favor testing cores after immersion in water for a period long enough to prevent the presence of moisture gradients. The advantage of this condition is that it is much more reproducible than a so-called dry condition, where the degree of dryness is uncertain, so that accidental variations in conditions can arise. In this connection, we can note that European Standard BS EN 12504-1 says: "If it is required to test the specimen in a saturated condition, soak in water at 20 ± 2°C (68 ± 4°F) for at least 40 hours before testing."[3]

What else affects core strength?

It has been observed that when drilling is performed in the horizontal direction, the resulting strength is lower than for cores drilled vertically downward. A likely reason for this is that, whereas drilling downward allows the drill to be fixed and held firmly in position, drilling horizontally almost inevitably permits a slight movement of the drilling barrel, even when holding-down bolts locate the drill. The consequences of such movement during drilling may be seen as shallow corrugations on the surface of the core.

Another possible reason for the difference between cores drilled vertically and horizontally can be sought in the presence of bleed water. Such water would have no effect when the load applied in the compression test is normal to the lenses of water trapped underneath coarse aggregate particles, but the lenses of bleed water parallel to the axis of the core might result in some weakening.

A reduction in strength of 8% of the value for vertical cores has been suggested:[9] this is probably the best correction factor that we have, but it cannot be assumed to be universally valid. In fact, Bartlett and MacGregor found that the direction of the core axis relative to the direction of casting has no effect on the strength of cores taken from high-strength concrete.[14] Some support of my doubts about the general use of a correction factor for horizontally drilled cores is given by the recent European Standard BS EN 12504-1: 2000, which makes no reference whatsoever to the direction of drilling.[3] This is in contrast to British Guide BS 6089: 1981, which enhances the strength of horizontal cores by 8.7%.[4]

Drilling cores in a horizontal direction introduces another factor influencing the strength of the core. This factor is not related to coring but (as already mentioned) is the consequence of the inherent variation in strength of concrete in the vertical direction of a concrete member or a lift, caused by an increase in the actual water-cement ratio in the uppermost layers of concrete. Such differences should not be "corrected" but should be recognized as a fact of life. A usual way to avoid this effect is not to take cores in the uppermost part of the concrete member.

What happens in slipformed concrete?

Slipformed concrete presents a special case. In my opinion, in such concrete there is no increase in the water-cement ratio with height because the apparent upward movement of water occurs continuously and steadily, owing to the slow progress of placing of concrete. For this reason, I would not expect the formation of lenses of trapped bleed water. A consequence of this situation is that there would be no lowering of the apparent strength of concrete in cores drilled horizontally.

This direction of drilling of cores is inevitable in a silo in service. However, I encountered an exception when a slipformed silo was demolished, and parts of the silo walls were laid horizontally on the ground. Cores were then taken by drilling horizontally, that is, in the direction of placing of concrete. In consequence, although the core is drilled horizontally, the core axis is parallel to the direction of placing of concrete. It is therefore difficult to say whether such a core should be classified as horizontal or vertical.

What is the effect of steel?

Another potential factor influencing the testing of cores and their apparent strength is the presence of reinforcing steel within the core. It is obvious that reinforcement should be avoided wherever possible; on the other hand, in some structures, the density of reinforcement is so great that it may be impossible to avoid it. If this is the case, there is a serious risk of adversely affecting the structural integrity of the given structural member by coring through the steel. I believe that, in such a situation, coring should not be resorted to, and nondestructive methods of testing should be used.

If, for whatever reason, cores cut through steel bars, then the cores can be tested for compressive strength, but only if there are no steel bars parallel to the axis of the core; this requirement is laid down in European Standard BS EN 12504-1: 2000.[3] British Guide BS 6089: 1981[4] and also the Concrete Society Report[15] give correction factors for the presence of one or more reinforcing bars in the core. The British Guide adds that "the in-situ strengths estimated from the [given expressions] cannot be equated to standard cube strengths."[4] I view all this as militating against the use of cores containing reinforcing steel.

How to test for splitting tensile strength?

ASTM C 42-99 also provides for testing cores to determine the tensile splitting strength. The requirements with respect to the cores themselves are the same as for compression testing, with the exception that capping is not to be applied. It follows thus that the value of L/D must not be greater than 2 and must not be smaller than 1. However, no correction for L/D is required. I am stating these requirements because I have seen results on tests on cores with a value of L/D as low as 0.5. Clearly such tests for the splitting tensile strength cannot be considered to conform to ASTM C 42-99. I have discussed the broader question of arbitrary modification of ASTM standard tests in an earlier section.[16]

Why is it not easy to interpret core tests?

This assertion is included in the title of this section but, at this stage, the reader might be surprised: after all, I have discussed the various factors influencing the strength of cores and the relevant correction factors. It could, therefore, be thought that all that is needed is to apply the various correction factors and, hey presto, we arrive at the compressive strength of concrete.

What is the variability of strength?

First of all, there is a scatter in the strength of cores even when taken in close vicinity to one another. Admittedly, there is also scatter in the strength of molded cylinders, but the scatter of strength of cores is larger because coring itself introduces variability. It may be worth repeating that, however carefully performed, the coring operation causes some damage to the core: there may be a weakening of the bond of large particles of aggregate near the surface of the core, and there may be a dynamic effect of drilling, which induces some fine cracks. It is, therefore, best to test more than one core, but this may not be desirable or even possible. The coring operation is expensive, and it may disturb the structure or mar its appearance.

According to Concrete Society Report No. 54,[17] the 95% confidence limits on the estimate of actual strength of the cores are as follows: ±12% of the value determined on a single core, and ±6% of the average value of four cores. Overall, the confidence limits are inversely proportional to the square root of the number of cores tested. These figures mean that, when the value determined by test is X, there is a 95% probability that the "true" value lies within 12 and 6%, respectively, of the value X determined in the test. This situation should be borne in mind when someone tries to argue that, say, a value of 27 MPa determined on a single core indicates that the expected value of 30 MPa has not been satisfied.

The precision statement in ASTM C 42 says that the results of two tests (each being an average of two adjacent cores) performed by two laboratories should not differ from each other by more than 13% of their average. European Standard BS EN 12504-1 gives no precision statement but acknowledges a greater variability of cores than of molded specimens.[3]

Which strength?

The first question is: what is meant by strength? It is sometimes forgotten that the intrinsic or "true" strength of concrete cannot be measured. By "intrinsic" I mean the strength of concrete that is independent of the characteristics of the test specimen.

We know that the strength of molded test cylinders cured, treated, and tested under standard conditions (such as those of ASTM C 31) represents, at best, the strength of concrete under precisely those conditions. I am saying "at best" because the value of strength actually determined applies only to the type of specimen used. For other types of specimen, say a cube, the strength recorded would be different, whereas the intrinsic strength of concrete is, of course, the same.

In the case of cores, the situation is much more complicated. The cores are taken from concrete in situ, whose history is most unlikely to replicate the treatment of standard molded specimens, and the age of the cores is usually greater, even significantly greater, than 28 days.

So, in interpreting the test result obtained by compression testing of cores, we have to start with this question: which type of strength are we interested in? Broadly speaking, there are two types: potential strength and actual strength. We should know the answer to this question prior to testing; it cannot be overemphasized that taking cores and testing them is of value only if we can interpret the results in a meaningful and unequivocal manner. Otherwise, the dispute about the quality of concrete that led, in the first place, to the decision to take cores is simply shifted to the stage after the cores have been taken.

We should therefore heed the injunction in Concrete Society Report No. 54, published in 2000, that states in bold letters: "No measurements should be taken or tests carried out if it is not known what the results will be used for."[17] Here is my opportunity strongly to dispute the dictum of those who say that a single test is worth more than a thousand words. In my opinion, this is not correct.

This may be an appropriate place to define potential strength and actual strength. Potential strength is the strength determined on standard molded cylinders in accordance with a standard procedure, such as that of ASTM C 31. The cylinders will have been made as well as possible in terms of consolidation and cured under constant conditions of temperature and humidity. In consequence, the compressive strength recorded is the "best" that the given mixture can achieve. Clearly, I am excluding the possibility of steam curing or the application of pressure in making the cylinder.

On the other hand, the actual strength is the strength of concrete in place. Consequently, the consolidation has been less good than in a test cylinder: there are bound to be some excess voids. The curing has been less good because the concrete in the interior of the structure could not be kept continuously wet. Likewise, the temperature was not kept constant. It follows that, at the age of 28 days, the actual strength is lower than the potential strength; under some circumstances, the situation may later on be reversed.

We can thus see that the potential strength is of interest when we want to know that the mixture used in construction conforms to what is specified. The actual strength is of interest when we want to know how good the concrete in the structure is. This difference is crucial to the interpretation of the value of strength determined on the cores.

How not to use the correction factors?

It could be expected that the discussion of various factors influencing the strength of cores earlier on in this section would make it possible to apply various correction factors so as to "convert" the test result into a value of either potential or actual strength. Alas, this is not so.

First, let me deal with the various correction factors. I have tried to point out that the factors are, at best, average values; sometimes, they are a compromise value. It follows that a given correction factor may not be entirely appropriate in the case of the actual mixture used.

More than that: even when the factor is appropriate, it is an average value around which there is some scatter. Thus, it is possible that in a particular case the correction factor used is excessive. It is possible that another correction factor is also excessive. As I see it, applying several factors, one after another, is a risky operation that reduces the accuracy of the prediction. I am saying this because I have seen a situation where the value of compressive strength on a core was sequentially corrected for the value of L/D, then for the diameter of the core, and finally for the moisture condition; and yet, we know that the effect of moisture condition on strength is greater in smaller cores than in larger ones. Therefore, the two corrections should not be applied independently.

I am, therefore, firmly of the opinion that a wholesale application of correction factors is highly undesirable. If the objective of the test were to determine, say, the strength of a core with $L/D = 2$, a diameter of 4 in. (100 mm), and in a moist condition, then this is how the core should have been taken and tested. If this was physically impossible, then the outcome of the test should have been compared with a core from concrete with known properties and with characteristics similar to those of the test core. On occasion, this may not be possible, in which case cores are not an appropriate method of resolving the given problem.

Can we determine the potential strength?

In seeking to "convert" the strength established by testing the cores into the potential strength of the given mixture, attempts are sometimes made to allow for the age and curing history of the concrete in the structure. I believe that this is an uncertain operation because this history is rarely known reliably. Moreover, any possible incomplete compaction requires a correction as well, and this is not easy. I realize that excess voidage can be assessed visually and that the density of a concrete specimen is affected by the presence of voids, but correcting for these may induce a large error because 1% of voids reduces the compressive strength by 5%.[9]

Because of these uncertainties, a rough-and-ready approach is sometimes used; this may be no less reliable than fine tinkering with corrections for individual factors. Concrete Society Report No. 11 simply states that the estimated potential strength is 1.3 times larger than the estimated actual strength.[15] This is unlikely to be universally correct because, as ASTM C 42 says, "There is no universal relationship between the compressive strength of a core and the corresponding compressive strength of standard-cured molded specimens." To my way of thinking, it follows that there is no such universal relationship between the estimated strength of molded specimens and the core strength.

I find the approach of ACI 318-99[2] and of ACI 301-99[18] to be preferable. They both say that concrete shall be considered adequate (which I understand to mean "as specified") when "the average of three cores is equal to at least 85% of f'_c [specified strength] and no single core is less than 75% of f'_c."[2] No allowance for age or curing history is mentioned. This approach is robust and has the merit of simplicity. The Commentary on ACI 318-99 is persuasive; it says, "Core tests having an average of 85% of the specified strength are entirely realistic. To expect core tests to be equal to f'_c is not realistic, since differences in the size of specimens, conditions of obtaining samples, and procedures for curing, do not permit equal values to be obtained."[2] A difficulty that sometimes occurs is that fewer than three cores are available.

The value of 85% in ACI 318-99 is not very different from the value used in BS 6089: 1981, which requires the "estimated in-situ cube strength" to be 83% of the specified characteristic strength of concrete, to which the partial safety factor for design strength is applied.[4]

ACI 318-99 applies the "85% correction" to the compressive strength of cores but remains silent on the interpretation of tests on the splitting tensile strength of cores. In my opinion, the "85% correction" should be applied to the splitting tensile strength as well because the factors responsible for this correction apply equally in splitting tension and in compression. Moreover, without a parity of application of the correction, the ratio of splitting tensile strength to the compressive strength applicable to molded cylinders would become distorted in the case of cores; there is no rational reason for such a distortion.

Who decides?

The merits of this simple approach cannot be overemphasized, but a party to a dispute may seek to manipulate the test results to its advantage. Before it does so, it would do well to read Malhotra's paper, in which he says: "The available test data on cores are full of contradiction and confusion."[10] He also cites Bloem:[18] "Core tests made to check adequacy of strength in place must be interpreted with judgment. They cannot be translated to terms of standard cylinders strength with any degree of confidence, nor should they be expected necessarily to exceed the specified strength f'_c."[18]

Bloem's reference to "judgment" and the reference in the Commentary in ACI 318-99 to "judgment on the part of the building official and design engineer" lead me to emphasize that such judgment is not the province of the testing laboratory. Perhaps this approach is observed in the U.S., but elsewhere I have seen technicians who, after reporting the results of core tests, yield to the temptation to express views about structural adequacy; to do so, the expertise of the structural engineer is essential.

It must be re-emphasized that testing and interpretation of test results are distinct operations. My view is supported by ACI 301-99, which states: "Core test results will be evaluated by Architect/Engineer and will be valid only if tests have been conducted in accordance with specified procedures."[18] As for the next step, if necessary, BS 6089: 1981 says, "Action to be taken in respect of a structural member in which the in situ concrete is considered to fall below the level required has to be determined by the engineer."[4]

Can we reduce the use of cores?

It is well known that all testing involves errors due to chance; indeed, it is chance that is responsible for the stochastic distribution of strength of concrete test specimens about the mean value. This distribution is reflected in the coefficient of variation of test results. Errors are distinct from mistakes caused by action, or absence of action, by the operator. So, chance—or in common parlance, luck—plays a role in testing.

Careful performance of all tests, strict adherence to standard test methods, and soundly-based judgment minimize the element of luck. As Jack Nicklaus (who has never played on a concrete golf course like the one on Das Island in the Arabian Gulf) is supposed to have said: "People tell me I am lucky. The funny thing is the harder I practice the luckier I get." When we make concrete we should try harder, so that the need for tests on cores is rare.

Nevertheless, perfection is unachievable, and this section has considered the use of cores in cases where there are doubts about the quality of concrete in a structure. Such a situation arises from time to time, and taking and testing cores may be an effective way of resolving the problem. Nevertheless, as we strive to make better concrete and to ensure that it complies with the specification, we should reduce the need for core tests and, hopefully, reach the situation in Mark Twain's *Tom Sawyer Abroad* (admittedly, concerned not with concrete but with apples) in which "there ain't-a-going to *be* no core."

References

1. Bartlett, F. M., and MacGregor, J. G., "Statistical Analysis of the Compressive Strength of Concrete in Structures," *ACI Materials Journal*, V. 93, No. 2, Mar.-Apr. 1996, pp. 158-168.
2. ACI Committee 318, "Building Code Requirements for Structural Concrete (318-99) and Commentary (318R-99)," American Concrete Institute, Farmington Hills, Mich., 1999, 391 pp.
3. European Standard BS EN 12504-1:2000, "Testing Concrete in Structures — Part 1: Cored Specimens," 8 pp.
4. British Standards Institution, BS 6089:1981, "Guide to Assessment of Concrete Strength in Existing Structures," 11 pp.
5. Gonnerman, H. F., "Effect of Size and Shape of Test Specimen on Compressive Strength of Concrete," *Proceedings ASTM Part 2*, 1925, pp. 237-255.
6. Murdock, J. W., and Kesler, C. E., "Effect of Length to Diameter

Ratio of Specimen on the Apparent Compressive Strength of Concrete," *ASTM Bulletin*, Apr. 1957, pp. 68-73.

7. Bartlett, F. M., and MacGregor, J. G., "Effect of Core Length-to-Diameter Ratio on Concrete Core Strengths," *ACI Materials Journal*, V. 91, No. 4, July-Aug. 1994, pp. 339-348.

8. McIntyre, M., and Scanlon, A., "Interpretation and Application of Core Test Data in Strength Evaluation of Existing Concrete Bridge Structures," *Canadian Journal of Civil Engineering*, V. 17, 1990, pp. 471-480.

9. Neville, A. M., *Properties of Concrete*, Fourth Edition, Longman, London, 1995, and John Wiley, New York, 1996, 844 pp.

10. Malhotra, V. M., "Contract Strength Requirements — Cores Versus In Situ Evaluation," *ACI Journal*, Apr. 1977, pp. 163-172.

11. Indelicato, F., "A Statistical Method for the Assessment of Concrete Strength Through Microcores," *Materials and Structures*, V. 26, 1993, pp. 261-267.

12. Naik, T. R., "Evaluation of Factors Affecting High-Strength Concrete Cores," *Proceedings of the First Materials Engineering Congress Part I*, Denver, 1990, pp. 216-222.

13. Bartlett, F. M., and MacGregor, J. G., "Effect of Moisture Condition on Concrete Core Strengths," *ACI Materials Journal*, V. 91, No. 3, May-June 1994, pp. 227-236.

14. Bartlett, F. M., and MacGregor, J. G., "Cores from High-Performance Concrete Beams," *ACI Materials Journal*, V. 91, No. 6, Nov.-Dec. 1994, pp. 567-576.

15. Concrete Society, "Concrete Core Testing for Strength," Concrete Society *Technical Report No. 11*, 1976, 44 pp.; Addendum, 1987, pp. 45-59.

16. Neville, A. M., "Standard Test Methods: Avoid the Free-For-All!," *Concrete International*, V. 23, No. 5, May 2001, pp. 60-64. [Section 7.1]

17. Concrete Society, "Diagnosis of Deterioration in Concrete Structures," Concrete Society *Technical Report No. 54*, 2000, 68 pp.

18. ACI Committee 301, "Specifications for Structural Concrete (ACI 301-99)," American Concrete Institute, Farmington Hills, Mich., 1999, 43 pp.

19. Bloem, D. L., "Concrete Strength Measurement — Cores Versus Cylinders," *Proceedings ASTM*, V. 65, 1965, pp. 668-696.

Letter to the Editor No. 1
A question of strength

The article on "Core Tests: Easy to Perform, Not Easy to Interpret" by Adam Neville (*Concrete International*, V. 23, No. 11, November 2001) is extremely interesting. In the first two paragraphs, he sets the stage for the remainder of the article in which he describes, with references, some of the important variables that might be encountered. These are not imaginary variables. They are the type of variables that might be encountered whenever the question of strength is at issue. For want of a better and more reliable method of arriving at the "approximate" truth, he suggests that the structural engineer of record serve also as the judge of record. To this end, it is hereby further suggested that the engineer of judgment read this article thoroughly before reaching that judgment.

Unfortunately, cores will probably continue to be used to settle arguments and, as Dr. Neville so succinctly suggests, let's try to build it right in the first place and avoid the arguments. This procedure may not always come out as planned or specified, and the use of cores appears to be the only viable alternative. To dispel some of these pitfalls in testing and judgment, it is further suggested that the job specifications state clearly how the laboratory tests should be interpreted (in addition to code requirements) and that these specifications be reviewed before the first core is ever drilled.

With respect to core diameter, two references might be of interest on this subject in addition to the excellent list of references provided by Dr. Neville. The first of these is "Core and Cylinder Strengths of Natural and Lightweight Concrete," by Richard H. Campbell and R.E. Tobin, which appeared in the April 1967 ACI Journal *Proceedings* (V. 64). The second article, "Does Core Size Affect Strength Testing?" by Ahmed Ahmed, was published in the August 1999 issue of *Concrete International* (V. 21, No. 8). A request has been made to ACI Committee 318 to revise paragraph 5.6.4.4 with respect to acceptable strength values for various core diameters.

R.E. (Bob) Tobin, FACI
Structural Engineer
Arcadia, Calif.

Letter to the Editor No. 2
Variations in testing

In reviewing Dr. Neville's article on cores, I was concerned the reader may be misled as to what compliance means in regard to strength tests. It is my opinion that a compressive strength test value low at acceptance age (28 days) does not mean the concrete was not in compliance if the contract documents invoke the "Building Code Requirements for Structural Concrete (318-02) and Commentary (318R-02)" and the "Specifications for Structural Concrete (301-99)." As you know, variation in compression-test values can arise from three likely sources: variation in concrete-making materials, variation in batching of those materials, and testing technique. Testing technique can provide large variations, as the data in Table 7.2.3 show. Some practices may impact core test values as well – perhaps getting proper data is not as easy as it may seem.

ACI 318 expects that test results will often be less than the specified strength and provides guidance when they are:

5.6 – Evaluation and acceptance of concrete

Commentary – An effort has been made in the code to provide a clear-

Table 7.2.3—One single batch of six-sack, non-air-entrained concrete was sampled and tested

Proper curing and testing	Poor sampling, curing or testing	
6490 psi (44.8 MPa)	Poorly consolidated	2570 psi (17.7 MPa)
6190 psi (42.7 MPa)	Froze	3010 psi (20.8 MPa)
6090 psi (42.0 MPa)	100 °F dry curing	5290 psi (36.5 MPa)
6190 psi (42.7 MPa)	Sloped top	4950 psi (34.1 MPa)
6480 psi (44.7 MPa)	Vertical embedded No. 4 rebar	6580 psi (45.4 MPa)
6510 psi (44.9 MPa)		
5980 psi (41.2 MPa)		
6440 psi (44.4 MPa)		
6250 psi (43.1 MPa)		
6330 psi (43.7 MPa)		
6295 psi (43.4 MPa) Average		
530 psi (3.7 MPa) Range		
184 psi (1.27 MPa) Standard deviation		
2.9% Coefficient of variation		

cut basis for judging the acceptability of the concrete, as well as to indicate a course of action to be followed when the results of strength tests are not satisfactory.

5.6.3.3 – Strength level of an individual class of concrete shall be considered satisfactory if both of the following requirements are met:

(a) Every arithmetic average of any three consecutive strength tests equals or exceeds f_c';

(b) No individual strength test (average of two cylinders) falls below f_c' by more than 500 psi when f_c' is 5000 psi or less; or by more than 0.10 f_c' when f_c' is more than 5000 psi.

Commentary – Strength tests failing to meet these criteria will occur occasionally (probably about once in 100 tests) even though concrete strength and uniformity are satisfactory. Allowance should be made for such statistically expected variations in deciding whether the strength level being produced is adequate.

5.6.5 – Investigation of low-strength test results

5.6.5.1 – If any strength test (see 5.6.2.4) of laboratory-cured cylinders falls below specified value of f_c' by more than the values given in 5.6.3.3(b) or if tests of field-cured cylinders indicate deficiencies in protection and curing (see 5.6.4.4), steps shall be taken to assure that load-carrying capacity of the structure is not jeopardized.

5.6.5.2 – If the likelihood of low-strength concrete is confirmed and calculations indicate that load-carrying capacity is significantly reduced, tests of cores drilled from the area in question in accordance with "Method of Obtaining and Testing Drilled Cores and Sawed Beams of Concrete" (ASTM C 42) shall be permitted. In such cases, three cores shall be taken for each strength test that falls below the values given in 5.6.3.3(b).

5.6.5.4 – Concrete in an area represented by core tests shall be considered structurally adequate if the average of three cores is equal to at least 85% of f_c' and if no single core is less than 75% of f_c'. Additional testing of cores extracted from locations represented by erratic core strength results shall be permitted.

Commentary – Core tests having an average of 85% of the specified strength are entirely realistic. To expect core tests to be equal to f_c' is not realistic, since differences in the size of specimens, conditions of obtaining samples, and procedures for curing, do not permit equal values to be obtained.

5.6.5.5 – If criteria of 5.6.5.4 are not met and if the structural adequacy remains in doubt, the responsible authority shall be permitted to order a strength evaluation in accordance with Chapter 20 for the questionable portion of the structure, or take other appropriate action.

Regarding testing, two cylinders should be tested at acceptance age, but often are not. ACI 318 defines a strength test as the average of the strengths of two cylinders made from the same sample of concrete. Equally important, the two strength values can be used to assess the testing quality. Wide scatter in companion values indicates poor testing techniques.

Finally, I recommend casting a hold cylinder. If a low value is obtained at acceptance age, the hold cylinder could be cured further. The results from the hold cylinder would give the engineer more data in the evaluation of the member and could lead to less coring.

Thank you for your thoughtful article.

Richard D. Stehly, FACI
American Engineering Testing, Inc.
St. Paul, Minn.

Letter to the Editor No. 3
Easy to perform?

This is another superb and very informative contribution by Dr. Neville. I must take exception with one theme of the article, however, namely that the procurement of core specimens is "easy to perform," as stated in the title and further expressed in the opening paragraph. In the United States, contractors specialized in diamond-cutting often perform the procurement of test cores. The ability of most drill operators is quite high as a result of rigorous training promoted by the Concrete Sawing and Drilling Association. This notwithstanding, the fact is that most core drilling in concrete is for the purpose of "making a hole," usually to place a pipe or other insert through. Thus, the drill operators go about their work in a manner that will most efficiently make a hole.

Unfortunately, making a hole does not always result in a good-quality core. Although positive anchorage of the drill to the coring surface is imperative, some slight movement may be tolerable when making a hole. It's not acceptable when procuring a core for test evaluation, as grooves or corrugations in the core walls will result. This may well explain Dr. Neville's speculation about the consequences of coring when a drill is anchored to a vertical surface. This configuration produces horizontal cores typically having lower strength than if they had been taken vertically.

Another common problem is the procurement of cores from elements that are more than 1 ft (0.3 m) thick. Commonly used core bits are 14 in. (350 mm) long. They are only able to penetrate about 12 in. (300 mm), however, due to the thickness of the end cap, which connects to the power source. The standard procedure when making a hole greater than 12 in. deep is to withdraw the bit once it bottoms out, break off the core, and remove it from the hole. With an extension rod added to the bit, that sequence is repeated until the full thickness has been penetrated. Although not commonly employed, longer core bits are available, as are core bits with separate flush-threaded extension barrels; however, these must be special-ordered.

In an extreme instance of this limitation, I was retained several years ago to participate in the evaluation of a very large structure that was suffering distress while still under construction. Prior to my involvement, a core had been taken through the 5-ft-thick (1.5 m) web of a large beam. A coring contractor, hired by the materials laboratory retained by the design engineer, procured the core.

The laboratory had a technician on site during the coring who was afraid of heights. The lab technician reportedly told the core driller where to obtain the core, but did not witness its procurement. What resulted was a milk crate full of core pieces with no individual piece more than 8 in. (200 mm) long. This, in addition to the results of other substandard testing, led to considering demolition of the structure.

Under my supervision, a second core specimen, using continuous coring equipment and taken immediately adjacent to the original core, was procured. We removed the core in a single piece and found it consisted of high-quality concrete throughout.

Because of these potential problems, it is my opinion that the procurement of core specimens is not "easy to perform" and further, that a fully qualified professional should be on site to supervise any core acquisition. Fortunately, most core drillers are fully competent and will procure quality cores if only they are made aware of the core's intended use. This is equally applicable to both vertical and horizontal cores. It is extremely important to explain the significance of obtaining a good core specimen to the core driller prior to starting work.

James Warner
Consulting Engineer
Mariposa, Calif.

Neville's response

It is always pleasing to the author of an article when readers send in letters with their comments; it is even more pleasing when the comments are positive. Thus, I am grateful to Bob Tobin, Richard Stehly, and James Warner for their extensive contributions.

I am particularly taken by Bob Tobin's idea that specifications should state how the laboratory tests should be interpreted and that these specifications should be reviewed before the cores are taken. This very much echoes my comment to the effect that, if we are not definite about interpreting the core tests results in a meaningful and unequivocal way before the cores have been taken, then the dispute about the quality of the concrete in the structure is simply shifted to a dispute about the concrete in the cores.

Bob Tobin's reference to an article by him and Richard Campbell on core and cylinder strengths, published in the ACI Journal in April 1967 has enabled me to reread that article. A particularly apposite comment in that article is: "It should also be recognized that concrete strength determination is not a precise mathematical science and that trends are more significant than exact numerical values." This should be borne in mind by all readers.

Richard Stehly provides very useful information on the variability of test results on concrete specimens and on the causes of that variability. He refers to various subclauses of 5.6 in ACI 318 and the related Commentary. His suggestion of making a hold cylinder is well worth considering in some situations. Overall, I would like to thank Richard Stehly for his helpful extension of my article.

James Warner is justified in upbraiding me about the title of my article: I admit to a tendency to choose catchy titles for my articles so that people may be encouraged to read them. Having said that, I am of the opinion that performing core tests is far easier than interpreting them. James Warner says that "most core drillers are fully competent," which agrees with my saying that taking of cores is "usually undertaken by skilled specialist personnel." I fully agree with James Warner's views on the need for a professional person to supervise the taking of cores, but I doubt that it is practicable to supervise the taking of every core.

I tried to emphasize the importance of a professional engineer in deciding the location of cores and in interpreting the test results. Without the professional's presence, the tester may well usurp that role.

Turning to the occasional problems in taking very long cores, I too have come across sheer incompetence. The problem was not with the driller, but with the engineer who interpreted the results. The cores were broken every 12 in. (300 mm), when an extension rod had to be installed on the bit, and this expert interpreted the breaks as an indication of a zone of weakness in the parent concrete. The expert was highly knowledgeable about structural analysis, but seemingly lacked site experience. I hope that James Warner's comments will encourage a greater involvement of professional engineers in core tests, which may have decisive influence on the acceptance of a structure.

Adam Neville

Chapter 8

Research

As I said in the Preface, the goal of this book is to achieve an improvement in concrete construction. Clearly, research is one of the means of progress toward that goal, but what kind of research?

Given the practical nature of my goal, I am not concerned with what is sometimes called pure scientific research, which aims at improving understanding the various *basic* phenomena involved in the behavior of concrete. That kind of research is valuable in scientific—I am tempted to say, cultural—terms but it is not part and parcel of the construction industry. My concern is with applied research. Section 8.1 indicates that a considerable proportion of research undertaken in civil engineering departments in universities does not contribute, even indirectly, to achieving better concrete structures.

Reasons for this can be found in the nature of the Ph.D. degree, which makes for very small projects that have to be completed in a couple of years. Moreover, most of these projects are laboratory based and are performed on small specimens under a narrow range of conditions. Furthermore, many supervisors of these projects have limited practical experience of construction, or even none. It is not surprising, therefore, that these supervisors often extend their own Ph.D. investigations, which may be of limited value, or even none.

In some cases, the research involves a study of materials that are no longer used or even available. I am aware of extensive studies of gravels in England: their size, natural grading, surface texture, petrographic nature, shape, and bond properties. In the early days, when gravels were abundant in English rivers, the findings of the research were useful. But the research continued for many years into a period when gravel deposits were either exhausted or protected from dredging and excavating by ecological considerations enforced by the law of the land. It was only the researcher's retirement that put an end to his life-long project.

I accept the fascination of such a single and prolonged project, but I doubt that it is research "likely to improve concrete," to use the words in the title of Section 8.1. I can imagine a parallel project in the European Union of the twenty-first century where 12 countries have a single currency: a study of durability of the banknotes of the old currencies, different in each country. This would be economic history, if that, but not practical economics.

So, Section 8.1 pleads for research whose goal is achieving better concrete in practice. This plea is developed in Section 8.2, but there is no exhaustive list of "projects to be undertaken" presented there. Rather, I point out practical problem areas such as: the variability of water content in batching, variability of aggregate properties, variability in cement, inadequate curing, strength testing, choice of appropriate test parameters, and inappropriate testing. Clearly, this is not an exhaustive list, but rather a prod in what I believe to be the right direction.

Although this is not, strictly speaking, a research topic, the need for improved training and education of all those involved in concrete—from operatives to designers—is mentioned. I have to admit that this theme recurs in a number of sections.

Section 8.3 also deals with good research and not-very-useful research. In the latter category, I mention replicating other people's tests but using local materials. This may be of value, but it is hardly research. My point is: yes, such work is worth doing, but it should not pretend to be research leading to the accolade of a Ph.D. degree.

The main thrust of Section 8.3 is to recognize that research needs to be conducted both on a micro- and a macro-scale. My impression is that, because of the nature of the Ph.D. degree in our universities, more work is done on the micro-scale than on the macro-scale. A shift would be beneficial with respect to achieving better concrete structures.

Whereas research is about doing things, it is of use only if its outcome is communicated to others. The means of communication is by publications. But publications are not just the researcher's hand-written test results or computer print-outs of numbers or graphs. The data have to be not only processed but also interpreted.

It follows that the value of a published outcome of research is influenced by the quality of the publication. By quality I do not mean the standard of the language, let alone appearance, although a good quality of both of these helps and encourages the reader. What is crucial is reliability, veracity, and completeness in reporting the outcome of tests and studies. Research rarely starts with a blank sheet of paper or without pre-conceived notions: just doing tests and seeing what comes out of it is a naïve approach and likely to be a waste of money. More than that, prior to starting the experimental work, the researcher should have a clear idea of what she or he is looking for, and this is by way of a hypothesis.

Now, the formulation of a hypothesis is likely to be rooted in a study of existing knowledge, that is, a review of publications. For this review to be balanced and objective, *all* the relevant publications should be consulted and they should be listed at the end of the paper reporting the author's research. Most researchers are scrupulous in following this approach, but occasionally the desire to prove a point may lead someone to ignore, discard, or hide those publications that oppose the author's thesis.

This need for objectivity in citing references is discussed in Section 8.5. The original paper that forms that section was written at the invitation of the editor of the Spanish journal *Materiales de Construcción* for its Fiftieth Anniversary issue. In presenting my paper, Professor M. Fernández Cánovas wrote:

> Although the article written by Prof. Neville is entitled "Objectivity in citing references" it has much more to say than is indicated by the title. It is a very serious analysis—although always accompanied by the good sense of humor that characterizes the author—that obliges us to think seriously as investigators about the objectivity with which we present or investigations, whether these are by articles, monographs, lectures, or in conferences etc.
>
> The opinions expressed in this article by Prof. Neville oblige us to meditate upon and, I would say, thoroughly examine our conscience in relation with the objectivity that we use, since, as he indicates, it can be very dangerous, as often is common practice in some fields of research, to only say half the truth, as, sometimes, this can unintentionally turn into lies.

Still on the subject of publications, there is included in this chapter Section 8.4. This discusses how papers and articles submitted for publication in a journal, or major works intended to be published as a book, are reviewed by editors, editorial boards, or by persons selected on an *ad-hoc* basis. The process is sometimes referred to as a peer review, and it is interesting to ponder on who is a peer (and I am not referring to the nearly defunct Upper House of the British Parliament) and what criteria are used in the peer review.

Most people have very little knowledge of the peer review process, and they may find some useful pointers in Section 8.4 and in the letters to the Editor of *Concrete International* at the end of that section.

Chapter 8: Research

Section 8.1: Is Our Research Likely to Improve Concrete?

This is not a criticism of recent research in the broad field of concrete, but I have had occasion to review publications of the last 13 years, and I want to ask some fundamental questions. A discussion of these questions could well be beneficial to the users of concrete.

The reason for my literature review is that my book, *Properties of Concrete,* which had been first published in 1963, and was in its third edition, became out of date. In 1994, I was therefore writing a completely new edition, not just a bolt-on version.

Much has changed since the last revision in 1981: many new standards and recommended practices have been written, and thousands upon thousands of research papers have been published, but there has been only a modest number of novel and important practical developments. Alas, this situation contrasts sharply with the rate of change in fields such as communications. As I see it, the goal of ACI, as well as of researchers, is primarily to improve the quality and use of concrete. Few of the thousands of papers published contribute fully to that aim. However, before justifying this view, I must emphasize that I view basic research on cementitious materials as most valuable: an improvement in the understanding of these materials and of the processes involved in hydration is of great importance *per se*. This is a contribution to the stock of human knowledge, and, as such, is valuable. However, the bulk of research should have a practical application in sight, not necessarily immediately, but within a foreseeable future.

What then is my criticism of the papers to which I referred earlier? Many of them are the outcome of a single Ph.D. project, which of necessity was confined to studying the effect of changing one or two variables at a time, using a very limited range of concrete materials, and often conducted at a temperature of 20 °C (68 °F), which is comfortable for the researcher. Real-life conditions are not like that. It is very difficult to generalize from these individual projects, which often have no set of conditions in common. This is why, having read a number of papers concerned with a single property of concrete, I was not able to distill the facts reported into generalized statements of the kind that a book reader wants.

Worse than that, many papers give conclusions about the influence of x on y which are valid at the fixed values of the other parameters, sometimes chosen quite unrealistically. These conclusions are subsequently cited and become embedded in the literature as if they were generally valid; there is a real risk in this.

My further criticism of some papers is that, under the heading "Conclusions," they mix new findings with conclusions like "the modulus of elasticity increases with strength." Is there a real need for confirming what is known, sometimes even without admitting that a conclusion simply is a confirmation of well-established knowledge? Perhaps it is not even a question of "admitting" but rather of a lack of familiarity with the existing literature.

Am I just being critical? I don't think so. Although we make better concrete now than in the past (but not always), there have been few breakthroughs in the last half-century. Among those that I consider as worthy of that classification are: air entrainment (discovered accidentally), superplasticizers (with a 40-year delay between invention and modest use), and high-performance concrete.

A diagnosis without a remedy is only of limited value. Were a simple remedy available, it would have been applied by now. There are, however, two actions that may help. One action is controlling the money for research. In many cases, the sole criterion for providing funds for university research is the personal ability of the researcher (a condition necessary but not sufficient) without much consideration of the purpose and the potential usefulness of the project. The proposal is judged by other researchers—the so-called peer review—whose interests are academic, in both senses of the word. There are of course some honorable exceptions. It would improve the situation if funding were conditional not only on a positive academic evaluation but also on the judgment from an industrial standpoint, which could be expressed by a financial input.

To put it bluntly, we should not provide funds for projects where the researcher cannot suggest a possible practical value of his or her work; nor should we support projects which are very similar to past work and which differ only in that a local material is used or the test temperature is higher or lower than before. Such investigations are of practical use but they do not constitute research.

The second "remedy" is linked to the preceding criticism: the nature of the narrowly structured and limited-scope papers arises from the fact that they are really abstracts of Ph.D. theses. A Ph.D. project, by its nature, has to involve experimental work that can be completed in less than two years. Consequently, the scope must be limited because the doctoral candidate has to submit a dissertation based on work done personally. This is what university regulations require. Need it be that way? A Ph.D. study is a training in research methods and need not necessarily be a free-standing research project or "discovery." If university regulations allowed research students not only to be part of a large and well-coordinated project team with longer-term objectives, but would also abandon the requirement of an individual doctoral dissertation, trivial work would be done away with and coherent and useful knowledge would be gained. This may seem revolutionary but is certainly not impossible. In this connection, it may be of interest to note that several British universities have introduced an "engineering doctorate" with emphasis apart and away from the traditional Ph.D. dissertation.

Chapter 8: Research

What kind of research do I see is badly needed? I am limiting myself to concrete used in the majority of structures, from buildings to bridges, from pavements to marine works (specialized structures such as dams, let alone lunar concrete, are another matter). What is needed is the application of some existing technologies to improve the actual uniformity of concrete produced. The concrete mixture selected in the laboratory may well be perfect, but site concrete often suffers from excessive variability. It is either the workability that varies or, if the workability is maintained constant as it must be for the concrete to be placeable, the strength varies. A common reason for this is the variation from hour to hour, or even batch to batch, of the moisture content of the aggregate or of its gradation. With respect to moisture content, there exist means of determining it during batching, but they are not routinely used. Possibly, further development is required, but what is really needed is the will to use such a technology. Of course, this costs money, but the cost can be offset by savings arising from a reduced standard deviation of the strength of the concrete. Moreover, better production control would allow us to move from the almost obsessive dependence on the 28-day strength test, aptly described in French as the engineer's *bagage culturel*.[1]

Most aggregate used these days is crushed, and crushing produces dust. Variation in the dust content results in a variation in water demand, and hence in a variability of concrete. Admittedly, the aggregate should be washed, but this is difficult in countries where fresh water is in short supply or is expensive. Moreover, the dust adheres to the larger particles during rain so that the dust content varies with the weather. The aggregate varies also owing to other reasons. What is needed is a means of a full quantitative description of the gradation, shape, and texture of the aggregate at the instant of its being conveyed into the batcher.

Elements of the relevant technologies exist in the chemical and mining industries, and what is required is the development of an operational technique for concrete batching. This way, continuous adjustment of mixture proportions can be achieved, with a consequent reduction in the variability of the concrete as produced. It is worth repeating that a lower variability means a smaller margin between the minimum strength and the target strength; hence, a reduction in production cost is possible.

How is this development to come about? Help may well come from an unlikely source: in many countries, there is an environmental and political pressure to use recycled aggregate from demolition. This aggregate will need careful classification and beneficiation; the techniques which will have to be developed will be applicable to ordinary crushed aggregate as well. Also, a whole new automated batching plant with continuous adjustment is needed for producing concrete with a low variability.

Who will undertake this development? I do not see much hope of practical "discoveries" in universities, which are a refuge for some who are reluctant to make decisions for which they may be held responsible. As well, development requires risk money and marketing. But, as long as no one firm has such a plant, other firms do not feel pressurized to have it.

So it is only a large conglomerate operating quarries and ready-mixed concrete plants that might become the trend setter. I have argued for making better concrete through the use of knowledge and technology now available, or nearly so, rather than for things new and wonderful. This is how it should be with a cheap mass material. Research in materials should be governed by practical, or economic criteria. An excellent book on this subject was recently written by Ashby[2] but, because this book deals with mechanical design, it may not be known by those concerned with concrete. These are the consequences of our pigeon-holed lives.

The fundamental point is that the selection of materials from among all those that are appropriate is nearly always on the basis of cost of the product, and not of the cost of the material alone. There are two broad possibilities. With some products, for example, a contact lens, the material cost of the glass is tiny but the value added in manufacture is very high. Consequently, research concentrates on developing materials that will improve the performance of the lens, virtually regardless of the cost of the material.

At the other extreme, where the product is also made of glass, a major share of the production cost of fiberglass insulation is the cost of the material, which uses a great deal of energy. The situation is similar when cement and aggregate are converted into concrete, which in turn is converted into a structural member. The cost of concrete represents, on average, about one-half of the cost of the structural member. The value added in making the member is relatively low, but the market is huge. It follows that the cost of the materials is very important: they must be good enough but it would be nonsense to transport "the best" aggregate over long distances.

From the preceding discussion it is clear that, as far as concrete is concerned, what is economically viable is improvement in the processing of existing materials for the purpose of reducing cost or improving quality control and reliability. Such improvements were made in cement manufacture; now is the time to make improvements in the production of concrete itself.

Are we moving in the right direction? I doubt it. Numerous conferences are held to discuss advances in concrete technology and to peer into the next century. Sometimes, the conferences are highly specialized so that academic cement chemists speak only to academic cement chemists and do not hear what the concrete producer is looking for. At other times, those attending the conference represent a broader spectrum, but the operation of parallel sessions means that, in reality, attendance at sessions is by narrow specialisms. It is not surprising that the discussion during the coffee breaks also is in cliques.

It is obvious but it bears repeating: concrete is not just cement with aggregate thrown in to make the cement go further. Much of the present-day research is like laboratory-made experimental bricks that are useful for the purpose of making better building bricks, but the true objective is to make a better brick wall. What is needed is a critical evaluation, with no punches pulled, of research in terms of its relevance to practice. This is not thought to be a worthwhile activity by many academics; anyway, it cannot be undertaken by those who are not in practice for real, that is, for money. Such critical evaluation is not performed often enough, and that is why much research, once published, is lost forever. Sad to say, some deserves to be.

Something needs to be done to improve the situation. I do not pretend to offer a panacea and can only hope that my strong, but hopefully not strident, plea will encourage ACI to use its influence to deflect research toward the goal of achieving better concrete in practice.

References

1. Rossi, P., and Wu, X., "Comportement en Compression du Béton: Mécanismes Physiques et Modèlisation," *Bulletin liaison Laboratoires Ponts et Chaussées,* V.189, Jan.-Feb. 1994, pp. 89-94.

2. Ashby, M. F., *Materials Selection in Mechanical Design,* Pergamon Press, London, 1992.

Letter to the Editor

Re: "Is Our Research Likely to Improve Concrete?" by Adam Neville in the March issue of *Concrete International*—Amen!

Richard C. Elstner
Wiss, Janney, Elstner Associates
Honolulu, Hawaii

Section 8.2: Suggestions of Research Areas Likely to Improve Concrete

In the March 1995 issue of *Concrete International*, an article[1] was published under the provocative title "Is Our Research Likely to Improve Concrete?"; this is in Section 8.1. It is difficult to tell whether the disapproval of the article was so strong that voicing disagreement was thought to be unnecessary; or whether the approval was so universal that expressing it was thought to be superfluous. In the end, only one discussion letter was published.[2] It consisted of a single word: "Amen." If nothing else, this is the acme of brevity.

What I would like to do now is to provoke a discussion on what worthwhile research should be undertaken for the purpose of improving our concrete in practice. I should emphasize that I am not looking for great new ideas that would take many years before they bear fruit; rather, I am concerned with the use of existing knowledge and even technology for the purpose of making better concrete possibly as early as the end of the millennium. Nor am I looking for variations on the theme of the various relationships of concrete properties. As it happens, a new book discussing these has just been published.[3]

I propose to consider 11 areas; their order of presentation does not reflect their relative importance, technical or financial.

Portland cement

I shall not discuss the production of cement, with just one very minor exception. After all, production of cement is primarily the concern of cement producers who have effected enormous improvements in the process of manufacture. These improvements are in energy use, and therefore in production cost, and also in terms of environmental pollution because this is mandatory these days. In other words, the driving force is efficiency and hence profit, as must be the case with any manufactured product. Nevertheless, the changes in the production of cement did not have as an objective the achievement of a product that would make a better concrete. Indeed, modern cement does not produce a better concrete, and sometimes produces a worse concrete than was the case with the cement of the 1950s and 1960s.[4] In essence, we do not have portland cement that would enable us to make concrete which is better than in the past. The only exception to this is cement with rounded grains developed in Japan, but the cost of production, which is a batch process, is prohibitive, at least so far. However, one can readily imagine an improvement in rheological properties of fresh concrete made with such a cement.

The preceding paragraph might be thought to be an accusation in that the cement manufacturers have not improved their product in a way parallel to enormous improvements in, say, telecommunications. It is not an accusation because the question to ask is whether such improvements in cement are worthwhile or even necessary. Let me develop this a little further.

Two broad types of concrete

By far the largest part of concrete produced in the world is used in routine construction for which fairly ordinary properties of concrete are required. These properties are generally limited to compressive strength and durability, and they can be achieved using present-day cements. It would be uneconomical to produce more expensive cement whose superior properties would not be exploited in the majority of applications. The present portland cement, often with other cementitious materials, is satisfactory. Improvements in what we might call ordinary concrete lie in improvements in aggregate, in batching, and in curing. I will consider these further on.

The second type of concrete is what has become known as high-performance concrete. Here, a better cementitious material is needed, but this material can be produced by judicious blending of portland cement with fly ash, ground-granulated blast-furnace slag and, above all, silica fume. In addition, in most cases, superplasticizers are also needed. I propose to discuss high-performance concrete later on.

Ordinary concrete—training requirements

What is needed to make satisfactory ordinary concrete? There is no mystery or complexity about the selection of mixture proportions to achieve satisfactory routine properties. The selection requires experience, coupled with knowledge of the influence of various factors upon the properties of concrete; this knowledge must be based on an understanding of the behavior of concrete.[3] If this is so, the reader could ask why a significant proportion of concrete in service is not satisfactory. I strongly suspect that, all too often, those who make concrete have ample experience, possibly some knowledge of the factors involved, but they lack the underpinning understanding. The reason for this lies in the inadequate education of those involved in concrete making.

Inadequate knowledge of factors influencing the behavior of concrete has harmful consequences in the operations of manual and technical staff. It is not infrequently thought that any fool can make concrete, but concrete is not a foolproof material.

The lack of understanding has its worst consequences at the professional level of the person in the design office or in charge of construction. This situation exists because learning about concrete is considered almost below the dignity of the person undertaking sophisticated structural calculations. And yet, these calculations involve parameters such as modulus of elasticity or the entire stress-strain relation, shrinkage, creep, temperature changes, or thermal movement; these parameters can be significantly affected by changes in the composition or even in concreting operations.

So my suggestion for improvement in the future is to undertake research into new and attractive methods of better training and education of all those involved in concrete construction. This costs money, but it is money well spent. It is a sound investment in terms of improving the quality of concrete and therefore promoting its more positive image.

Variability of water content in batching

Inadequate training and lack of knowledge possessed by concretors, important though they are, are not the only, or even the main, reason for poor concrete performance. Even when an excellent concrete mixture has been selected, concrete is found from time to time to be unsatisfactory. This may be by way of incorrect workability at the time of placing; or, a few weeks later, of inadequate strength; or, much later, by way of a lack of durability. The cause of any one of these must be sought in the variability of the concrete actually batched. This statement may come as a surprise, given the high quality of modern batching equipment. Whereas it is true that the mass of each ingredient, that is, cementitious materials, admixtures, water added as a liquid, and the various aggregate fractions, is the same in each batch, there are variations in the total mass of water and in the actual properties of aggregate.

Let me consider water first. The aggregate, especially its fine fraction, in most cases contains absorbed water and some surface water; in other words, the condition of the aggregate is usually other than saturated-and-surface-dry, so beloved in the laboratory. There exist various methods of determining the moisture content of the aggregate, but most of them are used only sporadically, perhaps once or twice a day, and at best every hour. It is hardly necessary to explain that in an aggregate stockpile the moisture content varies vertically, varies with changes in the weather, and varies with the passage of time.

There exist electrical devices, measuring resistance or capacitance, which give instantaneous or continuous reading of the moisture content of the aggregate in the storage bin, and these devices could automatically control the amount of water added into the mixer. However, an accuracy greater than 1% cannot be achieved in practice and frequent calibration is also necessary. Microwave absorption meters have also been developed; they are accurate and stable but expensive. Instruments emitting fast neutrons that are thermalized by hydrogen atoms in the water are also available, but I am not aware of their performance or even of their widespread use. All these instruments need to be carefully positioned so that they give a representative reading and yet are not damaged or do not interfere with the flow of materials into the mixer.

My view is that the use of continuous measurement of moisture in the aggregate coupled with automatic adjustment of the amount of added water would greatly reduce the variability of the resulting concrete. A lower variability readily translates into economy of cement in that the difference between the mean value and the specified or the necessary minimum cement content would be reduced. I am therefore advocating the development of a practical and reliable continuous moisture meter as a highly desirable development that would pay for itself from economies in batching. Moreover, the mere fact that a particular concrete supplier can virtually guarantee a consistent quality of concrete and reliability of supply would result in a financial premium.

Variability of aggregate properties

I would now like to turn to aggregate, and I realize that my remarks may well prove controversial and may even be regarded as unfounded. In many countries, the supply of natural aggregates in comminuted form is very limited or non-existent. The riverine supplies have become exhausted and the opening of new pits is environmentally unacceptable: the demand for shallow lakes for pleasure sailing is limited. The next best source of aggregate is crushed rock. This can produce excellent coarse aggregate and even fine aggregate, the latter possibly blended with natural sand. But my main concern is with crushed coarse aggregate.

On the basis of my experience on a number of very major projects in several countries, I am concerned about the variable dust content of such aggregate. This variability has a definite effect on the water demand of a given mixture: as the dust content goes up, the workability goes down. But a constant workability is essential for the concrete to be properly compacted by the means used on a given project, so that the water content of the mixture has to be increased in order to satisfy the water demand. A consequence of that is an increased water-cement ratio and a reduced strength.

Problems in dust control exist in a number of countries. Sometimes, they are the consequence of unavailability of water for washing the aggregate, or of a very high price of the water; at other times, the execution of washing is variable or simply poor because of inadequate supervision or even due to the belief that washing the aggregate is not an important task. But even on a well-regulated project, there are problems. There may be several reasons for this.

For example, the extent of adhesion of the dust to the coarse aggregate particles varies with the weather. More importantly, the amount of the adhering dust varies with the shape and texture of the aggregate, and with the process of crushing. Adhering dust can become detached by abrasion during mixing.

This brings me to the problem of variability of the shape of coarse aggregate in consequence of the crushing operations. It is known that the type of crusher used and the sequence of crushing affect the aggregate shape. So does the wear of the crusher. I have seen a situation where a brand-new crusher and a well-worn one operated in parallel. The shape of the aggregate particles produced by the two crushers varied but, entirely unavoidably, the delivery pattern of aggregate was such that there was no telling from which particular crusher came the aggregate being batched into the mixer at any instant. In consequence, what was a well-graded and properly screened aggregate produced an apparently randomly variable concrete mixture in terms of its rheological properties.

You may well feel that this situation is unavoidable. If it is, then someone has to waste money by selecting a mixture on the assumption of the worst possible case of aggregate shape; this costs, or more correctly wastes, money if you are making ordinary concrete. If you are trying to make high-performance concrete, then it just cannot be achieved.

The corollary to my preceding remarks is that, to make concrete of consistent quality, we need to develop practical means of measuring the shape and surface properties of aggregate and also better means of controlling the dust content. If we had such means, we could adjust, on a continuous basis, the mixture proportions so as to ensure uniformity of the rheological properties at a constant water-cement ratio. Such development is unlikely to be very easy or very rapid but a large concrete supplier armed with such means could become extremely successful in terms of his market share. As against that it can be argued—indeed, in some countries it is argued—that, as long as nobody supplies concrete with a narrowly consistent quality, nobody need to do so, and there is even a financial disbenefit in being better than everybody else.

There is one additional reason why better aggregate classification will become important: the expected much-increased use of recycled concrete and other waste. The physical properties of recycled aggregate can be very variable, and they therefore need to be properly determined. The use of such aggregate is already being promoted by legislation and by fiscal means in some European countries, and it is bound to increase because of two-fold environmental benefits: a reduction in quarrying and a reduction in landfill.

Quality control for high-performance concrete

In the preceding paragraph, I quoted a view that there is no benefit in doing better than others. However, I can give an example disproving this negative and miserable proposition. The example arose from the use of high-performance concrete, for which the production of concrete within very narrow tolerances was clearly required.

One particular ready-mixed concrete supplier who took pride in his workmanship developed the means of producing concrete with consistent properties in terms of workability and an extremely low water-cement ratio, and he supplied such concrete for the construction of that part of the structure which required high-performance concrete. It is said that the supplier lost money on that particular concrete. But he gained something else. First of all, he also won the contract for the supply of ordinary concrete for the rest of that particular project, even though his price per cubic meter was a penny or two higher. But there was more to it than that: the benefits of superior production control in his ready-mix plant enabled him to reduce costs when supplying concrete to a traditional specification.

I would like to make one additional remark applicable in countries where there is much vertical integration in terms of the production of cement, aggregate, and ready-mixed concrete. Such integration makes improvements in the consistency of the quality of aggregate and quality of batching operations practicable and economically worthwhile.

Moist curing

Curing of concrete has been an integral part of concreting operations for so long that you would not expect it to come under the heading of needed research. I am not referring to the effect of curing on the microstructure of hardened cement paste, but to the methods of ensuring that concrete is cured adequately.

Curing affects primarily the concrete in the cover to reinforcement, and it affects the actual quality of the concrete rather than its potential quality. Moist curing is essential for the cement to hydrate as much as possible; it is worth adding that full hydration is not necessary and, at low water-cement ratios, is impossible.

Curing is invariability specified but it is rarely achieved. Like batching and mixing, curing requires close supervision, perhaps even more so because, in retrospect, it is extremely difficult to prove that proper curing had not been applied. So it is not the specification that is at issue but the development of means of ensuring good practice.

It may be worth giving some reasons, admittedly not excusable, for the prevailing bad practice of inadequate curing.

First, curing is an operation which follows the end of the concreting operations; in consequence, there is a not-surprising desire to move on to the next phase of work.

Second, curing is seen by many as a silly operation, as a non-job: just sprinkling water, with nothing to show for it at the end of the day. In some climates, it can be argued that watering is not necessary because, any minute now, it will rain. In other climates it can be argued that water evaporates as soon as it has been applied to concrete, so that nothing is, and nothing can be, gained. Thus for one reason or another, why bother?

The third reason for not curing is that most personnel on site, often including even supervisory staff, do not believe in their heart of hearts that curing serves a really useful purpose. Many of them have never applied, or supervised, proper curing in all their lives, and they have been successful; so why change?

The fourth reason is the rather unkind argument that curing does not show: Who will know tomorrow whether the concrete was subjected to curing today? Perhaps, they argue that what the eye does not see the heart does not grieve over.

The fifth and last reason is the most compelling one, but it is also one that points toward a remedy: curing is not paid for as a separate item. To ensure good curing it would be worthwhile to develop a method of payment for it. This could possibly be by the consumption of ear-marked water, but a more clever method needs to be developed.

If I emphasize ensuring curing as a worthwhile topic for research, it is because curing can make all the difference between having good concrete at the end of the placing operation which becomes good concrete in service on the one hand, and on the other having good concrete ruined by the lack of a small effort. The importance of curing is nowadays even greater than in the past for three reasons:

First, modern cements, with their higher rate of gain of strength have unwittingly allowed a worsening in the curing practice. The explanation is as follows. Because strengths adequate for the removal of formwork or for trafficking the surface of the concrete are reached very soon, there is an excuse to discontinue effective curing at a very early age.

The second reason is that lower water-cement ratios are used than was the case in the past, and to prevent self-desiccation, ingress of water into the concrete is necessary.

Third, modern mixtures often contain fly ash and ground-granulated blast-furnace slag. These materials, especially the latter, react over longer periods of time, and consequently need prolonged curing.

Determination of the variability of cement

Even though I said at the outset of this section that I would not discuss improvements in the properties of portland cement as such, I would like to say something about the exploitation of the properties of cement as actually produced. I realize that what I am going to say may be inapplicable in some countries.

My concern is with the representatives of cement samples. Cement manufacturers may well insist that cement samples are taken scrupulously so as to be fully representative of the totality of the cement produced. Of course, but which totality? This is what I want to consider.

It is inevitable that there is a variation in the strength of cement from a given plant in consequence of the variability in the raw materials even within a single pit or quarry. Furthermore, differences in details of the process of manufacture, even though modern cement plants are highly sophisticated, and, above all, differences in the ash content and in the sulfate content of the coal used to fire the kiln, also contribute to the variability of the cement. This need not present problems *per se*, but a large variability prevents exploitation of, say, a high strength of cement because the possibility of an occasional lower strength must be allowed for.

The variability of cement from a single source is measured in the United States and in some other countries by ASTM C 917-91a, which uses mortar cubes and relies on a moving average of five grab samples, called also spot samples. It is the nature of sampling that is of crucial importance, and I shall shortly come back to it. However, let me first put the variability in actual terms: tests[5] at 87 United States cement plants performed in 1991 showed that 81% of them had a standard deviation of the seven-day strength lower than 2.10 MPa (305 psi); at the age of 28 days, only 43% of the plants had a standard deviation lower than 2.10 MPa (305 psi). The increase in standard deviation with age is typical of American cements[6] but not necessarily of cements made in other countries.

My view of grab samples is as follows. Clearly, a single grab sample would not be representative and would also be unduly affected by testing errors. It is therefore logical to use composite samples which are obtained by putting together subsamples from production over a certain period of time. However, if the composite sample represents production during 24 hours, the smoothing out of strength values is excessive. What the result of tests on the composite sample represents is the average properties of many thousands of tons of cement. Inevitably, there are variations within that bulk of cement, but only a very small part of it is used in a given part of a project. Thus, the properties of the composite sample are uninformative with respect to a given shipment of cement; even the cement manufacturer generally has no knowledge of the properties of any particular shipment of cement which represents less than 24 hours' production.

Cement manufacturers can shrug their shoulders and say: That's how we do it, and that's that! But it need not be. In the United States, some very large cement users such as State Highway Authorities have, in a few cases, negotiated contracts for cement such that the properties of cement were reported on a shipment-by-shipment basis. So it is possible for a large purchaser to lever the desired information when there is keen competition to sell. On the other hand, a monopoly or cartel situation can exist and result in a blank refusal to cooperate by all cement suppliers in a given area. The manufacturers' commercial strength lies in the fact that, for transportation cost reasons, cement generally does not travel far except by ship.

There is one other aspect of what I consider a lack of correspondence between the test sample results and the actual shipment which I view to be of importance, and this applies in research. The researcher may rely on the cement manufacturer's test results for parameters such as chemical composition or fineness. If these are not applicable to the cement used in the experiments, the researcher may find spurious correlations with whatever it is that he or she measures; alternatively, he or she may fail to find real correlations. More than that, the researcher is often not even aware of the fact that the manufacturer's certificate applies to average properties over 24 hours. The results of the research become part of the received wisdom, and yet they may be, to put it bluntly, wrong. Surely, this is in nobody's interest.

To demonstrate that I am not looking for non-existent problems, I can refer to personal experience.[7]

Strength testing of concrete

Continuing with the topic of testing, it may be appropriate to offer comments, admittedly not for the first time, about the standard 28-day cylinder test for concrete strength. My emphasis is on the age of 28 days. There is nothing sacrosanct about this age. A multiple of seven was chosen so as to avoid testing on Sundays, and the multiplier of four was chosen because the coarse-grained cement with a high content of dicalcium silicate which was produced three-quarters of a century ago gained strength only slowly. All this is well known, but we are still stuck with testing at the age of 28 days. As everybody knows, by then it is too late to remedy the situation if the strength was too low, and it is also too late to recoup a financial loss if the strength was unnecessarily high—hence, the search for an accelerated-curing test.

This search has continued for 40 years, and we are still talking about correlating the accelerated-curing strength and the 28-day strength for the purpose of predicting the latter from the former. And yet the actual strength of concrete in the structure is influenced not only by the concrete mixture used but also by the degree of compaction, bleeding, segregation, and curing. Thus, the 28-day strength of standard-cured test specimens is no more representative of the strength of concrete in the structure than is the strength of the specimens subjected to the accelerated-curing test.

My point is that what we really need is a direct use of the accelerated-curing strength test in quality control of the production of concrete. As far back as 1963, Smith and Chojnacki[8] wrote: "A suitable accelerated curing procedure can offer a more convenient and realistic way of ascertaining if the concrete will satisfy the purpose for which it was designed." I am convinced that the accelerated-curing test is superior as a quality control test as well as a compliance test. And yet, the best that is sometimes specified is that concrete is provisionally accepted on the basis of the accelerated-curing test and is subsequently expected also to pass the usual 28-day test. This is far too stringent a set of requirements because, for a given variability of concrete, the probability of passing two tests is smaller than that of passing either test alone.

We are hampered by the fact that the accelerated-curing test gives lower strength numbers than the 28-day test, but none of these numbers represents the true strength of concrete in the structure. We continue to be victims of what has aptly been described in French as the engineer's *bagage culturel*[9] and very heavy baggage it is, too. I am not recommending further development of accelerated-strength testing methods because quite a number of them are already in existence. What is needed is a convincing, fairly large-scale demonstration that the accelerated-curing test is safe both for quality control and for compliance purposes. This demonstration would require parallel but independent testing by the two methods over a period of time, coupled with education of influential engineers. I am suggesting that this be done on a large project involving several concrete mixtures. This would greatly help the cause of economic construction in concrete, and this is surely in our long-term interest.

Testing wrong parameters

I have some more criticism of testing, and that is the occasional testing of irrelevant parameters just because a test is available. One example concerns durability requirements, and I should confess that my views arise from my considerable involvement in concrete in the extremely aggressive environment in the coastal areas in the Middle East. Many problems arise from the corrosion of steel reinforcement induced by chloride ions which migrate into the concrete by ingress of seawater. The mechanisms of ingress have been discussed elsewhere;[3] here it suffices to say that the main relevant mechanism is sorption by capillary action. There is no generally agreed test for sorption but there exist several well-established tests for permeability of concrete: the best known ones are the German test of DIN 1048-1991 and the American test of the Bureau of Reclamation Procedure 4913-1992.

Permeability is measured under a head of water which forces it to travel through, or at least to penetrate into, the concrete. This head of water, that is the pressure, is high so that the natural state of concrete may be altered by rupture of some pores; also, blocking of some pores by silting can occur. Moreover, during the progress of the test, hydration of the hitherto unhydrated cement can take place so that the value of the coefficient of permeability varies with time and is not representative of the state of the concrete prior to the test. As an example, the value of the pressure specified in the United States Bureau of Reclamation test is 2.76 MPa (400 psi), which corresponds to a head of water of 282 m (925 ft). Such a head is relevant in high concrete dams but not in coastal structures or in structures in water-logged ground.

If an inappropriate test is used, the test results are at best irrelevant and at worst misleading. For example, there exist various waterproofing admixtures and other admixtures whose function is to reduce the permeability of concrete or, more specifically the ingress of water, which may be saline, into the concrete. The driving force may be rain, which probably represents a head of 200 mm (8 in.), or the head of water on the outside of a basement wall, which may be 2 or 3 m (7 to 10 ft). Often, the water pressure arises from waves or splash, but in very many cases there is no external pressure but only capillary action. There is a considerable market for products which would reduce ingress of water into concrete in structures in the coastal areas in the Middle East, in the Far East, and also in Australia. These, by the way, are the parts of the world where most construction can be expected in the foreseeable future. I have seen what you might call official tests by independent organizations evaluating such products by the use of comparative tests "with and without" the admixture where the discriminating test was the permeability test under a high pressure. For reasons which I have already given, the permeability test measures the wrong parameter and cannot therefore properly assess the effects of a given admixture. It is in everybody's interest that research establishing appropriateness of various test methods be undertaken.

Inappropriate testing

I would like to make one more remark about testing with respect to durability. Such tests invariably have to be accelerated in character; in other words, the deleterious action has to be exaggerated, be it by a higher or a lower temperature or by an increase in concentration of the attacking medium. One example of accelerated testing is the measurement of carbonation. The normal concentration of carbon dioxide in city air is about 0.3% by volume, with a highly polluted atmosphere containing about 1%. Accelerated tests are usually conducted at 4 or 5% and the relative humidity is maintained at 60 to 70%. I understand that in Switzerland a concentration of carbon dioxide of 100% is used so as to achieve a very rapid acceleration.[10]

There is a serious danger of distortion at very high levels of concentration of carbon dioxide, and the use of such levels in comparative tests may result in a totally false picture. Let me give an example. In tests at a concentration of 2% of carbon dioxide by volume, it was found[11] that the depth of carbonation of well-cured concrete containing fly ash or ground-granulated blast-furnace slag is greater by a factor of at least two than when only portland cement is present. When the concentration of carbon dioxide was 0.03%—a value typical of rural air—there was no such increase. A likely explanation of this difference in behavior at the two concentrations of carbon dioxide is that, at high concentrations, carbonation affected not only calcium hydroxide but also the calcium silicate hydrate phase. I am mentioning this because, if you want to use various cementitious materials in concrete, tests on their liability to carbonation, if conducted at a very high concentration of carbon dioxide, may wrongly discriminate against a particular mixture. It would be well worthwhile to conduct research to ensure that proper and valid tests are developed and that only such tests are used.

I would like to pursue the topic of inappropriate testing, but in a very different context. The majority of tests on the various properties of cement and of concrete were developed in the days when a water-cement ratio of 0.45 or higher was fairly typical. Nowadays, for many purposes, we use somewhat lower values of water-cement ratio. Moreover, to make high-performance concrete, we use water-cement ratios in the range of 0.22 to 0.30, and at these values the behavior of cement may be quite different. Of particular importance is the behavior in the presence of superplasticizers, which are an essential ingredient of high-performance concrete. This problem has been studied extensively at the Université de Sherbrooke in Canada by Aïtcin and co-workers[12] who point out that standard tests made on neat cement paste or on mortar do not reflect the behavior of concrete in which a superplasticizer is present.

With the increasing concern about the durability of concrete, and not only its strength, mixtures with low water-cement ratios, and therefore containing superplasticizers, are likely to become more widely used. The compatibility of portland cement and superplasticizers will become of wider importance, and this is a worthwhile field of research.

Calcium sulfate in cement

The problem of testing at an appropriate water-cement ratio is even more acute with respect to the role of calcium sulfate in the cement. The reason for this is that admixtures interact not only with the different phases of cement, but they also affect the solubility rates of the calcium sulfate added to the cement.[13] The term calcium sulfate covers the three possible forms, namely, gypsum, dihydrate, and anhydrite. A related problem is the presence of sulfate originating from sulfur in the coal, so that the amount of calcium sulfate added during grinding is smaller than in the past. Aïtcin[13] believes that the sulfate in the clinker is less soluble than that added. And yet, the standards for cement limit the total content of sulfate without distinguishing between its different sources.

This situation has serious implications for the hydration of tricalcium aluminate. It is at low water-cement ratios that the amount of sulfate and calcium ions is affected. This indicates the need for updating the standards so that they do not distort the behavior of mixtures used in practice. Cement manufacturers have a considerable interest in this as they do not want their cements to be incompatible with some superplasticizers. In fact, the need is not just to update the standards but, in the first place, to understand the behavior of mixtures at very low water-cement ratios in the presence of superplasticizers and with various forms of calcium sulfate and to establish practical criteria for compatibility.

Concluding remarks

There is no doubt that research leading to new knowledge and better understanding of the behavior of concrete must continue. However, this section is concerned with research aimed at improving concrete in practice, mostly by using existing knowledge and underlying technology. The main suggested areas for research are:
1. Better use of existing knowledge through improved training and education at all levels.
2. Control of variability of water content in the mixer.
3. Control of variability of aggregate shape and dust content.
4. Quality control in ready-mixed concrete
5. Ensuring adequate moist curing.
6. Method of reporting the properties and variability of cement.
7. Use of accelerated-curing strength test for quality control and compliance.
8. Testing ingress of salt water into concrete.

9. Accelerated testing of concrete for durability without undue distortion.
10. Testing cement and mortar at an appropriate water-cement ratio.
11. Sources of sulfates in cement and their properties.

What I have tried to identify are some problems in cement, aggregate, and concrete that hamper the future development of concrete. Let us admit it: concrete is not loved by the public at large. It is therefore in the interest of manufacturers, suppliers, designers, and contractors to contribute to a better concrete in the future. It is also in their interest to make it possible to produce good concrete more economically so that the client and owner never hesitate to consider concrete as the preferred material. This will be good for all those involved in concrete.

References

1. Neville, A., "Is Our Research Likely to Improve Concrete?," *Concrete International*, V. 17, No. 3, Mar. 1995, pp. 45-47. [Section 8.1]

2. Elstner, R. C., "Concrete Research," Letter to the Editor, *Concrete International*, V. 17, No. 7, July 1995, p. 8.

3. Neville, A. M., *Properties of Concrete*, Fourth Edition, Longman, London, 1995, and John Wiley, New York, 1996, 844 pp.

4. Neville, A. M., "Why We Have Concrete Durability Problems," *Concrete Durability: Katharine and Bryant Mather International Conference,* SP-100, V. 1, 1987, pp. 21-30.

5. Gaynor, R. D., "Cement Strength Data for 1991," ASTM Committee C-1 on Cement, 1993, 4 pp.

6. Poole, T. S., "Summary of Statistical Analyses of Specification Cube Mortar Test Results From Various Cement Suppliers, Including Four Types of Cement Approved for Corps of Engineers Projects," *Uniformity of Cement Strength*, ASTM STP No. 961, 1986, pp. 14-21.

7. Neville, A. M., "Cement and Concrete: Their Interrelation in Practice," *Advances in Cement and Concrete*, ASCE, 1994, pp. 1-14. [Section 4.1]

8. Smith, P., and Chojnacki, B., "Accelerated Strength Testing of Concrete Cylinders," *Proc. ASTM*, V. 83, 1963, pp. 1079-1101.

9. Rossi, P., and Wu, X., "Comportement en Compression du Béton: Mécanismes Physiques et Modèlisation," *Bulletin Liaison Laboratoires Ponts et Chaussées*, No. 189, Jan.-Feb. 1994, pp. 89-94.

10. Bulletin du Ciment, "Détermination Rapide de la Carbonatation du Béton," *Service de Recherches et Conseils Techniques de l'Industrie Suisse du Ciment*, V. 56, No. 8, Wildegg, Switzerland, 1988, 8 pp.

11. Bier, T. A., "Influence of Type of Cement and Curing on Carbonation Progress and Pore Structure of Hydrated Cement Paste," *Materials Research Society Symposium*, No. 85, 1987, pp. 123-134.

12. Tagnit-Hamou, A., and Aïtcin, P.-C., "Cement and Superplasticizer Compatibility," *World Cement*, V. 24, No. 8, Aug. 1993, pp. 38-42.

13. Aïtcin, P.-C., "Admixtures: Key Components of Modern Concrete," Onoda Research Center, Tokyo, August 1995, 14 pp.

Section 8.3: Concrete Research on a Micro- and a Macro-Scale

Types of research

This paper is about research in the field of concrete, and, as is usual for anyone who does not know how to start a paper, I shall begin by a classification. The easiest divisions are between useful research and useless research, and between good research and bad research.

In my opinion, there is no useless research, provided it is also good research. Anything which increases or improves our knowledge is worth doing, even if no practical application or benefit is perceived. Not only is increasing knowledge a laudable pursuit *per se* but there exist also many examples of even most esoteric discoveries being exploited in practice, even if many years later.

The distinction between good and bad research is another thing. I shall not attempt to define bad research but I can give a few examples.

The first one is misapplication of statistics. I have seen papers that give a multivariate regression analysis of the results of massive testing with practically every parameter of concrete or cement considered. The conclusions show which parameter is significant but give no thought to the underlying physical relation between it and the dependent variable tested. If it is not possible to explain the relation between the cause and effect, then a statistical correlation alone does not demonstrate that there is in fact such a relation. This distinction is of course well known.

I would also like to mention an unsound derivation of analytical models for the behavior of concrete. They are an excellent means of generalizing experimental data on the response to stress, strain, or environmental factors. What unfortunately sometimes happens is that the models that have been derived from a given set of data are subsequently tested against the very same data and thus proved to be reliable; no comment on this is necessary.

I have one further criticism of some models: occasionally, they include physical concepts that are far removed from the actual behavior of concrete. Professor K. H. Gerstle recently cited the example of plasticity-based formulations to describe brittle failure of concrete under combined compression and tension, and suggested that a consideration of actual physical modes of failure is necessary before the appropriateness of such models can be accepted. Much the same applies to some creep models.

A further example of bad research is in the not-invented-there category, known as NIH. On one occasion, when I challenged this type of research, I was solemnly told: "we cannot base *our* design code on foreign results." Yet Isaac Newton worked only in Europe.

Yet another example, which is perhaps a variant on the preceding one, is a paper describing in detail the properties of a local material just because it is local. There is no *a priori* hypothesis, simply a catalog of results.

My last example is a research paper that starts with a theoretical model, reports an experimental investigation, and observes that the two do not agree with one another. Instead of going back to the drawing board, the author, possibly because he needs a publication to his name just then, simply concludes that further work is required.

Readers of the present paper might doubt that such research papers actually get published. But they do. At one time, I was simultaneously a member of the editorial boards of four journals, and I remember seeing the same paper re-submitted to one journal after another.

Research on a micro-scale

I would now like to indulge in the classification of concrete research on the basis used in the opening paragraph of this section, namely on micro-scale and macro-scale. My primary concern is not so much with the scale of observation—nanometers or meters—but rather with the scale of application.

Let me put this in a rough historical context. Cement was invented as a practical material which "works," regardless of whether we date it from classical times, from John Smeaton or from Joseph Aspdin. Thereafter, improvements were effected by scientific research into the chemical and physical properties of the raw meal, the clinker, and the cement. All this work was necessary and beneficial.

The beneficiaries were the users and the manufacturers, and there is a distinction there. I remember, almost 40 years ago, in the course of my own research into the influence of the chemical composition of cement upon creep, finding as a sideline that the alkalies in cement had an influence on the rate of development of strength of mortar. Feeling excited, as a young man, that I have stumbled upon something, I proceeded to discuss this with the cement manufacturers' research people, only to be told that the relationship between the alkali content and strength was known to them, but it was "undesirable" to publish it because users might want to specify a particular alkali content; I was even shown an internal paper on the subject, in confidence of course.

There is thus a difference between truly independent research, I might say fearless research, and research in support of commercial interests. Until not so long ago, the attitude of one research organization of cement manufacturers toward admixtures was: the best admixture is more cement in the mixture. There is absolutely nothing wrong with supporting commercial interests but research papers sometimes do not make it entirely clear that the author is closely associated with a manufacturer; in other spheres of activity, there is a term "insider trading."

Let me repeat: micro-scale research on cement, on admixtures, on hydrated cement paste, on various cementitious and supplementary materials, on hardened paste and mortar containing various materials was, and continues to be, highly valuable. There is no possibility of thinking that enough has been done so that no more effort is required. This research has contributed to our ability to make better concrete and no doubt will continue to do so.

Features of macro-scale research

The preceding subsection dealt with research on the micro-scale, but in actual construction, achievement of satisfactory concrete requires also an input on the macro-scale. By this I mean concrete as it is actually batched, mixed, transported, placed, compacted, cured, and possibly subjected to some special treatment. Not only do these procedures require knowledge additional to that gleaned from the micro-scale work but, in some cases, the findings on the micro-scale cannot be properly applied to actual concrete.

Let me give an example. I have seen, more than once, papers with a title containing the word "concrete" and yet describing experiments on neat cement paste specimens, a few millimeters in size. Under these conditions, there is no influence of the inhomogenity arising from the presence of aggregate, no effect of the interface between the aggregate and the hydrated cement paste, no differential thermal effect of the two phases, and no thermal gradients that may exist in full-size concrete. The work described may well be of great value as a contribution to knowledge but this is not the same as a contribution to making good concrete.

Good concrete is not a commodity found in abundance. After all, concrete is considered by many to be a simple material—nothing but a mixture of cement, aggregate and water—which any fool can make; the trouble is that he or she does. In consequence, a not insignificant share of the concrete placed is indifferent, some of it is bad, and yet all of it could be so much better. What is needed is a means of showing those concerned with actual concreting how to do better. This, to my mind, comes under the heading of research on the macro-scale.

Purists may object to my applying the word research to something so mundane as improvement in practice. To fend off this accusation let me consider the purpose of research. As I see it, the purpose is to increase scientific knowledge or to improve practice. Often the former leads to the latter, sometimes the latter leads to the former but, even if there is no interaction, improvement in practice alone is a worthwhile objective. Actually, it is an objective of value to the community at large because the lives of all of us, and this is no exaggeration, are affected by the quality of the concrete that surrounds us. We may not love it, but we cannot live without roads and bridges, dams and power stations, buildings and airports, and so on.

Macro-scale problems

On reflection, it is not just the case that we should be doing better in building concrete structures: in fact, in some respects, we have been doing worse than in the past. I think it is generally accepted that post-World War II concrete, up to the mid-1960s, or even a little later, performed satisfactorily for many years in service; in other words, there were few problems with durability, excepting the cases where calcium chloride was deliberately introduced into the mixture.

Later on, the durability of concrete worsened significantly. One of the principal reasons for this was the fact that concrete continued to be specified by its compressive strength with, in most cases, no additional criterion of a minimum cement content or an effective maximum water-cement ratio. Thus the apparent specification for concrete remained the same but the actual composition changed substantially. The change was due to the fact that the newer cements were more finely ground and had a higher content of C_3S. It was thus possible to achieve the same 28-day strength at a higher water-cement ratio than previously, with the inevitable consequence of a higher permeability of the resulting concrete. Under aggressive conditions of exposure, be they the chlorides present under maritime or coastal conditions, be they carbon dioxide in the air of our cities, the passivity of the reinforcing steel became compromised, corrosion of reinforcement occurred, and damage of the structure took place.

I am not suggesting for a moment that research on a micro-scale is to blame. Why should it be? The researchers were improving the quality of cement and indeed improving its manufacture. It was not their business to foresee, or even to investigate, the consequences of these changes on construction practice. And at the concreting end, there was no relevant research, possibly because it was not in anybody's interest to finance such research. Structural designers thought that they were continuing to specify the same concrete as before: after all, they still required the same compressive strength. Ready-mix suppliers were satisfying the same specifications, and indeed were doing so at a lower cost, if not at a lower price. So nobody had reason to look into the matter and nobody did. That is, until some 15 years later when a large number of structures were found to be in need of repair—if not worse.

Nobody is to blame, and I am not trying to blame anybody. My point is that research was necessary, was not done, and there was no obvious body that should have done it.

Macro-scale research and design

I keep coming back to research on a macro-scale, and perhaps it is appropriate for me to look, in a broad sense, at the relevance of that kind of research to design. I wrote on this subject in the British journal *Concrete* in July 1978, and some of what I wrote then is still relevant.

Research is vital to nourish design: without research we could not design anything which has not been designed before; conversely, without new designs in the offing there would be no need to do research. But, as I said in 1978, this statement is too simplistic, and I should be clear about what I mean by research and by design.

Looking at design, I can quote from a typical British code of practice: "the purpose of design is the achievement of acceptable probabilities that the structure being designed will not become unfit for use for which it is required, i.e., that it will not reach a limit state." So safety and serviceability are the objective, although of course a satisfactory design has to fulfill also the requirements of economy and, preferably, aesthetics. I believe it was Nevil Shute who said that an engineer can design and build for a dollar what any fool can build for ten dollars.

To ensure safety and serviceability we use the tool of structural analysis, which enables us to prove that our design is satisfactory, that is that the structure will not fail. What is interesting and important is that, in addition to an actual, that is physical failure, there exists another type of failure, called by Professor E. H. Brown a designated failure. This designated failure is our limit of understanding of structural action or of behavior of the material.

An example of the latter is relevant to the present discussion. We may determine experimentally the tensile strength of a plain concrete specimen, and the result obtained depends of course on the test method, but we do not know what is the value of this strength in a reinforced concrete beam. In the design of flexural members, we have traditionally taken this value as zero, this being the limit of our understanding. In other words, we have considered design in which this limit is exceeded in a strength calculation as design that has failed. This then is a designated failure, although in reality tension stiffening occurs.

In fact, this restriction has been alleviated by research; for example, we can design partially prestressed concrete beams, and we can take tension stiffening into account in the calculation of serviceability of a reinforced concrete beam. This type of research, however, is much less concerned with the behavior of concrete than with structural action.

Let me now return to the topic of more conventional failure; that is, to the situation when a structure ceases to perform satisfactorily. As far as the ultimate limit state is concerned, failures in service are extremely rare; there are occasional failures during construction but that is another story. Absence of failure is predicated on two conditions: satisfactory structural analysis and actual construction conforming to the minimum properties of materials assumed in the analysis.

Serviceability problems

In fact, our design from the standpoint of the ability of the structure to carry the imposed loading, as well as self-weight of course, is very good. As far as the execution of construction is concerned, the quality of the concrete is, in some cases and in some respects, inferior to that specified. In consequence, already at the time of completion of the construction, the concrete can be said to have failed the requirements of the specification. At a later stage, the quality of reinforced concrete, again in some structures, deteriorates to the point where the structure becomes unserviceable or nearly so.

The preceding paragraph contains a number of qualified statements which require amplification. First of all, I have referred to the quality of concrete "in some respects." The property of concrete which is tested most frequently, nay, always, is its compressive strength. This, in an overwhelming majority of cases, is satisfactory, but a satisfactory strength is not very informative. Of course, the concrete has to be strong enough to resist the imposed stresses; once upon a time, it was possible to infer also that a strong enough concrete also has a low enough permeability and possibly even a low enough resistance to abrasion. This is no longer true as it is possible to produce concrete with an "adequate" strength but with a composition which guarantees none of the other properties required; I referred to the reasons for the change earlier in this paper.

So it is not the compressive strength of the concrete test specimens (that is, the potential strength of the concrete in the structure) which is inferior to that specified. A common fault is the quality of the concrete in the zone of cover to the reinforcement. The reason for this is the lack of curing. This is a topic for research to which I shall come back later.

As far as continuing serviceability is concerned, probably the biggest culprit is the lack of durability of reinforced concrete. The reasons for this include sulfate attack, alkali-aggregate reaction, and corrosion of reinforcement, to mention just the principal ones.

Research on a micro-scale has led to a good understanding of sulfate reactions, to the development of appropriate cements, and has resulted in construction where sulfate attack can be contained and harmful effects on the structure can be avoided.

The understanding of the alkali-aggregate reactions has also been greatly improved by research on a micro-scale but the application of this knowledge in construction has been less successful. What the engineer wants to know *a priori* is whether a given aggregate is likely to result in deleterious reactions with portland cement of a given type. Tests to answer this question often err on the safe side, and from the practical point of view this is not good enough. We cannot afford to reject an aggregate just to be on the safe side because the economic consequences of such action are large. The consequences may also be undesirable from the environmental point of view if the rejection of some sources of aggregate leads to an unnecessary exploitation of unspoiled areas of the country.

Corrosion of reinforcement is a large topic. Research on the micro-scale has led to an enormous improvement in our knowledge of the physical phenomena involved. Numerous papers describe the influence on corrosion of the hydroxyl ion concentration, of the chloride ion concentration, of the ratio of the two, of binding of chlorides by C_3A, of the release of bound chlorides, and so on. I have no doubt that all the test data as reported are valid and that the generalizations derived from the tests are valid in so far as the laboratory test conditions and the range of the parameters involved are concerned. But all this is of very little value in practice, at least so far.

This is a very sweeping statement which could be viewed as an unjustified condemnation. To fend off this accusation I should explain the reason for my views. For the best part of a decade, I have been involved in the investigation of failures of structures in service. A large proportion of these have suffered from the corrosion of reinforcement. The investigations have been supported by tests on various properties of concrete. The question is: how do the observed phenomena link with the knowledge obtained on the micro-scale?

The answer, sadly, is: not very well. What I mean is that, yes, we know that corrosion at a certain location can be ascribed to the presence of chloride ions at the surface of the reinforcing steel, but we do not know why, at some other location, there is no significant corrosion even though chloride ions are present. To put it in terms which an engineer requires: what is the threshold level of chlorides for corrosion to take place?

The mystery is likely to be related to the difference between the physical conditions in the actual structure and in the laboratory. In the latter, the conditions are controlled or known, and are usually steady at a constant value or with a well-defined cycle. In the actual structure, the conditions per force vary: the temperature changes not only with time but also from place to place depending on insolation. We still do not know enough about the temperature of concrete at its surface under given conditions of radiation or about the variation in temperature with depth. For that matter, we lack reliable quantitative knowledge of the influence of temperature on the rate of corrosion of steel in a reinforced concrete member.

Possibly more importantly than the variation in temperature, in an actual structure, the cycle of wetting and drying varies. We have known for a long time that cyclic wetting and drying is particularly conducive to corrosion so that concrete between the high- and low-water tide marks, and in the splash zone, was considered especially vulnerable. In such a situation, the drying period is short, at most 11 hours. It is only very recently that we observed the effects of a cycle in which the drying period is much longer and the wetting period is very short; this is a situation which obtains on horizontal surfaces where wetting occurs by occasional spray, by wash-down with seawater, by coiling of wet ropes, or by a deposition of dew; yes, there is a lot of dew in the dry coastal areas of the Middle East but you observe it only if you get up at 5 a.m.

The consequences of corrosion on the performance and life of many structures run into billions of dollars. They have spawned a new industry: concrete repairs. But we still do not know *a priori* which concrete will be durable under conditions likely to obtain at a given location, nor do we know which concrete repair will be durable; the next new industry will be the repair of repairs industry.

What is needed is a massive but well coordinated research on a macro-scale on corrosion problems in reinforced concrete. Such research would clearly rely on and build upon the micro-scale research, but it would include parameters and factors which cannot be studied on the micro-scale.

This brings me back to curing, which is one such factor. Imbibition of water by concrete or prevention of loss of water from concrete affects almost exclusively the outer layer of concrete, no more than the cover zone. And it is this zone alone that determines the ingress of chlorides under given conditions of exposure; hence the great importance of curing.

For this reason, I believe curing to be an important area for research on a macro-scale. Personally, I think proper curing of concrete should be enforced, but alas it is difficult to say today whether the concrete was properly cured yesterday. Perhaps an effort should be made to develop an *ex-post facto* wet-curing-measurement device. There may well be other possibilities such as integral indicators of the presence of free water in the cover zone concrete.

To readers who may judge this paper to drop to an unduly low level of what is considered research, I would say that, from the economic standpoint, the consequences of the lack of durability of reinforced concrete structures are far more important than any other problem with such structures. I am not advocating only dollar-led research, but the converse of ignoring the dollar sign will not serve the research community well in the longer term.

Other construction problems

From curing, I can generalize to other aspects of construction practice. Most problems in making good concrete arise not from the lack of fundamental knowledge about cement, aggregate, or the rheology of fresh concrete. They lie in a lack of proper application, often in respects which are thought to be almost trivial.

Let me give just two examples, both based on my experience. They both relate to batching. The influence of the water content of concrete on its properties is well known; for this reason, the total quantity of water being put in the mixer has to be controlled. However, that part of the mixing water which comes from the aggregate is often not measured in a continuous and reliable manner. This problem can be solved only by research on a macro-scale. Unfortunately, such research is not viewed as glamorous and does not attract would-be researchers. And yet, a solution which would ensure a constant water content of the mixture, and therefore a constant water-cement ratio and also a constant workability, would save large sums of money in obviating the production of non-conforming or substandard concrete. The need for large-scale testing of compressive strength might also be reduced.

I think it is unnecessary to give more examples of needed research on a macro-scale. The need for such research is clearly there, and what I am trying to do is to persuade both the researchers and those who allocate research funds that such research needs to be done.

A difficulty lies in the increasing specialization of researchers in recent years. The field of expertise of many has become narrower and narrower, and they sometimes fail to appreciate, or even notice, seemingly extraneous features of their tests. Unfortunately, in macro-scale work it is necessary to look more widely than solely at the parameters of the input in a laboratory-only-based study.

At the same time, I have to acknowledge the benefits of the work of those who look not just at a specific material but at a broader class of geomaterials. They will give new insights which, in turn, can be applied by researchers on a macro-scale.

Concluding remarks

The purpose of this paper has been to point out that practical problems in concrete construction and, above all, in the durability of concrete require research on both the micro- and the macro-scale. Work on the former scale has been extensive and is far ahead of the achievements on the macro-scale. Catching up by the latter is of great practical and economic importance and, even if less glamorous, should proceed without delay.

Section 8.4: Reviewing Publications: An Insider's View

It may not be immediately obvious why readers of *Concrete International*, who are concerned with the real, or concrete, world, should be bothered about procedures of reviewing papers and articles prior to their publication in a journal or a magazine, or even of book proposals. And yet, there are several reasons for airing this topic.

What are reasons for reviewing publications?

Reasons for publication review include establishing the reliability that the reader can attach to the printed word: A century ago, some people would say, "It must be true, I read it in a book." What is the situation today? A subset of reliability is to distinguish between a technical opinion about, say, a product, and a cleverly couched advertisement of that product by its manufacturer.

At the other extreme is the responsibility that may be carried by the author of a publication, which can possibly arise if the author advises a method or a material whose use under some circumstances may happen to be injurious, economically or physically, to the reader. We need no reminder that, alas, on both sides of the Atlantic (and probably around other oceans, too) we live in an increasingly litigious society.

The relevance of litigation to the review of publications arises also in establishing the credentials of an expert witness: the probative value of his or her published views may depend upon whether the publication was subject to previous review.

The importance of previous review of publications arises also in academic circles: Such publications are called "peer reviewed," and publishing in them carries much greater weight in deciding on the promotion of a faculty member than would the paper's appearance in other publications. A number of readers of *Concrete International*, as well as contributors to the magazine, are academics, so the topic of peer review is of burning importance to them. I hasten to add that I no longer have any academic ambitions and, consequently, do not care whether, or how, this article is reviewed. What I wish to achieve is to share my views with the concrete community at large and to generate some light on the process of review, rather than the heat that often accompanies it.

What is a peer?

The meaning of the term "peer" may not be obvious to everybody. I am clearly not alluding to "Peers of the Realm" in the United Kingdom. Parenthetically, until a year ago, they were members of the Upper House of Parliament, but British constitutional reform has put them in limbo.

The term "peer" is of Latin origin, meaning "equal," and it is useful to consider the more precise meaning of the word. *The New Shorter Oxford Dictionary*, published in 1993, offers the following relevant definition: "A person of the same standing or rank as the person in question; a person or thing of the same effectiveness as the one(s) in question; an equal." Turning to *Webster's Third New International Dictionary*, 1962, we find a peer defined as "one that is of the same or equal standing (as in law, rank, quality, age, ability) with another."

Interestingly, *The New Shorter Oxford Dictionary* offers also a definition of the combined term "peer review": "The evaluation by (other) experts of a research project for which a grant is sought, a paper received for publication in a learned journal...."

The preceding paragraphs imply that the judgment is exerted by the author's equals. This begs at least two questions. First, who are "the equals" as distinct from those with a better or higher knowledge? Second, what is the basis of their evaluation or judgment?

Who is your peer?

For many years now, I have realized that the world is full of "my betters" who are therefore not my peers. At the same time, there are some people whom I would like, rightly or wrongly, to consider my inferiors. There is thus a genuine problem of an objective definition of a peer.

In academic circles, the review of a publication by an assistant professor is likely to be undertaken by a full professor; clearly, these two are not peers. In the opposite situation, the review of a paper written by a full professor is unlikely to be performed by an assistant professor. Indeed, an assistant professor is not a natural choice for a reviewer.

The previous remarks point out that the term "peer review" is, at best, a bit of a misnomer, and at worst, a patronizing distortion of what happens in reality.

What are the criteria of peer review?

The number of reviewers of a given paper is usually two or three, but sometimes more. Given that, as I see it, the review is, in reality, often performed by people senior in knowledge and experience to the author, or better known than the author, I cannot help thinking that the process involved is that of judgment of the publication being reviewed. Judgment should not be capricious or arbitrary, but rather it ought to be based on specified criteria.

In the formal sense, these criteria are usually specified on the review form provided by the publisher of the journal. In reality, however, there is great scope for applying the reviewer's personal views and convictions, as well as preferences, if not likes and dislikes.

For example, there are two methods of determining the total chloride content in hardened concrete. The paper being considered uses method A and attempts to develop it further; on the other hand, the reviewer is

Chapter 8: Research

convinced that method A is much inferior to method B, which the reviewer has developed. It is, therefore, not surprising if the reviewer is unfavorably disposed to the paper, and this attitude may tip the balance against its acceptance.

Such an approach might be considered to be biased, but the bias is not deliberate or even conscious. Before World War II, Anita Loos wrote a well-known book titled *Gentlemen Prefer Blondes*; evidently some do. I think they are wrong, but I do not blame them or accuse them of unfairness. What I am trying to say is that the peer review process is not, and cannot be, a truly objective system of evaluation.

Consequently, the publication of papers by Author C does not *necessarily* mean that this author is a better researcher than Author D whose papers are published less frequently in refereed journals. I emphasize the word *necessarily* because the use of the peer review process is certainly better than no review at all.

The range and criteria applied to the review process depend on the extent and detail of the peer review. Some reviewers "improve" the paper or article by changing the style, and sometimes they verify meticulously as many statements and calculations as they can.

My approach is quite different: I do not see myself as a ghost writer or hidden co-author. I look for inconsistencies, for errors in the treatment of physical phenomena, for mistakes, for a lack of clarity, for tedious repetition. I also look for relevance to the avowed title and for the author's familiarity with the existing knowledge, so that wheels are not repeatedly invented. Lastly, if I, chosen to be a reviewer because of my expertise, cannot understand what the author is trying to say, I assume that some readers will have even greater difficulties. If they cannot understand the article, of what use is its publication except for academic promotion?

Overall, the peer review process is to ensure that the paper is not manifestly wrong. I feel very strongly that the one thing the peer review does *not* do is to provide a "certificate" to the effect that the views expressed in the paper are correct to the exclusion of other views. All that the review does, in my opinion, is to confirm that the content of the paper is not indubitably wrong.

What other factors influence publication?

The ultimate decision to publish or not to publish a paper lies with the journal editor or with an editorial board. The editor or the board are obviously strongly influenced by the reviewers' comments, but they also take into account other factors.

Some of these factors may militate against publication; others may have a positive influence. The situation is as follows. Each journal has a limited annual capacity of so many pages, which translates, broadly speaking, into a fixed number of papers.

In case someone thinks that a paper may be accepted with the notion that it will be published in the distant future, I should point out that such an approach is really quite unsatisfactory. The author of a paper wants it to be published with a minimum of delay so that the world at large becomes aware of the content of the paper. And even if the author is willing to wait patiently, this rebounds unsatisfactorily on future authors whose papers, however excellent, then have to wait in a long queue. This is why some journals openly practice the following approach: if the reviewers recommend the publication of a paper but classify it as less than "very good," but rather as "good" or "fair," the journal editor is empowered to decline publication of that paper if there is no space for it in the near future.

It follows that, had the "better" papers not been in the pipeline just then, the paper in question could have been published in the preferred journal. Conversely, when journal "copy" is in short supply and the journal has to fill its quota of pages, a "less good" paper has a better chance of publication.

The point I am trying to make is that the failure of a paper to appear in the preferred, or particularly prestigious, journal may be affected by totally extraneous factors and by a set of circumstances outside the given reviewers' control.

How are conference proceedings reviewed?

In my experience of assessing the research performance of faculty members in the United Kingdom, Australia, and Canada, the "powers that be" distinguish not only between publications in journals that use the peer review, known as refereed journals, and non-refereed journals, but also between journals and conference proceedings. These days, there are so many conferences, often international, that there is hardly a week in which there are not two or more conferences held dealing with concrete.

Most conferences publish a volume, or several volumes, of proceedings, but the presentations included in the proceedings may or may not be peer-reviewed. However, even when they are, many times the situation is not quite the same as in the case of regular journals.

Generally, a given conference has to attract a sufficient audience to be viable. This means that there has to be an adequate number of people who find the event attractive enough to attend. Authors of papers are expected to attend the conference so that the acceptance of an adequate number of papers will thus ensure the feasibility of a conference. It can follow that tough peer reviews, with a consequent acceptance of a relatively small number of papers, might injure the viability of the conference. This is often the reason why it is easier to have a paper published in conference proceedings than in a journal.

The preceding paragraph will help explain why publications in conference proceedings may rank low in the assessment of a faculty member. Thus not all peer-reviewed publications carry equal weight in the eyes of the faculty-member evaluators, and rightly so.

What I have described is sometimes circumvented. For some conferences, the peer review is conducted fairly stringently so that the papers published in the proceedings are of high quality. At the same time, the remaining papers are taken care of by the publishing of a supplementary volume of conference proceedings. This is fine, except that it is not explicitly stated, and therefore not generally known, that the standing of papers in the main volumes of proceedings and in the supplementary volumes is not the same. I am not criticizing this approach, but I do wish to point out that the interpretation of peer review with respect to conference proceedings is not that simple.

How are book proposals reviewed?

Authors of technical and scientific books will know, but probably not many other people, that proposals for a book are reviewed before a publisher decides whether or not to publish it. The reviews are usually undertaken by authors of existing books, or by experts on the subject. Thus, book proposals undergo true peer review, but this is only the first hurdle, which is a condition necessary but not sufficient.

The second hurdle is financial; indeed, financial considerations are dominant. Depending on the type of publisher, a book might be published only if it is felt that it will sell a sufficient number of copies to cover, among others, the cost of production, including that of the numerous and varied staff in the publisher's office, and will produce an acceptable profit (not necessarily so with University Presses). It may be useful to point out that the price charged by the publisher to the bookseller is discounted to between two-thirds and one-half of what the ultimate book purchaser, that is, the reader, pays. Actually, very recently, the pattern of selling books has been changing with the advent of Internet selling. As for the author, except for very successful books selling many thousands of copies, the author's royalties are only a small proportion of the book's selling price.

Because of these important financial considerations, the reviewer of the book proposal is expected seriously to assess, or perhaps more correctly to guess, the future sales and the potential market for the book, taking into consideration any existing competition. Of course, the reviewers themselves may be competitors of the proposed book; whether that influences their "judgment" is anybody's guess.

The overwhelming influence of financial aspects of book publishing means that a mediocre book with a large potential market may well be more likely to find a publisher than an erudite, scholarly, but abstruse proposal having a very small readership. Thus claiming a book as a feather in one's cap may not always be so wonderful; in case this is thought to be a deprecating remark, I would defend myself by declaring myself to be the author or co-author of eight books, two of which have run into at least four editions.

What I have described is not the whole picture: occasionally, but increasingly rarely, publishers will produce a book by a very eminent author, even if it will sell only a small number of copies, because the author's name adds luster to the reputation of the publishing house and that company's list of books.

I should add that when I refer to books, I mean books actually written by one or two, maybe even three, people. Many so-called books have the names of an editor or editors on the title page, with the book contents being in reality a collection of chapters, each written by a different person, and often with little cohesion and consistency between chapters. In my view, such books are of lesser value to the reader because the inter-relationships of the various topics are perforce less clearly treated. Alas, the majority of recent books are of the latter type, probably because the effort required to present an integrated view of a broad technological or scientific topic is really great.

Are reviews consistent?

The author of a paper rarely sees the entire reports by all the reviewers (though this is not the case for papers submitted to ACI periodicals). Usually, what is transmitted are recommendations for changes and improvements. Because there are two or three reviewers, it is not surprising that their views are often not coincident. A decision to publish or not to publish a paper is made on the basis of the individual recommendations taken all together. The opinions of the various reviewers may be given unequal weight depending on their detailed familiarity with the topic; sometimes, they are asked to state their own confidence level.

What is interesting to know is how divergent the views of the various reviewers are; alas, such information is not readily available (again, ACI periodicals are excluded; such information is made available to the authors, I am told). However, to give a clue about variability of the different reviewers' views, I can give some information obtained in a kindred field. This is the review of research proposals where the detailed opinions of several individual reviewers have subsequently been collated. Table 8.4.1 shows the scores, on a five-point scale, given by reviewers, whose ratings were between 2 and 5, in 20 cases of specific criteria being judged. In the majority of cases, the relative agreement among the reviewers is good; it is less so in cases 6, 8, 10, and 13; it is only in case 9 that the spread of scores runs from 2 to 5, and could be worrisome.

Table 8.4.1 illustrates the fact that there was a chance factor in the peer review: the particular reviewers who happened to be chosen. Let me give an example. At one time, I was simultaneously a member of editorial boards of four journals. At meetings of the various boards, we saw the reviews of all the papers submitted for publication. At the board of the first journal, a given paper was reported as being recommended for rejection; we agreed. A short time later, the same paper appeared before the board of the second journal; again, it was rejected. Later still, the same paper was considered at the board of the third journal, and again it was rejected. However, at the fourth board, the

Table 8.4.1—Scores given by reviewers*

Case number	A	B	C	D	E
1	2	4	4	(1)	—
2	4	4	4	(1)	—
3	4	4	4	(1)	—
4	3	3	4	(4)	—
5	3	3	4	(3)	—
6	4	3	(5)	4	5
7	4	5	(5)	4	4
8	4	3	(4)	3	5
9	4	4	(5)	2	5
10	4	3	(4)	3	5
11	(1)	2	—	—	—
12	(4)	3	—	—	—
13	(3)	1	—	—	—
14	(1)	2	—	—	—
15	(2)	3	—	—	—
16	(4)	4	5	5	—
17	(4)	5	4	5	—
18	(4)	4	4	4	—
19	(4)	5	4	4	—
20	(4)	4	5	5	—

*The circled scores were given by the author.

reviewers, except me, were different and perhaps kinder than the reviewers for the other journals: the paper was recommended for publication, and indeed, it was published in this refereed journal. I should add that I did not pass on my knowledge of the previous fate of the paper from journal to journal.

What is the significance of peer-reviewed publications in litigation?

When acting as an expert witness in litigation and arbitration, I found occasionally that much importance is placed on a given expert's publication list: how many peer-reviewed papers has the expert written? Even though I have written something like 130 such papers, I deplore this approach, which smacks of a beauty contest. Moreover, such comparisons across disciplines are almost meaningless: chemists can easily write numerous short papers, each based on one set of experiments; engineers write fewer papers, which are founded on more extensive study or experience; and those in humanities write only a few papers rooted in prolonged research.

Having written numerous peer-reviewed papers does not make one a useful expert in concrete or in construction. In my view, what is more important is the expert's knowledge of, and experience in, design and construction, including the actual use of concrete. Research experience is important in the academic world, but much less so in the legal realm in trying to ascertain what actually happened in the field, and this is what litigation is about.

Why disclaimers in publications?

We all learn from information provided by other people, and most often this is by way of the printed word. As I stated at the outset of this section, the immediate question is how reliable is this "word." This leads to the consideration of the responsibility for the printed content in publications.

With very few exceptions of some commercial publications, we should assume that all authors present facts and opinions that they believe to be correct and true. But in matters technical, there may be mistakes, and with a consequence that what appears in a published paper or book may result in economic loss to a third party. Who, if anyone, bears the responsibility? The answer too often is a matter for lawyers to help determine.

However, there is one aspect of responsibility that I would like to consider. Let me start with *Concrete International*. Every issue carries the words: "The Institute (ACI) is not responsible for the statements or opinions expressed in its publications. Institute publications are not able to, nor intended to supplant individual training, responsibility, or judgment of the user, or the supplier, of the information presented." Of relevance to this article is that ACI is not responsible for the statement or opinions of the supplier, that is, the author of any article in *Concrete International*. I, for one, would not wish the Institute to be responsible for me because this would entail censorship: any new ideas, suggestions, proposals, or interpretations might be suppressed, if only to be on the safe side. This would stifle progress and development. Moreover, when there is a discussion of an article by way of published letters that express views contrary to the original article, which point of view would ACI be responsible for?

ACI uses the same approach with its other publications. For example, in Monograph No. 6, titled *Hardened Concrete: Physical and Mechanical Aspects*, written by me, and published in 1971, there is a statement: "The Institute is not responsible, as a body, for the statements and opinions advanced in this publication."

With an increasing propensity of the populace at large to pursue before the courts anyone in sight to remedy a loss, the use of disclaimers continues to widen. At the most successful ACI Spring 2001 Convention in Philadelphia, I noticed that my book titled *Properties of Concrete*, on sale at the publications desk, bore a sticker to the effect that the book had not been reviewed by ACI and that ACI was not responsible for its contents. A similar sticker was attached to all books that had not been published by ACI. This is absolutely right and proper. Publications by ACI itself carry within them the same disclaimer as *Concrete International*.

The use of disclaimers is a consequence of social and legal attitudes in the present times, but it is no reflection on the quality of the material published. I hold this opinion very firmly, even if it is contrary to allegations sometimes made, for example, to discredit an expert witness on the grounds that a disclaimer implies that the opinion published by the expert cannot be trusted.

What is the relevance of this article to ACI?

Some readers of *Concrete International* may feel that my discussion of the review of publications is outside their interest. I would like to submit that this is not so. We all need relentlessly and continually to improve our knowledge; this includes reading journals and magazines, and we should know the standing of what has been published. We also have a responsibility to share our knowledge with others, and this includes writing about and publishing our experiences, good or bad, so as to improve the use of concrete. The mission of ACI is to improve the use of concrete; one way in which ACI members can contribute in this respect is to publish articles and papers. Some engineering institutions in the United Kingdom place a moral obligation, formally stated, on their members to publish a paper in the given institution's journal.

If ACI members are to publish articles and papers, it is useful for them to understand the system of peer review. I have tried to explain the operation of the system and to point out that the peer review process and its resulting recommendations are not an exact science. This is not surprising, but it should be borne in mind so that undue weight is not placed on peer-reviewed publications. I want to emphasize that I am not arguing against the use of peer review. Imperfect as it may be, it is the best system we have. I have also tried to point out how factors other than peer review play a role in decisions for publication, mainly journal space availability and financial considerations.

I very much hope that a better understanding of the process of reviewing publications will have the positive effect of encouraging ACI members to share their knowledge and experiences with the membership at large. Let the papers (and articles) roll in!

Letter to the Editor No. 1
Sharing knowledge about concrete

Adam Neville's comprehensive treatment of "peer review" and related matters in the September 2001 issue of *CI* is interesting and informative. I agree with his remarks at the end that it should encourage ACI members to write and share their experiences with the rest of us. Professor Neville emphasized that the peer review process was imperfect but the best we have. I agree, but I think ACI members should not infer that nearly all the things said in published peer-reviewed papers is so or, indeed, that the author was of the opinion that it was true when he wrote it.

What one gets from reading the journals, as they appear, is a chance to find out what's being published quite soon after it enters the concrete technology database. I recommend going to the five volumes of the *Manual of Concrete Practice*, which is updated annually. The thing about the contents of the *MCP* is that it has all been voted on—not by the reviewers, but by a whole committee of experts—and reviewed by a group of superexperts, the ACI Technical Activities Committee. This is better review than "peer review," but the product produced is the state of knowledge a few years old.

Bryant Mather
ERDC
Vicksburg, Miss.

Letter to the Editor No. 2

In his comprehensive article on reviewing publications, Adam Neville reminds us that we need to improve our knowledge relentlessly and continually. He continues, "We also have a responsibility to share our knowledge with others, and this includes writing about and publishing our experiences, good or bad, so as to improve the use of concrete." In the same issue, I believe the article by Roy Keck entitled "Improving the Durability of Concrete with Cementitious Materials" fails to meet the criteria of sharing both the good and the bad.

Mr. Keck rightly gives the caveat that Class F fly ash may need more air-entraining agent. He does not, however, pass along the known detrimental effects of Class C fly ash with a high calcium oxide content. In many instances, the addition of this Class C fly ash will make concrete more susceptible to sulfate attack. It can also exacerbate the alkali-silica reaction with reactive aggregates. These effects will make the concrete less durable.

If these caveats had been included, this informative article would have lived up to the mission of ACI, as stated by Neville, which is "…to improve the use of concrete."

John Wojakowski
Concrete Research Engineer
Kansas Department of Transportation
Topeka, Kan.

Neville's response

I am pleased that my exhortation to readers to contribute to the ACI Journals has produced two letters. It is gratifying that Mr. Wojakowski subscribes to my view that bad experiences, as well as good ones, should be shared. He then refers to another article in the same issue of *CI*, written by a different author, that did not mention some negative effects of the inclusion of Class C fly ash in concrete. It is useful to ponder his message. Occasionally, authors are carried away in reporting good news through personal enthusiasm or commercial interest.

Bryant Mather is right to say that what has been published in a peer-reviewed journal is not necessarily correct. As I see it, an editor's decision to publish indicates that the publication is not manifestly wrong, but not necessarily that it is wholly right. A call-out in my article says, "...the one thing the peer case review does not do is provide a 'certificate' to the effect that the views expressed in the paper are correct to the exclusion of other views."

An editor should have considerable independence and be free from political interference. There was a sad case a couple of years ago when the Journal of the American Medical Association fired its editor for publishing a survey unpalatable to some people. At the other extreme, when I reviewed a paper for an engineering journal and suggested that the paper was more appropriate for a physics journal, the editor wrote back to me saying that it was "flattering" for an engineering journal to publish such a paper. Personally, I do not share the view that there is a pecking order for physics and engineering.

Mather is right to point out that the content of the *Manual of Concrete Practice* is a few years old. For example, the gestation period of the ACI Guide on the Use of Silica Fume in Concrete was 9 years. This is why we need up-to-date articles and papers; in this respect, *CI* publishes the material submitted with commendable speed. I am less sure about his view on the greater validity of committee reports. My experience of writing design codes taught me that some equations and coefficients are really compromise values, and not necessarily those derived direct from tests and experience. They may be just what designers want, but publication of actual experience is also most valuable.

Finally, I would like to thank both contributors for taking the trouble to comment on my article. It is gratifying that they agree with me, but even a strong disagreement would be preferable to deafening silence, which could mean that one's article is of no interest whatsoever. We need to read more letters and articles: with so much and so varied concrete construction worldwide, there must be much that should be shared via *CI*.

Adam Neville

Section 8.5: Objectivity in Citing References

When I was invited to write a paper for the special fiftieth anniversary issue of the Spanish journal *Materiales de Construcción*, I asked myself what kind of paper? One possibility is a report on an experimental project. Such reports are published in numerous journals in prodigious numbers, and they often present detailed data on obscure tests, which are new, but which do not enrich our knowledge. A most apposite comment was made by Treval Powers—the great man who explained the behavior of concrete in physical terms—who wrote in 1968: "Sometimes it is questionable, I think, to assume that more laboratory experimentation is the best next step toward that goal [making satisfactory concrete]. Perhaps the time has come to give priority to analysis of information already available."

The early, good analyses have now largely given way to state-of-the-art reports, often followed a few years later by the "another-look" variety. I do not think that I should contribute to this proliferation of redigested material.

So I have decided to write something quite different: my personal judgment of objectivity in selecting references in some publications. Mine may seem to be a dangerous endeavor, but I do not expect anyone to imagine that my criticism could possibly apply to him or her; *amour propre* would not allow this. So I can rest secure in the assumption that every reader will consider anything that I write as applying only to others.

How objective can we be?

Before commenting on objectivity, or lack of it, I wish to say that I am not naïve enough to believe that absolute objectivity is attainable, even in science. When conducting an experimental investigation, it is not sensible to use an arbitrary set of physical conditions, conduct a test, and see what comes out as a result; the colloquial English description is "suck it and see." Such an approach hardly merits the label "research." We must have a hypothesis, and this implies an outcome that we are looking for. It is not surprising, therefore, that, while scrupulously honest, we look for what we expect with perhaps greater zeal than we observe other outcomes. I believe that this is human nature: you are more likely to discern traits of genius in your own child than in someone else's.

Serious problems arise when research results are assessed from the standpoint of commercial considerations. I will give anecdotal evidence of such an approach. In the 1950s, I was testing concrete made with high-alumina cement to establish the influence of alternating temperature with various lengths of temperature cycle upon the so-called "conversion" of the hydrated cement and the associated loss of strength of concrete. I found that the effect of temperature higher than normal was cumulative and irreversible; the consequences of this for concrete in service are significant. Being young (well, relatively young) I was proud of my discovery and told the manufacturer of high-alumina cement of my finding. His reaction deflated my ego: "we have known this for some time," he said, "but it is not in our interest to publicize this situation." Naively, I asked whether it would not be a good idea to state on each cement bag that the cement should not be used in construction likely to experience periods of exposure to temperature above, say, 30 or 35 °C. The answer surprised me, but I am sure it was commercially sound: potential users might think that if the cement is unsuitable at higher temperatures, it may not be all that good even at lower temperatures.

The corollary of this anecdote is that researchers working for a commercial enterprise may be prevented from publishing results harmful to their employer. This is a fact of life, and I am not ingenuous enough to quarrel with such a situation. However, what is worrying when commercial research is published only in part, anything "undesirable" being screened out. Partial knowledge may lead a reader to misinterpret the true situation because he or she would read the truth but not the whole truth. To revert to my previous example of high-alumina cement, if results of tests at, say, 25 °C are reported (and this fact is stated) the reader might not stop to think that a somewhat higher temperature might result in a significantly different behavior. With increasing age, I have learned to look at the nature of the author's employment: university, research institute, a trade association, or a manufacturer.

Selection of references to oneself only

Nobody undertakes research without a background of work done earlier, and this is alluded to by way of "references" at the end of the paper. Occasionally all, or nearly all, references are to the current author's earlier work; nobody else's publications are mentioned. I am not pointing a finger at any one in particular, and it is possible that I, too, committed that sin in the past; if so, let me quickly say, *mea culpa*.

My concern is, not so much with the egocentric attitude of the author, but with the consequent lack of objectivity. The work of other researchers may contain somewhat different findings and may throw a different light on the same subject. The moral is: beware of accepting uncritically the findings in papers that do not include any mention of work of other researchers.

Omissions of contradictory references

Still on the subject of selective use of references, I have come across a different situation. What I wish to describe concerns, once again, high-alumina cement. This is not my *idée fixe* but I did, at one time, undertake a considerable amount of research in that field. Indeed, in 1963, I wrote a paper[1] containing a warning about the consequences of "conversion" of hydrated cement upon structural integrity, and in 1975, after the failure of some structures in England, I published a book under the title *High Alumina Cement Concrete.*[2]

There is a second reason for my referring to high-alumina cement concrete. This material was used in construction in Spain (as well as in some other European countries) and to this day, some structural problems are being remedied. So people in Spain are aware of the fact that high-alumina cement in construction may be problematic.

The example of a selective use of references that I wish to discuss is a chapter in *Lea's Chemistry of Cement and Concrete*,[3] which bears the title "Calcium aluminate cements." This is a new name for high-alumina cement, which dissociates (no pun intended) it from a certain ill-repute of high-alumina cement.

The chapter is 70 pages long and contains 193 references, going back more than a century. A whole range of aspects of high-alumina cement is discussed, including conversion. There is not a single reference to any of my work. One reason for this could be that my work is of poor quality and cannot be relied upon: I am not qualified to dispute such a point of view. It is also possible that my work, which was concerned with the deleterious effects of conversion of hydrated high-alumina cement, is not palatable to authors working for probably the largest manufacturer of that cement.

To consider the second possibility, I have looked at the original book *The Chemistry of Cement and Concrete* by Sir Frederick Lea,[4] formerly Director of Building Research Station, a British government body. The third edition was published in 1970, and has inevitably become out of date; hence, the new book, published in 1998,[3] containing various chapters written by different authors, but of course nothing by the late Sir Frederick Lea. Lea's book[4] makes reference to my 1963 paper. He cites a contradictory view as well, and says: "there has been a considerable controversy on the long-term effect of storage of concrete in water temperatures around 18 degrees and the data are somewhat conflicting;"[4] this is what the reader should be told.

Interestingly enough, three items of my work are cited in a book, published as far back as 1962: *High-alumina cements and concretes*, by T. D. Robson,[5] who was the director of the company manufacturing high-alumina cement in the United Kingdom; the manufacture ceased some time ago.

The most recent book on the chemistry of cement edited by Bensted and Barnes[6] contains a chapter on calcium aluminate cements, in which there are numerous references to my work.

Were I to elaborate the excision of my name in Reference 3 (thus becoming an un-person in George Orwell's parlance) I could be accused of egocentricity, and rightly so. My point is general: an author should cite those references that he or she disagrees with and explain why. If the author is right, the readers will fully accept his or her views.

Objectivity and uncertainty

My plea is for objectivity but this does not guarantee certainty about the findings of research. Heisenberg's Principle of Uncertainty states that it is impossible to specify or determine simultaneously both the position and velocity of a particle. In other words, in making observations, we "upset" that which we measure.

Consequently, we know that there are inherent limitations to the precision with which we can observe natural phenomena. We have to live with this situation, but we should not gratuitously obscure the reader's understanding of our research. Likewise, we should scrupulously avoid a distortion of the physical background to our research by an innocent selection of references based on what the great philosopher Bertrand Russell called "a delusive support of subjective certainty."

Conclusions

This brief section is concerned with just a couple of aspects of objectivity in technical and scientific literature; my examples may seem unimportant. However, I believe that the broad issue of scientific reliability has recently very much come to the fore. The views of some scientists have become strongly polarized; for example, those who express the opinion that genetically modified plants are beneficial to mankind are opposed by those who see nothing but evil in interference with nature. Few scientists are willing to occupy the middle ground, which would recognize three millennia of man's modification of natural plants. The debate between the two camps of committed protagonists generates much heat but not light. Some scientists are employed by single-issue protest organizations and simply mouth their employer's propaganda; others work for enterprises developing and selling new plants and may not be able to engage in unhindered debate. All this is not good for science, and it has cost the scientists much of their former reputation and glory.

In this journal, we are concerned with construction materials. These materials involve commercial interests, and this we have to recognize. Nevertheless, there should be a clear distinction between commercial promotion on the one hand and, on the other, technical, scientific, and research publications. If this distinction is blurred, the users or the various materials, their specifiers, and the entire research and technical community may well become disoriented or even misled. Such a situation would not serve the society at large well. Nor would it serve those involved in research who would lose the respect of the society and might even be treated with contempt as no more than "hired guns."

We are nowhere near such a catastrophic situation but, bearing in mind what has happened in the realm of genetically modified food, we should not start on the slippery slope. The least we can do is to be scrupulous in reviewing literature, regardless of whether or not it is favorable to the thesis of the research project; we should also acknowledge the relevant work of earlier researchers. In other words, at all times, let us be objective. This will be rewarded by the esteem of others as well as by self-esteem.

References

1. Neville, A. M., "A Study of Deterioration of Structural Concrete Made With High-Alumina Cement," *Proceedings of Institution of Civil Engineers,* Vol. 25, July 1963, pp. 287-324.

2. Neville, A. M., *High Alumina Cement Concrete*, The Construction Press, 1975, 201 pp.

3. Scrivener, K. L., and Capmas, A., *Calcium Aluminate Cements,* Chapter 13, P. C. Hewlett, ed., *Lea's Chemistry of Cement and Concrete*, Arnold, 1998, pp. 709-778.

4. Lea, F. M., *The Chemistry of Cement and Concrete*, Third Edition, Arnold, 1970, 727 pp.

5. Robson, T. D., *High-Alumina Cement and Concrete,* Contractors Record & John Wiley, 1962, 263 pp.

6. Bensted, J., and Barnes, P., eds., *Structure and Performance of Cements,* Second Edition, E&FN Spon Press, 2001.

Chapter 9

Litigation

It is a pity to end a book about the most successful construction material, concrete, on a negative note, but the topic of this chapter is a reality. There is a considerable amount of litigation arising from construction contracts, especially in the United States and the United Kingdom, as well as in some Commonwealth countries. In my opinion, the amount of litigation is excessive.

What are the reasons for this? I realize that I must not stray into any legal issues, but my strong impression is that the adversarial legal system in common law jurisdictions is conducive to litigation. More than that, the system leads to prolonged lawsuits with numerous expert witnesses. In fact, it is my personal experience as an expert witness that has led to my writing Sections 9.1, 9.2, and 9.3.

The situation is aggravated by the number of parties involved in construction. In addition to the owner, the designer (sometimes both an engineer and an architect), and the contractor, there are often numerous subcontractors and even sub-subcontractors. The entities that may be sued in contract or in tort, or which can sue for non-payment, are parties such as installers, materials specialists, slipforming subcontractors, ready-mixed concrete suppliers, as well as suppliers of numerous materials and products, and those engaged in steel bending or fixing. Each party may have its experts, usually several, and indeed it needs to have them because expert evidence presented by one party and not challenged by other parties becomes accepted by court to the detriment of the "silent" party.

While this book is not a forum for a discussion of legal systems, I have to say that my experience of litigation in Spain and in France is that the inquisitorial legal system results in less highly controversial and less cut-throat procedures.

All of Chapter 9 deals with problems in litigation in courts in various countries, all using the common law system. While most engineers, architects, and contractors hope not to be drawn into litigation, the danger of being sucked in without any malfeasance or culpability lurks in the background. Alas, suing everyone in sight on a contingency fee basis is well established in several parts of the world. Being forewarned may be of help to the innocent.

Section 9.1 discusses litigation in construction in general, and this is followed in Section 9.2 by comments about expert witnesses. I am sure that most of them are honest and well-meaning, but there are some who see themselves as advocates of the party that pays them. It is even worse when they give opinions outside their field of competence, and do so with apparent confidence. All this is contrary to the expert's duty to court, and fortunately, experienced judges and shrewd jurors can see through biased, tendentious, and selective evidence. Alas, sometimes they do not.

The duties and performance of expert witnesses are discussed also in Section 9.3, but with particular relevance to my experience in the United Kingdom. Nevertheless, Section 9.3 is likely to be of interest to readers in other countries as well.

Not to end the book on a depressing note, I would like to say something positive about litigation concerning construction. My early involvement in litigation (prior to the mid-1990s) before I saw some of its more cut-throat aspects, led me to recognize that litigation can have a positive effect by way of input to research. I wrote about this in the *Magazine of Concrete Research* in 1989, in a Guest Editorial, and this is included in the present book as Section 9.4.

From my subsequent personal experience I can testify to studies of problems identified in lawsuits. First of all, I had to conduct a thorough search of the available literature in order to formulate my opinions and to confirm or refute (and sometimes even simply to understand) the opinions of other expert witnesses. In some cases, I found a need for an extensive study in depth. A number of my investigations led me to write articles for *Concrete International* and papers for other journals; many of these are included in the present book.

My articles on litigation and expert witnesses provoked some letters from readers (reproduced after the relevant Sections, 9.1 and 9.2) and even a comment from the Editor-in-Chief of *Concrete International* at the beginning of the March 2000 issue. This reads as follows.

> "Litigation is an expensive waste," says Adam Neville, in his piece on page 64 in *CI* this month. He continues with, " …and something should be done to reverse this 'growing concrete industry,'" However, Neville admits this growth is not true

worldwide but certainly is felt in the United States, and his native United Kingdom. Though his article focuses on litigation in construction, he admits early on that legal disputes are endemic through such prone societies: "It almost seems that any mishap or suffering can be converted into hard cash." But centering on concrete construction, he analyzes why this segment of industry is particularly prone to disputes. Among a number of reasons, he offers observations on the nature of specifications and their intent, the roles in the litigation mix of the owner and the owner's engineer, of the contractor, the designer, and of course, the lawyer. He outlines the influences propelling the individuals into court. In his concluding remarks, Neville – who has served as a legal expert worldwide – offers some "more specific and practicable conclusions." While these may not eliminate the problems, their realization can at least help mitigate the impact on the true losers in construction litigation: "…the owners of all future buildings," as the author defines them.

I hope that this book will be of benefit not only to expert witnesses but to all those concerned with achieving good concrete structures.

Section 9.1: Litigation—A Growing Concrete Industry

For a number of years, I have been heavily involved in litigation concerning construction, mainly concrete. Because in the United Kingdom most construction disputes are resolved by arbitration, I shall subsume the term arbitration in the term litigation. The main difference between litigation in a court of law and arbitration is that, whereas the proceedings and outcome of the former are, or become, public knowledge, arbitration is a private procedure shrouded in secrecy so that its outcome cannot serve as a lesson to others.

Dispute resolution

To present a complete picture, I should add that, in addition to litigation and arbitration, there exist several other methods of resolving construction disputes: mediation; adjudication; conciliation; and also the so-called pendulum arbitration. Some of these procedures are binding on the parties involved, and others are attempts to resolve the problem so as to avoid proceeding to litigation or arbitration; in other words, they are attempts to persuade the parties to come to an agreement voluntarily. This is not easy when their positions have become entrenched and their relations inimical. However, this section is not about methods of resolving disputes, but about disputes themselves.

The title of this section suggests that litigation is on the increase. This is not true worldwide, but it is the case, for example, in the United States and in the United Kingdom. Indeed, *Concrete International* finds it useful to run a periodic column titled "Concrete Legal Notes" and so does the British journal, *The Structural Engineer*. Why is this so? One answer is that we are living in an increasingly litigious society; for instance, an unsuccessful surgical intervention is often followed by a multimillion dollar lawsuit, even when no medical negligence is alleged. It almost seems that any mishap or suffering can be converted into hard cash.

Reasons for disputes

Construction is particularly prone to disputes, perhaps because there are clearly two or more parties bound by a contract that provides for an artifact in exchange for cash. This, in itself, is not unusual, but the artifact acquired is not one "off the shelf" like a motor car, but it is rather "made to measure" like a tailored suit or a portrait. The person painted may feel that the portrait is "not good," perhaps because it is considered by the subject not to be a good likeness, possibly with a surfeit of "warts." Objective assessment is not easy.

In the case of a concrete structure, a part of the contract documents is the specification. However, the specification cannot, by its very nature, describe everything in full detail; inevitably, it uses phrases such as "workmanlike manner" and uses terms that have a customary meaning but are not fully quantifiable: for example, "thoroughly mixed" or "well compacted."

Even codes are sometimes, and inevitably so, not fully explicit and precise. For example, the Uniform Building Code, 1985, when dealing with materials used in masonry, says: "... (material) quality shall be based upon generally accepted good practice, subject to the approval of the building official." Without goodwill, such wording may cause problems because what is "generally accepted good practice" is a matter of judgment or opinion. A builder whose "practice" is poor may feel that his is "good practice." In litigation, this can be disputed, and the resolution of the dispute can take up much time and a great deal of money.

I should note in passing that, generally, "the approval of the building official" does not exonerate the parties from the requirement to comply with the code, and it is this compliance that may be the subject of dispute or litigation.

The unavoidable absence of full and absolute detail need not, of itself, give rise to a serious dispute, provided there is an honest understanding of what is required and what can be provided. Such an attitude requires good will between the parties, and that is not always present.

Sometimes, the lack of good will arises through a tactless or overbearing attitude of the engineer representing the designer and, indirectly, the owner of the structure, towards the contractor. At other times, the contractor undertakes the construction with the object of making claims for payment for variation orders from the specification. I have seen cases where the tender price is unrealistically low because the contractor bids with the intention of making money by making claims for additional payment. I once saw a contractor's copy of the specification with a note in the margin: "This is good for a claim."

In another case involving a contractor from a non-English speaking country, there was a margin note in the other language: "They are fools—we shall get them on this point." Unfortunately for the contractor, I could read that language. What I am trying to say is that, with that kind of attitude already prior to the commencement of construction, nothing but strife can result during its progress.

Of course, the fault can lie also with the designer. In that case, it is less probable to be an explicit desire to influence the total cost of construction, but it is more likely to be a consequence of carelessness or even incompetence. The most common example is a poor specification. This is a topic upon which I have touched in an earlier article.[1]

Process of litigation

All of the preceding indicates how and why disputes arise: In the absence of good will —an attribute not particularly abundant in mankind—

they lead to litigation. Litigation means that the parties to the dispute leave it to a third, independent party—a judge, the jury, or the arbitrator—to determine the issue. There is nothing wrong with this by itself.

One problem is that, invariably, the parties do not present their case themselves, but do so through their lawyers. Again, there is nothing wrong in that by itself. Indeed, the lawyers' help is essential because only they know the contract law and the legal procedures; it has been said that a person who is his or her own lawyer has a fool for an adviser.

On the other hand, with honorable exceptions, lawyers are not technically versed and only rarely are fully specialized in the particular technical field around which the legal case revolves. Nevertheless, I am not suggesting for a moment that we should do without lawyers, or even that they are a necessary evil. They clarify and categorize the issues and they are able to formulate the pleadings in a well-defined and manageable manner. Most are very capable and genial, and nowadays, very many of them are women; I am not suggesting any cause-and-effect link between the two parts of the preceding phrase.

Trouble begins when a grim determination to get money begins to run a lawsuit; a crude and somewhat unfair word to describe this is greed. The greed may be on the part of the plaintiff (claimant) who is determined to squeeze the last ounce of blood out of the defendant (respondent). The greed may also be on the part of the lawyer. This may, but of course need not, be the case when attorneys are retained on the basis of contingency fees. This system means that their fees are contingent on "victory." Again, this in itself need not be bad. However, when a large share of the money that changes hands in consequence of the court decision goes into the pocket of the law firm, there is a strong temptation to fight to the death.

Because, as its name states, *Concrete International* is an international magazine, a comment from the United Kingdom may be in order. With litigation costs of individuals, such as homeowners or small subcontractors, paid for by the state, their lawyers have an interest to fight even cases that are doomed to be lost. With respect to this so-called "state legal aid" in general (and not in construction), Holcombe[2] wrote: "A change is drastically needed in the whole culture of our litigation process where lawyers are motivated not by justice, but more often just pure greed, and this is surely wrong."

I must make it clear that I am not critical of contingency fees as such. I am conscious of the fact, and I want to emphasize it, that for the "little man" without much money and without state support, there is a need to undertake litigation without a risk to all his savings, if not bankruptcy.

Evidence of expert witnesses

In construction disputes, the wish to fight on regardless, and to win by fair means or foul, may overshadow the need to establish the technical truth. In such a case, expert witnesses may be questioned less with the intention of hearing their technical knowledge, but more so for the purpose of discrediting them. This may be done, for example, by demonstrating inconsistency in their opinions.

Clearly, if an expert witness ascribes a certain type of cracking of concrete to, say, drying shrinkage in one case, and in another case, under closely similar circumstances, opines that the cracks are due to, for example, overloading, then the lawyer for the opposing side is justified in accusing the witness of being less than truthful. But there may be good reasons for inconsistency: What the witness believed to be correct 20 years ago, he or she may hold to be incorrect now.

An example of this situation is the publication of papers and, even more so, of books, over a period of time. I am potentially vulnerable in this respect because the first edition of my book, *Properties of Concrete*,[3] appeared in 1963 and the fourth edition in 1996. Specifically, in the older editions, I, like many others, held the view that strong concrete is automatically durable concrete; of course, now I am no longer of that opinion. I admitted my "error" in an article in *Concrete International*.[4]

In some cases, fortunately rare, lawyers may try to misrepresent a change in opinions in consequence of new and better knowledge, and to accuse the expert witness of shaping his or her evidence to suit his or her client's interests. The lawyers may also be very selective in representing the witness's published opinions, they may cite those opinions out of context, or even distort those opinions by citing "often" instead of "very often" or "high" instead of "very high." I also have experienced words added to a quotation, rather than omitted.

None of this contributes to establishing the true facts; all of this puts a great strain on the witness. I propose to discuss the position of an expert witness in a future article in *Concrete International* (see Section 9.2).

Cost of litigation

Litigation, good or bad, costs money, usually a lot of money. What is rarely considered is the question: whose money is it? In the first instance, the winning party thinks it is the loser's money. Suppose the owner of a building is paid damages by the contractor. The contractor carries insurance, so the money does not come out of its pocket. However, the insurer cannot, in the long run, be out of pocket because its pocket is really only a balancing vessel: those insured who suffer no loss put money into the vessel to pay for those who do suffer. So all the money is ultimately contributed by those insured, and not by the insurer.

If there is a large number of losses in a certain category, the insurance rates go up, and the contractor pays a higher premium. But this does not come out of the contractor's pocket: the premium is part of the operating costs so that, with a higher premium, the price of construction goes up. So who pays in the end? The owner. Of course, it is not the owner who won the lawsuit who pays, but the owner of the next building to be constructed, and indeed the owners of all future buildings.

A similar situation has arisen in medical claims: the cost of insurance paid by medical practitioners has rocketed, but it is actually paid for by the patients' fees.

Conclusions

The purpose of this section is not to describe the existing situation as I see it and just deplore it. Litigation is an expensive waste, and something should be done to reverse this "growing concrete industry," as described in the title of this section.

It would be easy to end this section by pious remarks such as: we should change the culture of dispute; we should abandon our adversarial, sometimes viciously so, attitudes; and we should seek to establish the truth with integrity and impartiality. Of course we should, but such exhortations are not enough.

In trying to end with some more specific and practicable conclusions, let me be more definite. As for contractors' claims for extra payment, there should be fuller discussions at the tender stage; the contractor should give a binding assurance that he or she understands fully the specification and does not propose to claim variation items except for changed quantities. Contractors that break that assurance should not be allowed to tender in the future.

In appropriate cases, contracts should be replaced by team working based on interdependence underlain by mutual trust. In other cases, negotiated contracts might be used, offering continuity of work. In some fields of endeavor, such as major retailers, continuity of purchase from a given manufacturer is guaranteed but, in construction, procurement is rarely integrated with execution. Partnering is the "in-word."

With respect to claims by owners, I would plead for a more realistic appreciation of the nature of concrete. Concrete is a man-made material consisting both of manufactured and natural ingredients, partly processed in a semi-industrial manner. Construction in concrete is often achieved under difficult conditions. It is therefore unrealistic to expect perfection in every detail and at every location.

To consider every crack as a fatal flaw is not sensible; concrete is a brittle material subject to shrinkage. For example, with reference to nonstructural floors, ACI 302.1 says: "Since drying shrinkage is an inherent characteristic of portland cement concrete, it is normal to experience some curling and cracking on every project."[5]

For the owner to expect the best, and not just that which is good enough for the required purpose, is commercially not sensible. The engineer's task is to produce a structure good enough for the purpose and not the best structure that money can buy. To give an extreme example, if a farmer needs a bridge to take a herd of cows over a railroad, it would be a sheer waste of money to build a bridge capable of carrying a Sherman tank.

Many similar statements can be made to emphasize realistic expectations of concrete structures that can be achieved at a reasonable price. Perhaps a document establishing this could be produced by ACI for the benefit of future owners of concrete structures, as well as of engineers and contractors. Would this also be of benefit to lawyers? To those who see their profession as serving the cause of justice, yes, it would be.

References

1. Neville, A., "What Everyone Who is 'In' Concrete Should Know About Concrete," *Concrete International,* V. 21, No. 4, Apr. 1999, pp. 57-61. [Section 6.2]

2. Holcombe, C., Letter to *The Times,* London, Apr. 28, 1999.

3. Neville, A. M., *Properties of Concrete*, Fourth Edition, Longman, London, 1995, and John Wiley, New York, 1996, 844 pp.

4. Neville, A., "Maintenance and Durability of Concrete Structures," *Concrete International,* V. 19, No. 11, Nov. 1997, pp. 52-56. [Section 3.2]

5. ACI Committee 302, "Guide for Concrete Floor and Slab Construction (ACI 302.1R-89)," American Concrete Institute, Farmington Hills, Mich., 1989, 45 pp.

Letter to the Editor No. 1
Litigation and the concrete industry

Adam Neville's article "Litigation—A Growing Concrete Industry," in the March 2000 issue of *Concrete International,* provides some of the reasons why there appears to be an "open season for lawsuits in the concrete industry." As he so clearly states, the reason in many cases is simply greed. Underlying this reason to a great degree, in the opinion of many, is the legal fraternity. It may still be true that for a relatively small fee and a mean disposition, one can file a lawsuit against anyone. Many of these lawsuits are literally frivolous, but the attorneys for the plaintiff frequently convince the courts that they are extremely valid. In fact, their validity is often measured by the amount of money stated in the initial claim. In most cases, this might be classified as "wishful thinking," but in many cases Mr. Neville's use of the word "greed" seems much more definitive.

A number of ideas have been proposed to correct these legal inequities…the lawyers in Great Britain, as well as other countries, have introduced their own punishment that makes it illegal to practice "barritry." This is the practice that lawyers might follow by working for no compensation until the final judgment is made in favor of their client. If a settlement is awarded to the plaintiff, it is divided between the attorney and the client. If the attorney does not prevail, he or she receives nothing. Most important, this practice of "barritry" can lead to loss of license as a barrister. By contrast, the practice of barritry is permissible in the U.S., but it is known as payment by "contingency." Repealing the contingency system following similar practice used in Great Britain might substantially reduce the number of lawsuits in the U.S. courts. Those unable to properly compensate a lawyer might obtain legal counsel "pro bono" (free) through their local bar association.

Another proposal that could substantially reduce construction litigation as well as other forms of "deep pocket" activity would be similar to the practice of posting a bond, which is commonly (and legally) used on many construction contracts. In this application, it is suggested that the plaintiff post a bond in the full amount of the expected judgment plus an estimated amount of the plaintiff's and the defendant's attorney fees and all court costs when the lawsuit is first filed. Then if the plaintiff loses the case, he or she does not initiate any expenditure for an otherwise innocent defender. Therefore, if greed is a motive, the plaintiff gets nowhere. It is debatable if this proposal would ever be adopted as a law by any of the States or by the Federal Government. However, it might serve to answer some of the questions raised by Mr. Neville in his most interesting article on litigation.

R. E. (Bob) Tobin, FACI
Structural Engineer Concrete Consultant
Arcadia, Calif.

Letter to the Editor No. 2

I have read the article by Adam Neville entitled "Litigation—a Growing Concrete Industry" with interest. This information is timely because of the large volume of litigation in progress in Canada and the U.S. At the Fifth CANMET/ACI International Conference on Durability of Concrete, held June 2000 in Barcelona, Spain, there was a special session dealing with the issues raised by Mr. Neville. My introductory remarks to the session were entitled "Have We Been Taken to the Cleaners by the Lawyers, or Are We a Willing Party?" Several other papers on the subject were presented.

Mr. Neville stated that the bulk of the money in litigation goes to the lawyers. This is true, but he does not mention that a number of so-called "expert witnesses"…have also benefited by participating in the process. In fact, some of our colleagues very aggressively seek this role and therein lies the problem. Unfortunately, in doing so, some of them have lost their reputations, and others are in the process of doing so.

In my opinion, the only way we can slow down the growth in litigation is to discourage both owners and contractors from resorting to this route. Instead, new mechanisms should be developed to resolve construction disputes. These could be arbitration and mediation, and should be overseen by committees established by professional organizations including ACI, ASCE, and others. This will considerably reduce the costs associated with dispute resolution.

V. M. Malhotra, ACI Honorary Member
ICON/CANMET
Ottawa, Ontario, Canada

Neville's response

The letters by Bob Tobin and Mohan Malhotra are not only very important but could readily become stand-alone comments. In the case of Dr. Malhotra, he refers to his presentation at the CANMET/ACI International Conference on Durability of Concrete, at which he spoke in an impressive and persuasive manner about expert witnesses, some of whom, as he writes, "have lost their reputations." Dr. Malhotra's letter mentioned other papers dealing with expert evidence at the Barcelona Conference. Alas, at that event, three papers were not presented; one supposedly because of an appeal in litigation. I found this influence on a scientific paper somewhat surprising and asked which "equation" applies: Truth equals truth, or Truth is a function of litigation? Indeed, the relation between truth and litigation is the nub of the matter.

Dr. Malhotra deplores the fact that "some of our colleagues very aggressively seek this (the expert witness) role." I fully agree with him, but if the plaintiffs use such people, the defense needs expert witnesses, too. An expert who does not trim his sails to the (attorney's) wind must be equally willing to be retained by either party; I am pleased to say that I have followed this rule on both sides of the Atlantic. For whichever party the expert works, charges should be billed for his or her time; as St Luke wrote, "a laborer is worthy of his hire." Indeed, an expert working for free, let alone on a contingency-fee basis, could be suspected of having a hidden axe to grind.

Dr. Malhotra is absolutely right in his comments about some expert witnesses. Some of them become professional witnesses and deliver a polished performance: I am referring to the quality of their presentation, and not to the technical content. Bickley[1] quotes Taylor on Evidence: "Perhaps the testimony which least deserves credit with a jury is that of skilled witnesses. These witnesses are usually required to speak, not to facts, but to opinions..." Others, when engaged by a party to litigation, are able to modify testing procedures: I observed tests to determine the rebound number of hardened concrete, in which the tester took six readings but recorded only five or even four. When I questioned his procedure, he replied that he was discarding the outliers, which thus never appeared on paper. And yet, ASTM C 805 clearly states that only a reading differing from the average of 10 readings by more than 7 units should be discarded. Discarding before the 10 readings have been taken could be interpreted as discarding values that are too high for the tester's liking, that is, higher than what he wishes to establish.

Another example of tailoring testing to an expert's objective is using established ASTM test methods for purposes for which they are not intended. I have seen the use of ASTM C 1202 to assess the permeability of concrete to water whereas ASTM clearly states that the "test method covers the determination of the electrical conductance of concrete to provide a rapid indication of its resistance to the penetration of chloride ions." It would require a jury of scientists or engineers to appreciate the difference. It would require a like jury to realize that ASTM E 96 is not intended to be used on concrete, as the test method is intended to "cover the determination of water vapor transmission of materials...such as paper, plastic films, other sheet materials, fiber boards, gypsum and plaster products, wood products, and plastics." It is difficult to see how concrete fits in this list, and yet I saw an expert witness apply the test method to concrete. Taking the most charitable view, it is possible that the witness was inexperienced, or naïve, or unduly influenced, but these are not appropriate qualities of an expert. Unfortunately, misuse of test methods, without expressly considering the scientific basis for doing so, is on the slippery path to the so-called junk science haunting the courts and clearly discussed by Matson.[2]

Therefore, Dr. Malhotra is right in his criticism of those whom he describes as "so-called expert witnesses." There are far too many of them in various lawsuits, and they cost a lot of money. They also give concrete an undeservedly bad name by portraying spurious bad concrete or, rather, "virtual bad concrete."

The problem of expert witnesses has now been tackled in England; as this took place after my article had been written, I would like to refer to this topic. Given that *Concrete International* is an international journal, procedures in other countries rooted in common law and in the adversarial approach should be of interest to a wide range of readers.

The Civil Procedure Rules,[3] which came into force in 1999, are said to represent the most far-reaching change in 100 years. The rules give the judge a great power to manage the case, removing much of the control of litigation from the parties and their attorneys. Rule 35.1 states: "Expert evidence shall be restricted to that which is reasonably required to resolve the proceedings." Specifically, Rule 35.3(1) says: "No party may call an expert or put in evidence an expert's report without the court's permission." Moreover, Rule 35.4(4) states: "The court may limit the amount of expert's fees and expenses that the party who wishes to rely on the expert may recover from any other party." Irrecoverable costs and the risk of displeasing a judge should prove to be great deterrents.

These rules alone should satisfy Dr. Malhotra's concerns: expert witnesses could not claim excessive fees on the assumption that the employing party will recover them from other parties; nor could numerous experts in the same field be employed. Sometimes they are hired to deny the opposing party the availability of experts in a given field. As expenses are limited as well, a party in the U.S. could not and would not need or want to employ an "imported" witness like me!

Although agreement between the experts of opposing parties was usually sought in England (in contrast to the cases in which I was involved in California where there was no contact whatever) the new Rule 35.12 expressly states that the court may direct a discussion between experts requiring them (and not the lawyers) to "identify the issues in the proceedings; and where possible, reach an agreement on an issue." Such a discussion is privileged, that is, it may not be referred to in court.

The court may direct the experts to "prepare a statement for the court showing those issues on which they agree; and those issues on which they disagree and a summary of their reasons for disagreeing." Such an agreement does not bind the parties, so that the experts cannot usurp the lawyers' powers. I view this approach as very valuable because it concentrates the experts' minds and prevents a repetition of my experience in a case under English law. I did not agree to another expert's suggestion because I thought it to be scientifically flawed. Later on, he disagreed with me on another matter, presumably for scientific reasons as well. However, he offered to change his mind in return for my giving in on the earlier issue. This comes back to my question about truth being a function of litigation and subject to bargaining.

I have discussed the new English Rules at some length because they intend to improve the operation of the adversarial system but not to do away with what is enshrined in England as much as in the U.S.

The above comments do not deny that there may be honest and sincere disagreement between experts that is independent of their respective paymasters. Total objectivity is elusive; for example, one expert may prefer a particular design method, if only because he is familiar with it. But the expert must not omit to consider material facts that could detract from his concluded opinion, for example, that other design methods have been used and have proved to be successful. What is needed in an expert is the willingness to listen, to argue rationally, and to be prepared to change his or her mind on the basis of technical arguments.

Dr. Malhotra refers to the comment in my article about some cases in which "the bulk of the money in litigation goes to the lawyers." Mr. Tobin blames "the open season for lawsuits" on lawyers. The solutions to this problem proposed by both Mr. Tobin and Dr. Malhotra are laudable but they require a fuller discussion than is possible here. All I wish to do is thank them both for their valuable contributions.

Adam Neville

References
1. Bickley, J. A., "The Engineer and Litigation," *The International Construction Law Review,* V. 8, Part 3, July 1991, pp. 398-405.

2. Matson, J. V., *Effective Expert Witnessing,* Third Edition, CRC Press, 1998.

3. *The Civil Procedure Rules 1998*, Statutory Instrument No. 3132, London, England.

Section 9.2: An Expert Witness—As Reliable as Concrete?

In the last issue of *Concrete International* there appeared my article on litigation in concrete construction.[1] Such litigation almost invariably involves expert witnesses who have an important role in influencing the outcome of the dispute. Clearly, percipient witnesses, that is, witnesses of fact, also have an important role, but they are not the subject of the present section.

The writing of this section has also been prompted by two letters in *Concrete International* concerning the role of expert witnesses and the veracity of their statements. In a nutshell, McDonald said that the best chemists and petrographers are in the unfortunate situation of having to defend their positions.[2] In the second letter, Idorn said that the expert investigator must be truthful.[3] What Idorn says is right but what McDonald says is sometimes true. It follows that the topic is worthy of consideration.

Having been an expert witness in numerous disputes concerning a wide range of structures in England, the United States, Hong Kong, Spain, Singapore, France, and the Middle East over the last 20 years, I would like to say something that may be of interest not only to American readers but also to those elsewhere.

Duties of an expert witness

It is obvious that a witness has to tell the truth. When he or she is sworn in, he (for brevity, "she" is subsumed in "he") undertakes to tell the truth, the whole truth, and nothing but the truth. I do not think it necessary to start with St. John's record of Pilate's words: "What is truth?" Nevertheless, the triple reference to truth in the oath requires some consideration.

A witness can tell the truth but withhold some vital and crucial facts. He would thus fail to tell the whole truth. A very simple example of the above is as follows. Imagine that a witness said: I performed the tests on absorption of coarse lightweight aggregate according to ASTM C 127. This was the truth, and the jury might believe that the testing was done as it should be. But it was not the whole truth because the scope of ASTM C 127 explicitly excludes lightweight aggregates.

It is also possible for a witness to tell the whole truth and to add some statements that are not true. By doing so he would not comply with the undertaking to tell nothing but the truth. An example of failing to tell nothing but the truth is as follows. Imagine that a witness said: The tests that I performed were according to ASTM C 642. This would give credence to his tests. While he performed all the tests prescribed by ASTM C 642, he performed also some tests not covered by that standard but devised by him for the purpose. Thus he did not satisfy the requirement to tell nothing but the truth.

All this is obvious, but occasionally an expert witness cites those published references that support his argument but ignores those that undermine, or even refute, his opinion in support of his client. In the United States, the United Kingdom, and in many other countries, but not, for example, in France, each party to the dispute appoints its experts. It would be naive not to recognize that each party seeks an expert who is likely to be sympathetic to its cause, and may not engage one whose views are likely to be harmful to the party in question. However, truth must prevail at all cost.

Role of an expert witness

Many construction disputes involve technical matters, sometimes complex, which are not immediately understood by the jury or the judge. The phenomena involved may require an explanation that, in turn, involves testing; hence the need for experts who can unravel the complexities and present their opinions to the judge and jury. This requires knowledge of materials and material behavior, and structural design and analysis, so as to elucidate and explain physical phenomena and events. All this will be used by the attorneys for the two parties to the dispute and enable the judge and jury to establish or apportion responsibility and the financial consequences. We should not lose sight of the reality that a dispute is not so much about who is right as about who pays.

In an earlier article,[1] I commented on the fact that we live in an increasingly adversarial society and we work in an adversarial legal system which, by its nature, leads to taking extreme positions. This contrasts with an inquisitorial system of finding solutions to disputes such as in France. In that country, the tribunal often appoints its own expert. Sometimes, that expert tenders his opinion to the judge in private. In other cases, the tribunal expert is there in addition to the parties' experts; he helps the judge to interpret, reconcile, or choose from the conflicting opinions.

I am not competent to comment on which system is preferable, and anyway this is not a forum for seeking change, if change indeed be needed. What matters is that the expert witness honestly contributes to establishing the truth. It may be valuable to quote from the summary of a presentation before the British Parliamentary and Scientific Committee (of which I am a member): "The role of the expert witness is to clarify and explain the specialist evidence brought before the court, yet the system pits specialist against specialist rather than engaging them in reasoned debate, producing a consensus opinion leading to an understandable explanation to the jury of often complex issues."[4]

Attributes of an expert witness

From the preceding discussion it could be concluded that the situation is straightforward and there should be no doubts about the role of the expert witness. Of course, most experts are scrupulously honest and truthful, and they can give a positive answer to the question once asked by a judge:

"Would you hold the same views if you were retained by the other side to the dispute?"

Alas, some other witnesses are partisan, selective in the evidence they use, and indulge in advocacy of their client's cause. Advocacy is exclusively the attorneys' domain. Even an intrinsically honest witness is not serving the cause of justice well if he espouses his client's case. By the same token, a doctor who is emotionally involved with his or her patient may well become distracted from the best medical path.

An extreme case was quoted by Haley and Shaw.[5] A witness for the defense was being particularly partisan, but counsel for the plaintiff had uncovered an article written by the witness some five years previously, in which he wrote: "How should the expert avoid becoming partisan in a process that makes no pretense of determining the truth but seeks only to weigh the persuasive effect of arguments deployed by one adversary or the other?" He then referred to the sleight-of-hand in the "three-card trick" and continued: "If by an analogous sleight-of-mind an expert witness is able to present the data so that they seem to suggest an interpretation favorable to the side instructing him, that is, it seems to me, within the rules of our particular game."

The castigation by the trial judge is worthy of being quoted:

> The function of a court of law is to discover the truth relating to the issues before it. In doing that it has to assess the evidence adduced by the parties. The judge is not a rustic who has chosen to play a game of "three-card trick." He is not fair game. Nor is the truth. That some witnesses of fact, driven by a desire to achieve a particular outcome to the litigation, feel it necessary to sacrifice truth in pursuit of victory is a fact of life. The court tries to discover when it happens. But in the case of expert witnesses the court is likely to lower its guard. Of course the court will be aware that a party is likely to choose as its expert someone whose view is most sympathetic to its position. Subject to that caveat, the court is likely to assume that the expert witness is more interested in being honest and right than in ensuring that one side or another wins. An expert should not consider that it is his job to stand shoulder-to-shoulder through thick and thin with the side which is paying his bill.

These words emphasize the particular requirements of honesty on the part of an expert witness, a professional man or woman with a duty to the public. So why do some expert witnesses behave otherwise? I suppose the answer lies in human frailty: The temptation of a high remuneration is great. The expert witness is paid for the time he spends on the case. The hourly rate is generally higher than for work in the design office or on site, although nowhere near the lawyer's rate, at least in England.

I am not being particularly critical or sanctimonious. An expert witness's work is onerous and stressful. He must rapidly think on his feet and he cannot afford to give a sloppy, let alone wrong, answer. All this while the cross-examining attorney may be out to wrong-foot or trip the witness. So I must remain sympathetic to, and understanding of, the witness's task, even if he occasionally yields to shortcuts that seem to offer him protection. However, I am critical of the notion of expert witnesses retained on the basis of contingency fees, overtly or covertly. It is one thing for an advocate to be remunerated on that basis; it is another for an expert witness who is seeking the truth to find a conflict between the truth and money.

The actual situation

It would be unrealistic to expect the expert witness to be infallible. There are cases when he simply does not know, and he should say so. Likewise, there are cases where the evidence is conflicting; he should say so, too. And there are cases where he is sure of his ground but has not yet heard all the evidence; he should then say: if there is any further evidence, I may reconsider my opinion.

It is realistic to accept that, when there are "shades of gray" in a situation, the expert witness may not be entirely unaffected by his client's interests. But he must not stray from the "gray" into black or white. And he must not be "economical with the truth." A particularly objectionable example of this is quoting from published work in a selective manner. I have seen a quotation to the effect that shrinkage cracking occurs in slabs but omitting the following sentence: "However, this can be prevented by a uniform restraint." Such an omission not only creates a false impression, but is also stupid: it is all too easy for the opposing party to check the source of the quotation. Obvious, isn't it? But covert omission is sometimes practiced, not only by experts but also by advocates.

Having said all that, I believe that very few experts are deliberate liars, but some occasionally overstep their area of expertise and venture into areas of which they have only limited knowledge. This may happen when an eminent researcher in the field of cement and concrete, but without practical experience in construction, without knowledge of concrete on the job, and without knowledge of structural behavior, says "I will now put on my engineering hat" and opines on structural safety. This is as if I "put on my legal hat" and opined on a question of law (even though I qualified as an arbitrator in Scotland and England). But then the world is full of salon lawyers and generals.

A professional expert witness?

Who becomes an expert witness? It is obvious that an expert witness must be, first and foremost, an expert in the field in which he offers his opinions. If he has no experience with using concrete on the job site, he should not opine about good and bad practice. It may be easy to pontificate on the basis of book knowledge or laboratory experience, but there is a danger of holding unrealistic expectations of what can be achieved on the job. I cannot resist mentioning that, on one occasion, a graduate student performing routine testing on a structure subject to litigation, confided in me that he was planning to become an expert witness as his first job.

But even with experience under the belt, one should not become a professional expert witness to the exclusion of other professional activity. Without such activity, there is a danger of becoming out of touch with practice and changes in materials, procedures, and even standards. Also, if someone sees nothing but "structures in trouble" his views may become jaundiced and he may cease to take a balanced view of engineering practice. A man who conducts nothing but *post mortem* examinations may cease to be a good physician.

Who should act as an expert witness?

So to become an expert witness one should, first of all, be an expert practitioner. As to how to become an expert witness, there is a view that one should follow courses and learn the tricks of the trade. In my view, training is not necessary: Much can be achieved by experience and study.

Another pertinent question is: should a witness invariably act for the owner of a structure or alternatively always for the contractor? I am of the opinion that acting sometimes for one party and sometimes for the other enables an expert witness to maintain a balanced view of life. Otherwise, he might convince himself that the contractor is always out to grab money and to do down the owner, or else that the owner will use any trick not to pay the contractor his due.

The preceding point has a bearing on relations between witnesses. If witnesses alternate, not necessarily equally, between working for the plaintiff and for the defendant, then different witnesses may occasionally work alongside one another and at other times for opposing parties. I have experience of being engaged, at the same time, alongside Witness X in one case, and against him in another case. This system makes for better professional, as well as personal, relations between the experts. Thus they are less likely to forget that, after all, they have a common goal: establishing the truth.

Good relations between the witnesses make it easier to achieve agreed statements. In many cases, the judge orders the experts from the two sides to agree on "facts as facts" and "figures as figures." The witnesses are also expected to reach agreement on theories and test methods, wherever possible, and to produce a list of unresolved items. This reduces time spent in court and the cost of the litigation. In an earlier article,[1] I pointed out that it is not just the loser who pays; in the long run, the cost is passed on to future owners.

If relations between the witnesses for the two sides have become hardened into permanent animosity, progress is hampered. If a party goes so far as to forbid its experts to communicate with the other party's experts (and I have encountered that), then achieving agreement outlined in the preceding paragraph becomes impossible. He who is imbued with enmity to other experts should not act as a witness.

Conclusions

Resolution of construction disputes requires a judge and jury as well as attorneys to plead their clients' cases. All of the above need technical help in elucidating the issues and in finding valid explanations of the phenomena involved. Such help is provided by expert witnesses. If the object of the litigation is justice, as it should be, then there is a heavy burden on the witness: he should be competent, honest, and independent. The expert witness should be as reliable as concrete.

Being competent, honest, and independent is onerous. There is a story about a communist party leader in the bad old days. He boasted that his party members were honest, loyal, and intelligent. When questioned more closely, he admitted that those members who were loyal and intelligent were not honest; those who were honest and intelligent were not loyal; and those who were honest and loyal were not intelligent. The leader's boast was the truth. But was it the whole truth? Was it nothing but the truth? Or shades of truth?

I hope that those who contemplate acting as expert witnesses will find this section helpful. By the same token, those less interested in truth than in personal gain might also find this section useful in that it will deflect them to other pastures.

References

1. Neville, A., "Litigation—A Growing Concrete Industry," *Concrete International,* V. 22, No. 3, Mar. 2000, pp. 64-66. [Section 9.1]

2. McDonald, D. B., "Letters," *Concrete International,* V. 19, No. 7, July 1997, p. 9.

3. Idorn, G. M., "Letters," *Concrete International,* V. 20, No. 1, Jan. 1998, p. 8.

4. Parliamentary and Scientific Committee, "Scientific Evidence in Courts of Law," London, Jan. 26, 1998.

5. Haley, G., and Shaw, G., "Expert Evidence and the Truth," *The New Gazette,* Hong Kong, Dec., 1995, p. 50.

Letter to the Editor No. 1
The expert witness re-examined

In the article by Adam Neville, "An Expert Witness—As Reliable as Concrete?" [*Concrete International,* V. 22, No. 4, April 2000, pp. 59-61], the author has shared many insights relative to the expert witness industry, and quite correctly pointed out that it is a huge industry indeed. His accounts of experts and "not-so-expert experts" are both factual as well as somewhat entertaining. Having served as an expert witness on many litigation matters over the last three decades, I was particularly interested in the author's reflections. His observations are consistent with my experience, although I believe further discussion on some of his points is in order.

His observation that many players are drawn to become expert witnesses as "The hourly rate is generally higher than for work in the design office or on site" is factually correct; however, that charge is usually limited to actual time spent on a particular matter. True experts expend an enormous amount of both time and money in maintaining their expertise, which is generally not directly billable to a client. This includes subscribing to and reading many journals and other technical literature, as well as participating in technical organizations, of which ACI is but one. Additionally, they will often attend highly specialized technical conferences, held in many places around the world. The hourly rate charged for services must be sufficient to cover these non-billable expenses, and thus the real compensation is not nearly as great as it first appears. The degree of expertness that has been achieved is the direct result of these non-billable technical activities and independent research, and thus the ones who are most active are usually the most expert. They must charge accordingly, and often will receive the highest hourly rates. I believe this is as it should be.

As credentials, some will submit long lists of organizations to which they hold membership; however, membership in itself does not provide much knowledge or expertise. It is attendance at the meetings and active participation in the technical activities of such organizations that result in real knowledge. The personal contacts one makes through such activity are also of great value, for it is here that one develops a network of real experts with whom technical matters can be discussed, much to the benefit of all.

I must wholeheartedly agree with the author that "training is not necessary: much can be achieved by experience and study," and find this a noteworthy observation. It is not unusual to find experts touting their academic degrees. While basic education and an understanding of fundamentals are certainly pertinent, one does not become an expert on them alone. It is not until an individual has obtained many years of practical experience, beyond that formal education, that one can become a true expert. It is this writer's view that what one has done to further one's knowledge following formal education is of far more significance than whatever degree one once obtained.

Neville comments, "Few experts are deliberate liars, but some occasionally overstep their area of expertise and venture into areas of which they have only limited knowledge." My experience has been that self-proclaimed "experts" very often venture into areas in which they have little real expertise and, not infrequently, into areas in which they have virtually no real experience! I would suggest that the first qualification of an expert is to "know what one knows not," for certainly no one knows all there is to know about all the aspects of concrete, or for that matter, any other material or subject.

Is it reasonable to expect a construction expert to understand the details of design analysis or cement chemistry and paste microstructure? To develop expertness in construction, one would have devoted most of one's time and energy on site. Likewise, one should not expect chemists to possess a detailed understanding of site work or design, for their primary activity is in the laboratory, not in the design office or on site during construction.

Yes, the true experts "know what they know not." They are honest with both themselves and their clients, and admit their limitations. They will have developed relationships with other experts whom they can recommend to fill voids in their own knowledge. And when this is so, the expert will be able to be "competent, honest, and independent" and "be as reliable as concrete."

James Warner
Consulting Engineer
Mariposa, Calif.

Letter to the Editor No. 2

Adam Neville made a very good point when he said, "A dispute is not so much about who is right as about who pays." This has created the "deep pockets concept"—bringing into litigation any and all parties with the ability to pay, regardless of their possible contribution to the cause of defective construction.

Years ago, I was one of the expert witnesses in a case of a fire in the main building of a hotel. A separate cabana building was spared by the fire; during litigation, it was used to store the luggage of the guests after the fire, and the plans, specifications, and related case documents available for examination by lawyers and expert witnesses. The engineer who only designed the air conditioning of the cabana building was sued also, and could only get out of the case after considerable time and legal expense.

After a long career in engineering that has included field engineering, design, teaching and engineering research, construction contract administration, troubleshooting consulting, and expert witness work, I have found that in most litigation cases there is more than one cause for defective construction.

When, in my opinion, my client's performance has substantially contributed to the construction defects, I explain the basis of my opinion and advise the client to propose a settlement of the case. So far my advice has been followed and my clients have avoided the costly time devoted to litigation, and saved the legal and expert witness fees. In this way, an expert witness can comply with the "three truths" and be competent, honest, and independent, and remain loyal to the client.

Ignacio Martín
Consulting Engineer
San Juan, Puerto Rico

Neville's response

It is very gratifying that two very eminent engineers with extensive experience in all aspects of engineering practice as well as in expert witness work have written letters. This is the essential link between the witness work and engineering experience that I tried to emphasize in my article, and it is the putting oneself forward as an expert, in the absence of practical experience, that I decried. With appropriate background and experience, the expert's fees should be commensurately high, as James Warner rightly points out; what I was criticizing is high fees charged by junior academics without professional experience. Indeed, it is highly doubtful that someone with a Ph.D. and several years' employment as an assistant professor, but no design and no construction site experience, is qualified to testify on what's wrong with an actual structure; and yet, this is what I have encountered. Mr. Warner is absolutely right to emphasize the vital importance of "many years of practical experience" rather than "whatever degree one once obtained." Alas, the titles of "doctor" and "professor" may be used to impress the jury.

Mr. Warner rightly asks: "Is it reasonable to expect a construction expert to understand the details of design analysis or cement chemistry and paste microstructure?" The answer is: of course not. Conversely, we could ask: is it reasonable to expect a cement chemist versed in paste microstructure to understand construction methods and structural design? The answer, in Mr. Warner's words—"One should not expect chemists to possess a detailed understanding of site work or design, for their primary activity is in the laboratory, not in the design office or on site during construction"—should be written on tablets of stone (or concrete).

Yet, I have heard of experts who are highly successful in academic laboratory-based research, but have no engineering qualifications and no engineering experience. Sometimes such "experts" testified on how a structure should have been built. Views on the "correct" water-cement ratio, on the "required" slump, on details of finishing techniques, and on what the designer "should" have specified were very confidently stated. These opinions were just that: opinions; they were not based on code requirements or on accepted practice. But if the opinions are stated suavely and with charm, then members of a jury who are not technically fully knowledgeable might easily mistake the form for the substance of meaningful technical content.

At the same time, the last thing that the court needs is witnesses who have been engaged by a party to litigation because they sing a tune pleasing to that party, and not because they have well-trained voices.

Mr. Warner is absolutely right in pointing out that "membership [of organizations] in itself does not provide much knowledge or expertise"; yet, I have heard of witnesses without construction or concrete design experience who feel qualified to recommend repair methods and claim such expertise on the basis of membership of a repair organization.

Ignacio Martín's comments, coming as they are from someone with a great breadth of practical experience as well as of engineering research, and a Past President of ACI on top of all that, are extremely valuable. I am pleased he has not disagreed with me, and of course, I fully agree with him. All I can say is to repeat the final words of his letter, which are the same as the final words of Mr. Warner's letter, and are also the words in my article to the effect that to dare to put himself or herself forward as an expert, the person should be—nay, must be—"competent, honest, and independent."

I am thus most grateful to Messrs. Martin and Warner for complementing and amplifying my article; let us all hope that their words will be heeded.

Adam Neville

Section 9.3: An Expert Witness as Seen by an Expert Witness

I have acted as an expert witness in a large number of cases and in several jurisdictions: England, Hong Kong, Singapore, California, France, Middle East, and Spain. While legal systems and legal procedures vary, the role of the expert witness is seemingly simple: by clarifying and explaining scientific and technical evidences, to help the tribunal to establish the truth of the technical issues.

But what is truth? I am not echoing the words of Pontius Pilate, but only looking at the witness's oath: To tell the truth, the whole truth, and nothing but the truth. At first sight, this should ensure that the truth is readily established: alas, this is not necessarily the case. Let me give some examples of what I have witnessed (no pun intended).

First, about the whole truth. A witness said: "I performed my tests according to a certain British standard." This was the truth, but it was not the whole truth. The standard says that it applies to normal weight concrete. It does not say that it does not apply to lightweight concrete, but by implication such concrete is not covered by the given standard. Now the witness's concrete was lightweight, but he did not say so. He was not asked, and thus a part of the truth was excluded.

How to tell nothing but the truth seems obvious, but I have seen a report saying: "I have tested according to a given standard." Yes, the witness did so, but she also performed some tests not covered by the standard and included the results in the same report without saying that those tests were not British standard tests.

It seems to me that the oath is not watertight enough for, shall I say, an ingenious witness. Perhaps it would be better to require the witness to swear that he or she will be honest, which I think is more than being truthful. I realize, however, that a legal definition of the term honest may prove very difficult.

A non-concrete example of telling the truth, but not being honest, is afforded by a story about an international sports competition in the bad old days of the cold war. It so happened that all participant countries but two dropped out, and only the U.S. and the Soviet Union took part in the competition. The Americans won. One of the Soviet newspapers reported: "Russians runners-up"; another paper wrote: "Americans second from the bottom." These statements were the truth. But were they the whole truth? Were they nothing but the truth? Or were they shades of truth?

In addition to being honest, the witness has to be competent and independent. Being independent is somewhat difficult because the witness is paid by one party which is in litigation or arbitration against another party. Nevertheless, I view being independent as being of vital importance. At an early stage in the proceedings, I may find that my views do not

Fig. 9.3.1—Reprinted from "The Best of Queen's Counsel" by A. Williams and G. Defires, with permission from HarperCollins Publishers, Ltd

support "my" party's case; if so, I should clearly inform my party. An intelligent party will realize that any weaknesses in its case that I have discovered will be also discovered by the witnesses of the opposing party. To underestimate your opponent is a cardinal mistake in warfare or "lawfare." Knowing the strengths and weaknesses of its case, a party will decide how to settle the dispute, usually by some sort of compromise.

In my experience, no party, however strong appears its case, is 100% right. Likewise, hardly ever is a party totally culpable: the concrete may be "bad" but the specification may be ambiguous, if not inconsistent, or the design may be wanting. I have always given my independent views at an early stage in the proceedings, and only once did "my" party not want to know, and sacked me. I do not regret this because I was saved from appearing partisan.

My remarks may seem perhaps naïve, but in some jurisdictions lawyers work on the basis of contingency fees, which means that their being paid depends on winning the case. The witness is not supposed to be paid according to the outcome of the case, but sometimes this seems to be so, if only because a witness who is instrumental in winning is likely to be engaged again by the same lawyer; a "loser" will be dropped. With contingency fees in some form or other being considered in Great Britain, the situation is of interest; the cartoon in Fig. 9.3.1 is relevant.

Most witnesses are scrupulously honest and are competent in the field in which they testify, but I have encountered those who are willing to opine well outside their field of competence; in one case, a professor of chemistry who had never worked on a construction site said: "I will now put on my engineering hat," and expressed "definite" views on methods of placing reinforcement. Another pure scientist opined confidently on forces induced by earthquake. This problem is less likely to occur in the future, given the new Civil Procedure Rules introduced by Lord Woolf, whereby the tribunal has large powers in approving witnesses, but the rules do not apply in Scotland.

But even when a witness stays within his field of competence and the judge controls the witnesses, machinations are possible. In one case in which I was involved, the witnesses from the two sides were ordered to meet and agree facts and figures as much as they could, so as to save time in formal proceedings. At one point, I proposed that, on the basis of calculations, we agree on what was the maximum temperature of the concrete; the witness for the other side refused to do so. Much later, he proposed agreement on some other point, but I felt he was wrong and did not agree. He then suggested that, if I agreed to his proposal, he would agree to my views on the temperature. I find such behavior shocking: Either something is true or it is not; the truth of one physical phenomenon being conditional on agreement about something totally different is more like horse trading.

Why do some witnesses behave like this? I am afraid that the answer is human frailty, which usually means money or sheer greed. Expert witnesses are paid a high hourly rate, much higher than in a university or on a construction site. So it should be, because the witness ought to spend many hours of unpaid time to maintain his or her expertise. A friend of mine said that, in the days of professional fees in guineas, the shillings were for doing the work and the pounds for knowing how to do it. [A guinea was one pound and one shilling.]

If I have given the impression of being unduly critical of expert witnesses, then I hasten to add: only of some of them, and, as St. Luke says,

Fig. 9.3.2—Professor Adam Neville (right) received Honorary Membership of The Concrete Society from Professor Ted Kay (photo courtesy of UPPA Commercial Photographers)

not of the "ninety and nine persons, which need no repentance." I realize that no expert is a paragon of virtue who is infallible. Sometimes, he errs; sometimes he simply does not know; if so, he should just say so. Much more importantly, he should say when evidence is conflicting. Specifically, an expert witness may prefer one design method, if only because he is familiar with it. But the witness must not omit to say that another method has also been successfully used.

I realize that, when there are shades of gray, the witness may well be affected by his party's interests, but he must not stray into black or white. And he must not be economical with truth, which will be uncovered in cross-examination.

In my view, a witness should have a very considerable relevant professional experience and just be "natural." Training, which is offered by professional "witness trainers," is not necessary or even desirable. It is not a good thing to tell a witness how to use his or her voice so as to appear plausible; it is not the tone but the words that matter. When I was about to appear for the first time in a court in California, I asked my lawyer what I should do; he answered: "Just tell the truth."

In my opinion, someone who is a professional expert and who does not practice engineering is somewhat unsatisfactory as such a person has lost contact with developments in his or her field. I would not like to be examined by a physician who does nothing but post mortem examinations.

This section highlights problems with some experts, but experts are needed if the object of litigation or arbitration is justice, as should be the case, even though disputes are often not about who is right but about who pays. That is a matter for the parties to the lawsuit or the arbitration. The expert must remain above this, even though such a stance is onerous and hard to maintain.

Section 9.4: The Influence of Litigation on Research

Research is motivated by many forces: curiosity, inspiration of the brilliant person, desire for personal promotion of the more pedestrian one, or commercial and industrial need. In recent years, research within the purview of this Magazine [of Concrete Research] has also occasionally been affected by litigation.

In this context, litigation should be taken to include not only action in a court of the land but also other, perhaps even more common forms of settling disputes, such as arbitration. Many disputes concern the performance of concrete or its adequacy for the intended purpose. These are more difficult to resolve than the issues of conformance to specification as they involve an assessment of the characteristics of the actual concrete against the characteristics necessary for the desired performance. An example of this is the durability of concrete structures in a hostile environment, often involving corrosion of the reinforcement.

Early in the dispute, each of the parties involved needs to establish why the responsibility for the "problem" lies with the other party or parties. In the example of durability, it may be necessary to show that the fault lies not with the concrete (which, it is claimed, conforms to the specification) but with the particular mechanism of transport of the aggressive medium through the concrete. The back-up of this requires a study of the relevant, possibly newly postulated, mechanism.

Such a study inevitability starts with a state-of-the-art review, which must be full and complete, and cannot be selectively biased because any report by one party is subject to fierce scrutiny by the other. Therefore, it can be argued that the degree of objectivity finally achieved is greater than would be the case with an individual researcher, for however high is one's honesty and probity, no individual is free from prejudices, preferences, or bias, possibly hidden even from oneself.

An experimental work is equally rigorously vetted. In some cases, for instance patent infringement disputes, experimental work may be supervised jointly by expert witnesses for both parties concerned so that "facts can be established as facts" and thus the need to lead evidence in court is obviated.

These activities clearly constitute research. It not only helps to solve the problem in hand but provides information of more general applicability. The research is performed rapidly, being driven by the court or arbitration timetable, and it is not starved of funds as the sums in the claim frequently run into many millions, or even many tens of millions, of pounds. As the outcome of the tests and studies has to be communicated to the opposite party, the research does not remain secret and, anyway, once the dispute has been settled, as it must be, publication of the scientific and technical work is generally permitted.

An editorial is no place for a description of specific examples of litigation-driven research, and individual cases obviously must not be mentioned, but perhaps it is appropriate to refer to two important areas where considerable research has been prompted, if not forced, by litigation. One of these is the stability of precast concrete structures. The investigation of an actual failure was followed by extensive analytical and experimental work, which resulted in remedial work ahead of any possible further failures, in better subsequent structures, and in a better design code.

The other area of improvement is the resistance of concrete to the penetration of salts which leads to corrosion of the reinforcement. This started as a problem in a hot, coastal environment but also applies in a temperate climate, although the rate of attack there is considerably slower. Our understanding of the phenomena involved has improved enormously and this has lead to an improvement in mixture design as well as in the development of new methods of measuring the ability of concrete to allow the movement of salts through it: the latter development is still far from complete.

An important feature of research arising from litigation is that much of it has been performed, not by the traditional research organizations but by those for whom research does not represent a daily activity: testing houses, contractors, and designers. Their work has increased the total amount of research output and is thus of benefit to the construction world at large.

In the past, exigencies of war, be it preparation for it or for its avoidance, contributed to the development of many areas of technology, often of the high-tech variety. Today we hope that a war-driven world will become a thing of the past. Litigation may provide a corresponding drive for research on concrete and concrete structures, although hopefully not so pervasive that it also induces a war-like situation. So, while we fear litigation, we have to acknowledge that it is good not only for the lawyers but also for concrete engineers.

Epilogue

As its title implies, this book presents Neville's views on concrete. Of course, my views are not one person's ideas conceived in isolation: the input has been from various sources and various activities.

I am saying this because some people like to think of themselves as having only one role in life: a researcher, a designer, a concrete materials specialist, a ready-mix supplier. Looking at concrete from any single standpoint does not give a balanced view, and certainly not a complete one.

At the other extreme, it is very difficult for one person to be, or to have played, all those roles. And yet, considering concrete from all these points of view is essential in order to present a balanced and all-inclusive picture.

It helps therefore if the author of a book has a broader, if not a broad, experience, but this is not enough either. Published information needs to be consulted, digested, and taken into account. I have tried to do all this, and this is why the present book does not make too many categorical statements that do not admit dissent. In some instances, I have made definitive statements because I felt that the situation was clear-cut. In other instances, I have presented more than one point of view, but I also expressed my own opinion, believing that it prevailed. Not to do so, but simply to say "it could be like this, it could be like that" would not be helpful to the reader who needs to make decisions about concrete in practice. Whether my approach is right, only time will tell, that is, the readers' reaction will give the answer.

I would like to make an irreverent aside. In 1974, the government of the People's Republic of China made first contacts with British non-government bodies and chose the Concrete Society as a politically neutral body. At that time, I was President of the Society and, together with three other officers, I was invited to visit China. While there, I was given a "vest pocket edition" of a book titled *Quotations from Chairman Mao Tsetung*, which has a bright red cover (the famous "red book"), exactly the same color as my book, *Properties of Concrete*.

In his book, Mao quotes Lenin as saying: "Concrete analysis of concrete conditions is the most essential thing in Marxism, the living soul of Marxism." I do not know what these philosophers and leaders meant by this saying but, in the context of the present book, I subscribe to the essential nature of "concrete analysis of concrete conditions." By the way, vest pocket books are no more and the red book, too, is likely to be consigned to history.

As I have said more than once, the goal of this book is to contribute to the use of better concrete and to a better use of concrete. We all know that concrete is an excellent construction material and that excellent concrete structures can be built and operated. The term "operated" reflects the need for maintenance and for adequate durability; it is the consideration of durability in design that needs increased attention in the future.

Still on the subject of design, I hold quite strong views on the internationalization of concrete, to coin a clumsy phrase. Because I have worked in a wide range of countries, I am aware of the fact that designs produced in one country are used in a second country, the construction being executed by a contractor from a third country, and sometimes the unskilled work force is from a fourth country. This is all to the good, but there exist problems that need not be there. These problems arise from units of measurement, from standards, and from design codes.

Taking the first of these, in almost all countries, the system of units of measurement is the so-called SI system, based on the meter as a unit of length, Newton as a unit of force, and kilogram as a unit mass. The only significant exception is the United States. I am aware of the fact that metric units of measurement have been legal in the United States for a long time. I am also aware of the metrication policy of ACI. These are all good intentions but, as one could say, the road to hell is paved with prestressed concrete.

In practice, on many jobs, it is the good old Imperial units that "rule," albeit under the modernized name of American units. There is a curious twist to this, for me personally. In my younger days I was educated on a diet of pounds and feet, including concepts of pound-mass and pound-force. However, in the United Kingdom, in 1968, all these were replaced by the SI units of measurement. Not surprisingly, there was little enthusiasm among engineers to learn new tricks.

The change-over was given a tremendous push by the British government, which decided that all design projects for which the government was the client had to be in SI units. The consultants who wanted these jobs were obliged to "re-tool" themselves. Having done that, they preferred to use SI units for private clients as well, because to have two systems in one office is to court disaster. In consequence, the Imperial units quickly went out the window, and the contractors and the building suppliers had to change, too. Admittedly, this was not a painless change, but it is not unusual, even for change for the better, to be painful.

Like everybody else, I became fully operational in SI units, and I consigned the Imperial units to the same place to which the Empire had been consigned: history, proud but passé. My trouble came when I became involved in technical problems in the United States in the 1990s. I still remembered enough about the Imperial units to know how many inches to a foot, although the American units of volume of fluids were different from the British units. I was also able to convert mentally feet into millimeters and the like. What I had lost was the "feel" for sacks per cubic yard (as against kilograms per cubic meter) and, worse still, for pressures in pounds per square foot (as against Newtons per square meter). I simply had to relearn.

Let me turn to the effect of the dual system of units of measurement on international construction. The world leader in standards for materials and for testing methods and procedures is ASTM. That organization has been highly proactive in adopting the SI system of units, but I suspect that full success requires dropping the old units altogether. Those of us who live and work outside the United States have a strong interest in ASTM standards because there are numerous areas where British, and now European, standards do not exist, so that we turn to ASTM.

The ASTM standards are updated much more rapidly and frequently than European standards and, in my opinion, it would be beneficial all round if ASTM standards became worldwide standards. I can see difficulties in forming standard-writing committees with, say, European membership, if only because of the distance between the two continents. Nevertheless, this is not impossible: in Europe, we have found a way of operating from Portugal to Greece, and from Finland to Italy, and do so despite different languages in the various countries.

I hope someone will give the idea of truly international standards for materials and for testing a strong push. This needs a powerful and influential person: my suggestion is a successful, committed, and important engineer as a President of the United States. Long enough time has elapsed since Herbert Hoover, once again to elect an engineer (but I don't have a vote).

The third area in need of internationalization is the design codes. Of course, the underlying physical phenomena are the same the world-over. The principles of consideration of loading, of a probabilistic approach to loads, stresses, and the material properties are also the same, and so are the concepts of strength, robustness, and serviceability. However, the details and values of load factors and of partial safety factors differ. The outcome design is usually the same, and so it should be, but some codes apply higher factors to, say, the live load but lower partial safety factors for materials. This is why, as is well-known, one must not use selectively parts of different codes on a single project. And yet, sometimes we need codes from other countries for particular purposes. As an aside, it has been said that the only common international factor of safety used by the designer is the number of years to retirement plus two.

Quite recently, I came across this situation in connection with the design of circular silos. There exists an ACI code for silo design, also an Australian code, and a German code, but there is no British code. In the particular case in which I was involved, the general design code was the British code, but it was necessary to supplement it with a "foreign" code, and the ACI code was particularly useful and convenient. This use of two codes required great vigilance with respect to the various factors, as mentioned earlier, and the interpretation of the "combined" use of the two codes was not always unambiguous.

In Europe, there exist already some European design codes. In my view, it would be all-round beneficial if we could achieve a harmonization and, eventually, unification of design codes between Europe and the United States.

Some readers might feel that I have moved away from concrete construction and digressed into broad issues. Let me justify myself. As we are striving for better concrete structures, we should make the path to that objective as free from obstacles and difficulties as possible. I submit that a more uniform approach to measurements, testing methods, material properties, and to design would help by providing better information and guidance to all concerned, by decreasing uncertainty, and by eliminating sources of disagreement and thus reducing problems.

Mine may be "The voice of one crying in the wilderness …" but my aim is to make the "paths [of concrete] straight." I hope that *Neville on Concrete* will contribute to making better concrete; to make the book more readable I have tried to follow the words of W. S. Gilbert in "Yeomen of the Guard" who wrote:

"For he who'd make his fellow creatures wise,
Should always gild the philosophic pill."

And, finally, may concrete, and all who work in concrete, prosper.

Appendix

Details of Original Publications

The various sections of this book were first published as articles, papers, and letters to the Editor as listed below.

CHAPTER 1: CEMENTITIOUS MATERIALS
1.1—Neville, A., "Cementitious Materials—A Different Viewpoint," *Concrete International*, V. 16, No. 7, July 1994, pp. 32-33.
1.2—Neville, A., "There is More to Concrete than Cement," *Concrete International*, V. 22, No. 1, Jan. 2000, pp. 61-63.
1.3—Neville, A., "Silica Fume in a Concrete Specification: Prescribed, Permitted or Omitted?," *Concrete International*, V. 23, No. 3, Mar. 2001, pp. 73-77.
1.4—Neville, A., "Whither Expansive Cement?," *Concrete International*, V. 16, No. 9, Sept. 1994, pp. 34-35.
1.5—Neville, A., "A 'New' Look at High-Alumina Cement," *Concrete International*, V. 20, No. 8, Aug. 1998, pp. 51-55.
Letters to the Editor, *Concrete International*, V. 21, No. 3, Mar. 1999, pp. 7-10.

CHAPTER 2: WATER AND CONCRETE
2.1—Neville, A., "Water and Concrete: A Love-Hate Relationship," *Concrete International*, V. 22, No. 12, Dec. 2000, pp. 64-68.
2.2—Neville, A., "Water—Cinderella Ingredient of Concrete," *Concrete International*, V. 22, No. 9, Sept. 2000, pp. 66-71.
2.3—Neville, A., "Effect of Cement Paste on Drinking Water," *Materials and Structures*, V. 34, No. 240, July 2001, pp. 367-372.

2.4—Neville, A., "Seawater in the Mixture," *Concrete International*, V. 23, No. 1, Jan. 2001, pp. 48-51.

Letters to the Editor, *Concrete International*, V. 23, No. 7, July 2001, pp. 8-11.

2.5—Neville, A., "How Useful is the Water-Cement Ratio?," *Concrete International*, V. 21, No. 9, Sept. 1999, pp. 69-70.

Letters to the Editor, *Concrete International*, V. 23, No. 1, Jan. 2000, pp. 11-15.

Letters to the Editor, *Concrete International*, V. 23, No. 3, Mar. 2000, pp. 7-9.

CHAPTER 3: DURABILITY

3.1—Neville, A., "Consideration of Durability of Concrete Structures: Past, Present, and Future," *Materials and Structures*, V. 34, No. 236, Mar. 2001, pp. 114-118.

3.2—Neville, A., "Maintenance and Durability of Concrete Structures," *Concrete International*, V. 19, No. 11, Nov. 1997, pp. 52-56.

3.3—Neville, A., "The Question of Concrete Durability: We Can Make Good Concrete Today," *Concrete International*, V. 22, No. 7, July 2000, pp. 61-63.

3.4—Neville, A., "Good Reinforced Concrete in the Arabian Gulf," *Materials and Structures*, V. 33, No. 234, Dec. 2000, pp. 655-664.

3.5—Neville, A., "Chloride Attack of Reinforced Concrete: An Overview," *Materials and Structures*, V. 28, No. 176, Mar. 1995, pp. 63-70.

CHAPTER 4: SPECIAL ASPECTS

4.1—Neville, A., "Cement and Concrete: Their Interaction in Practice," *Advances in Cement and Concrete*, Proceedings of an Engineering Foundation Conference, University of New Hampshire, American Society of Civil Engineers, 1994, pp. 1-14.

4.2—Aïtcin, P.-C., and Neville, A., "High-Performance Concrete Demystified," *Concrete International*, V. 18, No. 1, Jan. 1993, pp. 21-26.

4.3—Neville, A., and Aïtcin, P.-C., "High Performance Concrete—An Overview," *Materials and Structures*, V. 31, No. 206, Mar. 1998, pp. 111-121.

4.4—Neville, A. M., "Aggregate Bond and Modulus of Elasticity," *ACI Materials Journal*, V. 94, No. 1, Jan.-Feb. 1997, pp. 71-74.

Discussion, *ACI Materials Journal*, V. 94, No. 6, Nov.-Dec. 1997, pp. 612-613.

4.5—Aïtcin, P.-C., Neville, A., and Acker, P., "Integrated View of Shrinkage Deformation," *Concrete International*, V. 19, No. 9, Sept. 1997, pp. 35-41.

CHAPTER 5: CONSTRUCTION

5.1—Neville, A., "Concrete Cover to Reinforcement or Cover-Up?," *Concrete International*, V. 20, No. 11, Nov. 1998, pp. 25-29.

Letters to the Editor, *Concrete International*, V. 21, No. 5, May 1999, pp. 8-9.

5.2—Neville, A., "Specifying Concrete for Slipforming," *Concrete International*, V. 21, No. 11, Nov. 1999, pp. 61-63.

5.3—Neville, A., "Efflorescence—Surface Blemish or Internal Problem? Part 1: The Knowledge," *Concrete International*, V. 24, No. 8, Aug. 2002, pp. 86-90.

5.4—Neville, A., "Efflorescence—Surface Blemish or Internal Problem? Part 2: Situation in Practice," *Concrete International*, V. 24, No. 9, Sept. 2002, pp. 85-88.

5.5—Neville, A., "Autogenous Healing—A Concrete Miracle?," *Concrete International*, V. 24, No. 11, Nov. 2002, pp. 76-82.

Letters to the Editor, *Concrete International*, V. 25, No. 2, Feb. 2002, pp. 18-20.

CHAPTER 6: TECHNOLOGY AND DESIGN

6.1—Neville, A., "Concrete Technology—An Essential Element of Structural Design," *Concrete International*, V. 20, No. 7, July 1998, pp. 39-41.

Letters to the Editor, *Concrete International*, V. 20, No. 9, Sept. 1998, p. 7.

6.2—Neville, A., "What Everyone Who is *In* Concrete Should Know *About* Concrete," *Concrete International*, V. 21, No. 4, Apr. 1999, pp. 57-61.

6.3—Neville, A., "A Challenge to Concretors," *Concrete International*, V. 24, No. 1, Jan. 2002, pp. 59-64.

6.4—Neville, A., "Concrete Technology and Design—The Twin Supports of Structures," *Concrete International*, V. 24, No. 4, Apr. 2002, pp. 52-58.

Letters to the Editor, *Concrete International*, V. 24, No. 11, Nov. 2002, pp. 10-11.

6.5—Neville, A., "Creep of Concrete and Behavior of Structures: Part I: Problems," *Concrete International*, V. 24, No. 5, May 2002, pp. 59-66.

Letters to the Editor, *Concrete International*, V. 24, No. 11, Nov. 2002, p. 12.

6.6—Neville, A., "Creep of Concrete and Behavior of Structures: Part II: Dealing with Problems," *Concrete International*, V. 24, No. 6, June 2002, pp. 52-55.

CHAPTER 7: TESTING

7.1—Neville, A. M., "Standard Test Methods: Avoid the Free-For-All," *Concrete International*, V. 23, No. 5, May 2001, pp. 60-64.

Letters to the Editor, *Concrete International*, V. 23, No. 10, Oct. 2001, pp. 10-18.

7.2—Neville, A., "Core Tests: Easy to Perform, Not Easy to Interpret," *Concrete International*, V. 23, No. 11, Nov. 2001, pp. 56-65.

Letters to the Editor, *Concrete International*, V. 24, No. 5, May 2002, pp. 8-11.

CHAPTER 8: RESEARCH

8.1—Neville, A., "Is Our Research Likely to Improve Concrete?," *Concrete International*, V. 17, No. 3, Mar. 1995, pp. 45-47.

Letters to the Editor, *Concrete International*, V. 17, No. 7, July 1995, p. 8.

8.2—Neville, A., "Suggestions of Research Areas Likely to Improve Concrete," *Concrete International*, V. 18, No. 5, May 1996, pp. 44-49.

8.3—Neville, A., "Concrete Research on a Micro- and a Macro-Scale," *Cement and Concrete Research*, V. 22, No. 6, pp. 1067-1076.

8.4—Neville, A., "Reviewing Publications: An Insider's View," *Concrete International*, V. 23, No. 9, Sept. 2001, pp. 56-61.

Letters to the Editor, *Concrete International*, V. 24, No. 4, Apr. 2002, pp. 6-7.

8.5—Neville, A., "Objectivity in Citing References," *Materiales de Construcción*, V. 51, Nos. 263-264, 2001, pp. 5-11.

CHAPTER 9: LITIGATION

9.1—Neville, A., "Litigation—A Growing Concrete Industry," *Concrete International*, V. 22, No. 3, Mar. 2000, pp. 64-66.

Letters to the Editor, *Concrete International*, V. 22, No. 11, Nov. 2000, pp. 8-9.

9.2—Neville, A., "An Expert Witness—As Reliable as Concrete?," *Concrete International*, V. 22, No. 4, Apr. 2000, pp. 59-61.

Letters to the Editor, *Concrete International*, V. 22, No. 12, Dec. 2000, pp. 6-7.

9.3—Neville, A., "An Expert Witness as Seen by an Expert Witness," *Concrete*, Jan. 2001, pp. 73-77.

9.4—Neville, A., "The Influence of Litigation on Research," Guest Editorial, *Magazine of Concrete Research*, V. 19, No. 148, 1989, p. 123.

Index

A

Abrasion, 3-63, 4-38, 8-28
    of aggregate, 8-12
Absorbed water in aggregate, 8-11
Absorption of chlorides, 3-56
Absorption of electromagnetic
    radiation, 2-59
Abutments and cover, 5-12
Accelerated curing test, 8-17
Accelerator, 2-38
Accuracy, 6-35, 6-38, 6-39
ACI Building Code, 5-8
Acid rain, 2-8
Active fine powders, 1-13
Addition, 1-15
Additive, 1-15
Adhesion bonds, 4-18
Adiabatic conditions, 4-59
Adjudication, 9-4
Admixtures, 1-4
    chemical, 4-4
    and chlorides, 3-55
    compatibility of, 3-23
    in curing water, 2-23
    and drinking water, 2-29, 2-33
    and durability, 3-10
    mineral, 4-3
    in shrinkage-compensating
        concrete, 1-22
    in slipforming, 5-20
Adversarial system, 9-8, 9-12, 9-15
Advocacy by witness, 3-16
Advocates, 9-2
Aeration of seawater, 2-40
Aerosol, seawater, 5-7
Aesthetic aspects, 4-63, 5-5, 5-30, 5-31
Age and creep, 6-5
Aggregate,
    bond, 2-21, 4-23
    classification, 6-28, 8-6, 8-13,
    and coefficient of thermal
        expansion, 4-5
    contaminated, 3-15, 3-55
    content and durability, 3-9
    content and modulus of
        elasticity, 4-43
    cooling, 2-37, 3-37
    crushed, 4-24, 4-32, 5-20

crushing, 3-46
and durability, 3-46
grading, 4-23, 6-26
grading, final, 4-39
grading and workability, 6-27
in high-performance concrete, 4-30, 4-31
and interface zone, 1-11, 4-23
local, 6-34
moisture content, 6-19, 6-26
and modulus of elasticity, 4-5
natural, 3-47
pre-cooled, 4-67
recycled, 6-20, 6-27
rounded, 5-21
from seashore, 2-37
from seashore and efflorescence, 5-25
shape, 4-23, 6-28
shape, influence on bleeding, 2-5
size, influence on bleeding, 2-5
for slipforming, 5-20
specification for, 6-5, 6-6
stockpiles, 6-19
strength, 4-18, 4-23
surface water, 2-14
texture, 4-23, 6-28
thermal properties, 4-59
and thermal strains, 4-4
variability, 4-4, 6-26, 8-6, 8-12
and workability, 4-4, 4-5
Aggressive agents
and bleeding, 2-5
and cementitious materials, 3-20
and durability, 2-51
penetration by, 3-5, 3-8, 3-10
and silica fume, 3-21
Aggressive water, 2-8
Air currents and chlorides, 3-38
Air entrainment,
by algae, 2-21
and hardness of water, 2-22
and high-performance concrete, 4-19

Air permeability, 5-43
Algae, 2-20
and air entrainment, 2-21
Alkali
attack and high-alumina cement, 1-31
bicarbonates, 2-17
carbonates, 2-17, 2-23
content in cement, 2-11, 4-9, 4-30
in curing water, 2-23
oxides, 4-34
sulfates, 3-24, 4-31, 4-34, 4-35, 4-56, 5-27
soluble, 4-35, 5-25
and strength, 8-24
in wash water, 2-22
Alkali-activated slag cement, 5-33
Alkali-aggregate reaction, 2-16, 2-17, 3-5
and corrosion, 3-64
and cracking, 3-14
and mineral waters, 2-17
and seawater in mixture, 2-38, 2-43
and silica fume, 3-21
and water, 2-10, 3-36
Alkali-carbonate reaction, 3-5
Alkali-reactive aggregate, *see Alkali-aggregate reaction*
Alkali-silica reaction, *see Alkali-aggregate reaction*
Alkalinity and cover, 5-5
Alumina and silica fume, 3-21
Aluminate leaching, 2-34
Aluminum in drinking water, 2-9, 2-29, 2-32, 2-33, 2-34
Ammonia, 2-15
Analysis of idealized structures, 6-35
Analysis, structural, 6-5
Anhydrite, 3-24, 4-34, 8-20
Anode, 3-53, 3-54, 3-61, 5-41, 5-44, 5-48
and surface barriers, 3-63
Appearance, 3-1
and efflorescence, 5-23, 5-33

Arabian Gulf
   and concrete quality, 3-34
   exposure conditions, 2-43, 3-2
   and seawater, 2-48
Arbitration, 9-4, 9-10, 9-28
   pendulum, 9-4
Arbitrator, 9-6
Arch, 5-30
Arid countries, 2-37, 2-43
Assessment of structures, 3-40
Attorney, 9-19
Autoclaved concrete, 5-49
Autogenous healing, 2-8, 5-2, 5-36, 5-46
Autogenous shrinkage, 2-7, 4-54, 4-55, 4-60
   and durability, 3-44
   in high-performance concrete, 4-36
   and w/c, 3-31
   and wet curing, 3-50
Average strength, 7-5, *see also* f_{cr}'

B

Backfill, 4-40
Bacteria in water, 2-22
Basements and efflorescence, 5-32
Basic creep, 6-54
Batching, 6-21, 8-30
   automated, 8-6
   plant, 3-29, 6-26, 8-11
Beam, 4-66
   built-in, 6-35
   and creep, 6-56
   and deck construction, 6-8
   encastré, 6-35
Bias, 8-34
Bill of quantities, 3-18, 3-31
Binder, 1-15
Bleed water, 3-47, 3-48
   description, 2-4
   significance of, 1-10
Bleeding, 7-34
   effect of aggregate, 5-21

   in high-performance concrete, 4-35
   and silica fume, 3-45
   and w/c, 2-52
Blemish, 5-23
Blended cement, 1-5, 6-18, 8-10
   and drinking water, 2-23
Blisters, 3-42
Block, paving, 5-33
Blocking of cracks by corrosion products, 5-44
Bloom, 5-23
Bond of aggregate, 1-8, 2-21, 4-2, 4-33, 4-42, 4-53
Bond failure, 4-45
Bond microcracking, 4-43, 4-45, 4-47, 4-62
Bonds,
   adhesion, 4-14
   chemical, 4-18
   cohesion, 4-18
   and strength, 4-23
Book publishing, 8-36, 8-37
Box girder, 6-10
   and creep, 6-56
Brackish groundwater, 3-56
Brackish water,
   and chlorides, 2-8
   as mixing water, 2-44
Bricks, 5-32
Brickwork and efflorescence, 5-2, 5-24, 5-31
Bridge, 4-15, 6-10
   and cover, 5-12, 5-13
   and efflorescence, 5-30
Bridge piers, 6-7
   and creep, 6-48
   and slipforming, 5-17
Briquette, 7-9
Brittle failure, 8-23
Brittleness, 4-48
Brucite, 3-39
Buckling, 7-11
Buildability, 5-14
Building code, 6-15

Building and creep, 6-50
Building official
    and cores, 7-39
    and testing, 7-27
Building stone and efflorescence, 5-24
Bulk cement paste, 1-10
Burlap, *see Hessian*

C

Calcium aluminate cement, 1-25,
    1-27, 3-25, 8-45, *see also*
    High-alumina cement
Calcium aluminate hydrate, attack
    on, 5-26
Calcium carbonate, 5-25
    and autogenous healing, 5-37,
        5-38
    structure, 5-47
    volume of, 3-39
    in water, 2-32
Calcium chloride, 3-60, 6-8
Calcium chloroaluminate, 3-58,
    3-59
Calcium chloroferrite, 3-58
Calcium hydrogen carbonate, 5-38
Calcium hydroxide
    and drinking water, 2-29
    in interface zone, 1-9, 4-44
    leaching, 3-36, 5-32
    and silica fume, 3-21
    solubility, 5-25
    structure, 5-46
Calcium monosulfoaluminate, 4-56
Calcium nitrite and corrosion, 3-36
Calcium silicate hydrate,
    attack of, 5-26
    and silica fume, 3-21, 4-58
Calcium sulfate, 4-34, 5-25
    in cement, 4-20, 4-56
    in expansive cement, 1-22
    reactivity, 3-24
    solubility, 4-34
    synthetic, 3-24
Calcium sulfoaluminate, 3-59, 4-56
Calcium sulfoferrite, 4-56

Camber, 6-56, 6-58
Cantilever, 5-4
Cantilever box girder, 6-11
Capillary action 8-18
Capillary pores, 4-57
    discontinuous, 3-21
    and durability, 3-8
    evaporation from, 4-61
    and free water, 2-4
    menisci, 4-36
    open, 3-56
    and w/c, 2-48
Capping, 7-21, 7-31
Carbon dioxide,
    attack by, 3-10
    and autogenous healing, 5-37,
        5-39
    solubility, 5-25
Carbon in fly ash, 4-8
Cabonate alkalinity, *see Water*
    hardness
    in drinking water, 2-33
Carbonates in curing water, 2-23
Carbonation
    and autogenous healing, 5-38
    and chlorides, 3-59
    and cover, 3-28, 5-5
    and corrosion, 3-15, 4-38
    of high-alumina cement, 1-40
    influence on porosity, 3-39
    rate of, 3-39
    and relative humidity, 3-39,
        6-13
    shrinkage, 4-61
    and strength, 3-20
    testing, 8-19
    and water, 2-10
    and w/c, 3-19, 4-10
Carbonic acid and rain, 5-31
Cathode, 3-54, 3-56, 5-41, 5-43,
    5-49
    and oxygen, 3-61
    and surface barriers, 3-63
Cathodic protection, 3-34, 3-64
Cavitation, 2-11, 3-39

Cement
    chemists, 8-7
    composite sample, 4-6
    composition, 4-7
    composition and durability, 4-11
    and concrete interaction, 4-1, 4-3
    effect on water, 2-26
    grab sample, 4-6
    manufacturers, 6-18
    manufacturers' certificate, 4-7
    mortar test, 4-6, 4-20
    "new," 3-7, 3-19
    "new," strength development, 3-8
    replacement, 1-15
    strength, 4-6
    superplasticizer compatibility, 4-19
    type and seawater, 4-11
    unhydrated, 3-50, 4-58
    variability, 4-6
Cement content
    in high-performance concrete, 4-31, 4-40
    specified, 3-9
    variability, 2-59
    and w/c, 3-19
    and workability, 3-20
Cementitious materials, 1-4, 1-6, 1-7
    and corrosion, 3-60
    description, 4-4
    and durability, 3-12, 3-45
    in high-performance concrete, 4-31
    in hot climate, 3-61
    and strength, 4-7
    terminology, 1-1, 1-6
    and thermal problems, 4-12
    and w/c, 2-50, 2-51, 2-53, 3-20
Cementitious properties, latent, 1-6, 3-23
CEN, 7-22
Certainty, 8-46
Chairs, 3-29, 5-10, 5-14
Chance variations, 6-39

Change of use, 7-11, 7-27
Characteristic strength, 3-19, 6-24, 6-25, 6-26
Chemical
    admixtures, 1-4
    attack and cracking, 3-14
    bonds, 4-18
    reactions, rate of, 3-39
Chimneys, 5-17
Chlorides
    absorption, 3-56
    acid soluble, 2-39, 3-55
    airborne, 3-55
    in aggregate, 2-20
    attack, 3-53
    binding, 8-28, 3-58
    binding by aluminates, 3-59
    bound, 3-55, 6-14
    bound and slag, 6-14
    and carbonation, 3-59
    chemically bound, 3-58
    content, 2-39
    content, permissible, 3-55
    content, threshold, 3-58
    content total, 3-55
    and curing, 2-43
    diffusion, 3-56, 7-7
    diffusion test, 7-18
    and efflorescence, 5-31
    free, 3-55, 3-58, 3-60
    ingress, 3-55
    ingress and carbonation, 3-59
    ingress and curing, 8-30
    limits, 2-39
    in mixture, 3-54
    origin of, 3-41
    penetration, 7-6, 7-7, 7-14
    penetration test, 6-14
    physically bound, 3-58
    in portland cement, 2-40
    profile, 3-15, 3-41, 3-57
    in seawater, 2-37
    and silica fume, 3-21
    in slag, 3-55
    and sulfate attack, 4-11

10-14 Index

total, 3-55
transport, 5-44, 5-45
and Type V cement, 3-59
in wash water, 2-21
in water, 2-20
water-soluble, 2-39, 2-44, 3-55
wind-borne, 3-38
Chloroaluminates, 2-39, 4-11
Chrysotile, 4-59
"Ciment fondu," 3-25
Civil Procedure Rules, 9-12, 9-26
Cladding, 6-7, 6-16
Claimant, 9-6
Claims, 9-5
Clay in water, 2-20
Climate and durability, 3-36
Clinker, 4-34
and calcium sulfate, 1-22
grinding, 3-8
Coastal areas, 8-18
Codes, 3-50, 5-8, 6-15, 6-20, 6-24, 8-27, 10-2, 10-3
and creep, 6-47, 6-58, 6-60
European, 10-4
Coefficient of thermal expansion, 4-60
of aggregate, 4-61, 6-17
of cement, 6-51
differential, 3-5
of repair material, 3-17
Coefficient of variation, 7-40
Cohesion, 5-19, 5-20
bonds, 4-18
loss of, 5-25
Cold joints, 3-26
in slipforming, 5-18
Collapse of structure, 1-27, 1-36, 6-13
Colored concrete and efflorescence, 5-33
Columns, 4-37, 4-64, 4-65
and cover, 5-13
and creep, 6-7, 6-16, 6-49, 6-54
differential shortening, 6-55
Committee reports, 8-42

Common law, 9-1, 9-12
Compaction, 2-53, 3-6, 4-16
and cover, 3-29
determination of, 3-29
and workability, 3-27
Compatibility, 6-35
of repair materials, 3-17
of superplasticizer and cement, 4-33, 4-35
Compliance, 2-53, 7-11, 7-26, 8-17, 9-5
Composite cement, 6-18
Composite hard material, 4-42
Composite material, 1-1, 6-63
Composite soft material, 4-24, 4-42
Composites, 6-31
Compressive strength, 6-7
and mixing water, 2-19
Conciliation, 9-4
Concrete
masonry, 5-8
plain, 5-8, 6-37
quality and cover, 5-13
teaching in universities, 6-8
technologists, 3-26
under-reinforced, 6-37
unreinforced, 5-8, 6-37
Concretor, 3-11
Conduits, 2-9, 2-26
Conference proceedings, 8-35
Confidence limits, 6-62, 7-36
Consolidation, *see Compaction*
Construction, 5-1, 6-29
joints, 5-17
methods and specification, 3-26
monocoque, 6-31
speed, 5-17
Consumption of concrete, 2-12, 4-41
Consumption of steel, 6-30
Consumption of water, 2-12
Containment vessels
and creep, 6-60
and slipforming, 5-17
Contingency fees, 9-6, 9-9, 9-11, 9-17

Continuous beams and creep, 6-48, 6-55
Continuous slabs, 6-55
Contraction joints, 6-17
Contractors, 6-28
    and testing, 7-28
Controlled permeability formwork, 3-48
Conversion, 1-25, 3-25, 8-43, 8-44
    consequences of, 1-30
    and exposure conditions, 1-34
    influence on durability, 1-34
    influence on strength, 1-34, 1-40, 2-6
Cooling aggregate, 4-67
Cooling concrete, 4-67
Coral aggregate, 2-41
Cores
    horizontal, 7-34, 7-45
    interpretation, 7-28
    location, 7-28
    moisture condition, 7-31, 7-32, 7-33
    size, 7-31, 7-32, 7-42
    and strength, 2-53
    tests, 7-26
    tests, corrections, 6-40, 7-30
    vertical, 7-34
Coring, 6-39
Correction factors, 7-29, 7-30
Correlation, 8-23
Corrosion, 8-28
    and autogenous healing, 5-36, 5-43, 5-49
    and blended cements, 3-60
    and brackish water, 3-61
    and calcium chloride, 2-38
    and cementitious materials, 3-60
    and chlorides, 2-20, 3-53
    and cover, 5-5
    and cracking, 3-14, 3-42
    and curing, 3-61
    and curing water, 2-23
    and depassivation, 3-15
    effect of oxygen, 33-54
    effect of relative humidity, 3-54
    effect of temperature, 3-62
    with high-alumina cement, 1-31
    inhibitors, 2-33, 3-63
    initiation, 3-37, 3-40, 3-61
    initiation and curing, 3-61
    prevention, 3-63
    of prestressing steel, 1-40
    protection, 2-34
    and seawater, 2-2
    and seawater in mixture, 2-37
    and tricalcium aluminate, 4-11
    and water, 2-8
Cost, 8-7
Courts, 9-6, 9-9, 9-22
Cover, 3-27, 3-48
    actual, 5-8
    and autogenous healing, 5-41
    as built, 5-6
    and carbonation, 4-39
    and coral aggregate, 2-41
    and corrosion, 5-11
    and curing, 3-18, 4-38, 6-19
    distribution of, 5-12
    excessive, 5-6
    "negative," 5-8, 5-9
    nominal, 5-8
    and penetrability, 3-58
    positive, 5-8
    purposes of, 3-28, 5-4
    and quality of concrete, 5-12, 5-13
    tolerances, 3-28, 5-8, 5-15
    verification, 5-9
Covermeter, 6-41
Cracks
    and attack, 3-39
    and autogenous healing, 5-37, 5-47
    causes of, 3-14, 3-41, 6-34
    clogging, 5-38
    control, 6-37
    and cover, 5-6, 5-7
    and design, 3-61

10-16 Index

and filler, 5-37, 5-39
filling of, 3-15, 5-37
flexural, 3-49, 5-7
initiators, 3-50
live, 3-42
old, 3-15
in reinforced concrete, 5-41
reversible, 3-15
and shrinkage, 3-15, 3-49, 3-50
in slabs, 4-12
stable, 3-15, 3-42
thermal, 3-15
Crazing, 3-50, 4-61
Creep, 6-7, 6-46
and age, 6-47
basic, 2-7, 6-47, 6-54
in beams, 6-7, 6-9, 6-10, 6-47
and behavior of structures, 6-46, 6-59
beneficial effects of, 6-55
in buildings, 6-50
and cement content, 3-35
and cladding, 6-49
coefficient, 6-10, 6-56, 6-60
in columns, 6-48, 6-54
and construction, 6-59
and deflection, 6-6, 6-9, 6-10
differential, 6-53
drying, 2-7, 6-47, 6-54
expressions, 6-47, 6-62
extrinsic factors, 6-47
of high alumina cement, 6-62
influence of strength, 4-7
intrinsic factors, 6-46
lower bound, 6-55
of plain concrete, 6-46
in prestressed concrete, 6-46
problems, 6-46
recovery, 6-24, 6-47, 6-60
in reinforced concrete, 6-46
and speed of construction, 6-55
and strength, 4-67
upper bound, 6-55
and water, 2-6

Crushed rock, 8-12
Crushing aggregate, 8-12
Cryptoflorescence, 5-24
Crystalline deposit, 5-24
C-S-H, 4-58, *see also Calcium silicate hydrate*
Cubes, 7-29, 7-36
Curing
adiabatic, 4-59
and carbonation, 6-19
and chloride ingress, 6-19
and cracking, 2-55
and drinking water, 2-34
and fly ash, 3-31, 4-33
high-performance concrete, 4-35, 4-62, 4-64
inadequate, 3-30
influence on cooling, 4-63
isothermal, 4-59
membrane, 3-49, 4-36, 4-63, 5-21
membrane, effect on surface, 3-8
and modern cements, 8-15
by seawater, 2-23, 2-43, 2-44, 2-48
and shrinkage, 4-63
and slag, 3-31
in slipforming, 5-21
water suitability, 2-5
wet, 2-14, 3-18, 3-49, 4-63
wet, effect on surface, 3-8
wet, and temperature, 2-5
and wind, 5-21
Curling, 4-67, 9-8
"Custom-built concrete," 6-3, 6-16
Cylinder strength, 8-17
and concrete quality, 2-55
field cured, 7-43

D

Dampness, 2-20
and seawater in mixture, 2-38
permanent, 5-32
Dams and permeability, 5-25, 7-8

D-cracking, 3-42
Decalcification, 5-25
Decomposition of cement paste, 5-25
Deductive approach, 6-2, 6-62
Defendant, 9-6, 9-18
Deflocculation, 3-23
Deformation of concrete, 6-6
Deicing agents, 3-5, 3-13
    and cover, 5-5
Deionized water and cement, 2-35
Delamination, 3-54, 4-39, 5-9
Delayed ettringite formation, 6-12
Deleterious reactions
    and aggregate, 3-10
    and efflorescence, 5-33
    progress, 3-38
    rate of, 3-38
    and steel, 3-10
Demolition, 3-16
    material, 6-27
Density of concrete, 4-43, 6-36
Density of matrix, 4-18, 4-23
Depassivation, 3-15
Deposits on surface, 5-23
    crystalline, 5-23
Desalinated water, 2-5
Desalinating concrete, 3-64
Design, 10-2
    and construction, 6-22
    and creep data, 6-48, 6-59
    and material properties, 1-38, 6-6, 6-29, 6-63
    office, 5-14
    and technology, 6-1, 6-33
"Designer concrete," 6-3, 6-16
Detailing, 3-50, 5-11, 5-14, 6-35
    and cracking, 3-62
    of reinforcement, 3-29, 3-42, 3-43
Detergents in water, 2-21
Deterioration of structures, 3-5, 3-41
    causes of, 3-5, 3-44
    and exposure conditions, 3-4
Dew, 5-7, 8-29
Dicalcium silicate content, 4-9

Differential
    creep, 6-53, 6-59
    settlement, 6-49, 6-59
    shortening, 6-52, 6-54
    shrinkage, 6-10
Diffusion
    in autogenous healing, 5-38
    of carbon dioxide, 3-39
    of chlorides, 3-56
    definition, 3-6
    and durability, 3-44
    of vapor, 7-7
Dihydrate, 8-20
Direct tensile strength, 6-40
Direct tension, 5-40, 6-40, 7-8, 7-9
Disclaimers, 8-39
Discoloration, 6-12
    by curing water, 2-23
Discs, 7-10
Dispute resolution, 7-11
Distilled water, 2-5
Dolomitic limestone, 4-22, 4-24
Domestic sewage, 2-3
Drainage, 6-15, 6-29
Drinking water and leaching, 2-2, 2-9
Drive joints, 6-17
Drying creep, 6-54
Drying shrinkage, *see Shrinkage*
    and repair material, 3-17
    and water, 2-6
    and water content, 2-54
Ducts, 2-28
Duplex film, 4-44
Durability, 3-1, 3-22
    attitudes to, 3-4, 3-13
    and cement content, 8-26
    and cementitious materials, 3-22
    classes, 3-1
    and cover, 5-13
    definition, 3-4, 3-22
    effect of aggregate, 3-46
    of high-alumina cement, 1-31, 1-37

of high-performance concrete, 3-24
importance of, 3-10
and lime, 2-21
of "new" cements, 8-26
and shape of structure, 3-18, 3-45, 6-8
specifying for, 3-27
and strength, 3-6, 3-12, 3-13, 4-38
and water content, 2-54
and w/c, 2-58, 3-6, 8-26
Dust
in aggregate, 3-47, 4-31, 8-6, 8-12, 8-13
airborne, 3-55
salt-laden, 5-7
and water content, 8-12
Dynamic loading, 7-5

E

Earthquake, 3-46, 6-36
and creep, 6-54
Earth-retaining walls, 5-33
Ecological considerations, 6-27
Economy, 6-24, 6-27
Eddies in air, 3-38, 3-56
Editor, 8-34, 8-37, 8-42
Editorial board, 8-34, 8-37
Education, 6-4, 6-17, 6-28, 8-11
Efflorescence, 2-8, 5-2, 5-23, 5-30
and damage, 5-33
minimizing, 5-32
mock, 5-31
and seawater, 2-20, 2-38
tests for, 5-32
Elastic strain, 6-47
Electrical
charge on cement, 4-34
properties of high-alumina cement, 3-25
resistivity, 2-12
Electrochemical
cell and corrosion, 2-39, 2-40, 3-53, 5-43
potential and exposure, 3-54

Electrolytic conduction, 2-12
Elements, structural, 6-13
Elevators, 6-52
Elongated particles, 3-47, 4-24, 4-32, 5-20, 6-26
Embedded steel and corrosion, 5-5
Environmental aspects, 6-8, 8-9, 8-13
of aggregate, 8-6
Epoxy coating, 3-64
Equilibrium, structural, 6-35
Erosion, 2-11
Ettringite
composition, 1-21
delayed formation, 3-14
European Union and drinking water, 2-28, 2-31, 2-34
Evaporation, 4-11
Evaporative surface, 5-27
"Everyman's concrete," 6-12, 6-13, 6-16
Excess voids, 7-37
Exothermic reaction, 4-56
Expansive cements, 1-20
Expansive compounds, 5-26
Expected life, 3-22
Expert witness, 9-6, 9-10, 9-11, 9-14, 9-15, 9-17, 9-23
and publications, 8-38
Explosive loading, 7-22
Exposure conditions
extreme, 3-34
and seawater in mixture, 2-41
severe, 5-6
variable, 3-38
Extenders, 11-4

F

f'_c, 7-5, 7-39
f'_{cr}, 7-5, 7-6, 7-9
Façade, 6-13, 6-50
Factor of safety, 6-36
Failure, 1-30, 6-11, 8-27
designated, 8-27
investigation, 6-9
False set, 1-28

Fast neutrons emission, 8-11
Filler in cracks, 5-44
Fillers, 4-21
    calcareous, 3-45
    in cement, 1-2
    and chlorides, 3-61
    and corrosion, 3-61
    and durability, 3-23
    in high-performance concrete, 4-40
    and w/c, 2-50
Fibers in concrete, 6-31
Fineness
    of cement, 3-8
    modulus, 4-32
    of silica fume, 1-13
Finish screening, 3-46
Finishing, 3-49
    and cementitious materials, 2-62
Finite element method, 6-36
Fire, 7-4
    and cover, 5-5
    endurance, 5-5
    and high-performance concrete, 4-38
    protection, 3-28
    rating, 5-5
    resistance, 5-5
    resistance and cover, 5-15
    tests, 7-19, 7-23
    and water, 2-8
Fitness for purpose, 3-22, 6-16, 6-20, 6-29
Flaky particles, 3-46, 3-47, 4-24, 4-32, 5-20, 6-26
    and bleeding, 2-5
Flame penetration, 5-5
Flat roofs and creep, 6-11
Flat slabs, 6-63
Flatwork, 6-17
Flaws, 4-30
Flexural strength, 7-9
Flocculation, 4-16
Floors, garage, 5-27
Flow through cracks, 5-42

Flow and durability, 3-44
Flowing water, damage by, 2-11
Fly ash
    and durability, 3-22
    and heat development, 4-12
    and high-performance concrete, 3-21
    in interface zone, 1-10
    specific gravity, 2-50, 2-62
    and strength development, 2-62
    and w/c, 2-50
Fluorescence microscopy, 7-3
Fog misting, 4-63
Footings, 6-15, 6-37
Form-release oil, 3-48
Formwork
    controlled permeability and efflorescence, 5-32
    friction in slipforming, 5-20
    plywood, 4-65
    release agents and drinking water, 2-34
    removal, 3-8, 3-19, 3-31, 3-50, 4-9, 4-10
    for slipforming, 3-26, 5-17, 5-18
    steel, 4-65
Foundations, 6-13, 6-15
Fracture, 4-23
Free water and thermal insulation, 2-12
Freezing and thawing, 2-10, 3-5, 3-13, 6-13
    in high-performance concrete, 4-18, 4-38
    and self-desiccation, 4-38
    and strength, 3-20
Fresh concrete, action of frost, 2-10
Friedel's salt, 3-58
Frost, 2-10

G

Galvanic coupling, 5-44
Gel pores, 4-60
Gel water, 2-4, 4-57
Generalists, 6-3, 6-23

Geomaterials, 8-31
Geotextiles, 4-36
Glass microbeads, 6-31
Glass-reinforced cement, 6-51
"Good practice," 9-5
Grab samples, 8-15
Granite, 4-24
Gravel, 8-1
Green concrete and autogenous healing, 5-36
Grinding aids and drinking water, 2-34
Ground granulated blast-furnace slag, *see Slag*
Groundwater, sulfates in, 5-26
Guides, 6-38
Gutters, 5-31, 5-32
Gypsum, 4-20, 8-20
    solubility, 4-34
    types, 3-24

H

Handbook, 6-48
Hardened cement paste composition, 4-17
Heat of hydration, 3-35, 3-36, 4-30, 4-55
    with fly ash, 4-32
    with slag, 4-32
Heavy metals and leaching, 2-33
Hemihydrate, 3-24, 4-20, 4-34
Hessian, 3-49, 4-36
Heterogeneity of concrete, 4-50, 4-51
High-alumina cement, 1-2
    and chlorides, 3-61
    collapse of structure, 1-27
    and drinking water, 2-2, 2-9, 2-34
    and efflorescence, 5-32
    erroneous use, 6-8
    inspection, 3-17
    manufacturer, 8-44
    non-structural use, 3-31
    problems, 1-30
    review, 1-25
    and seawater in mixture, 2-39
    structural use, 1-25, 1-27, 1-28
    use of, 3-25
    and wet conditions, 1-39
    wet curing, 2-6
High-performance concrete, 8-10
    and aggregate size, 4-29
    and aggregate strength, 4-18
    and autogenous shrinkage, 6-7
    and cementitious materials, 4-16, 4-29
    definition, 3-24
    density, 4-29
    description, 4-1, 4-15, 4-29
    and "designer concrete," 6-16
    and durability, 2-54
    mixtures, 4-22, 4-24, 4-25
    modulus of elasticity, 4-29
    overview, 4-29
    permeability, 4-29
    quality control, 8-13
    rheological properties, 4-19
    and superplasticizers, 4-16
    and water, 2-7
High-range water-reducing admixtures, *see Superplasticizers*
High-rise buildings, 4-15, 6-7
High-strength concrete, 4-2, 4-15, 4-29, 4-37, 4-45
    and cores, 7-34
    and lightweight aggregate, 4-48, 4-53
Highway construction, 4-12, 6-28
"Hired gun," 8-46
Holistic approach, 6-30
Homogeneity of mixture, 3-29
Hot climate
    and corrosion, 3-62
    dry, 4-10
    and high-alumina cement, 1-39
HPC, *see High-performance concrete*
Humidity,
    influence on strength of high-alumina cement, 1-26

Humins, 2-15
Hydrated cement paste, 1-8
Hydration
    and curing, 4-10
    full, 3-30
    and heat, 4-55, 4-58
    kinetics, 4-20
    maximum, 4-17
    and strength, 4-55, 4-58
    and w/c, 2-65, 4-17
    and w/c determination, 2-52
Hydraulic cement, 1-5, 1-16, 2-4, 5-25
Hydraulic properties, 1-7
Hydrochloric acid, 3-56
Hydrophobic admixtures, 2-9
Hydroxyl ion, 3-64, 5-43, 8-28
Hygral movement, 3-37
Hypothesis, 8-3, 8-24, 8-43
Hysteresis, 4-24

I

Ice, 4-67
    crushed, 2-14
    shavings, 2-14, 3-37
Imbibition, 8-30
Incinerator ash, 6-27
Indeterminacy, 6-36
Inductive approach, 6-2, 6-62
Industrial waste water, 2-3
Inference, 7-6
Inhomogeneity, 4-4, 4-5
Initial set and fly ash, 2-62
Inquisitorial legal system, 9-2, 9-15
In-situ strength, 7-33
Insolation, 2-40, 8-29
    and diffusion, 3-62
    and efflorescence, 5-24
Inspection, 3-14, 3-17, 3-40, 3-51
Insurance, 9-7
Interface, 4-30, 4-49, 8-25
    influence on stress-strain relation, 4-44
    porosity, 4-44

Interface zone, 1-1, 3-45, 4-2, 4-5, 4-43, 4-44, 7-8
    description, 1-8, 1-9, 1-10
    in high-performance concrete, 4-33
    influence on durability, 4-5
    influence on permeability, 4-5
    stress concentration at, 4-48
    and unhydrated cement, 3-50
Internationalization, 10-2
Interstitial phase, 4-20
Investigators, 3-43
Ironworkers, 5-4
Irrigation, 6-29
ISO, 7-22
Isothermal conditions, 4-56
Iteration, 6-36

J

Japan, 5-31
Jetty, 6-8
    and corrosion, 2-41
    shape and durability, 3-38
Joints,
    formed, 6-17
    sawn, 6-17
Judges, 9-2, 9-6, 9-15, 9-19
Judgment, 6-41
Junk science, 9-12
Jury, 9-2, 9-6, 9-12, 9-15, 9-19, 9-22
Justice, 9-9, 9-19, 9-28

K

Kiln and sulfur, 3-24
Kinetics of creep and shrinkage, 6-62
Kinetics of hydration, 4-20

L

Lateral constraint, 7-5
Lawyers, 9-9
Leachates, 2-30
Leaching, 2-8, 2-18, 2-29, 2-30, 2-31, 2-32, 5-23, 5-27
    and autogenous healing, 5-47

of calcium hydroxide, 3-36
    by flowing water, 2-23
    of heavy metals, 2-33
    by pure water, 2-26
Leakage and construction joints, 5-17
Le Chatelier's expression, 4-57, 4-58
Length-diameter ratio, 7-29
Life cycle, 3-17
    cost, 4-41
Life of structure, 1-10, 3-16, 3-38, 3-43, 4-37, 6-14, 6-20, 8-29
Lightweight aggregate, 4-43
    and absorption, 4-46
    bond, 4-46
    and hydration, 4-46
    surface texture, 4-46
    and w/c, 2-56
Lightweight concrete, 4-23
    and insulation, 2-11
    tests, 7-5
Lime bloom, 5-23
Lime-saturated water, 5-71
    and autogenous healing, 5-41
Lime weeping, 5-23, 5-33
Limestone, 4-24
    ground, 4-21
Limit state, 8-27
Limiting cracking, 6-37
Linear materials, 6-36
Links, 5-8
Liquid nitrogen, 3-37
Litigation, 9-1, 9-4, 9-13
    cost, 9-7
    and publications, 8-32
    and research, 9-30
    research benefits, 9-29
    and testing, 7-1
Load
    -carrying capacity, 3-54, 7-4, 7-7, 7-27, 7-43
    dead, 6-11
    factor, 10-3
    live, 6-36, 10-4
    self-weight, 6-36
    tests, 7-4

Loading system
    actual, 7-27
    designed, 7-27
Long building, 6-17
Loss of prestress, 6-60
    and age at loading, 6-56
Low-heat portland cement, 4-67

M

Macro-scale research, 8-23, 8-25
    and design, 8-26
Magnesium sulfate, attack by, 3-39
Maintenance, 3-13, 4-41
    corrective, 3-12, 3-40
    need for, 3-11,
    preventive, 3-2, 3-12, 3-14, 3-17, 3-41
    procedures, 3-40
    regular, 3-14
    routine, 3-17
    systematic, 3-2
Making good, 3-41
Manual staff, 8-10
Marine environment, 2-37
Masonry
    bridges, 7-4
    construction, 6-9
    efflorescence, 5-24, 5-30, 5-31
Mass concrete, 6-9
Massive construction, 6-9
Material parameters, 6-7
Materials scientists, 3-26, 6-24, 6-43
Materials specialists, 3-32
Maturity meter, 4-56
Maximum aggregate size, 4-24
    and cement content, 3-9, 3-20
    and cores, 7-31, 7-32
    in high-performance concrete, 4-30, 4-62
    and strength, 4-45
    and testing, 7-21
Mechanization, 6-4, 6-28, 6-31
Mediation, 9-4, 9-10

Metakaolin, 6-14
　in high-performance concrete, 4-40
　and *w/c*, 2-50
　and *w/cm*, 2-62
Microbial growth and drinking water, 2-34
Microclimate, 3-20, 3-27, 6-13, 6-15
Microcracking, 4-5, 4-30
　bond, 4-50
Micro-scale research, 8-23, 8-24
Microsilica, 1-16, *see also Silica fume*
Microstructure, 2-50, 3-7, 8-14
　of interface zone, 4-43
Microwave absorption, 2-59
　meter, 8-11
Microwave oven test, 2-55
Middle East, 5-7, 5-9
　and cover, 5-13
Mineral
　addition, 1-15
　admixtures, 1-7, 1-15, 3-23
　waters as mixing water, 2-17
Mirabilite, 5-27, 5-33
Mix, *see Mixture*
Mixers, 6-19, 6-25
Mixing time, 3-29, 6-19, 6-21
　and output, 6-21
　with silica fume, 6-19, 6-21
Mixing water, 2-14
　quality, 2-3
　standards, 2-14, 2-16
Mixture
　design and specific gravity, 2-62
　in high-performance concrete, 4-30, 4-31
　ingredients, 6-17
　proportions, adjustments, 8-6
　proportions from charts, 3-20
　selection, 3-19
　selection, incompatibility in, 5-20
　for slipforming, 5-20
　without specification, 6-15

Mobility, 5-19
Models, 8-23
Modulus of elasticity, 4-15, 6-7
　and aggregate, 4-23, 6-44
　of aggregate, 4-42, 4-46, 4-47, 6-44
　of aggregate and of cement paste, 4-51, 4-52
　and autogenous healing, 5-36, 5-42
　and bond, 4-42
　of high-performance concrete, 3-24, 4-29
　of high-strength concrete, 4-46
　of lightweight aggregate, 4-48
　of matrix, 4-42
　of repair material, 3-17
　and strength, 4-43
Modulus of rupture, 7-9, 7-10
Moist curing, 8-14
Moisture
　changes, 3-14
　condition of cores, 7-38
　content of aggregate, by capacitance, 8-11
　content of aggregate, continuous measurement, 8-11, 8-12
　content of aggregate by resistance, 8-11
　gradient in core, 7-33
　movement and autogenous healing, 5-40
Moment-carrying capacity, 3-28
Moment of resistance, 5-6
Monocoque construction, 6-31
Monolithic behavior, 4-42, 4-45, 4-46, 4-48, 4-53
Mortar,
　hard lime, 5-31
　lining of pipes, 2-26, 2-27, 2-32, 2-34, 5-40
　lining of pipes and microbial growth, 2-34

in masonry, 5-24
soft lime, 5-31
strength, 7-9
and sulfate attack, 5-27
Multiphase material, 4-56

N

Naphthalene sulfonates, 4-20
Natural sand, 8-12
Neat cement paste, 4-13
Non-compliance, 7-2
Nondestructive tests, 7-19
Nonelastic deformation, 6-46
Nonlinearity, 6-36
Nuclear power stations, 6-60

O

Objectivity, 8-3, 8-43, 8-45, 9-29
Oil platforms and slipforming, 5-17
Operatives, 5-10, 6-28
Ordinary concrete, 6-20
Organic
 admixtures, 2-34
 fibers, 2-34
 plasticizers, 2-34
Outliers, 7-20
Overloading, 5-6, 7-27, 9-7
Oxygen, dissolved in water, 2-44

P

Packing
 in interface zone, 1-9, 1-10
 and silica fume, 3-45, 4-5, 4-33, 4-44
Paint and efflorescence, 5-32
Panels, 6-50
Parasitic bending moments, 6-49, 6-55, 6-59
Parasitic shearing forces, 6-49, 6-55, 6-59
Passivation, 3-28, 3-53, 3-54, 5-5
Patio and joints, 6-17
Paving slabs, 5-27
Pedestal, 7-5

Peer review, 8-3, 8-5, 8-32, 8-33, 8-34, 8-37, 8-40
 and litigation, 8-39
Penalty payment, 7-27
Penetrability
 to chlorides, 3-63
 and durability, 2-51, 3-44
 of high-performance concrete, 3-21
 with "new" cements, 3-7
 and pore structure, 1-9
 tests, 3-63
Penetration resistance, 3-43
Performance specification, 2-58, 3-27, 6-6, 6-14, 6-20, 6-61
Permeability
 and carbonation, 4-10
 by chloride test, 7-7
 of dams, 5-25
 definition, 3-5, 3-6
 and durability, 2-51, 4-10, 6-10
 effect of aggregate, 4-5
 and fire resistance, 4-38
 influence on chemical attack, 3-10
 and porosity, 2-9
 and pressure gradient, 7-15
 test, 7-7
 and w/c, 2-58, 3-19
Permeable concrete and efflorescence, 5-32
Persian Gulf, 2-48, *see also* Arabian Gulf
Personnel, qualifications, 3-11
Petrographic examination, 2-63, 5-43, 5-49, 7-3, 7-26
Petrography, 2-64, 2-65, 6-43
pH
 of acid rain, 2-8
 and autogenous healing, 5-39
 of drinking water, 2-29, 2-31, 2-32, 2-33
 of drinking water and high-alumina cement, 2-34
 of high-alumina cement, 1-31
 of pore water, 3-60

of Type K cement, 1-23
of water, 2-20
of water in conduits, 2-9
Phase, 1-1
Ph.D., 8-1, 8-2, 8-4, 8-5
Physical salt attack, 5-27
Piers and cover, 5-12
Piles, 5-36
Pipe and autogenous healing, 5-36, 5-40, 5-41, 5-42
Pipe lining, 2-26, 5-43
and corrosion, 5-43
Pipelines, 2-27, 2-28
Pits, 8-12
Pitting corrosion, 3-62
Pitting of steel, 3-5
Pitting by water, 2-11
Plain concrete and chlorides, 5-31
Plaintiff, 9-6, 9-18
Plaster and efflorescence, 5-32
Plastic additives and drinking water, 2-34
Plastic shrinkage, 2-5, 4-62
and bleeding, 4-62
Plastic shrinkage cracking, 2-5, 3-5, 3-44
in high-performance concrete, 4-35
Pleadings, 9-6
Pockmarks, 3-48
Poisson's ratio, 4-5, 4-45
Polymer concrete, 2-9
Polymers in concrete, 6-30
Ponding, 4-63
Pop-outs, 3-42, 7-21
Pore
blocking, 8-8
connectivity, 2-51, 3-8
silting, 8-18
size, 6-14, 6-33
size and permeability, 7-22
structure and durability, 3-27
structure and strength, 6-18
water, alkalinity, 5-5
water and chlorides, 3-59

water and corrosion, 3-54
water as electrolyte, 3-53
water, pH, 3-60
Porosity
and carbonation, 3-39
and durability, 2-51
of high-alumina cement, 1-26
of high-performance concrete, 4-18
of interface zone, 1-9, 4-44
and permeability, 2-9
and w/c, 2-60, 2-64
Portland cement
and cementitious materials, 6-18
compounds in, 4-56
disadvantages, 3-35
patent for, 1-2
terminology, 1-5
and w/c, 2-50
Portland composite cement, 1-5
Portlandite, 4-56, *see also Calcium hydroxide*
Potassium
and drinking water, 2-29
Potassium carbonate, 5-25
Potassium sulfate, 5-25
solubility, 2-16
Powers' expression, 4-57, 4-58
Pozzolans
and efflorescence, 5-32
in high-performance concrete, 4-40
Precast concrete, 6-50
and autogenous healing, 5-36
and cover, 5-11
and efflorescence, 5-31
Precision, 6-35, 6-36, 6-37, 6-38, 6-39, 6-40, 8-46
Precision of derived quantity, 6-39
Prestressed concrete
and chlorides, 2-20, 2-39, 3-55, 6-8
and corrosion, 3-62
and cover, 5-12

and creep, 6-56, 6-62
pipes, 2-28
and seawater, 2-23
Prestressing steel in high-alumina cement, 1-40
Proctor needle, 5-18
Production control, 8-13
Production, vertical integration, 8-14
Publications, 8-32
Pure water, attack by, 2-8
Principle of uncertainty, 8-46

Q

Quality of concrete, 1-10, 5-14, 5-16
  actual, 3-30
  and cover, 5-12, 8-28
  and penetrability, 3-62
  potential, 3-30
Quality control, 4-39, 6-31, 8-17
  for high-performance concrete, 8-13
Quenching, 2-8

R

Rainfall and water properties, 2-19
Rainwater and efflorescence, 5-27, 5-30
Raveling, 6-17
Reactive powders, 3-45
Reactive silica, 4-32
Ready-mixed concrete, 6-25, 6-26, 8-26
  and mixture selection, 6-15
  and specification, 3-9, 6-13
Rebound hammer, 3-43
Recovery of strength, 5-42, 5-43
Recycled aggregate, 6-27, 8-6
Recycled concrete, 8-13
Redistribution of stresses, 6-58
Re-entrant angles, 6-17
Refereed journal, 8-35
References, 8-43
  to oneself, 8-44
  omission, 8-44
Refractory concrete, 1-2, 1-32, 3-25

Rehabilitation, 3-17, 3-48
  expenditure on, 3-13
Reinforced concrete
  and chlorides, 2-39
  and seawater in mixture, 2-37
Reinforcement
  bending, 5-11
  cage, 5-10
  and creep, 6-47, 6-54
  cranking, 5-11
  detailing, 3-42, 3-43, 6-35
  epoxy coated, 3-48
  galvanized, 3-48
  lapping, 5-11
  position, 5-1
  in slipforming, 5-19
Relative humidity
  and autogenous healing, 5-39, 5-42
  effect of wind, 3-37
Relaxation methods, 6-36
Relaxation of stress, 6-24
Reliability, 8-46
Remedial measures, 3-15
Repair materials, 3-17, 3-43
Repairs, 3-2, 3-16, 3-17, 4-40, 8-29
  of cracks, 3-14
  expenditure on, 3-13
  of repairs, 3-41, 4-40
Repeatability, 6-39
Replacement materials, 1-4, 3-23, 6-18
Replacement of structure, 3-43
Reproducibility, 2-33
Research, 6-22, 8-1
  bad, 8-23
  basic, 8-4
  commercial, 8-44, 8-46
  and design, 8-26
  and litigation, 9-29, 9-30
  on portland cement, 8-9
  proposal, 8-38
  and testing, 7-6
  usefulness, 8-4, 8-9
Reservoir, 2-29

Resistivity of cement paste, 3-58
Resistivity of concrete, 3-54
Respondent, 9-6
Restraint and cracking, 3-42
Restraint of deformation, 3-14
Retarder in slipforming, 5-20
Reviewer, 8-35, 8-36
Rheological properties, 4-19, 8-9, 8-13, 8-30
Rice husks, 4-40, 6-14
RILEM, 7-22
Risk, 1-30
Robots, 6-4, 6-28, 6-31
Robustness, 6-5
Roof, flat, 3-16
Rounded cement, 8-9
Rust, 3-50

S

Safety factors, 6-24
    partial, 6-24, 6-25, 10-3
Safety of personnel, 3-17, 3-40
Safety, responsibility for, 6-9
Sag, 6-6, 6-57
St. Eustatius and seawater in mixture, 2-41, 2-42
Salinity of seawater, 2-38
Salt
    crystal whiskers, 5-31
    crystallisation, 5-24, 5-27
    deposits on concrete, 5-25
    hydration distress, 5-28
    ingress, 3-57
    weathering, 3-42, 3-44
Sampling plan, 7-28
Saturated-and-surface-dry aggregate, 6-19
Seashore and efflorescence, 5-31
Sea spray and corrosion, 2-40
Seawater
    airborne, 3-55
    and autogenous healing, 5-40
    and chlorides, 2-8
    and corrosion, 2-23
    and curing, 2-5, 2-43
    as mixing water and efflorescence, 2-20, 5-31
    salinity, 2-38
    salt composition, 2-48
Sedimentation, 2-5
Segregation, 7-26
    and workmanship, 3-11
Self-desiccation, 3-31, 4-18, 4-36, 8-15
    and vapor pressure, 4-57
Self-stressing concrete, 1-20
Self-weight, 6-36
Service attachments and creep, 6-52
Service life, 6-9
Service load, 6-36
Serviceability, 3-16, 3-22, 6-5, 7-4, 7-16, 8-27, 8-28
Setting
    and cementitious material, 2-62
    effect of lime, 2-21
    effect of sulfates, 2-21
    of mortar, 2-16
Setting time, 5-18
    influence of mixing water, 2-18
    influence of sugar, 2-17
Severe exposure, 5-6
Shaft lining and slipforming, 5-17
Shear walls, 5-17, 6-56
Ship docks, 6-9
Shipment of cement, 8-16
Shrinkage
    of aggregate, 6-7, 6-17
    autogenous, 4-54, 4-63, 4-64
    and bridge piers, 6-48
    carbonation, 4-54
    chemical, 4-54
    -compensating concrete, 1-22
    compensation, 1-20
    consequences of, 1-20
    of high-performance concrete, 4-35, 4-64
    and hydration, 4-63
    reinforcement, 3-63, 5-10, 6-37
    restrained, 4-62, 6-17
    and self-desiccation, 4-54

thermal, 4-54
total, 4-54, 4-62
types, 4-54
Shrinkage cracking, 3-35, 3-44, 4-38, 6-37
and autogenous healing, 5-39
and chloride ingress, 2-40
and cover, 3-28
and water, 2-6
SI system, 10-2
Silica fume, 1-2, 1-12
and autogenous healing, 5-39
cost, 4-32
and durability, 3-21, 3-22
and efflorescence, 5-32
form of, 1-17
in high-performance concrete, 4-19, 4-32, 4-33
influence on bleeding, 3-45
influence on permeability, 3-21
influence on plastic shrinkage, 3-45
and interface zone, 1-10, 4-44
and mixing time, 3-30, 6-19, 6-21
and packing, 4-44
and self-desiccation, 4-36
slurry, 1-18
in specification, 1-12
specifying use of, 1-16
standards for, 1-14
and strength, 4-32
and w/c, 2-50
and workability, 5-20
Silos
design, 10-4
and slipforming, 5-17
Silt in water, 2-20
Silting of cracks, 5-37
Site, 6-28
experience, 5-11
operatives, 5-14
Skid resistance, 6-17
Slab, 6-37
on grade, 3-22, 6-15, 6-37
suspended, 6-15

Slag, 1-2
and drinking water, 2-33
and durability, 3-22
and heat development, 4-12
and strength development, 2-62
and w/c, 2-50
and workability, 5-20
Slate waste, 6-27
Slipformed concrete, 3-26
cores, 7-34
Slipforming, 5-2, 5-17
and cold joints, 5-18
Slump loss, 4-32
in slipforming, 5-2
Sodium
in drinking water, 2-29
in salt, 2-38
Sodium carbonate, 5-25
Sodium chloride, 3-60
dry, 2-8
in seawater, 2-38
and corrosion, 3-63
Sodium sulfate, 5-26
deposits, 5-33
from soil, 5-27
solubility, 2-16
Soffits and cover, 5-12
Solar radiation, 2-40
Sorption, 7-7
and durability, 3-44
South Africa and seawater in mixture, 2-43
Spacers, 3-29, 5-10, 5-14
Spalling, 3-40, 3-54, 4-39, 5-9, 5-30, 6-9, 6-16
Specialists, 6-3, 6-22, 6-23, 8-31
Specific gravity of cement, 2-54
Specific surface of cement, 4-56
Specification, 3-50, 6-13, 9-25
content of, 6-13
for cover, 5-1, 5-8
incompatibility in, 3-32
and litigation, 9-3, 9-4
performance-type, 2-58, 3-27, 6-6, 6-14, 6-20, 6-61

prescriptive, 3-27, 6-14
and serviceability, 8-27
and silica fume, 1-12
for slipforming, 5-17
writing, 3-26
Specified strength, 7-5, *see also* f'_c
Specifying silica fume in concrete, 1-16
Spillway, 5-30
Splash zone, 8-29
and salt ingress, 3-57
Splitting tensile strength, 7-9, 7-10, 7-14, 7-35, 7-39
Spot samples, 8-15
Spray, 8-29
Staining by curing water, 2-23
Staircase, 6-41
Stalactites, 5-30, 5-31
Standard deviation, 6-25, 6-39
Standard test methods, 7-4, 7-17
Standards, international, 10-2, 10-3
Standards for silica fume, 1-14
State-of-the-art, 8-43
Statistical significance, 6-38
Steam curing, 7-37
Steel
    bending shop, 5-14
    coating, 3-34
    construction, 6-5
    consumption, 6-30
    in cores, 7-35
    embedded, 5-8
    fixer, 5-4, 5-16
    mesh, 5-15
Stiffening of concrete, 5-2
Stiffening time in slipforming, 5-17, 5-18, 5-19
Stirrups, 5-8
Stockpile, 8-11
Strain capacity, 5-5, 5-37
Strategic Highway Research Program, 1-30
Strength
    actual, 4-8, 6-26, 7-37

of cement and of concrete, 4-5, 4-6
and cement content, 3-27
characteristic, 6-24, 6-25, 6-26
development, 4-9
and durability, 4-10
and heat, 4-55, 4-58
high-early, 4-2
high long-term 4-2
and hydration, 4-55, 4-58
increase in, 3-7, 3-20
intrinsic, 7-36
mean, 4-39, 6-25
minimum, 4-39
of "new" cements, 4-8, 4-9
and permeability, 4-10
potential, 6-26, 7-37, 7-38, 8-28
precision of, 6-40
scatter, 6-25, 6-26
specified, 3-7, 4-39, 7-42
testing, 8-16
"true," 7-36
variability, 6-25
and void ratio, 6-33
and *w/c*, 6-33
Stress-strain relation, 4-42, 4-43, 4-44, 4-45, 4-47, 4-49, 4-51, 6-35
    in high-strength concrete, 4-46
    and lightweight aggregate, 4-47
Structural action
    and cores, 7-27
    and cracking, 3-42
Structural adequacy, 6-37, 7-26, 7-39
Structural assessment, 6-9
Structural behavior, 6-24, 6-34, 9-17
Structural design
    and concrete technology, 6-5
    and durability, 6-5, 6-7
    and properties of concrete, 6-7
    and *w/c*, 3-9
Structural designer, 6-10
Structural engineering, 6-22

Structural form, 6-34
    and efflorescence, 5-32
Structural function, 6-41
Structural safety, 9-17
Structural use of high-alumina cement, 11-25, 1-27, 1-28
Structures,
    idealized, 6-35, 6-36
    rounded, 6-31
    shape and exposure, 3-38, 6-31
Submerged concrete, 2-43, 2-44
Substrate, 4-67
    incompatibility, 3-17
Sugar in water, 2-17
Sulfate
    in coal, 4-34, 8-15
    in portland cement, 3-23
    in wash water, 2-21
    in water, 2-20
Sulfate attack, 2-8
    and efflorescence, 5-23, 5-33
    effect of silica fume, 3-21
    reactions, 5-26
    and seawater in mixture, 2-43
Sulfate-resisting cement and chlorides, 3-55
Sulfur
    in coal, 4-34
    in kiln, 4-34
    in oil, 4-34
Sun, exposure to and salts ingress, 3-57
Superplasticizers, 6-30
    action of, 4-34
    in cement paste, 8-19
    compatibility, 3-24, 3-31, 8-20
    and corrosion, 3-60
    and drinking water, 2-33
    in high-performance concrete, 4-31
    in slipforming, 5-19, 5-20
    and tricalcium aluminate, 4-34
    and water content, 2-54
    and w/c, 2-50, 2-58
Supersaturated solution, 5-28

Supervision, 3-18, 3-21, 3-51, 5-14, 6-43, 8-14
Supplementary cementing materials and w/cm, 2-61
Supplementary materials, 1-15, 3-23, 4-21, 6-18
Surface
    barriers, 3-63
    blemish, 3-48, 5-28
    deposits on concrete, 5-26
    scaling, 5-33
    texture of lightweight aggregate, 4-46
    -volume ratio, 4-61, 4-66
    -volume ratio and cladding, 6-59
    -volume ratio and creep, 6-47, 6-48
Suspension bridge, 6-49
Sustained stress, 6-46
Swelling, 4-55, 4-60
Symbiosis of design and materials, 6-29, 6-30, 6-31

T

Taiwan and water supply, 2-28
Tall buildings, 6-56
Tanks, 2-29
    and autogenous healing, 5-36
    and slipforming, 5-17
Teaching concrete, 6-32, 6-33, 6-42
Technical staff, 8-10
Technologists, 6-33
Technology and design, 6-1, 6-5, 6-32
Temperature changes, 4-37, 8-29
    and autogenous healing, 5-40
    and cracking, 3-14
Temperature of concrete,
    effect of curing, 2-6
    effect of evaporation, 2-5
Temperature cracks, 3-42
Temperature cycling, 4-47, 4-48
    and efflorescence, 5-33
Temperature and deterioration, 3-5
Temperature differential, 3-5

Temperature gradient, 3-36, 4-32
   in a column, 4-65
   and cracking, 3-35
   and curing, 2-6
   and hydration, 4-59
   and mixture proportions, 3-26
   in slab, 4-67
   and slipforming, 5-18, 5-20
Temperature influence
   on high-alumina cement, 1-25
   on plastic shrinkage cracking, 3-37
   on reactions, 3-37
   on stiffening, 3-36
Temperature reinforcement, 3-63, 4-2
Tensile force, 5-4, 5-6
Tensile strain, 4-66
   capacity, 5-37
Tensile strength, 4-37, 4-61, 4-62, 4-65, 5-42, 6-16, 6-40, 7-9, 8-27
   of rock, 7-1
   tests, 7-8, 7-12
Tensile stress, 6-47
Tension stiffening, 8-27
Ternary, 4-23
Test cylinders, 4-8, 6-33
Tester, 6-37, 7-46
Testing, 7-1
   errors, 8-16
   inappropriate, 8-19
   machine, 7-29
   plan, 7-27
   for research, 7-16
   and safety, 7-28
   and specification, 7-28
   wrong parameters, 8-18
   at wrong w/c, 8-19
Thenardite, 5-27, 5-33
Thermal
   behavior of high-performance concrete, 6-9
   conductivity, 2-11
   cracking, 3-35, 3-36, 4-32
   cracking and setting, 2-5
   effects and autogenous healing, 5-40
   effects and cracking, 3-14
   expansion, 6-51
   gradients, 8-25
   gradients and cracking, 3-44
   insulation, 2-11
   movement, 6-17, 6-46, 6-52
   properties and slipforming, 5-17
   shrinkage, 4-58
   strains and aggregate, 4-4
   stresses and curing, 2-6
   stresses in slabs, 4-12
Tidal zone and salts ingress, 3-57
Tide, 8-29
Tide and corrosion, 2-40
Ties, 3-29
Tolerances on reinforcement, 5-1
Towers, 5-17
   and creep, 6-49
   and shrinkage, 6-50
Tradesmen, 6-28
Trafficking and curing, 3-31
Training, 6-4, 8-11
Transient loads, 6-54
Transition zone, 3-21, *see also* Interface zone
Trapped air, 3-48
Trial mixes, 4-21
Triaxial compression, 6-46
Tricalcium aluminate
   and chloride binding, 6-14
   and chlorides, 3-58
   and heat evolution, 3-59
   reactivity, 4-20
   and sulfate attack, 3-59
   and sulfates, 8-20
Tricalcium silicate
   content, 3-8, 4-9
   in "new" cements, 3-19
Truth, 9-14, 9-16, 9-17, 9-24
Tunnel lining, 6-9
Turbo-alternator, 6-48
Turbulent flow, 5-43
   and oxygen, 2-44

10-32 Index

Turkey and water supply, 2-28
Type I cement, 4-33
Type K cement, 1-21, 1-22, 1-23
Type M cement, 1-21, 1-23
Type O cement, 1-22
Type S cement 1-21, 1-23

U

Ultimate limit state, 8-27
Ultimate strength, 5-6
Ultrasonic pulse velocity, 6-38, 6-43
Uncertainty, 8-45
Unhydrated cement, 4-58
    and autogenous healing, 5-37, 5-39
Uniaxial compression, 7-5
Uniformity of concrete, 8-6
Units of measurement, 10-2
Unreinforced concrete, 5-14
UVP, *see Ultrasonic pulse velocity*

V

Validity of claims, 9-9
Validity of tests, 2-33, 7-5
Vapor transmission, 7-8, 7-15
Variability, 8-2
    of aggregate, 8-12
    of cement, 8-15
    of concrete, 6-24
    of strength, 7-36
Variation orders, 9-5, 9-8
Very fine powders and durability, 3-23
Viruses in water, 2-22
Voids and durability, 2-51
Volume-surface ratio, 4-61, 4-66
    and cladding, 6-59
    and creep, 6-47, 6-48
Volumetric proportioning, 2-50, 2-54

W

w/c, 2-50
    and autogenous healing, 5-42
    and creep, 6-47
    definition, 2-50
    determination, 2-51
    and durability, 2-51
    by fluorescence microscopy, 7-3
    of fresh concrete, 2-59
    of hardened concrete, 2-52
    in high-performance concrete, 4-40
    influence on strength of high-alumina cement, 1-29
    maximum, 1-28
    optimum, 4-18
    of original mixture, 2-51, 2-53
    and position, 7-34
    specified, 3-9
    standard deviation of, 2-59
    and strength, 2-5
    by volume, 2-54
w/cm, 2-2, 2-55, 4-22, 4-23
    during production, 2-59
Wall effect, 4-5, 4-44
Walls and cover, 5-13
Warm and humid environment, 2-43
Warping, 4-67, 9-8
Wash down with seawater, 2-40, 8-29
Wash water, 2-3, 2-14, 2-21
    guide for, 2-15
Washing aggregate, 2-14, 3-47, 8-12
Waste as aggregate, 8-13
Waste products, 6-27
Water,
    acid, 2-16
    adsorbed, 2-6
    on aggregate, 6-19
    algae in, 2-20, 2-21
    alkalinity, *see* pH
    attack by cement, 2-26
    bound, 4-18
    brackish as mixing water, 2-37, 2-38, 2-44, 3-15, 3-47
    chemically combined, 2-4, 4-57
    and concrete, 2-1, 2-3
    consumption, 2-12
    for cooking, 2-28

for curing, 2-23, 3-47
deionized and cement, 2-35
demand, 2-21
demand, influence of fine particles, 2-21
domestic waste, 2-22
drinking, 2-30, 2-32, 3-47
drinking quality, 2-26, 2-28, 3-47
evaporable, 2-4
for food, 2-28
free, 4-61
gel, 2-4, 4-57
hardness, 2-22, 2-29, 2-31
hardness and autogenous healing, 5-39
for human consumption, 2-28
of hydration, 2-4, 4-18
impurities in, 2-19
industrial, 2-20
loss, 4-60
mine waste, 2-20
mixing, 3-47
natural, 2-20
non-evaporable, 2-4, 4-57
penetration, 2-9
physically adsorbed, 2-4
potable, 2-26, 2-28, 3-47
pure, effect on concrete, 2-23, 2-26
pure and leaching, 2-18, 3-36
-reducing admixtures, 4-21
-reducing admixtures in slipforming, 5-20
re-used, 2-16
saline, 3-15
solids in, 2-20, 2-21
standards for, 2-21
tap, 3-47
wash, 2-21
waste as mixing water, 2-24
Water-cement ratio, *see w/c*
Water-cementitious material ratio, *see w/cm*
Water curing, 2-5
"Waterproof" concrete, 2-9
Waterproofing admixtures, 2-9, 8-18
Water-resistant concrete, 2-9, 6-29
Water-retaining structures, 2-10
Water content,
of aggregate, 2-52
in high-performance concrete, 4-2
and strength, 2-51
total, 1-29
variability, 8-11
and w/c, 2-54, 3-19
and workability, 2-3, 2-4, 2-51, 4-16
Watertightness, 2-10, 2-56
Weathering, 3-38, 5-2, 6-12
Wedge-splitting test, 7-16, 7-24
Weep holes, 5-33
Wet chemistry, 5-26
Wet curing, *see Moist curing*
Wetting and drying, 2-7, 3-56, 4-37, 4-61, 5-33, 6-13, 8-29
and salt ingress, 3-56, 3-57
White portland cement, 5-33
Wind, 3-57
and creep, 6-54
and durability, 3-37
loading, 6-36
and plastic shrinkage, 4-62
Wire mesh reinforcement, 6-15
Witness, 9-14, 9-24
Workability
and cement content, 3-20
retention, 5-20
in slipforming, 5-19
and superplasticizers, 3-23
and water content, 2-3, 2-51
and water demand, 6-26
and w/c, 2-51, 3-19
Worker training, 3-11
Workmanship, 3-50, 4-11, 4-38, 6-29, 9-5

X

X-ray diffraction, 5-26